# Fundamental Theories of Physics

Volume 214

The international monograph series "Fundamental Theories of Physics" aims to stretch the boundaries of mainstream physics by clarifying and developing the theoretical and conceptual framework of physics and by applying it to a wide range of interdisciplinary scientific fields. Original contributions in well-established fields such as Quantum Physics, Relativity Theory, Cosmology, Quantum Field Theory, Statistical Mechanics and Nonlinear Dynamics are welcome. The series also provides a forum for non-conventional approaches to these fields. Publications should present new and promising ideas, with prospects for their further development, and carefully show how they connect to conventional views of the topic. Although the aim of this series is to go beyond established mainstream physics, a high profile and open-minded Editorial Board will evaluate all contributions carefully to ensure a high scientific standard.

Debraj Das · Shamik Gupta

# Facets of Noise

## Effects in Classical and Quantum Systems

Springer

Debraj Das
Quantitative Life Sciences
The Abdus Salam International Centre
for Theoretical Physics (ICTP)
Trieste, Italy

Shamik Gupta
Department of Theoretical Physics
Tata Institute of Fundamental Research
Mumbai, India

ISSN 0168-1222 ISSN 2365-6425 (electronic)
Fundamental Theories of Physics
ISBN 978-3-031-45311-3 ISBN 978-3-031-45312-0 (eBook)
https://doi.org/10.1007/978-3-031-45312-0



This Springer imprint is published by the registered company Springer Nature Switzerland AG
The registered company address is: Gewerbestrasse 11, 6330 Cham, Switzerland

Paper in this product is recyclable.

*To my beloved parents,*
*Tripti Das and Dipak Kumar Das*

*Debraj Das*

*To my beloved parents,*
*Geetanjali Gupta and Late Subir Kumar Gupta*
*and*
*To*
*Stefano Ruffo*

*Shamik Gupta*

# Foreword

Probabilistic reasoning has become a main tool in several branches of physics and in many ways has a role today akin to analysis in the development of classical physics. Probability theory until a few decades ago was not part of the mathematical curriculum of a physicist: as a consequence there was a lack of introductory books in this discipline with applications in physics. One must also consider that probability reached a mature mathematical formulation and status only in the third decade of the twentieth century with the work of the famous Russian mathematician A. N. Kolmogorov.

In classical physics the only sector which made explicit reference to probability was statistical mechanics once the atomic structure of matter was established and accepted. However it was a "probability" created by physicists and largely based on kinetic theory, until the appearance of the book "*Mathematical Foundations of Statistical Mechanics*" by another well known Russian mathematician A. I. Khinchin. The book appeared in English in 1949. We quote from this book which contains also a brief historical sketch:

"*In the books on physics the formulation of the fundamental notions of the theory of probability as a rule is several decades behind the present scientific level, and the analytic apparatus of the theory of probability, mainly its limit theorems, which could be used to establish rigorously the formulas of statistical mechanics without any complicated special machinery, is completely ignored.*"

The theory of probability and stochastic processes developed considerably in the second half of the 20th century, both in mathematics and in physical applications. A rigorous statistical mechanics emerged providing an important motivation to extend probability theory to infinite dimensional spaces and becoming a wide chapter of functional analysis. With its help it was possible to found the rigorous theory of phase transitions in equilibrium states of thermodynamic systems and to construct the first models of quantum field theory via the identification of quantum fields with random variables in quantum mechanics at imaginary times. A particular interesting area was represented by stochastic differential equations whose prototype had been introduced by the physicist P. Langevin but whose rigorous theory required advanced tools of stochastic processes.

We now describe some of the features and highlights of the present book. It is divided into three parts: *Background, Applications, Appendices*. Each part is divided into chapters, each one preceded by a very useful abstract. The *Background* includes in particular an introduction to the theory of dynamical systems which had a considerable impact on the development of probability via the ergodic problem and the concept of deterministic chaos. *Applications* include topics which can be found mainly in research articles sometimes difficult to read. The last subject in this part is devoted to quantum systems. The interpretation of probability in quantum theory is not simple and the authors limit themselves to cases in which randomness is of external origin like in observations made at random times. Finally the *Appendices* contain all sorts of topics which are relevant for making the main text self-contained.

A merit of the present book is to provide a manual of intermediate difficulty introducing the reader to the language of modern probability and to several of its physical applications. The publication of this book is welcome: even if it cannot cover all possible subjects relating probability and physics, the authors contribute to filling a gap in the communication between these disciplines.

October 18, 2023

Professor Giovanni Jona-Lasinio
Sapienza—Università di Roma
Rome, Italy

# Preface

The word "noise" is often associated with an unpleasant and disruptive sound. Acoustically, noise consists of frequencies distributed across a wide spectrum with no particular correlations among them. This fact makes future predictions of noise much harder in stark contrast to that for a melodious tone. In a broader sense, noise refers to randomly fluctuating information that one encounters in a wide variety of domains, from stock-price fluctuations that make future price predictions a challenge to random variations of quantities in cellular biology that impact crucially gene expressions and cellular characteristics. While noise was traditionally seen as an undesirable quantity and largely a nuisance, it has transpired in recent years to play a rather constructive role in various systems of relevance. An example is offered by the physical phenomenon of stochastic resonance, in which, quite counterintuitively, it is the noise that serves the desirable role of amplifying weak signals.

The genesis of noise research may be traced back to 1827 with the Scottish botanist Robert Brown observing the irregular motion of pollen grains suspended in a fluid. The erratic motion of the pollen grains was later named the Brownian motion, and the particles exhibiting the motion came to be known as Brownian particles. This fundamental phenomenon of Brownian motion remained unexplained for roughly eight decades, until Albert Einstein, in one of his annus mirabilis papers, and, independently, Marian Smoluchowski provided an explanation of the phenomenon. Einstein offered a theoretical understanding of the Brownian motion and demonstrated both qualitatively and quantitatively that the underlying erratic movements arise due to random buffeting of the Brownian particles by the molecules of the surrounding fluid. His work resulted in concepts that eventually formed the cornerstones of the field of statistical mechanics, such as the fluctuation-dissipation relation and the diffusion equation, and also unveiled the remarkable connection between the mean-square displacement (characterized by the diffusion constant) of the Brownian particles and the resistive force (characterized by the friction coefficient) experienced by them as they wade their way through the surrounding fluid. He obtained an important relation known as the Stokes-Einstein relation that relates macroscopic measurable quantities (e.g., the diffusion constant and the friction coefficient) to microscopic thermal fluctuations characterized by the temperature of the surrounding fluid.

The pioneering work of Einstein led to the derivation of the so-called diffusion equation that describes the time evolution of the spatial probability distribution of the Brownian particles. The same form of the diffusion equation was independently derived earlier by Adolf Fick in 1855. However, the microscopic origin of diffusion was lacking in the literature until the work of Einstein. For a localized initial position of the Brownian particles, the solution to the diffusion equation is a Gaussian distribution, a result that was verified in a remarkable experiment by Jean Baptiste Perrin in 1908 for which he was awarded the Nobel prize in 1926. Perrin used Stokes's law of viscosity to obtain the frictional force on a small spherical particle moving through a viscous fluid. By estimating the diffusion constant from the trajectories of the particle, Perrin was able to obtain a highly accurate estimate of Avogadro's number, thereby contributing to the acceptance in the scientific community at large of such concepts as atoms and molecules as fundamental building blocks of matter. In 1908, Paul Langevin adopted a completely different approach to the study of the Brownian motion by formulating an equation of motion for the Brownian particles that incorporates it a dissipative viscous force and a random complementary force. Using the so-called equipartition theorem and the form of the friction constant obtained from Stokes's law, he derived the mean-square displacement of the Brownian particles, which conformed exactly to the result obtained by Einstein. The mathematical tools to quantitatively analyze noise were subsequently formulated by some of the greatest mathematicians of all times, including Paul Pierre Lévy, Norbert Wiener, Kiyosi Itô, Ruslan Leont'evich Stratonovich, and many others. Their work laid the foundation of the theory of stochastic differential equations, of which the Langevin equation is a prominent example, and which today forms the backbone of analysis in as disparate a field as quantitative finance dealing with stocks, derivatives, and options.

In modern times, study of noise has emerged as a relevant and a very active area of research across various scientific disciplines, including natural sciences such as physics, chemistry, biology, and mathematics, as well as applied fields such as economics, finance, and computer science. In biology, the research has largely focused on understanding how fluctuating gene expression levels affect cell transitions, the role that thermal fluctuations play in the motion of molecular motors, and the impact of negative feedback in gene networks on suppressing random fluctuations. In machine learning, introducing small noise in training data has been found to improve the extrapolation abilities of neural networks, while noise correction is found to enhance classifier accuracy. In financial markets, studies have explored the relationship between past trades and current stock prices, the connection between large price fluctuations and trading volume, and the diffusive nature of stock prices. Research in economics has examined seasonal fluctuations and their resemblance to business cycle fluctuations. In ecology, the effects of random environmental and demographic forces on population dynamics have been investigated, and the adverse impact of noise, such as traffic noise masking acoustic signals used by birds and animals, has been unveiled.

In this book, our primary focus is on noise in physical systems. Our aim is to discuss how noise impacts the static and dynamic properties of classical and quantum systems. To this end, we provide a broad exploration of the different facets of noise arising in a variety of physical contexts, including the noise that arises from interaction with the environment, the noise whose inclusion into a dynamical setup allows to extract valuable information about the corresponding noiseless dynamics, the noise resulting from sampling of initial conditions in finite-size interacting systems, the noise that is induced by the application of an external field. In our endeavor to make the book a comprehensive exploration of the topic, we categorize it into two parts, namely, (i) Background and (ii) Applications, with each consisting of five chapters. In the background part, we discuss a number of key fundamental concepts and principles that underpin the study of noise in physical systems. These ideas and tools are then borrowed and employed extensively in the second part in which we provide several practical applications in a wide range of physical phenomena emerging in the presence of noise. The phenomena we unveil include those of spontaneous synchronization, bifurcations, dynamics in long-range interacting systems, and peculiar characteristics of projective measurements in quantum systems.

Specifically, in the background part, we start in Chap. 1 with a discussion of random variables and probability distributions, and explore how these mathematical tools allow us to quantify and analyze uncertainty and variability in physical systems. Next, in Chap. 2, we delve into the realm of stochastic processes, in which we investigate the behavior of systems that evolve stochastically in time. From Markov processes to Brownian motion, we discuss how these mathematical models can capture and provide predictions in a statistical sense for the observable static and dynamic properties of stochastic processes. In Chap. 3, we turn to the kinetic theory approach to studying the collective dynamical behavior of systems comprising a large number of interacting degrees of freedom, and how they may relax to stationary states characterized by time-independent observable properties. Chapter 4 deals with the time-honored statistical mechanical approach to study of physical systems. Beginning with a recapitulation of basic thermodynamics, we illustrate in this chapter how the statistical mechanical theory à la Gibbs and Boltzmann achieved the remarkable feat of establishing a direct relationship between microscopic characterization and macroscopic properties of many-body systems. The background part of this book ends with a discussion on nonlinear dynamical systems in Chap. 5, wherein we discuss briefly the fundamental concepts of fixed points, bifurcations, limit cycles, and introduce and study a paradigmatic and illustrative model of nonlinear systems exhibiting the phenomenon of spontaneous synchronization, the Kuramoto model.

The application part of this book deals with the effects on physical systems of noise arising in situations pervading length and time scales of several orders of magnitude. We start in Chap. 6 with unveiling of the effects of noise that arises invariably in the dynamics of any physical system in interaction with its environment. Here, we focus on the noisy version of the Kuramoto model, wherein the effects of the environment are incorporated in the dynamics in terms of a Gaussian, white noise. We provide exact analytical expressions for the autocorrelation of the synchronization order parameter in the nonequilibrium stationary state that the dynamics attains at long times. The main result that emerges is that the stronger the noise in the dynamics, the shorter the time that it takes to decorrelate the system on a macroscopic scale.

Chapter 7 explores how noise may be usefully and efficiently used in extracting information about deterministic nonlinear dynamical systems. The chapter focuses on bifurcation, a phenomenon wherein the flow structure of a dynamical system changes qualitatively on varying the dynamical parameters. Introducing noise into the dynamics of a model nonlinear system, we demonstrate in this chapter how the conditional probability distribution of microscopic configurations, characterizing the likelihood of occurrence of configurations based on the occurrence of a previous set of configurations in time, provides an efficient tool in the limit of weak noise for extracting the bifurcation behavior of the corresponding noiseless dynamics.

In Chap. 8, the focus is on noise arising from the sampling of initial conditions and its interplay with environmental noise in a finite-size system of particles interacting via long-range interactions. As representative examples, we consider mean-field systems of classical Heisenberg spins and study how they relax to the Boltzmann-Gibbs equilibrium state in the course of dynamical evolution. We show that while the relaxation under deterministic dynamics proceeds through long-lived intermediate quasistationary states, fluctuations induced by environmental interactions lead to a fast relaxation to equilibrium, thereby wiping out signatures of quasistationarity observed in the deterministic dynamics.

Chapter 9 investigates noise in the form of fluctuations induced by a small external field applied to the same mean-field classical spin systems as considered in Chap. 8. We analyze the response of these systems to the external field and obtain the critical exponents characterizing the power-law behavior of the response near the critical point of phase transition that the systems exhibit in the absence of the field. The exponents are obtained in two different ways: a static analysis that assumes the long-time phase space distribution after the application of the field to be a priori of the same mathematical form as the one before the application of the field, and a dynamic analysis, in which one follows the exact evolution of the initial distribution under the application of the field. We show that the exponents obtained within the two approaches differ significantly, and demonstrate that the values derived following the dynamics are the ones actually observed on performing numerical integration of the equations of motion. Interestingly, we find that both the sets of values of the critical exponents satisfy the same scaling relation, which is incidentally the same as the one known for mean-field Hamiltonian particle systems that are quite different from the studied spin systems. This chapter offers to hint on the one hand at the universality of critical behavior for mean-field systems and cautions on the other hand against relying on a static approach, with no reference to the dynamical evolution, to extract values of critical exponents.

Chapter 10 explores the issue of non-unitary evolution of quantum systems subject to projective measurements at random times. Considering two different evolution schemes, we investigate the survival probability of an initial state after a certain number of measurements have been performed on the system. The interplay between classical noise associated with the random measurement times and the inherent quantum noise in the sense of unpredictability associated with measurements in quantum systems is examined. Through numerical and semi-analytical results, we reveal universal exponential decay of the survival probability under one scheme and a power-law decay under the other.

Our earnest endeavor in this book is to strike a fine balance between theoretical expositions and practical applications. Through this book, we embark on developing in the readers a deep appreciation of the multifaceted nature of noise effects, and on arming them with theoretical concepts and tools that would facilitate their future expedition into the kingdom of noise effects in physical systems. We hope that this book will serve as a valuable resource for students, researchers, and enthusiasts alike, besides fostering in them a sense of admiration for physical effects arising from noise and their mathematical modeling and applications.

This book is based on the Ph.D. work in Physics of one of the authors (DD) under the supervision of the other (SG). The thesis work was performed at the Ramakrishna Mission Vivekananda Educational and Research Institute, Belur, India. The thesis was defended and the degree was awarded on 30 September 2020. The application part of the book is based on the following joint publications of DD and SG: Debraj Das and Shamik Gupta, *Exact correlations in the nonequilibrium stationary state of the noisy Kuramoto model*, J. Phys. A: Math. Theor. **51**, 445003 (2018); Debraj Das, Sayan Roy, and Shamik Gupta, *Reconstructing bifurcation behavior of a nonlinear dynamical system by introducing weak noise*, Commun Nonlinear Sci Numer Simulat **72**, 575 (2019); Debraj Das and Shamik Gupta, *Relaxation to equilibrium in models of classical spins with long-range interactions*, J. Stat. Mech.: Theory Exp. 084007 (2019); Yoshiyuki Y. Yamaguchi, Debraj Das, and Shamik Gupta, *Critical exponents in mean-field classical spin systems*, Phys. Rev. E **100**, 032131 (2019); Debraj Das and Shamik Gupta, *Quantum random walk and tight-binding model subject to projective measurements at random times*, J. Stat. Mech.: Theory Exp. 033212 (2022).

It is a great pleasure to warmly thank a number of colleagues for fruitful and enjoyable discussion and collaboration on topics discussed in this monograph: Anish Acharya, Mustansir Barma, Tanmay Biswas, Alessandro Campa, Rohitashwa Chattopadhyay, Pierfrancesco di Cintio, Sushanta Dattagupta, Stefano Gherardini, Luca Giuggioli, Late Arun Jayannavar, Rainer Klages, Christian Maes, Rupak Majumder, David Mukamel, Soumya Kanti Pal, Édgar Roldán, Sayan Roy, Stefano Ruffo, Mrinal Sarkar, C. L. Sriram, and Yoshiyuki Y. Yamaguchi. We are particularly grateful to Anish Acharya, Rohitashwa Chattopadhyay, Sanjeev Kumar, Rupak Majumder, Soumya Kanti Pal, and C. L. Sriram for a careful reading of the manuscript.

SG acknowledges support from the Science and Engineering Research Board (SERB), India under SERB-CRG Grant CRG/2020/000596. SG also thanks ICTP, Trieste, Italy, for support under its Regular Associateship scheme. Finally, we acknowledge the kind support of Springer Nature Switzerland AG. We are especially thankful to Zachary Evenson and Naveen Kumar Madhan Mohan, and, in particular, to Angela Lahee, all from Springer, for their encouragement, assistance, and patience.

Trieste, Italy
Mumbai, India
August 2023

Debraj Das
Shamik Gupta

# Contents

# Part I
# Background

# Chapter 1
# Random Variables and Probability Distributions

**Abstract** In this chapter, after recapitulating the elements of probability theory, we introduce the notion of random variables and probability distribution of random variables. Next, we discuss in turn joint, marginal, and conditional distribution of random variables, as well as such attributes of a probability distribution of a random variable as mean, variance, standard deviation, raw, and central moments. We also discuss the concepts of covariance and correlation between random variables and that of moment and cumulant generating functions of a random variable. In the following section, we provide some familiar examples of probability distributions and then move on to discuss two central results related to sums of independent and identically-distributed (i.i.d.) random variables, the Law of Large Numbers, and the Central Limit Theorem. We then discuss stable distributions, which allow a generalization of the Central Limit theorem. This is followed by a brief overview of the elements of the extreme value theory that deals with extreme values of a sample of i.i.d. random variables. As an illustration of the concepts developed, we discuss in detail the paradigmatic example of a random walk in one dimension.

## 1.1 Elements of Probability Theory

We start this chapter with a brief reminder on probability theory. This chapter is based on materials found in standard references, e.g., Refs. [1, 2]. To set off, let us quickly summarize some familiar set-theoretic notations.

1. A set is a collection of elements, e.g., $A = \{a, c, d\}$, $B = \{1, 2\}$, $C = \{\text{Mumbai, London, Paris, New York}\}$. In a set, the ordering of its elements is irrelevant, in contrast to a sequence, a tuple, or a permutation of a set. Thus, $\{a, c, d\}$, $\{a, d, c\}$, $\{c, a, d\}$ all represent the same set $A$. The cardinality of a set is given by the number of its elements. For example, the set $A = \{a, c, d\}$ has a cardinality of 3.
2. The fact that an element "$a$" is in set $A$ and an element "2" is not in set $A$ is denoted respectively by $a \in A$ ($a$ belongs to $A$) and $2 \notin A$ (2 does not belong to $A$).



D. Das and S. Gupta, *Facets of Noise*, Fundamental Theories of Physics 214,
https://doi.org/10.1007/978-3-031-45312-0_1

3. Given two sets $A$ and $B$, their union is defined as $A \cup B \equiv \{x|x \in A \text{ or } x \in B\}$, while their intersection is defined as $A \cap B \equiv \{x|x \in A \text{ and } x \in B\}$. The difference is on the other hand defined as $A \backslash B \equiv \{x|x \in A \text{ and } x \notin B\}$. Here, the notation "$\{x|x \in A \text{ or } x \in B\}$" denotes set of $x$ values that satisfy the condition stated to the right of the symbol "$|$", namely, that $x$ has to belong to either of the two sets $A$ and $B$.
4. The empty or null set containing no element is denoted by $\emptyset$.
5. For two sets $C$ and $D$, if every element of $C$ is also an element of $D$, we say that $C$ is a subset of $D$ (denoted by $C \subset D$), or equivalently, $D$ is a superset of $C$ (denoted by $D \supset C$). The empty set is the subset of every set.
6. The power set of a set $A$ is the set of all subsets of $A$, which includes the empty set and the set $A$ itself. For a set with cardinality $n$, the cardinality of its power set is $2^n$.

**Example** Consider two sets $A = \{1, 2, 3\}$ and $B = \{2, 4, 6\}$. Then, we get $A \cup B = \{1, 2, 3, 4, 6\}$ and $A \cap B = \{2\}$. The set $C = \{1, 2\}$ is a subset of $A$, i.e., one has $C \subset A$. The power set of $A$ with cardinality 3 is the set $\{\emptyset, \{1\}, \{2\}, \{3\}, \{1, 2\}, \{1, 3\}, \{2, 3\}, \{1, 2, 3\}\}$ that contains $2^3 = 8$ elements.

With this background, let us now move on to discuss the fundamentals of probability theory. A random trial is an experiment whose outcome is random, i.e., one gets in general a different outcome every time the experiment is repeated under identical conditions. Consider the random trial in which one tosses a coin: every time it is tossed, one may get heads or tails, but a priori (i.e., before one has tossed it), there is no way to know with certainty whether one gets heads or tails.

Let $\omega$ denote an elementary outcome of the random trial and $\Omega$ denote the set of all possible $\omega$'s; thus, we have $\omega \in \Omega$. Then, $\Omega$ is called the sample space. An event $E$ is a subset of $\Omega$. The set of all events is the power set of $\Omega$.

**Example** Consider the random trial involving tossing a die: Here, we have $\omega = 1, 2, 3, 4, 5, 6$, while we have $\Omega = \{1, 2, 3, 4, 5, 6\}$. An example of an event is $E = \{2, 4, 6\}$ (the event that the face on which the die falls has an even number).

Given a sample space $\Omega$, a probability measure is a function that assigns to each event $E \subset \Omega$ a non-negative real number $P(E)$, such that the following axioms, introduced by the Russian mathematician Andrey Kolmogorov, are satisfied.

1. $P(E) \geq 0 \; \forall \; E \subset \Omega$,
2. $P(\emptyset) = 0$ (Probability of the event "nothing happens" is zero, since it cannot be that the random trial does not yield any outcome) and $P(\Omega) = 1$, and
3. For $E_1, E_2, \ldots \subset \Omega$, if $E_i \cap E_j = \emptyset \; \forall \; i \neq j$ (events that are mutually exclusive, meaning thereby that they cannot occur simultaneously), then $P(\cup_i E_i) \equiv P(E_1 \cup E_2 \cup E_3 \ldots) = \sum_i P(E_i)$.

**Example** Tossing a fair coin once. We have $\Omega = \{H, T\}$ and $P(\Omega) = P(\{H, T\}) = 1$. Now, the two events $E_1 = \{H\}$ and $E_2 = \{T\}$ are mutually exclusive, and so $P(\Omega) = P(E_1 \cup E_2) = P(E_1) + P(E_2) = 1$. The coin being fair implies that $P(E_1) = P(E_2)$, namely, the coin is equally likely to fall on heads as on tails. Combining everything, we get $P(E_1) = P(E_2) = 1/2$.

We now state a number of important and useful results.

1. Complement of an event: Let $E^c$ denote the complement of the event $E$, that is, $E^c = \Omega \backslash E$. Then, $E$ and $E^c$ being mutually exclusive, we have using the property 3 above that $P(E \cup E^c) = P(E) + P(E^c)$, while we have $P(\Omega) = 1 = P(E \cup E^c)$. We finally get $P(E^c) = 1 - P(E)$.
2. Inclusion-exclusion formula: Consider two events $E_1$ and $E_2$ that can intersect. The intersection consists of all outcomes that are in both $E_1$ and $E_2$. Then, we have by virtue of property 3 that $P(E_1 \cup E_2) \neq P(E_1) + P(E_2)$. Instead, we have

$$P(E_1 \cup E_2) = P(E_1) + P(E_2) - P(E_1 \cap E_2). \tag{1.1}$$

***Proof*** Let us define three mutually-exclusive events $C_1 \equiv E_1 \backslash E_2, C_2 \equiv E_1 \cap E_2$, and $C_3 \equiv E_2 \backslash E_1$, see Fig. 1.1 for Venn diagrams showing the three events. Next, we use property 3 given above, to get $P(E_1 \cup E_2) = P(C_1 \cup C_2 \cup C_3) = P(C_1) + P(C_2) + P(C_3)$. On the other hand, we have $P(E_1) = P(C_1 \cup C_2) = P(C_1) + P(C_2)$, $P(E_2) = P(C_2 \cup C_3) = P(C_2) + P(C_3)$, and $P(E_1 \cap E_2) = P(C_2)$. It then immediately follows that $P(E_1 \cup E_2) = P(E_1) + P(E_2) - P(E_1 \cap E_2)$.

3. Conditional probability: Given two events $E_1$ and $E_2$, with $P(E_2) > 0$, the conditional probability of $E_1$ given $E_2$, denoted by $P(E_1|E_2)$, is the probability of $E_1$ occurring if $E_2$ has or is assumed to have happened, and is defined as follows:

$$P(E_1|E_2) \equiv \frac{P(E_1 \cap E_2)}{P(E_2)}. \tag{1.2}$$

4. Bayes' theorem: For two events $E_1$ and $E_2$, with $P(E_2) > 0$, the theorem states the following:

$$P(E_1|E_2) = \frac{P(E_2|E_1)P(E_1)}{P(E_2)}. \tag{1.3}$$

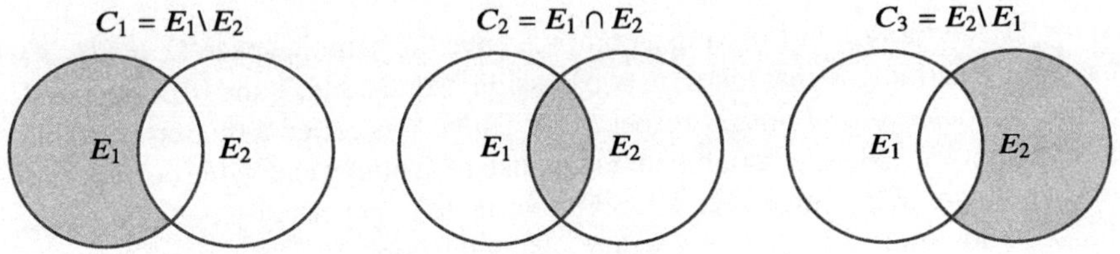


**Fig. 1.1 Venn diagram for proving Eq. (1.1), see text**

***Proof*** From the definition of conditional probability, Eq. (1.2), we have

$$P(E_1|E_2) \equiv \frac{P(E_1 \cap E_2)}{P(E_2)} \text{ if } P(E_2) > 0\,, \tag{1.4}$$

$$P(E_2|E_1) \equiv \frac{P(E_2 \cap E_1)}{P(E_1)} \text{ if } P(E_1) > 0\,. \tag{1.5}$$

Using Eq. (1.5) to solve for $P(E_2 \cap E_1)$, using $P(E_2 \cap E_1) = P(E_1 \cap E_2)$, and substituting the result Eq. (1.4), one obtains Eq. (1.3).

5. Independence of events: Two events $E_1$ and $E_2$ are said to be independent if knowing whether $E_1$ has happened or not does not influence the probability of $E_2$, that is,

$$P(E_2|E_1) = P(E_2)\,. \tag{1.6}$$

Using Eq. (1.2), it then follows the condition of independence of two events: Two events $E_1$ and $E_2$ are independent, provided

$$P(E_1 \cap E_2) = P(E_1)P(E_2)\,. \tag{1.7}$$

In general, for $n \geq 2$ events $E_1, E_2, \ldots, E_n$ to be independent, all of the following conditions have to hold:

$$\begin{aligned}
&P(E_i \cap E_j) = P(E_i)P(E_j) \text{ for all distinct } i, j \in \{1, 2, \ldots, n\}\,;\\
&P(E_i \cap E_j \cap E_k) = P(E_i)P(E_j)P(E_k) \text{ for all distinct } i, j, k \in \{1, 2, \ldots, n\}\,;\\
&\vdots\\
&P(E_1 \cap E_2 \cap E_3 \cap \ldots \cap E_n) = \prod_{i=1}^{n} P(E_i)\,.
\end{aligned} \tag{1.8}$$

## 1.2 Random Variables and Probability Distribution

Consider the random trial of tossing a coin; we have the sample space $\Omega = \{H, T\}$. Let us define a variable $X$ that takes on two possible values, say, 1 and 0, corresponding to the two outcomes $H$ and $T$, respectively. Then $X$ is called a random variable. Thus, a random variable is a variable that may take on a numerical value corresponding to each outcome in a given $\Omega$. A random variable is generically denoted by the symbol $X$, whose particular values are denoted by the corresponding lower-case letter $x$. Another popular example could be rolling a die once and having the random variable $X$ as denoting the number on the top face of the die. Then the possible values of $X$ are $x = 1, 2, 3, 4, 5, 6$.

There are two types of random variables: discrete and continuous. A discrete random variable is one that takes on only a countable number of distinct values. Consider tossing a coin three times in a row and the random variable $X$ denoting the number of times that the coin comes up heads. Then, $X$ is a discrete random variable whose possible values are only $0, 1, 2, 3$. In contrast to a discrete random variable, a continuous random variable is one that takes on an infinite number of possible values that are defined over an interval or a range. An example of a continuous random variable could be all real numbers on the interval $[-100, 100]$.

The probability distribution of a discrete random variable is specified by a list of probabilities associated with each of the possible values of the random variable. Consider a discrete random variable $X$ that takes on possible values $x_1, x_2, \ldots$. Let $P_X(x_i)$ denote the probability that $X$ takes on the value $x_i$; $i = 1, 2, \ldots$. Then, the quantities $P_X(x_i)$ must satisfy the following properties:

1. $0 \leq P_X(x_i) \leq 1 \; \forall \, i$, and
2. $\sum_i P_X(x_i) = 1$ (Normalization).

Plotting $P_X(x_i)$ against $x_i$ then yields the probability histogram. We will refer to $P_X(x_i)$ as the probability distribution of $X$.

For a continuous random variable $X$ that takes on values that are real numbers, the probability distribution $P_X(x)$ is defined such that $P_X(x)\mathrm{d}x$ is the probability that $X$ takes on any value in the infinitesimal interval $[x, x + \mathrm{d}x]$. Note that while the probability distribution of a discrete random variable provides probability values for the random variable taking specific values, the same for a continuous random variable provides probability values for the random variable taking values over a range. The function $P_X(x)$ must satisfy the following properties:

1. $P_X(x) \geq 0 \; \forall \, x$, and
2. $\int \mathrm{d}x \; P_X(x) = 1$, with the integration performed over the whole range of allowed values of $X$ (Normalization).

Plotting $P_X(x)$ against $x$ then yields the probability density curve, and the condition 2 above implies that the total area under the curve is unity.

The cumulative distribution function of a continuous random variable $X$, denoted by $F_X(x)$, is the probability that $X$ takes on a value less than or equal to $x$:

$$F_X(x) = P(X \leq x) \,, \tag{1.9}$$

and is thus given for real $X$ by

$$F_X(x) = \int_{-\infty}^{x} \mathrm{d}x \; P_X(x) \,. \tag{1.10}$$

It is evident that $F_X(x)$ satisfies

$$\lim_{x \to -\infty} F_X(x) = 0, \quad \lim_{x \to \infty} F_X(x) = 1\,, \tag{1.11}$$

and that

$$P_X(x) = \frac{\mathrm{d}F_X(x)}{\mathrm{d}x}\,. \tag{1.12}$$

For a discrete random variable, the corresponding equation for the cumulative distribution function is

$$F_X(x) = \sum_{\substack{i \\ x_i \le x}} P_X(x_i)\,. \tag{1.13}$$

## 1.3 Joint, Marginal, and Conditional Distribution

The joint distribution $P_{X,Y}(x_i, y_j)$ of two discrete random variables $X$ and $Y$ that respectively take on values $x_1, x_2, \ldots$ and $y_1, y_2, \ldots$ is the probability of simultaneous occurrence of the random variable $X$ taking on the value $x_i$ *and* the random variable $Y$ taking on the value $y_i$. The joint distribution satisfies $0 \le P_{X,Y}(x_i, y_j) \le 1 \,\forall\, x_i, y_j$ and is normalized: $\sum_i \sum_j P_{X,Y}(x_i, y_j) = 1$. The above definition of the joint distribution may be extended to any given number of random variables.

For continuous random variables, the joint distribution $P_{X,Y}(x, y)$ is defined such that $P_{X,Y}(x, y)\mathrm{d}x\mathrm{d}y$ is the probability of simultaneous occurrence of $X$ taking on any value in the interval $[x, x + \mathrm{d}x]$ *and* $Y$ taking on any value in the interval $[y, y + \mathrm{d}y]$; the function $P_{X,Y}(x, y)$ must satisfy the positivity condition $P_{X,Y}(x, y) \ge 0$ and the normalization condition $\int \mathrm{d}x \int \mathrm{d}y\; P_{X,Y}(x, y) = 1$. The integration is over all values assumed by $X$ and $Y$.

Two discrete random variables $X$ and $Y$ that respectively take on values $x_1, x_2, \ldots$ and $y_1, y_2, \ldots$ are said to be independent if we have

$$P_{X,Y}(x_i, y_j) = P_X(x_i)P_Y(y_j) \;\forall\; x_i, y_j\,. \tag{1.14}$$

Generalizing to more than two random variables, a finite set of $n$ random variables $\{X_1, X_2, \ldots, X_n\}$ is mutually independent if we have

$$P_{X_1,X_2,\ldots,X_n}(x_1, x_2, \ldots, x_n) = P_{X_1}(x_1)P_{X_2}(x_2)\ldots P_{X_n}(x_n)\,, \tag{1.15}$$

where $x_i$ denotes the possible values of the random variable $X_i$.

The above definition of independence may be straightforwardly extended to continuous random variables. Two continuous random variables $X$ and $Y$ are said to be independent if we have

$$P_{X,Y}(x, y) = P_X(x)P_Y(y) \tag{1.16}$$

for all values $x$ and $y$ of the random variables $X$ and $Y$.

Given two discrete random variables $X$ and $Y$ having a joint distribution $P_{X,Y}$, the marginal distribution of $X$ and $Y$ are defined respectively as

$$P_X(x_i) \equiv \sum_j P_{X,Y}(x_i, y_j), \;\; P_Y(y_j) \equiv \sum_i P_{X,Y}(x_i, y_j) \,. \tag{1.17}$$

Evidently, both $P_X(x_i)$ and $P_Y(y_j)$ are normalized to unity. If $X$ and $Y$ are continuous random variables, then the corresponding defining equations are

$$P_X(x) \equiv \int \mathrm{d}y \; P_{X,Y}(x, y), \;\; P_Y(y) \equiv \int \mathrm{d}x \; P_{X,Y}(x, y) \,. \tag{1.18}$$

Again, $P_X(x)$ and $P_Y(y)$ are both normalized to unity. A particular special case is when $X$ and $Y$ are random variables such that their sum is a constant: $X + Y = A$, where $A$ is a constant. In this case, we have

$$P_X^{(c)}(x) = \int \mathrm{d}y \; P_{X,Y}(x, y)\, \delta(y - (A - x)) = P_{X,Y}(x, A - x) \,, \tag{1.19}$$

with the superscript $c$ a reminder of the constraint $X + Y = A$. From the above equation, we see that $P_X^{(c)}(x)$ depends on a special value of $Y$, namely, the value $y = A - x$. As such, $P_X^{(c)}(x)$ is *not* normalized to unity; Had it been completely independent of any value that $Y$ assumes, $P_X^{(c)}(x)$ would have been normalized to unity.

For two discrete random variables $X$ and $Y$, the conditional probability distribution of $Y$ given $X$ is the probability distribution of $Y$ when $X$ is given to have a particular value. Thus, the conditional probability distribution of $Y$ given $X = x_i$ is given by

$$P(Y = y_j | X = x_i) = \frac{P_{X,Y}(x_i, y_j)}{P_X(x_i)} \,. \tag{1.20}$$

The above definition holds also for continuous random variables.

## 1.4 Mean, Variance, Standard Deviation, Raw and Central Moments

The mean value of a random variable $X$, usually denoted by the symbol $\mu$, is defined as

$$\mu \equiv \langle x \rangle \,. \tag{1.21}$$

Here, we have defined $\langle f(x) \rangle$ for any function $f$ of a random variable $X$ as

$$\langle f(x) \rangle = \begin{cases} \sum_i f(x_i) P_X(x_i) & \text{for discrete } X\,, \\ \int \mathrm{d}x\, f(x) P_X(x) & \text{for continuous } X\,. \end{cases} \tag{1.22}$$

The mean is also the first raw moment $\mu'_1$ of the underlying probability distribution. In general, the $n$-th raw moment of a probability distribution is defined as

$$\mu'_n \equiv \langle x^n \rangle; \quad n = 0, 1, 2, \ldots \,. \tag{1.23}$$

Note that we have

$$\mu'_0 = 1 \,. \tag{1.24}$$

By contrast, the central moments of a probability distribution are defined as

$$\mu_n \equiv \langle (x - \mu)^n \rangle; \quad n = 0, 1, 2, \ldots \,, \tag{1.25}$$

with

$$\mu_0 = 1, \;\; \mu_1 = 0 \,. \tag{1.26}$$

The raw moments may be expressed in terms of the central moments, e.g.,

$$\mu'_2 = \mu_2 + (\mu'_1)^2, \; \mu'_3 = \mu_3 + 3\mu_2\mu'_1 + (\mu'_1)^3 \,, \mu'_4 = \mu_4 + 4\mu_3\mu'_1 + 6\mu_2(\mu'_1)^2 + (\mu'_1)^4 \,. \tag{1.27}$$

The deviation of specific values that $X$ assumes from its average value $\mu$ is characterized in terms of the variance of $X$, defined as

$$\sigma^2 \equiv \langle (x - \mu)^2 \rangle = \mu_2 \,. \tag{1.28}$$

Expanding the right hand side, we obtain $\sigma^2$ in terms of the mean and the second raw moment $\mu'_2$, as

$$\sigma^2 = \mu'_2 - \mu^2 \,. \tag{1.29}$$

The quantity $\sigma$ is called the standard deviation, and which is a measure of the amount of variation of values that $X$ assumes about its mean value. A low standard deviation implies that the different values that $X$ assumes are close to the mean, while a high standard deviation indicates that the different values are more spread out with respect to the mean.

It is often useful to characterize a probability distribution using normalized and dimensionless central moments defined as

$$\gamma_n \equiv \frac{\mu_n}{\mu_2^{n/2}} = \frac{\mu_n}{\sigma^n}; \quad n = 0, 1, 2 \ldots . \tag{1.30}$$

One then readily observes that $\gamma_0 = 1$, $\gamma_1 = 0$, $\gamma_2 = 1$. The quantity $\gamma_3$ is known as the skewness of the distribution. The skewness vanishes when the distribution is symmetric about its mean. When the distribution is skewed to values smaller (respectively, higher) than the mean, one has $\gamma_3 < 0$ (respectively, $\gamma_3 > 0$).

The mode of a probability distribution is the value of the underlying random variable at which the probability distribution attains its maximum value. On the other hand, the median of a continuous probability distribution is the value of the underlying random variable at which the cumulative distribution function has the value $1/2$. In other words, the median is defined as any real number $m$ that satisfies the inequalities

$$F_X(x \leq m) \leq \frac{1}{2} \text{ and } F_X(x \geq m) \geq \frac{1}{2}, \tag{1.31}$$

where $F_X(x)$ is the cumulative distribution defined in Eq. (1.10).

## 1.5 Covariance and Correlation

For two random variables $X$ and $Y$ (either discrete or continuous) with respective mean $\mu_X$ and $\mu_Y$ and standard deviation $\sigma_X$ and $\sigma_Y$, their covariance and correlation are respectively defined as

$$\text{cov}_{XY} = \sigma_{XY} \equiv \langle (X - \mu_X)(Y - \mu_Y) \rangle, \tag{1.32}$$

$$\text{corr}_{XY} = \rho_{XY} \equiv \frac{\langle (X - \mu_X)(Y - \mu_Y) \rangle}{\sigma_X \sigma_Y}, \tag{1.33}$$

yielding

$$\rho_{XY} = \frac{\sigma_{XY}}{\sigma_X \sigma_Y}, \tag{1.34}$$

where, e.g., we have

$$\langle (X-\mu_X)(Y-\mu_Y)\rangle = \int \mathrm{d}x\mathrm{d}y\ P_{X,Y}(x,y)\,(x-\mu_X)\,(y-\mu_Y)\,, \tag{1.35}$$

for continuous $X, Y$. Here, $P_{X,Y}(x,y)$ is the joint distribution defined in Sect. 1.3. We observe that $\sigma_{XX} = \sigma_X^2$.

The above definition may be extended to $n$ random variables $X_1, X_2, \ldots, X_n$ (either discrete or continuous) with respective means $\mu_1, \mu_2, \ldots, \mu_n$, for which the covariance $\sigma_{X_iX_j}$ is defined as

$$\sigma_{X_iX_j} \equiv \langle (X_i-\mu_i)(X_j-\mu_j)\rangle\,, \tag{1.36}$$

while $\rho_{X_iX_j} = \sigma_{X_iX_j}/(\sigma_{X_i}\sigma_{X_j})$. Obtaining the covariance between every pair of variables $(X_i, X_j)$ generates a matrix of dimension $n \times n$ called the covariance matrix.

Covariance and hence correlation offers a measure of the amount of interdependence between a set of random variables. Indeed, when two random variables $X$ and $Y$ are independent, we have $\sigma_{XY} = 0$ (implying from Eq. (1.34) that $\rho_{XY} = 0$). Note that while $\sigma_{XY}$ ranges in values from $-\infty$ to $\infty$, the correlation $\rho_{XY}$, being nothing but scaled covariance, ranges in values from $-1$ to 1. If $\rho_{XY} = 1$ (respectively, $\rho_{XY} = -1$), then $X$ and $Y$ are said to be completely correlated (respectively, completely anti-correlated). Note that while statistically independent variables are always uncorrelated, i.e., $\rho_{XY} = 0$, the converse is not necessarily true. Indeed, a zero correlation between every pair in a set of random variables $\{X_1, X_2, \ldots, X_n\}$ implies that one has $P_{X_i,X_j}(x_i, x_j) = P_{X_i}(x_i)P_{X_j}(x_j)$ for every pair $(X_i, X_j)$. On the other hand, independence of this set of random variables demands a much more stringent condition to be satisfied, namely, condition (1.15).

## 1.6 Moment and Cumulant Generating Functions

The moment-generating function $M(t)$ of $X$ is defined as

$$M(t) \equiv \langle \mathrm{e}^{xt}\rangle\,, \tag{1.37}$$

provided there exists an $h > 0$ such that we have $|t| < h$. Expanding the exponential on the right hand side, one obtains

$$M(t) = \sum_{n=0}^{\infty} \mu'_n \frac{t^n}{n!}\,, \tag{1.38}$$

which makes it evident that $M(t)$ indeed generates the raw moments of the distribution $P_X(x)$. The moment-generating function is closely related to the characteristic function $\phi(t)$ of a random variable $X$ defined as the average of $\mathrm{e}^{\mathrm{i}tx}$:

$$\phi(t) \equiv \langle e^{itx} \rangle = \int_{-\infty}^{\infty} dx\, e^{itx} P_X(x)\,, \tag{1.39}$$

with $t$ being a real quantity. In other words, $\phi(t)$ is given by the Fourier transform of the probability distribution, which therefore is obtained as the inverse Fourier transform of the characteristic function:

$$P_X(x) = \frac{1}{2\pi} \int_{-\infty}^{\infty} dt\, e^{-itx} \phi(t)\,. \tag{1.40}$$

The characteristic function for the sum of two independent random variables is given by the product of the two corresponding characteristic functions. To show this, let us consider two independent random variables $X$ and $Y$ with the probability distribution $P_X(x)$ and $P_Y(y)$, respectively. As $X$ and $Y$ are independent, the joint probability distribution is given by $P_{X,Y}(x, y) = P_X(x)P_Y(y)$. For a random variable $Z(X, Y) = X + Y$, the probability distribution is then obtained as

$$\begin{aligned} P_Z(z) &= \int dxdy\, P_{X,Y}(x, y)\, \delta(Z(x, y) - z) \\ &= \int dxdy\, P_X(x)P_Y(y)\, \delta(Z(x, y) - z)\,. \end{aligned} \tag{1.41}$$

Using the Fourier integral representation of the Dirac delta function, namely,

$$\delta(Z(x, y) - z) = \frac{1}{2\pi} \int_{-\infty}^{\infty} dt\, e^{it(Z(x,y)-z)}\,, \tag{1.42}$$

we obtain from Eq. (1.41) that

$$\begin{aligned} P_Z(z) &= \frac{1}{2\pi} \int_{-\infty}^{\infty} dt\, e^{-itz} \int dx\, e^{itx} P_X(x) \int dy\, e^{ity} P_Y(y) \\ &= \frac{1}{2\pi} \int_{-\infty}^{\infty} dt\, e^{-itz}\, \phi_X(t)\, \phi_Y(t)\,. \end{aligned} \tag{1.43}$$

Now, using the above equation, one obtains the characteristic function $\phi_Z(t)$ of the random variable $Z$ as

$$\begin{aligned} \phi_Z(t) &= \int_{-\infty}^{\infty} dz\, e^{itz} P_Z(z) = \int_{-\infty}^{\infty} dt' \left( \frac{1}{2\pi} \int_{-\infty}^{\infty} dz\, e^{iz(t-t')} \right) \phi_X(t')\, \phi_Y(t') \\ &= \int_{-\infty}^{\infty} dt' \delta(t - t') \phi_X(t')\, \phi_Y(t') \\ &= \phi_X(t)\, \phi_Y(t)\,. \end{aligned} \tag{1.44}$$

The cumulants $\kappa_n$ of the probability distribution $P_X(x)$ are defined in terms of the cumulant-generating function $K(t)$ of $X$ as

$$K(t) \equiv \ln M(t) = \sum_{n=1}^{\infty} \kappa_n \frac{t^n}{n!}, \tag{1.45}$$

so that the $n$-th cumulant is obtained by differentiating the above expansion $n$ times and evaluating the result at zero:

$$\kappa_n = K^{(n)}(0); \quad n = 1, 2, \ldots . \tag{1.46}$$

The cumulants are related to the raw moments as

$$\kappa_1 = \mu_1', \ \kappa_2 = \mu_2' - (\mu_1')^2, \ \kappa_3 = 2(\mu_1')^3 - 3\mu_1'\mu_2' + \mu_3' . \tag{1.47}$$

The utility of the generating functions is that instead of computing each raw moment or each cumulant individually by performing separate integrals or sums, one has to compute the integral or the sum only once while computing the generating function. The latter then serves as the parent function, from which all the moments and all the cumulants can be obtained by taking suitable derivatives, which is a simpler task than performing multiple integrals and sums.

## 1.7 Examples of Probability Distributions

1. Discrete uniform distribution: Here, the discrete random variable $X$ takes on with equal probability integer values in an interval $[a, b]$, with $a$ and $b \geq a$ being integers. We have

$$P_X(k) = \frac{1}{(b - a + 1)}; \quad k \in [a, b] . \tag{1.48}$$

   The mean is $\mu = \frac{a+b}{2}$, while the variance is $\sigma^2 = \frac{(b-a+1)^2-1}{12}$.
2. Bernoulli distribution: Here, the discrete random variable $X$ takes values 1 or 0 with respective probability $p$ and $q = 1 - p$, with the real quantity $p$ satisfying $0 \leq p \leq 1$. Thus, we have

$$P_X(1) = p, \ P_X(0) = q . \tag{1.49}$$

   The mean is $\mu = p$, while the variance is $\sigma^2 = pq$.
3. Binomial distribution: Here, the discrete random variable $X$ equals the sum of $n$ independent Bernoulli trials, so, e.g., $X = k$ means the number of times that 1 occurs among the $n$ trials equals $k$. We thus have $k = 0, 1, 2, \ldots, n$, and

$$P_X(k) = \binom{n}{k} p^k q^{n-k}; \quad k = 0, 1, 2, \ldots, n . \tag{1.50}$$

The mean is $\mu = np$, while the variance is $\sigma^2 = npq$.

4. Poisson distribution: Here, the discrete random variable $X$ takes integer values $k = 0, 1, 2, \ldots$, with the probability distribution being

$$P_X(k) = \frac{\lambda^k e^{-\lambda}}{k!}; \quad k = 0, 1, 2, \ldots, \tag{1.51}$$

with the real parameter $\lambda > 0$ being both the mean and the variance: $\mu = \sigma^2 = \lambda$.

5. Discrete exponential distribution: Here, the discrete random variable $X$ takes integer values $k = 1, 2, \ldots$, with the probability distribution being

$$P_X(k) = r(1-r)^{k-1}; k = 1, 2, 3, \ldots, \tag{1.52}$$

with the real parameter $r$ satisfying $0 < r < 1$. The mean is $\mu = 1/r$, while the variance is $\sigma^2 = \frac{(1-r)}{r^2}$.

6. Discrete power-law distribution: Here, the discrete random variable $X$ takes integer values $k = 1, 2, \ldots$. The probability distribution has the form

$$P_X(k) = \frac{1}{\zeta(s)k^s}; \quad k = 1, 2, 3, \ldots, \tag{1.53}$$

with the real and positive quantity $s$ satisfying $s > 1$, and $\zeta(s) \equiv \sum_{n=1}^{\infty} 1/n^s$ being the Riemann zeta function. The mean is finite only for $s > 2$ and equals $\mu = \frac{\zeta(s-1)}{\zeta(s)}$, and is undefined otherwise. The variance on the other hand is finite only for $s > 3$, being given by $\sigma^2 = \frac{\left[\zeta(s)\zeta(s-2)-\zeta^2(s-1)\right]}{\zeta^2(s)}$, and is otherwise undefined.

7. Continuous uniform distribution: Here, the continuous random variable $X$ takes on with equal probability values in an interval $[a, b]$, with $a$ and $b$ being real quantities satisfying $-\infty < a < b < \infty$. We have

$$P_X(x) = \frac{1}{b-a}; \quad x \in [a, b], \tag{1.54}$$

and is zero otherwise. The mean is $\mu = (a+b)/2$, while the variance is $\sigma^2 = (b-a)^2/12$.

8. Exponential distribution: Here, the continuous random variable $X$ takes values in $[0, \infty)$. The probability distribution is

$$P_X(x) = \lambda e^{-\lambda x}; \quad x \in [0, \infty). \tag{1.55}$$

The parameter $\lambda > 0$ is called the rate parameter. The mean is $\mu = \frac{1}{\lambda}$, while the variance is $\sigma^2 = \frac{1}{\lambda^2}$.

9. Gaussian or Normal distribution: Here, the continuous random variable $x \in (-\infty, \infty)$ has the probability distribution

$$P_X(x) = \frac{1}{\sqrt{2\pi\sigma^2}} \exp\left(-\frac{(x-\mu)^2}{2\sigma^2}\right); \quad x \in (-\infty, \infty), \tag{1.56}$$

with the real parameter $\mu$ being the mean, and the real parameter $\sigma^2 > 0$ being the variance.

10. Cauchy or Lorentzian distribution: Here, the continuous random variable $x \in (-\infty, \infty)$ has the probability distribution

$$P_X(x) = \frac{1}{\pi\gamma\left[1 + \left(\frac{x-x_0}{\gamma}\right)^2\right]}; \quad x \in (-\infty, \infty), \tag{1.57}$$

with the real parameter $x_0$, called the location parameter, specifying the location of the peak of the distribution, while the real parameter $\gamma > 0$ called the scale parameter, specifies the half-width at half-maximum of the distribution. Both the mean and the variance are undefined.

11. Pareto (also called power-law) distribution: Here, the continuous random variable $x \in [x_{\text{ch}}, \infty)$ has the probability distribution

$$P_X(x) = \frac{\alpha}{x_{\text{ch}}\left(\frac{x}{x_{\text{ch}}}\right)^{\alpha+1}}; \quad x \in [x_{\text{ch}}, \infty), \tag{1.58}$$

with the real parameter $x_{\text{ch}} > 0$ being the lower cut-off, while $\alpha > 0$ is called the exponent of the power law. The mean is undefined for $\alpha \leq 1$ and is otherwise given by $\mu = \frac{\alpha x_{\text{ch}}}{\alpha-1}$. The variance is undefined for $\alpha \leq 2$ and is otherwise given by $\sigma^2 = \frac{x_{\text{ch}}^2 \alpha}{(\alpha-1)^2(\alpha-2)}$.

## 1.8 Law of Large Numbers and the Central Limit Theorem

In this section, we state without proof two central results related to sums of independent and identically-distributed random variables, the Law of Large Numbers and the Central Limit Theorem. For a more detailed discussion of the theorems as also for their derivation, the reader is referred to the classic text by Feller, Ref. [1].

Consider a set of $n$ independent and identically-distributed (i.i.d.) random variables $\{X_1, X_2, \ldots, X_n\}$. Each of them can take on values distributed according to a common probability distribution $P_X(x)$ with mean $\mu \equiv \int \mathrm{d}x\, x P_X(x)$. For example, a particular realization of the set $\{X_1, X_2, \ldots, X_n\}$ could be the set of values $\{x_1, x_2, \ldots, x_n\}$. Consider the sample average, given by

$$\overline{X}_n \equiv \frac{1}{n}\sum_{i=1}^{n} X_i, \tag{1.59}$$

which clearly takes on different values corresponding to different realizations of the set $\{X_1, X_2, \ldots, X_n\}$, and is thus itself a random variable. Since the $X_i$'s are independent random variables, it is easy to check using Eq. (1.59) and the defining equation of a mean, Eq. (1.21), that[1]:

$$\langle \overline{X}_n \rangle = \frac{1}{n} \sum_{i=1}^{n} \int \mathrm{d}x_1 \mathrm{d}x_2 \ldots \mathrm{d}x_n \; P_{X_1}(x_1) P_{X_2}(x_2) \ldots P_{X_n}(x_n) x_i = \mu \,, \tag{1.60}$$

where we have used the fact that $X_1, X_2, \ldots, X_n$ are i.i.d. random variables. The Law of Large Numbers states that the sample average $\overline{X}_n$ converges to its mean value $\langle \overline{X}_n \rangle$ as $n \to \infty$. Using Eq. (1.60), we therefore have

$$\overline{X}_n \to \mu \text{ as } n \to \infty \,. \tag{1.61}$$

There are two versions of the Law of Large Numbers. The Strong Law of Large Numbers states that the convergence in Eq. (1.61) happens almost surely, that is, we have

$$P\left( \lim_{n \to \infty} \overline{X}_n = \mu \right) = 1 \,. \tag{1.62}$$

On the other hand, the Weak Law of Large Numbers states that for any positive number $\epsilon$ however small, which may be interpreted as a measure of precision or tolerance, we have

$$\lim_{n \to \infty} P\left( |\overline{X}_n - \mu| < \epsilon \right) = 1 \,. \tag{1.63}$$

As is evident on comparing Eqs. (1.62) and (1.63), the difference between the strong and the weak version of the Law of Large Numbers is that while the former describes how a sequence of random variables behaves in the limit $n \to \infty$, the latter describes the convergence of a sequence of probabilities in the same limit. An important implication of the Law of Large Numbers is that while estimating the mean value of an observable in an experiment by repeating it a large number of times under identical conditions, the sample average of the results obtained from a large number of trials becomes closer to the mean value as more trials are performed.

While $\overline{X}_n \to \mu$ as $n \to \infty$, it is relevant to ask about the size and the distribution of the stochastic fluctuations $\overline{X}_n - \mu$ as $n \to \infty$. The Central Limit Theorem (CLT) states that provided the variance $\sigma^2 = \int \mathrm{d}x \, (x-\mu)^2 P_X(x)$ is finite: $\sigma^2 < \infty$, the distribution of the scaled random variable $\left(\overline{X}_n - \mu\right)/(\sigma/\sqrt{n})$ is a Gaussian distribution with mean equal to zero and variance equal to unity, as $n \to \infty$. Thus, one has a convergence in distribution of the probability of the random variable $\left(\overline{X}_n - \mu\right)/(\sigma/\sqrt{n})$ to a Gaussian distribution with zero mean and unit variance.

[1] In the case when the $X_i$'s are discrete random variables, one has to perform the appropriate summation instead of the integration in Eq. (1.60) to obtain $\langle \overline{X}_n \rangle$.

The remarkable and universal aspect of the theorem lies in the fact that the limiting distribution is a Gaussian even when the original variables $X_i$ themselves are not Gaussian-distributed. In mathematical terms, the CLT implies that for finite and real quantities $a$ and $b$, we have

$$\lim_{n\to\infty} P\left(a \leq \frac{\left(\overline{X}_n - \mu\right)}{(\sigma/\sqrt{n})} \leq b\right) = \int_a^b \mathrm{d}u\ \frac{1}{\sqrt{2\pi}} \mathrm{e}^{-u^2/2}. \tag{1.64}$$

The above equation implies that for $a, b$ fixed and finite, the probability that as $n \to \infty$, the random variable $\left(\overline{X}_n - \mu\right)/(\sigma/\sqrt{n})$ lies between $a$ and $b$ is given by a Gaussian distribution integrated between $a$ and $b$, even though the $X_i$'s that make up the quantity $\overline{X}_n$ are themselves not Gaussian distributed. It may be noted that when $n$ is finite, the probability that $\left(\overline{X}_n - \mu\right)/(\sigma/\sqrt{n})$ has a value smaller than $a$ or larger than $b$ is not given by a Gaussian distribution, but the point is that the weight of these non-Gaussian tails tends to zero as $n \to \infty$. In this sense, the CLT is concerned only with the central region, namely, values of $\overline{X}_n$ satisfying $\overline{X}_n - \mu \sim \mathcal{O}(1/\sqrt{n})$.

## 1.9 Stable Distributions

Random variables following an arbitrary distribution when added, the sum (also a random variable) follows in general a different distribution. However, for certain special distributions, called stable distributions, it just so happens that the distribution of the sum has exactly the same shape as the common distribution of the random variables whose sum is being considered [1].

Consider $X_1$ and $X_2$ to be independent realizations of a random variable $X$. Then $X$ is said to be stable if for any constants $a > 0$ and $b > 0$ the random variable $aX_1 + bX_2$ has the same distribution as $a_1 X + b_1$ for constants $a_1 > 0$ and $b_1$. If we have $b_1 = 0$, the distribution is said to be strictly stable. Stable distributions constitute a family of continuous probability distributions parametrized by four parameters, namely, the location parameter $\mu \in (-\infty, \infty)$, the scale parameter $c \in (0, \infty)$, and two shape parameters $\alpha \in (0, 2]$ and $\beta \in [-1, 1]$. While $\beta$, called the skewness parameter, gives a measure of asymmetry of the distribution, the parameter $\alpha$ is the so-called stability parameter. One cannot express analytically the probability distribution of a general stable distribution, although the corresponding characteristic function can be expressed analytically, as

$$\phi(t; \alpha, \beta, c, \mu) = \exp\left[\mathrm{i}t\mu - c^\alpha |t|^\alpha \left(1 - \mathrm{i}\beta\, \mathrm{sgn}(t)\, \Phi\right)\right], \tag{1.65}$$

where $\mathrm{sgn}(t)$ is the sign of $t$, and we have

$$\Phi \equiv \begin{cases} \tan\left(\frac{\pi\alpha}{2}\right); & \alpha \neq 1\,, \\ -\frac{2}{\pi}\log|t|; & \alpha = 1\,. \end{cases} \tag{1.66}$$

The corresponding stable distribution function is denoted by $P_X(x; \alpha, \beta, c, \mu)$, and is given on using Eq. (1.40) by

$$P_X(x; \alpha, \beta, c, \mu) = \frac{1}{2\pi}\int_{-\infty}^{\infty} \mathrm{d}t \; \mathrm{e}^{-itx}\phi(t; \alpha, \beta, c, \mu)\,. \tag{1.67}$$

It is evident from Eqs. (1.65) and (1.66) that

$$\phi(t; \alpha, -\beta, c, -\mu) = \phi(-t; \alpha, \beta, c, \mu)\,, \tag{1.68}$$

implying

$$P_X(-x; \alpha, -\beta, c, -\mu) = P_X(x; \alpha, \beta, c, \mu)\,. \tag{1.69}$$

That the characteristic function (1.65) yields a stable distribution may be appreciated from the fact that the characteristic function for the sum of two independent random variables is given by the product of the two corresponding characteristic functions, as shown in Eq. (1.44). From Eq. (1.65), we see that with $\beta = 0$, the characteristic function is a stretched exponential function, and then one has

$$P_X(x; \alpha, 0, c, \mu) = \frac{1}{2\pi}\int_{-\infty}^{\infty} \mathrm{d}t \; \mathrm{e}^{-it(x-\mu)-c^\alpha|t|^\alpha}\,. \tag{1.70}$$

The right hand side of the above equation is invariant under the transformation $(x - \mu) \to -(x - \mu)$. As a result, the distribution $P_X(x; \alpha, \beta, c, \mu)$ with $\beta = 0$ is symmetric about $\mu$ and is referred to as a (Lévy) symmetric alpha-stable distribution.

For $\alpha < 1$, the distribution $P_X(x; \alpha, \beta, c, \mu)$ is supported on $[\mu, \infty)$ and $(-\infty, \mu]$ for $\beta = 1$ and $\beta = -1$, respectively. For all other cases, $P_X(x; \alpha, \beta, c, \mu)$ is supported on $(-\infty, \infty)$. We give below analytical closed-form expression for the distribution function $P_X(x; \alpha, \beta, c, \mu)$ for a few cases of interest, corresponding to specific values of $\alpha$ and $\beta$.

1. Gaussian distribution ($\alpha = 2$ and $\beta$ any value in the range $[-1, 1]$): From Eqs. (1.65) and (1.66), it is clear that in this case with $\alpha = 2$ implying $\Phi = 0$, the skewness parameter $\beta$ has no effect, and one may therefore set $\beta = 0$. We then have

$$\phi(t; 2, 0, c, \mu) = \exp\left[it\mu - c^2t^2\right]\,, \tag{1.71}$$

$$P_X(x; 2, 0, c, \mu) = \frac{1}{\sqrt{4\pi c^2}}\exp\left[-\frac{(x-\mu)^2}{4c^2}\right]; \quad -\infty < x < \infty\,. \tag{1.72}$$

Note that in this case, the location parameter $\mu$ coincides with the mean of the distribution, while the scale parameter $c$ is related to the standard deviation $\sigma_X$ by $c = \sigma_X/\sqrt{2}$.

2. Cauchy or Lorentz distribution ($\alpha = 1$ and $\beta = 0$): In this case, we have

$$\phi(t; 1, 0, c, \mu) = \exp\left[it\mu - c|t|\right] , \tag{1.73}$$

$$P_X(x; 1, 0, c, \mu) = \frac{c}{\pi\left[(x-\mu)^2 + c^2\right]}; \quad -\infty < x < \infty . \tag{1.74}$$

Here, the parameter $\mu$ coincides with the location of the peak of the distribution, and the parameter $c$ specifies the half-width at half-maximum of the distribution.

3. Lévy distribution ($\alpha = 1/2$ and $\beta = 1$): In this case, we have

$$\phi(t; 1/2, 1, c, \mu) = \exp\left[it\mu - \sqrt{-2ict}\right] , \tag{1.75}$$

$$P_X(x; 1/2, 1, c, \mu) = \sqrt{\frac{c}{2\pi}} \frac{1}{(x-\mu)^{3/2}} \exp\left[-\frac{c}{2(x-\mu)}\right]; \quad \mu < x < \infty , \tag{1.76}$$

where in obtaining Eq. (1.75) we have used $\operatorname{sgn}(t) = t/|t|$ and $(|t| - it)^2 = -2it|t|$. Note that here the location parameter $\mu$ sets the lower limit of the support of the distribution. The mode of the distribution is related to both $\mu$ and $c$, and is given by $\mu + c/3$.

Given $P_X(x; \alpha, \beta, c, \mu)$ as the probability distribution of $X$, and with $Y$ defined as the sum of $n$ independent realizations of $X$:

$$Y \equiv \sum_{i=1}^{n} k_i (X_i - \mu) , \tag{1.77}$$

it follows that the probability distribution of $Y$ is also a stable one, for which the characteristic function is given by $\phi(t; \alpha, \overline{\beta}, \overline{c}, \overline{\mu})$ (see Eq. (1.65)) with [3]

$$\overline{\beta} \equiv \beta \frac{\sum_{j=1}^{n} |k_j|^\alpha \operatorname{sgn}(k_j)}{\sum_{j=1}^{n} |k_j|^\alpha} , \qquad \overline{c} \equiv c \left(\sum_{j=1}^{n} |k_j|^\alpha\right)^{1/\alpha} ,$$
$$\overline{\mu} \equiv \begin{cases} 0; & \alpha \neq 1 , \\ -\frac{2\beta c}{\pi} \sum_{j=1}^{n} k_j \log|k_j| ; & \alpha = 1 . \end{cases} \tag{1.78}$$

When $X_i$'s are sampled from a Gaussian distribution ($\alpha = 2$, $\beta = 0$) with mean $\mu_X$ and variance $\sigma_X^2$, one has the location parameter $\mu = \mu_X$ and the scale parameter $c = \sigma_X/\sqrt{2}$. In this case, we see from Eq. (1.78) that

$$\overline{\beta} = 0, \quad \overline{\mu} = 0, \quad \overline{c} = c\sqrt{\sum_{j=i}^{n} |k_j|^2}, \tag{1.79}$$

which implies that $Y$ is also a Gaussian random variable ($\alpha = 2, \overline{\beta} = 0$) with mean $\mu_Y = \overline{\mu} = 0$ and variance $\sigma_Y^2 = 2\,\overline{c}^2 = \sigma_X^2 \sum_{j=1}^n |k_j|^2$. In addition, for the particular case with $k_i = 1/n \; \forall \, i$, we see that the random variable $Y = \frac{1}{n}\sum_{j=1}^n X_i - \mu_X$ is distributed as a Gaussian with zero mean and variance $\sigma_X^2/n$. In other words, we may say that in this case, the scaled random variable $Y/(\sigma_X/\sqrt{n})$ is distributed as a Gaussian with zero mean and unit variance, which is consistent with the CLT discussed in Sect. 1.8.

Let us recall that the CLT discussed previously states that for sample average $\overline{X}_n$ of a set of $n$ i.i.d. random variables $\{X_1, X_2, \ldots, X_n\}$, see Eq. (1.59), the scaled random variable $\left(\overline{X}_n - \mu\right)/(\sigma/\sqrt{n})$ is distributed as a Gaussian with zero mean and unit variance as $n \to \infty$. This result holds independently of the form of the distribution $P_X(x)$ of the $X_i$'s, so long as the variance $\sigma^2 = \int \mathrm{d}x \; (x - \mu)^2 P_X(x)$ is finite, and in this sense, the Gaussian distribution is an attractor in the space of all possible $P_X(x)$'s with finite variance. We have also seen above that the Gaussian distribution is a stable distribution, and in fact, it is a stable distribution that has all its moments finite. One may ask: Are there non-Gaussian stable distributions that are also attractors in a similar sense?

### *1.9.1 Generalized Central Limit Theorem*

A generalization of the CLT due to Gnedenko and Kolmogorov [1, 4–6] states that provided we have

$$P_X(x) \sim \begin{cases} A_-|x|^{-(1+\alpha)} & \text{as } x \to -\infty, \\ A_+|x|^{-(1+\alpha)} & \text{as } x \to \infty, \end{cases} \tag{1.80}$$

then, for $\alpha \in (0, 2]$, the quantity $\widetilde{X}_n/n^{1/\alpha} \equiv (\sum_{i=1}^n X_i)/n^{1/\alpha}$ in the limit $n \to \infty$ converges in probability to a stable distribution with index $\alpha$ and skewness parameter $\beta \in [-1, 1]$ given by

$$\beta = \frac{A_+ - A_-}{A_+ + A_-}. \tag{1.81}$$

In the light of Eq. (1.80), it then follows that:

1. For $\alpha = 2$, when the variance $\sigma^2$ is infinite, the quantity $\widetilde{X}_n/\sqrt{n}$ converges in probability to a stable distribution with index $\alpha = 2$, which is nothing but a Gaussian (note that for a Gaussian, the actual value that $\beta$ takes in the range $[-1, 1]$ does not matter, see the discussion around Eq. (1.72));

2. For $\alpha > 2$, when the variance $\sigma^2$ is finite, the CLT implies that the quantity $\widetilde{X}_n/\sqrt{n}$ converges in probability to a Gaussian distribution.

Given that $\alpha$ can have an infinite number of possible values in the range $(0, 2]$, yielding an infinite number of stable distributions, it follows from above that each one of them is an attractor in the space of distribution functions $P_X(x)$.

In Eq. (1.80), we have taken the "left" and the "right" exponent to be equal. When the two exponents are unequal and are given by $\alpha_-$ and $\alpha_+$, respectively (with $\alpha_\pm \in (0, 2]$), then the smallest of the two wins, in that the quantity $\widetilde{X}_n/n^{1/\alpha}$ converges in probability to a stable distribution with index $\alpha = \min(\alpha_-, \alpha_+)$ and skewness parameter $\beta$ that is either equal to $-1$ if $\alpha_- < \alpha_+$, or, is equal to $+1$ if $\alpha_+ < \alpha_-$.

In Fig. 1.2, we show the Generalized CLT at work for three illustrative examples.

## 1.10 Extreme Value Theory

Consider the set $\{X_i\} \equiv \{X_1, X_2, \ldots, X_n\}$ of $n$ i.i.d. random variables. Let $P_X(x)$ denote the underlying common probability distribution, and let

$$M_n \equiv \max(X_1, X_2, \ldots, X_n) \tag{1.82}$$

denote the maximum of the given sequence. Clearly, $M_n$ is also a random variable, whose value varies between different realizations of the set $\{X_i\}$. The Extreme Value Theory (EVT) concerns itself with obtaining the probability distribution of $M_n$. The reader is referred to Refs. [7, 8] for a recent overview of the theory and its many facets. For the cumulative distribution of $M_n$, we have

$$\begin{aligned} Q_n(y) &\equiv P(M_n \le y) \\ &= P(X_1 \le y, X_2 \le y, \ldots, X_n \le y) \\ &= \int_{-\infty}^{y} \mathrm{d}x_1 \int_{-\infty}^{y} \mathrm{d}x_2 \ldots \int_{-\infty}^{y} \mathrm{d}x_n \; P_{X_1,X_2,\ldots,X_n}(x_1, x_2, \ldots, x_n), \end{aligned} \tag{1.83}$$

since indeed for arbitrary $y$, the event $M_n \le y$ corresponds to the joint probability that each of the $X_i$'s has a value less than or equal to $y$, and is thus given in terms of the joint probability distribution $P_{X_1,X_2,\ldots,X_n}(x_1, x_2, \ldots, x_n)$. Next, noting that the $X_i$'s are independent random variables, we have, on using Eq. (1.15), that

$$P_{X_1,X_2,\ldots,X_n}(x_1, x_2, \ldots, x_n) = \prod_{i=1}^{n} P_{X_i}(x_i), \tag{1.84}$$

thus yielding

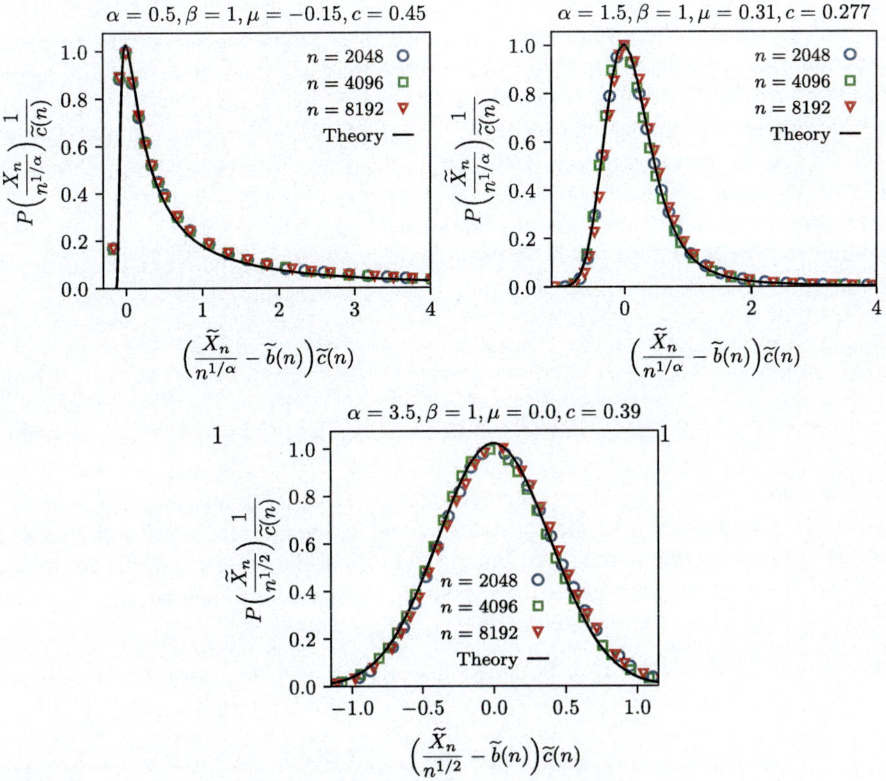


**Fig. 1.2 For $n$ random variables $X_i$ sampled independently from the distribution $P_X(x) = \alpha/[x_0\ (x/x_0)^{1+\alpha}]$, with $x \in [x_0, \infty)$, and with $\alpha = 0.5$ and 1.5, the first two panels show convergence in distribution of the quantity $\widetilde{X}_n/n^{1/\alpha}$; $\widetilde{X}_n \equiv \sum_{i=1}^{n} X_i$, to a stable distribution characterized by the four indices $\alpha, \beta, c, \mu$. Here, we have taken $x_0 = 1.0$. The convergence is in accordance with the generalized Central Limit Theorem. The third panel shows for $\alpha = 3.5$ the convergence of the quantity $\widetilde{X}_n/\sqrt{n}$ to a Gaussian distribution, in accordance with the Central Limit Theorem.** In all the panels, the numerically-obtained distributions have been suitably scaled by the quantities $\widetilde{b}(n)$ and $\widetilde{c}(n)$ that refer respectively to the value at which the unscaled distributed is peaked and the corresponding peak value of the distribution

$$Q_n(y) = \left(\int_{-\infty}^{y} \mathrm{d}x\ P_X(x)\right)^n = \left(1 - \int_{y}^{\infty} \mathrm{d}x\ P_X(x)\right)^n. \tag{1.85}$$

Once $Q_n(y)$ is computed using Eq. (1.85), the probability distribution of $M_n$ is obtained as

$$P_{M_n}(M) = \frac{\mathrm{d}Q_n(y)}{\mathrm{d}y}\bigg|_{y=M}. \tag{1.86}$$

The precise form of the probability distribution of $M_n$ then depends on the specific form of the parent distribution $P_X(x)$. Yet, quite remarkably, universal results emerge for $Q_n(y)$ and hence for $P_{M_n}(M)$ when one considers the limit $n \to \infty$. A parallel may be drawn with the sample mean $\overline{X}_n$ in Eq. (1.59) for which we have seen in Sect. 1.8 that universal results in terms of convergence to a Gaussian distribution emerges by virtue of the CLT, when one considers the limit $n \to \infty$.

Let us now summarize the mentioned universal results regarding the form of the cumulative distribution $Q_n(y)$. When both $n$ and $y$ are large, there exist sequences of constants $a_n > 0$ and real $b_n$, such that

$$Q_n(y) \xrightarrow[z \equiv (y-a_n)/b_n \text{ fixed and finite}]{y \to \infty,\, n \to \infty} F\left(\frac{y-a_n}{b_n}\right), \tag{1.87}$$

$$\text{i.e., } \lim_{n \to \infty} Q_n(a_n + b_n z) = F(z)\,. \tag{1.88}$$

In other words, $y$ when suitably-scaled converges in distribution to one that is independent of $n$, namely, $F(z)$. The constants $a_n$ and $b_n$ are non-universal in the sense that they depend on the parent distribution $P_X(x)$. On the other hand, the limiting distribution $F(z)$ can only be of three possible forms determined by the large-$x$ behavior of $P_X(x)$, as summarized below:

- Weibull law: When $P_X(x)$ has bounded tails: $P_X(x) \sim A\beta(x_0 - x)^{\beta-1}$; $\beta > 0$ as $x \to x_0^-$ and $P_X(x) = 0$ for $x > x_0$, one obtains from Eq. (1.85) as $y \to x_0^-$ that

$$Q_n(y) = \left[1 - A\,(x_0 - y)^{\beta}\right]^n\,. \tag{1.89}$$

  Using the transformation $y = a_n + b_n z$ with $a_n = x_0$ and $b_n = (An)^{-1/\beta}$, one gets $Q_n(y(z)) = (1 - (-z)^{\beta}/n)^n$, and then obtains from Eq. (1.87) that

$$F(z) = \begin{cases} e^{-(-z)^{\beta}}; & z < 0\,, \\ 1; & z \geq 0\,, \end{cases} \tag{1.90}$$

  which is the Weibull cumulative probability. Hence, the Weibull probability distribution is given by

$$F'(z) = \begin{cases} \beta(-z)^{\beta-1} e^{-(-z)^{\beta}}; & z < 0\,, \\ 0; & z \geq 0\,. \end{cases} \tag{1.91}$$

- Fréchet law: When $P_X(x)$ has a power-law decay: $P_X(x) \sim A\alpha x^{-(1+\alpha)}$; $\alpha > 0$ for large $x$, one obtains from Eq. (1.85) that

$$Q_n(y) = \left[1 - Ay^{-\alpha}\right]^n\,. \tag{1.92}$$

  Using the transformation $y = a_n + b_n z$ with $a_n = 0$ and $b_n = (An)^{1/\alpha}$, one gets $Q_n(y(z)) = (1 - z^{-\alpha}/n)^n$, and then obtains from Eq. (1.87) that

$$F(z) = \begin{cases} e^{-z^{-\alpha}}; & z > 0, \\ 0; & z \le 0, \end{cases} \tag{1.93}$$

which is the Fréchet cumulative probability. Hence, the Fréchet probability distribution is given by

$$F'(z) = \begin{cases} \alpha z^{-1-\alpha} e^{-z^{-\alpha}}; & z > 0, \\ 0 & z \le 0. \end{cases} \tag{1.94}$$

- Gumbel law: When $P_X(x)$ has unbounded tails that decay faster than a power law: $P_X(x) \sim A e^{-x^\delta}$; $\delta > 0$ for large $x$, one obtains from Eq. (1.85) that

$$Q_n(y) = \left[1 - \frac{A}{\delta}\Gamma\left(\frac{1}{\delta}, y^\delta\right)\right]^n = \exp\left[n \log\left\{1 - \frac{A}{\delta}\Gamma\left(\frac{1}{\delta}, y^\delta\right)\right\}\right], \tag{1.95}$$

where $\Gamma(a, z)$ is the incomplete Gamma function. For large $y$, using the known asymptotic behavior $\Gamma(s, x)/(x^{s-1} e^{-x}) \to 1$ as $x \to \infty$, Eq. (1.95) may be reduced to

$$Q_n(y) \approx \exp\left[n \log\left\{1 - A\frac{e^{-y^\delta}}{\delta y^{\delta-1}}\right\}\right] \approx \exp\left(-nA\frac{e^{-y^\delta}}{\delta y^{\delta-1}}\right) \approx \exp\left(-\frac{nA}{\delta} e^{-y^\delta}\right), \tag{1.96}$$

which using $y = a_n + b_n z$ yields

$$Q_n(y(z)) = \exp\left(-\frac{nA}{\delta} e^{-(a_n + b_n z)^\delta}\right). \tag{1.97}$$

Let us anticipate that $a_n \gg b_n$ (which we will eventually verify below), so that to leading order in $b_n/a_n$, we have

$$Q_n(y(z)) = \exp\left(-\frac{nA}{\delta} e^{-a_n^\delta - \delta a_n^{\delta-1} b_n z}\right). \tag{1.98}$$

Let us choose

$$\frac{nA e^{-a_n^\delta}}{\delta} = 1, \quad \text{and} \quad \delta a_n^{\delta-1} b_n = 1, \tag{1.99}$$

which yield on account of $n$ being large that $a_n = (\ln n)^{1/\delta}$ and $b_n = (1/\delta)(\ln n)^{1/\delta-1}$, verifying what we had anticipated earlier: $a_n \gg b_n$. With these definitions of $a_n$ and $b_n$, using Eqs. (1.98) and (1.87) gives

$$Q_n(y(z)) \approx e^{-e^{-z}} = F(z), \tag{1.100}$$

which is the Gumbel cumulative probability. Here, one has $-\infty < z < \infty$. The Gumbel probability distribution is given by

$$F'(z) = e^{-z-e^{-z}}; \quad z \in (-\infty, \infty). \tag{1.101}$$

Further comments may be made on the scaling of the maximum with $n$, by using Eq. (1.85). Large enough $y$ near the tail implying small value of the integral $\int_y^\infty dx\ P_X(x)$, we obtain on using Eq. (1.85) that

$$Q_n(y) \approx \exp(-n \int_y^\infty dx\ P_X(x)), \tag{1.102}$$

which for the Weibull and the Fréchet class is given by

$$Q_n(y) \approx \exp\left(-nA(x_0 - y)^\beta\right), \tag{1.103}$$

and

$$Q_n(y) \approx \exp\left(-nAy^{-\alpha}\right), \tag{1.104}$$

respectively. As the cumulative probability $Q_n(y)$ is of order one, the scaling with $n$ of the maximum for the Weibull and the Fréchet class is set by

$$(x_0 - y) \sim (nA)^{-1/\beta}, \quad \text{and} \quad y \sim (nA)^{1/\alpha}, \tag{1.105}$$

respectively. Since for the Weibull case the underlying distribution $P_X(x)$ is bounded from above, what the first result tells us is that with the increase of $n$, $y$ gets closer to the bound $x_0$. On the other hand, in the case of Fréchet, the distribution $P_X(x)$ being unbounded from above, the second result implies that $y \to \infty$ as $n \to \infty$. For the Gumbel case, one obtains from Eq. (1.96) that the scaling with $n$ of the maximum is given by

$$y \sim (\ln n)^{1/\delta}. \tag{1.106}$$

## 1.11 An Application: The Random Walk Problem in One Dimension

Here, as an illustration of the concepts developed in the preceding sections, we discuss the paradigmatic example of a random walker in one dimension [9]. A particle (the random walker) is moving on a one-dimensional lattice of lattice spacing equal to $a = 1$ and in discrete times $i = 0, 1, 2, \ldots$ (the time interval between two consecutive time steps is $\Delta t = 1$). At every time step, the particle tosses a coin afresh, with no memory of outcome of previous tosses. If the coin falls on the heads (respectively, on the tails), the particle moves one step to the right (respectively, one step to the left). The coin is a fair one, which implies that one has equal probability of $1/2$ for the coin to fall on the heads and on the tails. The particle starts from the center of the lattice, which is the origin, with the lattice extending from $-\infty$ to $\infty$.

### *1.11.1 Site Occupation Probability*

Let us now ask the question: what is the probability $W(m, N)$ that the particle is at position $m$ (i.e., at a distance $ma$ from the origin) at the $N$-th step? The particle is initially at the origin, that is, one has

$$W(m, 0) = \delta_{m,0} \,. \tag{1.107}$$

Clearly, while considering $W(m, N)$, we have $m \in [-N, N]$. Since the particle is equally likely to move to the left and to the right, the problem has a left-right symmetry. Therefore, we may without loss of generality choose $m > 0$. Let $N_\pm$ be the number of steps taken respectively to the right and to the left in reaching position $m$ in $N$ steps, in a given realization of the dynamics. Then, we have $N_+ + N_- = N$ and $N_+ - N_- = m$. Solving, one gets $N_+ = (N + m)/2$ and $N_- = (N - m)/2$. The probability of occurrence of such a realization is $(1/2)^{N_+}(1/2)^{N-N_+}$, and the number of such realizations is just $\binom{N}{N_+}$ (the different realizations differing only in the sequence in which left and right steps are taken). Since all such realizations happen as mutually-exclusive events, the probabilities of the different realizations would add up. Consequently, we obtain $W(m, N)$ as

$$W(m, N) = \binom{N}{N_+}\left(\frac{1}{2}\right)^{N_+}\left(\frac{1}{2}\right)^{N-N_+} = \frac{1}{2^N}\binom{N}{(N+m)/2} \,. \tag{1.108}$$

Note that $\sum_{m=-N}^{N} W(m, N) = \sum_{N_+=0}^{N} \binom{N}{N_+}\left(\frac{1}{2}\right)^{N_+}\left(\frac{1}{2}\right)^{N-N_+} = (1/2 + 1/2)^N = 1$, as required. One may write down how the probability distribution $W(m, N)$ updates in time. Noting that this probability gets contributions only from moves from nearest-neighbor sites, we have

$$W(m, N+1) = \frac{1}{2}W(m-1, N) + \frac{1}{2}W(m+1, N) \,. \tag{1.109}$$

The next question to ask is: what is the average displacement from the initial location at the $N$-th step, given by $\langle m\rangle_N = \sum_{m=-N}^{N} m W(m, N)$? Using $W(m, N) = W(-m, N)$ immediately implies that $\langle m\rangle_N = 0$, which is just a reflection of the fact that the random walker is at every time step equally likely to move to the left and to the right. What then is the mean squared displacement (MSD) from the initial location at the $N$-th step, given by the variance $\sigma_N^2 \equiv \langle m^2 \rangle_N = \sum_{m=-N}^{N} m^2 W(m, N)$? In order to obtain the variance, let us employ the technique of moment generating function discussed in Sect. 1.6, and compute the moment generating function at the $N$-th step:[2]

$$M(k, N) \equiv \langle \mathrm{e}^{mk} \rangle_N = \sum_{m=-N}^{N} \mathrm{e}^{mk} W(m, N) . \tag{1.110}$$

Using Eq. (1.109), we find that $M(k, N)$ satisfies

$$M(k, N+1) = \frac{1}{2}\mathrm{e}^{k} M(k, N) + \frac{1}{2}\mathrm{e}^{-k} M(k, N) . \tag{1.111}$$

We need to now compute the function $M(k, N)$ by solving Eq. (1.111). To this end, consider the function

$$\widetilde{M}(k, z) \equiv \sum_{N=0}^{\infty} M(k, N) z^N , \tag{1.112}$$

which using Eq. (1.111) is seen to satisfy

$$\widetilde{M}(k, z) - \mathrm{e}^{mk} W(m, 0) = z \cos k \, \widetilde{M}(k, z) . \tag{1.113}$$

Using the initial condition, Eq. (1.107), yields

$$\widetilde{M}(k, z) = \frac{1}{1 - z \cos k} = \sum_{N=0}^{\infty} [z \cos k]^N . \tag{1.114}$$

On comparing the last equation with Eq. (1.112), we arrive at our desired solution:

$$M(k, N) = (\cos k)^N . \tag{1.115}$$

Now that we have obtained the moment-generating function $M(k, N)$, we obtain the raw moments $\mu'_n \equiv \sum_{m=-N}^{N} m^n W(m, N)$ as

$$\mu'_n = \left. \frac{\partial^n M(k, N)}{\partial k^n} \right|_{k=0} . \tag{1.116}$$

In particular, we obtain the average displacement in $N$ steps:

$$\langle m \rangle_N = 0 , \tag{1.117}$$

as obtained earlier. On the other hand, the variance $\sigma_N^2$ is obtained as

[2] Note that we are using here the variable $k$ to stand for $t$ in Eq. (1.37).

$$\sigma_N^2 = N\,. \tag{1.118}$$

We thus conclude that the MSD scales linearly with $N$.

Consider the case of very large $N$ and $m \ll N$. Using Stirling's formula $\ln n! = (n + 1/2)\ln n - n + (1/2)\ln(2\pi) + \mathcal{O}(1/n)$ as $n \to \infty$, one obtains

$$\begin{aligned}\ln W(m, N) \simeq \left(N + \frac{1}{2}\right)\ln N - \frac{1}{2}(N + m + 1)\ln\left[\frac{N}{2}\left(1 + \frac{m}{N}\right)\right] \\ - \frac{1}{2}(N - m + 1)\ln\left[\frac{N}{2}\left(1 - \frac{m}{N}\right)\right] - \frac{1}{2}\ln(2\pi) - N\ln 2\,.\end{aligned} \tag{1.119}$$

Further, since we have $m \ll N$, using $\ln(1 \pm m/N) = \pm(m/N) - m^2/(2N^2) + \mathcal{O}(m^3/N^3)$, we obtain

$$W(m, N) \approx \sqrt{\frac{2}{\pi N}}\exp\left(-\frac{m^2}{2N}\right);\quad N \text{ large}, m \ll N\,. \tag{1.120}$$

Note that the expression on the right is not normalized to unity, since it is valid only for values of $m \ll N$ and not for the full range of values of $m$, namely, $m \in [-N, N]$.

Let us now move on to discuss the continuous space-time limit of the random walk problem. Denoting by $x \equiv ma$ the displacement of the particle at time $t \equiv N\Delta t$ from the origin, the continuous space-time limit of the random walk corresponds to taking the simultaneous limits $a \to 0$, $\Delta t \to 0$, $N \to \infty$, in such a way that the quantities $x, t$, and $D \equiv a^2/(2\Delta t)$ remain fixed and finite. In this limit, the probability density $\widetilde{W}(x, t)$ of finding the walker within a small range $[x, x + \Delta x]$ at time $t$ is evidently given by

$$\widetilde{W}(x, t)\Delta x = \sum_{x \le ma \le x+\Delta x} W(m, N)\,. \tag{1.121}$$

Noting that $\Delta x$ is small, one may rewrite the above equation as

$$\widetilde{W}(x, t)\Delta x = W(m, N)\frac{\Delta x}{2a}\,, \tag{1.122}$$

where the factor of two on the right hand side arises due to the fact that only half of the sites within $\Delta x$ are accessible. Indeed, depending on whether $N$ is even or odd, only the even and the odd sites within $\Delta x$, respectively, are accessible by the random walker. We remark that to keep $D$ fixed and finite, one needs to take the limits $a \to 0$ and $\Delta t \to 0$ in a way that the quantities $a^2$ and $\Delta t$ approach zero in a similar fashion, i.e., $a$ approaches zero much slowly than $\Delta t$. Now, to keep both $x$ and $t$ fixed and finite, one has to have the scaling $m \sim 1/a$ and $N \sim 1/\Delta t$, so given the slower approach of $a$ to zero with respect to $\Delta t$, we necessarily have $m \ll N$. Then, we may use Eq. (1.120) in Eq. (1.122), to obtain

$$\widetilde{W}(x,t) = \frac{1}{\sqrt{4\pi Dt}} \exp\left(-\frac{x^2}{4Dt}\right). \tag{1.123}$$

From Eqs. (1.121) and (1.123), we may write

$$\sum_{x_1=m_1a\le ma\le x_2=m_2a} W(m,N) = \int_{x_1}^{x_2} \mathrm{d}x\, \widetilde{W}(x,t) = \int_{x_1}^{x_2} \frac{\mathrm{d}x}{\sqrt{4\pi Dt}} \exp\left(-\frac{x^2}{4Dt}\right). \tag{1.124}$$

To show the emergence of the CLT in the context of the random walk problem, we now demonstrate how Eq. (1.124) may be expressed in a form similar to Eq. (1.64). Let $\Delta_i$; $i = 1, 2, 3, \ldots$ denote the random variable for the outcome of the coin toss at the $i$-th time step with $\Delta_i = 1$ for the heads and $\Delta_i = -1$ for the tails. Clearly, we have $\Delta_i = \pm 1$ with equal probability of $1/2$. Since coin toss at every step is performed independently, $\Delta_i\ \forall\, i$ is a random variable sampled independently from a discrete probability distribution

$$P_{\Delta_i}(+1) = \frac{1}{2}, \quad P_{\Delta_i}(-1) = \frac{1}{2}. \tag{1.125}$$

It then follows that the mean and variance of the random variable $\Delta_i$ are

$$\mu = \langle \Delta_i \rangle = 0, \quad \sigma^2 = \langle \Delta_i^2 \rangle = 1, \tag{1.126}$$

respectively. From the dynamical rules of evolution of the random walker, we may express the displacement $m$ from the origin at the $N$-th time step in terms of $\Delta_i$ as

$$m = \sum_{i=1}^{N} \Delta_i; \quad i = 1, 2, 3, \ldots. \tag{1.127}$$

Thus, $m$ is a sum of independent and identically-distributed random variables $\Delta_i$ with a finite variance, providing conditions ideal for an application of the CLT. To this end, let us first define the random variable

$$\overline{m} \equiv \frac{1}{N} \sum_{i=1}^{N} \Delta_i = \frac{m}{N}. \tag{1.128}$$

In the aforementioned continuous space-time limit of the random walk, one may write $(\overline{m} - \mu)/(\sigma/\sqrt{N}) = \sqrt{N}\,\overline{m} = m/\sqrt{N} = x/\sqrt{2Dt}$. In this limit, using the one-to-one mapping between the random variables $m$ and $m/\sqrt{N}$, we get from Eq. (1.124) that

$$P\left(A \leq \frac{(\overline{m} - \mu)}{(\sigma/\sqrt{N})} \leq B\right) = \int_A^B \mathrm{d}u \, \frac{1}{\sqrt{2\pi}} \mathrm{e}^{-u^2/2}, \tag{1.129}$$

with

$$A \equiv \frac{x_1}{\sqrt{2Dt}}, \quad B \equiv \frac{x_2}{\sqrt{2Dt}}. \tag{1.130}$$

Comparing Eq. (1.129) with Eq. (1.64), we thus see the emergence of the CLT in the continuous space-time limit of the random walk. It may be noted that the latter limit of the random walk defines the paradigmatic Brownian motion, which we will discuss in detail in Chap. 2.

## References


1. W. Feller, *An Introduction to Probability Theory and its Applications. Vol. 1 and 2* (Wiley, New York, 2009)
2. A. Papoulis, S.U. Pillai, *Probability, Random Variables, and Stochastic Processes*, 4th edn. (McGraw-Hill, Boston, 2002)
3. J.P. Nolan, Univariate Stable Distributions: Models for Heavy Tailed Data (Springer. Cham (2020). https://doi.org/10.1007/978-3-030-52915-4
4. B.V. Gnedenko, A.N. Kolmogorov, *Limit Distributions for Sums of Independent Random Variables* (Addison-Wesley, Massachusetts, 1954)
5. J.P. Bouchaud, M. Potters, *Theory of Financial Risk and Derivative Pricing: From Statistical Physics to Risk Management*, 2nd edn. (Cambridge University Press, Cambridge, 2003). 10.1017/CBO9780511753893
6. H.E. Stanley, *Introduction to Phase Transitions and Critical Phenomena* (Oxford University Press, New York, 1987)
7. S.N. Majumdar, A. Pal, G. Schehr, Phys. Rep. **840**, 1 (2020). https://doi.org/10.1016/j.physrep.2019.10.005
8. A. Hansen, Front. Phys. **8**, 604053 (2020). https://doi.org/10.3389/fphy.2020.604053
9. S. Chandrasekhar, Rev. Mod. Phys. **15**(1), 1 (1943). https://doi.org/10.1103/RevModPhys.15.1

# Chapter 2
# Stochastic Processes

**Abstract** In this chapter, we lay the foundation of stochastic processes. We start with a general definition of a stochastic process, and then specialize to its many variants: stationary, homogeneous, and Markovian stochastic processes. We next discuss stationary Markov processes, and how they give rise to homogeneous Markov processes. For the latter, we introduce the notion of stationary state and condition of detailed balance. Next, we discuss how one studies analytically a homogeneous Markov process via the so-called Fokker-Planck equation for the conditional probability distribution of the relevant random variables. As an example of the formalism developed, we discuss the case of the ubiquitous Brownian motion, focusing in particular on two complementary approaches à la Langevin and Einstein to characterize the motion. While the approach due to Langevin relies on a phenomenological stochastic differential equation of motion for the Brownian particle, the one due to Einstein is instead based on a differential equation for the probability distribution of the velocity of the Brownian particle. In the following two sections, we discuss generalizations of the Langevin equation, and its derivation from first principles for a model system.

## 2.1 General Definition

The content of this chapter is based on materials found in standard references on stochastic processes, e.g., Refs. [1–7]. The reader is particularly encouraged to consult Ref. [8], which collects together six classic papers on stochastic processes.

We start with defining a stochastic process. By the word process, one generally means a series of changes in time. In the context of a physical system, a process involves dynamical evolution of configurations of the system while starting from an initial configuration. We all are familiar with deterministic processes, the most prominent example of which is perhaps a process evolving according to the Newton's laws of motion. In a deterministic process, a given initial configuration evolves to yield a unique configuration at a given instant of time. By contrast, a stochastic process involves a dynamical evolution that for a given initial configuration may yield many different configurations with varying probabilities at a given instant of time.



D. Das and S. Gupta, *Facets of Noise*, Fundamental Theories of Physics 214,
https://doi.org/10.1007/978-3-031-45312-0_2

In order to fix our ideas, let us first consider a discrete-time stochastic process, wherein time is labeled by the index $t$ that takes on integer values: $t = 0, 1, 2, \ldots$ (so the time step is $\Delta t = 1$), and which involves a finite number of configurations denoted by $\mathcal{M}$. The configuration of the system at time $t$, denoted by $C_t$, is then a stochastic process, since $C_t$ may take on one of $\mathcal{M}$ possible options. To characterize the process, we may define the one-point probability $P_1(C, t)$ as the probability to observe the configuration $C$ at time $t$, the two-point probability $P_2(C_1, t_1; C_2, t_2)$ as the joint probability to observe the configuration $C_1$ at time $t_1$ and the configuration $C_2$ at time $t_2$, the three-point probability $P_3(C_1, t_1; C_2, t_2; C_3, t_3)$ as the joint probability to observe the configuration $C_1$ at time $t_1$ and the configuration $C_2$ at time $t_2$ and the configuration $C_3$ at time $t_3$. In this way, one may define quite generally the $n$-point probability $P_n(C_1, t_1; C_2, t_2; \ldots; C_n, t_n)$, which is the joint probability to observe $n$ given configurations $C_1, C_2, \ldots, C_n$ at respective times $t_1, t_2, \ldots, t_n$, with $n \in \mathbb{Z}^+$. These joint probabilities satisfy the obvious condition $0 \leq P_n \leq 1 \; \forall \; n$ and the normalization condition

$$\sum_{C_1, C_2, \ldots, C_n} P_n(C_1, t_1; C_2, t_2; \ldots; C_n, t_n) = 1 \; \forall \; n \,. \tag{2.1}$$

Moreover, $P_n(C_1, t_1; C_2, t_2; \ldots; C_n, t_n)$ does not change on interchanging two pairs $(C_i, t_i)$ and $(C_j, t_j)$, and that $P_m(C_1, t_1; C_2, t_2; \ldots; C_m, t_m)$ is obtained as the marginal probability with respect to $P_n(C_1, t_1; C_2, t_2; \ldots; C_n, t_n)$ for $n > m$, as follows

$$P_m(C_1, t_1; C_2, t_2; \ldots; C_m, t_m) = \sum_{C_{m+1}, C_{m+2}, \ldots, C_n} P_n(C_1, t_1; C_2, t_2; \ldots; C_m, t_m; C_{m+1}, t_{m+1}; C_{m+2}, t_{m+2}; \ldots; C_n, t_n) \,. \tag{2.2}$$

In order to characterize completely the stochastic process $C_t$, one would need to specify the functional form of the infinite set of joint probabilities $\{P_1(C, t), P_2(C_1, t_1; C_2, t_2), P_3(C_1, t_1; C_2, t_2; C_3, t_3), \ldots, P_n(C_1, t_1; C_2, t_2; \ldots; C_n, t_n), \ldots\}$.

Let us now introduce the concept of conditional probability in the context of stochastic processes. The quantity $\mathcal{P}(C_n, t_n | C_1, t_1; C_2, t_2; \ldots; C_{n-1}, t_{n-1})$ gives the probability to observe the configuration $C_n$ at time $t_n$, while conditioned on having observed configurations $C_1, C_2, \ldots, C_{n-1}$ at respective times $t_1, t_2, \ldots, t_{n-1}$, with $t_1 < t_2 < t_3 < \ldots < t_{n-1} < t_n$. One has the obvious normalization

$$\sum_{C_n} \mathcal{P}(C_n, t_n | C_1, t_1; C_2, t_2; \ldots; C_{n-1}, t_{n-1}) = 1 \,. \tag{2.3}$$

Suppose now we ask: what is the probability of observing the sequence of configurations $C_1, C_2, \ldots, C_{n-1}, C_n$ at respective times $t_1 < t_2 < \ldots < t_{n-1} < t_n$? Using the result on conditional probability $P(A|B) = P(A; B)/P(B)$, with $P(A; B)$ being the joint probability to observe events $A$ and $B$ (see Eq. (1.20)), we have

$$
\begin{aligned}
&P_n(C_1, t_1; C_2, t_2; \ldots; C_{n-1}, t_{n-1}; C_n, t_n) \\
&= \mathcal{P}(C_n, t_n | C_1, t_1; C_2, t_2; \ldots; C_{n-1}, t_{n-1}) P_{n-1}(C_1, t_1; C_2, t_2; \ldots; C_{n-1}, t_{n-1}) \\
&= \mathcal{P}(C_n, t_n | C_1, t_1; C_2, t_2; \ldots; C_{n-1}, t_{n-1}) \\
&\times \mathcal{P}(C_{n-1}, t_{n-1} | C_1, t_1; C_2, t_2; \ldots; C_{n-2}, t_{n-2}) P_{n-2}(C_1, t_1; C_2, t_2; \ldots; C_{n-2}, t_{n-2}) .
\end{aligned}
\tag{2.4}
$$

Proceeding as in the second step, we eventually obtain

$$
\begin{aligned}
&P_n(C_1, t_1; C_2, t_2; \ldots; C_{n-1}, t_{n-1}; C_n, t_n) \\
&= \mathcal{P}(C_n, t_n | C_1, t_1; C_2, t_2; \ldots; C_{n-1}, t_{n-1}) \\
&\times \mathcal{P}(C_{n-1}, t_{n-1} | C_1, t_1; C_2, t_2; \ldots; C_{n-2}, t_{n-2}) \\
&\times \mathcal{P}(C_{n-2}, t_{n-2} | C_1, t_1; C_2, t_2; \ldots; C_{n-3}, t_{n-3}) \\
&\times \mathcal{P}(C_{n-3}, t_{n-3} | C_1, t_1; C_2, t_2; \ldots; C_{n-4}, t_{n-4}) \\
&\times \ldots \times \mathcal{P}(C_2, t_2 | C_1, t_1) P_1(C_1, t_1) .
\end{aligned}
\tag{2.5}
$$

### *2.1.1 Stationary Stochastic Processes*

A stochastic process is said to be stationary when the moments

$$
\langle C_{t_1} C_{t_2} \ldots C_{t_n} \rangle \equiv \sum_{C_1, C_2, \ldots, C_n} C_1 C_2 \ldots C_n \; P_n(C_1, t_1; C_2, t_2; \ldots; C_n, t_n) \tag{2.6}
$$

are unaffected by a shift in time by an amount $\tau$, for all $n$, all $\tau$, and all $t_1, t_2, \ldots, t_n$. In other words, we have

$$
\langle C_{t_1+\tau} C_{t_2+\tau} \ldots C_{t_n+\tau} \rangle = \langle C_{t_1} C_{t_2} \ldots C_{t_n} \rangle . \tag{2.7}
$$

In particular, $\langle C_{t+\tau} \rangle = \langle C_t \rangle$ for arbitrary $\tau$ implies that $\langle C_t \rangle$ is independent of time. It then follows that the average of outcomes of a one-time measurement at time $t$ of any observable $O \equiv O(C_t)$, captured by the quantity $\langle O(C_t) \rangle$, will be independent of time. Put it simply, when stationary, the average value of the observable does not change with time, while the two-point correlation, given by $\langle O(C_{t_1}) O(C_{t_2}) \rangle$, is a function of the time difference $|t_1 - t_2|$ and not of individual times $t_1, t_2$.

Equation (2.7) implies that all joint probabilities of a stationary stochastic process are invariant under a shift in time:

$$
P_n(C_1, t_1 + \tau; C_2, t_2 + \tau; \ldots; C_n, t_n + \tau) = P_n(C_1, t_1; C_2, t_2; \ldots; C_n, t_n) \; \forall \, n , \tag{2.8}
$$

that is, $P_n(C_1, t_1; C_2, t_2; \ldots; C_n, t_n)$ is independent of the choice of origin of the time axis with respect to which the times $t_1, t_2, \ldots, t_n$ are measured. In particular, the one-point probability $P_1(C, t)$ is independent of time (and is denoted by $P_1(C)$), while the two-point probability $P_2(C_1, t_1; C_2, t_2)$ is a function of only the time difference $|t_1 - t_2|$. It then follows on using $P(C_1, t_1; C_2, t_2) = \mathcal{P}(C_2, t_2|C_1, t_1)P_1(C_1)$ for $t_1 < t_2$ that the two-point probability being a function of only the time difference $|t_1 - t_2|$ implies that the conditional probability $\mathcal{P}(C_2, t_2|C_1, t_1)$ is also a function of the time difference:

$$\mathcal{P}(C_2, t_2|C_1, t_1) = f_{t_2-t_1}(C_2|C_1)\,. \tag{2.9}$$

### 2.1.2 *Power-Spectrum, Correlations, and White Noise*

Now that we have seen what a stationary stochastic process is, we may ask: what is its signature as observed by an experimentalist? Let $x_t$ denotes a stochastic process in time, and let $x(t)$ denote the possible values of $x_t$. The Fourier transform of the signal $x(t)$ and its inverse Fourier transform are respectively given by

$$X_\omega = \int_{-\infty}^{\infty} \mathrm{d}t\; x(t)\mathrm{e}^{-\mathrm{i}\omega t}\,, \tag{2.10}$$

$$x(t) = \frac{1}{2\pi}\int_{-\infty}^{\infty} \mathrm{d}\omega\; X_\omega \mathrm{e}^{\mathrm{i}\omega t}\,. \tag{2.11}$$

Since $x(t)$ is real, we have $X_{-\omega} = X_\omega^\star$, where $\star$ denotes complex conjugation. Given the data for $x(t)$, one of the first things that may be computed is the correlation function $\langle x(t_1)x(t_2)\rangle$, which gives a measure of how correlated are the values of the random variable $x_t$ at two different times. The angular brackets correspond to averaging over the underlying stochasticity. Since we have assumed the underlying stochastic process to be stationary, following the discussion in Sect. 2.1.1, we conclude that $\langle x(t_1)x(t_2)\rangle$ is a function of only the time difference $|t_1 - t_2|$. We have

$$\langle x(t_1)x(t_2)\rangle = \int_{-\infty}^{\infty} \frac{\mathrm{d}\omega\mathrm{d}\omega'}{(2\pi)^2}\langle X_\omega X_{\omega'}\rangle \mathrm{e}^{\mathrm{i}(\omega t_1+\omega' t_2)}\,. \tag{2.12}$$

Writing $t_1 = \mathcal{T} + \tau/2$ and $t_2 = \mathcal{T} - \tau/2$, we have

$$\langle x(t_1)x(t_2)\rangle = \int_{-\infty}^{\infty} \frac{\mathrm{d}\omega\mathrm{d}\omega'}{(2\pi)^2}\langle X_\omega X_{\omega'}\rangle \mathrm{e}^{\mathrm{i}(\omega+\omega')\mathcal{T}+\mathrm{i}(\omega-\omega')\tau/2}\,, \tag{2.13}$$

so that the condition of stationarity imposes that we must have $\langle X_\omega X_{\omega'}\rangle = 0$ unless one has $\omega + \omega' = 0$. In other words, we must have $\langle X_\omega X_{\omega'}\rangle = 2\pi\, S(\omega)\delta(\omega + \omega')$.

Here, the quantity $S(\omega)$ is what is known as the power spectrum of $x(t)$. We thus obtain

$$\langle x(t_1)x(t_2)\rangle = \int_{-\infty}^{\infty} \frac{\mathrm{d}\omega}{2\pi} S(\omega)\mathrm{e}^{\mathrm{i}\omega(t_1-t_2)}\,, \tag{2.14}$$

or, equivalently, that

$$\langle x(t)x(t+\tau)\rangle = \int_{-\infty}^{\infty} \frac{\mathrm{d}\omega}{2\pi} S(\omega)\mathrm{e}^{\mathrm{i}\omega\tau}\,, \tag{2.15}$$

with $\langle x(t)x(t+\tau)\rangle$ being a function of only $|\tau|$. Inverse Fourier transform then yields

$$S(\omega) = \int_{-\infty}^{\infty} \mathrm{d}\tau\ \langle x(t)x(t+\tau)\rangle \mathrm{e}^{-\mathrm{i}\omega\tau}\,. \tag{2.16}$$

We thus find that the power spectrum and the correlation function are Fourier transform of one another, a relation that goes by the name of Wiener-Khinchin theorem. Note that the quantity $\langle x(t)x(t+\tau)\rangle$ being a function of only $|\tau|$, we may rewrite Eq. (2.16) as

$$S(\omega) = 2\int_{0}^{\infty} \mathrm{d}\tau\ \langle x(t)x(t+\tau)\rangle \cos(\omega\tau)\,. \tag{2.17}$$

A particular case of the power spectrum is when $S(\omega)$ is independent of $\omega$, which implies that the stochastic process $x_t$ is completely uncorrelated in time. Such a signal is what is known as the white noise. It is not an overstatement to say that one of the simplest stationary stochastic processes is a white noise. Any behavior other than $S(\omega)$ being equal for all $\omega$ implies that $x_t$ is correlated in time.

Let us work out S($\omega$) for the case of $x_t$ having zero mean and correlations given by

$$\langle x(t)x(t+\tau)\rangle = C\,\mathrm{e}^{-\gamma|\tau|}\,, \tag{2.18}$$

where $C > 0$ is a constant, while the constant $\gamma > 0$ is the inverse correlation time. Using Eq. (2.16), one gets

$$S(\omega) = \frac{2C\gamma}{\omega^2+\gamma^2}\,, \tag{2.19}$$

which resembles the form of a Lorentzian distribution. The white-noise limit corresponds to the limit of large $\gamma$ in which the quantity $S(\omega)$ becomes independent of $\omega$.

### 2.1.3 Homogeneous Stochastic Processes

A stochastic process is said to be homogeneous when the conditional probability $\mathcal{P}(C_2, t_2|C_1, t_1)$ with $t_1 < t_2$ is a function of only the time difference $= t_2 - t_1$:

$$\mathcal{P}(C_2, t_2|C_1, t_1) = g_{t_2-t_1}(C_2|C_1)\,. \tag{2.20}$$

Note that Eq. (2.20) is of the same form as Eq. (2.9). However, Eq. (2.9) is a consequence of the underlying stochastic process being stationary, while it is Eq. (2.20) that defines a homogeneous stochastic process. Indeed, Eq. (2.9) follows from the fact that for a stationary stochastic process the two-point probability is invariant under a time-shift. On the other hand, Eq. (2.20) does not imply that the two-point probability is invariant under a time-shift:

$$\begin{aligned} P_2(C_2, t_2+\tau; C_1, t_1+\tau) &= \mathcal{P}(C_2, t_2+\tau|C_1, t_1+\tau)P_1(C_1, t_1+\tau) \\ &= g_{t_2-t_1}(C_2|C_1)P_1(C_1, t_1+\tau) \\ &\neq P_2(C_1, t_1; C_2, t_2)\,. \end{aligned} \tag{2.21}$$

Thus, a stationary process is of course homogeneous. By contrast, a homogeneous process is not necessarily stationary.

### 2.1.4 Markovian Stochastic Processes

A stochastic process is said to be Markovian when the conditional probability $\mathcal{P}(C_n, t_n|C_1, t_1; C_2, t_2; \ldots; C_{n-1}, t_{n-1})$ to observe $C_n$ at time $t_n$, with $t_1 < t_2 < t_3 \ldots < t_{n-2} < t_{n-1} < t_n$, depends only on the configuration $C_{n-1}$ at the previous time instant $t_{n-1}$ and not on configurations that have been observed at earlier instants $t_1,\ t_2,\ t_3, \ldots, t_{n-2}$. In this sense, a Markovian stochastic process (in short, a Markov process) is memoryless: what will happen at the next instant depends on the information available at the current instant and not on the information up to the current instant. Thus, for a Markov process, we have

$$\mathcal{P}(C_n, t_n|C_1, t_1; C_2, t_2; \ldots; C_{n-1}, t_{n-1}) = \mathcal{P}(C_n, t_n|C_{n-1}, t_{n-1})\,. \tag{2.22}$$

One has the normalization

$$\sum_{C_n} \mathcal{P}(C_n, t_n|C_{n-1}, t_{n-1}) = 1\,, \tag{2.23}$$

as follows from Eq. (2.3). Using Eq. (2.5), we obtain for a Markov process that

$$
\begin{aligned}
&P_n(C_1, t_1; C_2, t_2; \ldots; C_{n-1}, t_{n-1}; C_n, t_n) \\
&\quad = \mathcal{P}(C_n, t_n | C_{n-1}, t_{n-1}) \mathcal{P}(C_{n-1}, t_{n-1} | C_{n-2}, t_{n-2}) \mathcal{P}(C_{n-2}, t_{n-2} | C_{n-3}, t_{n-3}) \\
&\quad \times \ldots \times \mathcal{P}(C_2, t_2 | C_1, t_1) P_1(C_1, t_1) . \qquad (2.24)
\end{aligned}
$$

We thus see that $P_n$ for any $n \geq 2$ can be uniquely expressed in terms of two functions, the one-point probability $P_1(C, t)$ and the conditional probability $\mathcal{P}(C, t|C', t')$ with $t > t'$, and so a Markov process is completely specified by these two functions. In this sense, the Markovian nature offers a huge simplification with respect to a general stochastic process, for which, as mentioned earlier, a complete specification requires knowing an infinite set of joint probabilities.

Consider the case $n = 3$, when we have for $t_1 < t_2 < t_3$ that

$$
\sum_{C_2} P_3(C_1, t_1; C_2, t_2; C_3, t_3) = \sum_{C_2} \mathcal{P}(C_3, t_3|C_2, t_2) \mathcal{P}(C_2, t_2|C_1, t_1) P_1(C_1, t_1) . \qquad (2.25)
$$

On the other hand, using Eq. (2.2), we get

$$
\sum_{C_2} P_3(C_1, t_1; C_2, t_2; C_3, t_3) = P_2(C_1, t_1; C_3, t_3) = \mathcal{P}(C_3, t_3|C_1, t_1) P_1(C_1, t_1) . \qquad (2.26)
$$

Comparing Eqs. (2.25) and (2.26), we arrive at the so-called Chapman-Kolmogorov equation given by

$$
\mathcal{P}(C_3, t_3|C_1, t_1) = \sum_{C_2} \mathcal{P}(C_3, t_3|C_2, t_2) \mathcal{P}(C_2, t_2|C_1, t_1) . \qquad (2.27)
$$

All of the above treatment may be generalized to the case of number of configurations being infinite by considering the limit $\mathcal{M} \to \infty$. We now move on to discuss stationary Markov processes.

## 2.2 Stationary Markov Processes: The Master Equation

From the discussions of the previous section, it follows that a stationary Markov process will be characterized by the two functions, the one-point probability $P_1(C)$ and the two-point probability $P_2(C_1, t_1; C_2, t_2)$ that is invariant under a time shift: $P_2(C_1, t_1 + \tau; C_2, t_2 + \tau) = P_2(C_1, t_1; C_2, t_2)$ for all $\tau$. We have already seen in Sect. (2.1.1) that stationarity demands that we must have the conditional probability $\mathcal{P}(C_2, t_2|C_1, t_1)$ a function of only the time difference:

$$
\mathcal{P}(C_2, t_2|C_1, t_1) = T_\tau(C_2|C_1); \quad \tau = t_2 - t_1 . \qquad (2.28)
$$

Note that by virtue of Eq. (2.23), we have $\sum_{C_2} T_\tau(C_2|C_1) = 1$.

For a stationary Markov process, the Chapman-Kolmogorov equation (2.27) reads as

$$T_{t_3-t_1}(C_3|C_1) = \sum_{C_2} T_{t_3-t_2}(C_3|C_2)T_{t_2-t_1}(C_2|C_1)\,, \tag{2.29}$$

or, defining $\tau \equiv t_2 - t_1$, $\tau' \equiv t_3 - t_2$, we have

$$T_{\tau+\tau'}(C_3|C_1) = \sum_{C_2} T_{\tau'}(C_3|C_2)T_\tau(C_2|C_1)\,. \tag{2.30}$$

Having discussed thus far the case of discrete-time evolution, we now treat the case of continuous time, which is obtained by considering the time step $\Delta t$ to be infinitesimally small ($= \mathrm{d}t \to 0$) instead of unity. In this case, the Chapman-Kolmogorov equation (2.30) converts to a differential equation on considering $\tau \to 0$ and $\tau' \to 0$, as we now show. To this end, let us first investigate the form of $T_{\tau'}(C_3|C_2)$ as $\tau' \to 0$. Starting from $C_2$, the system in the ensuing small time interval $\tau' \to 0$ may (i) make a transition to a configuration $C_3 \neq C_2$, or, (ii) make no transition and therefore remain in $C_2$. To proceed, we now introduce a $\tau'$-independent quantity $W(C_3|C_2) \geq 0$ that denotes the transition probability per unit time (commonly referred to as the transition rate) from a configuration $C_2$ to another configuration $C_3$ such that in the considered limit of $\tau' \to 0$ one has, for a given $C_2$, the normalization $\sum_{C_3} T_{\tau'}(C_3|C_2) = \sum_{C_3} W(C_3|C_2)\tau' = 1$. Then, the probability for (i), i.e., the probability for making a transition from $C_2$ to $C_3 \neq C_2$ in time interval $\tau' \to 0$ is given by $W(C_3|C_2)\tau'$ with $C_3 \neq C_2$. On the other hand, the probability for (ii), i.e., the probability to remain at $C_2$ in the time interval $\tau' \to 0$ is given by $[1 - \sum_{C_3 \neq C_2} W(C_3|C_2)\tau'] \equiv [1 - a(C_2)\tau']$ with $a(C_2) = \sum_{C_3 \neq C_2} W(C_3|C_2)$. Note that in writing the probabilities we indeed have respected the normalization $\sum_{C_3} W(C_3|C_2)\tau' = a(C_2)\tau' + W(C_2|C_2)\tau' = 1$. Now, considering both (i) and (ii), we may write, for small $\tau' \to 0$, a single equation for $T_{\tau'}(C_3|C_2)$, namely,

$$T_{\tau'}(C_3|C_2) = (1 - \delta_{C_2,C_3})W(C_3|C_2)\tau' + (1 - a(C_2)\tau')\delta_{C_2,C_3}\,, \tag{2.31}$$

which is valid for all $C_3$. Let us for later purposes define a quantity $\kappa(C_2)$ as

$$\kappa(C_2) \equiv \sum_{C_3} W(C_3|C_2) = a(C_2) + W(C_2|C_2)\,. \tag{2.32}$$

Note that the $W$'s being rates and not probabilities, $\kappa(C_2)$ is not required to be unity.

Using the result (2.31) in Eq. (2.30), we get

$$T_{\tau+\tau'}(C_3|C_1) = \tau' \sum_{C_2 \neq C_3} T_\tau(C_2|C_1)W(C_3|C_2) + (1 - a(C_3)\tau')T_\tau(C_3|C_1)\,. \tag{2.33}$$

Dividing throughout by $\tau'$ and taking the limit $\tau' \to 0$, we get

$$\frac{\partial}{\partial \tau} T_\tau(C_3|C_1) = \sum_{C_2 \neq C_3} \left[ W(C_3|C_2) T_\tau(C_2|C_1) - W(C_2|C_3) T_\tau(C_3|C_1) \right] . \quad (2.34)$$

The above differential form of the Chapman-Kolmogorov equation (2.30) is what is known as the master equation valid for a stationary Markov process.

Given a homogeneous Markov process, one may go over the arguments leading from Eq. (2.29) to Eq. (2.34) to realize that they all remain valid, and so the master equation (2.34) also holds for a homogeneous Markov process.

### 2.2.1 *Extraction of a Subensemble: How Homogeneous Markov Processes May Arise*

We have already seen that for a stationary Markov process, the origin of the time axis does not matter. On the other hand, in a typical experiment (be that a numerical experiment performed on a computer or one that is performed in a laboratory), one starts taking observations at a fixed time instant, and so for the experimentalist, this time instant of start of the experiment forms the origin of the time axis. The question then is: given that the experimental system is described by a stationary Markov process, how does the experimentalist adapt the formulation of stationary Markov processes to discuss the aforementioned experimental situation that clearly violates the stationarity condition through having a particular origin for the time axis? We now answer this question in the following.

Let us be more precise and denote by $t_0$ the time instant of the start of the experiment that necessarily involves preparing the initial condition; one starts with either a specified initial configuration, or, more generally, with an ensemble of initial configurations sampled according to a specified probability $p(C_0)$, namely, the probability for the event $C_{t_0} = C_0$. The underlying stochastic process being stationary and Markovian, one has for subsequent times the one-point and the conditional probability given by

$$P_1(C_1, t_1) = \sum_{C_0} T_{t_1 - t_0}(C_1|C_0) p(C_0); \quad t_0 < t_1 , \quad (2.35)$$

$$P_2(C_2, t_2|C_1, t_1) = T_{t_2 - t_1}(C_2|C_1); \quad t_0 < t_1 < t_2 . \quad (2.36)$$

Then, the Markovian stochastic process $C_{t \geq t_0}$, with $C_{t_0} = C_0$ distributed according to $p(C_0)$, is non-stationary, as we have singled out a special time $t_0$ as the origin of the time axis, and yet the transition probability of this Markov process depends on the time difference. Such non-stationary Markov processes for which the transition probability depends on the time difference alone are homogeneous Markov processes, according to the discussion in Sect. 2.1.3, and which arise as subensembles of stationary Markov processes as described in the foregoing. Note that here, it is not being claimed that a homogeneous Markov process can necessarily be embedded in a stationary Markov process, but only that a subensemble of a stationary Markov process is a homogeneous Markov process.

In a subensemble, one expects that after a long time, the system relaxes to a stationary state, whereby one has

$$P_1(C_1, t_1) \to P_1(C_1) \text{ as } t_1 \to \infty\,. \tag{2.37}$$

As this has to be true for arbitrary preparation protocol for the initial condition contained in the function $p(C_0)$, one must then have on using equation Eq. (2.35) that

$$T_{t_1-t_0}(C_1|C_0) \to P_1(C_1) \text{ as } t_1 \to \infty\,. \tag{2.38}$$

In such a subensemble, we rewrite the master equation (2.34) as

$$\frac{\partial}{\partial \tau} T_{t-t_0}(C|C_0) = \sum_{C' \neq C} \left[ W(C|C') T_{t-t_0}(C'|C_0) - W(C'|C) T_{t-t_0}(C|C_0) \right], \tag{2.39}$$

leading to the more familiar form of the master equation:

$$\frac{\partial}{\partial t} P(C,t) = \sum_{C' \neq C} \left[ W(C|C') P(C',t) - W(C'|C) P(C,t) \right]; \quad t \geq t_0\,, \tag{2.40}$$

which has to be interpreted thus: the solution of the above equation subject to the initial condition $P(C, t_0) = \delta_{C,C_0}$ is the conditional probability $T_{t-t_0}(C|C_0)\ \forall\ C_0$. It must be emphasized that $P(C, t)$ in Eq. (2.40) refers to a conditional probability and should not be confused with the one-point probability $P_1(C, t)$.

The master equation (2.40) has the form of a gain-loss equation for the probability $P(C, t)$ for configuration $C$. The first term on the right hand side, given by $\sum_{C' \neq C} W(C|C') P(C', t)$, represents ways by which $P(C, t)$ increases in time through transitions from other configurations $C'$ into configuration $C$. On the other hand, the second term on the right hand side, given by $\sum_{C' \neq C} W(C'|C) P(C, t)$, represents ways by which $P(C, t)$ decreases in time through transitions from configuration $C$ into other configurations $C'$.

The master equation (2.40) may be put in a particularly appealing and useful form in the following manner. Making use of the definition (2.32), one may rewrite the equation as one involving unconstrained summation:

$$\frac{\partial}{\partial t} P(C,t) = \sum_{C'} \left( W(C|C') - \kappa(C) \delta_{C,C'} \right) P(C',t); \quad t \geq t_0\,, \tag{2.41}$$

Next, one defines a column vector $|P(t)\rangle$ constituted by the set of probabilities $\{P(C,t)\}$, and whose dimension is given by the total number number of configurations, i.e., the total number of $C$'s. One also defines a matrix $\mathbb{W}$, whose elements are given by

$$\mathbb{W}_{CC'} \equiv W(C|C') - \kappa(C)\delta_{C,C'} \,. \tag{2.42}$$

It then follows on using the definition of $\kappa(C)$ in Eq. (2.32) that the sum of elements in any column of the matrix $\mathbb{W}$ equals zero, i.e., one has $\sum_C \mathbb{W}_{CC'} = \sum_C W(C|C') - \kappa(C') = 0$. Also, the diagonal element in any column is given by the negative of the sum of all off-diagonal elements in the same column, i.e., we have $\mathbb{W}_{C'C'} = W(C'|C') - \kappa(C') = -\sum_{C\neq C'} W(C|C')$. In terms of the matrix $\mathbb{W}$ and the column vector $|P(t)\rangle$, Eq. (2.41) takes the form

$$\frac{\partial|P(t)\rangle}{\partial t} = \mathbb{W}|P(t)\rangle \,, \tag{2.43}$$

which, drawing an analogy with the Schrödinger equation in quantum mechanics, may be interpreted as having the same form as the Schrödinger equation in imaginary time and with the Planck's constant $\hbar$ set to unity. Of course, the "Hamiltonian" for our case, whose role is played by $\mathbb{W}$, is not Hermitian. Indeed, all the elements of the matrix $\mathbb{W}$ are real numbers. Note that $\mathbb{W}$ is a matrix of dimension $\mathcal{M} \times \mathcal{M}$, while $|P(t)\rangle$ is a column vector of dimension $\mathcal{M} \times 1$; recall that $\mathcal{M}$ is the total number of possible configurations $C$.

Now, given that the trace of a matrix equals sum of its eigenvalues and the fact the diagonal elements of $\mathbb{W}$ sum up to a negative real number, it then follows that the sum of eigenvalues of $\mathbb{W}$ is a negative real number. The sum of the elements in each column of $\mathbb{W}$ being zero implies that it has zero determinant. Since the product of eigenvalues of a matrix equals its determinant, it then follows that zero is an eigenvalue of $\mathbb{W}$. As the elements of $\mathbb{W}$ are real, all complex eigenvalues occur in conjugate pairs. The sum of the eigenvalues being real negative implies that all non-zero eigenvalues are either real negative or complex with negative real parts. We will show below that the real parts of all non-zero eigenvalues are negative.

Equation (2.43) has the formal solution

$$|P(t)\rangle = \mathrm{e}^{\mathbb{W}t}|P(0)\rangle \,. \tag{2.44}$$

As shown in Appendix A, one may decompose any general square matrix in terms of its left and right eigenvectors, see Eq. (A.9). The left and right eigenvectors need not be the same when the matrix is not symmetric, but they correspond to the same eigenvalue. Moreover, the left eigenvector corresponding to a given eigenvalue is orthogonal to the right eigenvectors corresponding to all the other eigenvalues. Denoting by $|\lambda_L^{(\alpha)}\rangle$ and $|\lambda_R^{(\alpha)}\rangle$ the left and the right eigenvectors of $\mathbb{W}$, respectively, with the corresponding eigenvalue being $\lambda^{(\alpha)}$, and the normalization $\langle\lambda_L^{(\alpha)}|\lambda_R^{(\alpha)}\rangle = 1$, one has the decomposition

$$\mathbb{W} = \sum_{\alpha} \lambda^{(\alpha)} |\lambda_R^{(\alpha)}\rangle\langle\lambda_L^{(\alpha)}| . \tag{2.45}$$

Substituting the above expansion in Eq. (2.44) and then expanding the exponential, one then straightforwardly obtains

$$|P(t)\rangle = \sum_{\alpha} e^{\lambda^{(\alpha)} t} |\lambda_R^{(\alpha)}\rangle\langle\lambda_L^{(\alpha)}|P(0)\rangle . \tag{2.46}$$

Let us now prove that the real parts of all non-zero eigenvalues are negative. To this end, we adapt the so-called Gershgorin circle theorem to our problem; for the theorem, see, e.g., Ref. [9]. Let us consider the $i$-th column of the matrix $\mathbb{W}$,[1] and mark in the complex plane the diagonal element $-|\mathbb{W}_{ii}|$, which will lie on the negative real axis at a distance $|\mathbb{W}_{ii}|$ from the origin. Next, centered on the marked point, we draw a disc of radius $R_i \equiv \sum_{j \neq i} |\mathbb{W}_{ji}|$, which by virtue of the fact that the column sum of $\mathbb{W}$ is zero will pass through the origin on its right with the imaginary axis as the tangent, see Fig. 2.1. Such a disc is called a Gershgorin disc. Corresponding to different columns of the matrix $\mathbb{W}$, one thus has a set of Gershgorin discs, all of which pass through the origin on their right with the imaginary axis as the tangent. Consider the left eigenvector $\langle\lambda_L^{(\alpha)}|$, and denote by $[\lambda_L^{(\alpha)}]_{i_m}$ the largest in absolute value of all the elements of $\langle\lambda_L^{(\alpha)}|$. Consider the $i_m$-th component of the eigenvalue equation $\langle\lambda_L^{(\alpha)}|\mathbb{W} = \lambda^{(\alpha)}\langle\lambda_L^{(\alpha)}|$, given by $\sum_j [\lambda_L^{(\alpha)}]_j \mathbb{W}_{ji_m} = \lambda^{(\alpha)}[\lambda_L^{(\alpha)}]_{i_m}$. Rearranging, we get $\sum_{j \neq i_m} [\lambda_L^{(\alpha)}]_j \mathbb{W}_{ji_m} = (\lambda^{(\alpha)} - \mathbb{W}_{i_m i_m})[\lambda_L^{(\alpha)}]_{i_m}$, which gives $|(\lambda^{(\alpha)} - \mathbb{W}_{i_m i_m})| = |\sum_{j \neq i_m} \mathbb{W}_{ji_m}[\lambda_L^{(\alpha)}]_j/[\lambda_L^{(\alpha)}]_{i_m}| \leq \sum_{j \neq i_m} |\mathbb{W}_{ji_m}||[\lambda_L^{(\alpha)}]_j/[\lambda_L^{(\alpha)}]_{i_m}| \leq \sum_{j \neq i_m} |\mathbb{W}_{ji_m}| = R_{i_m}$. Here, we have used the fact that for complex numbers $z_1, z_2, \ldots$, we have $|z_1 + z_2 + \ldots| \leq |z_1| + |z_2| + \ldots$, and the fact that by definition, we have $|[\lambda_L^{(\alpha)}]_j/[\lambda_L^{(\alpha)}]_{i_m}| \leq 1$. We conclude then that every eigenvalue of $\mathbb{W}$ lies within or on the boundary of at least one of the Gershgorin discs. Since all the discs pass through the origin on their right with the imaginary axis as the tangent, the real parts of all non-zero eigenvalues are negative. Hence, for the matrix $\mathbb{W}$, it is the eigenvalue zero that has the largest possible real part, which itself is zero.

Since we have just proved that among all the eigenvalues $\lambda^{(\alpha)}$, the one with the largest possible real part is $\lambda^{(\alpha,\max)} \equiv 0$, while the remaining $\lambda^{(\alpha)}$'s are either real negative or complex with negative real parts, it then follows from Eq. (2.46) that in the $t \to \infty$, one has the result

$$\lim_{t\to\infty} |P(t)\rangle = |\lambda_R^{(\alpha,\max)}\rangle\langle\lambda_L^{(\alpha,\max)}|P(0)\rangle . \tag{2.47}$$

[1] The index $i$ labels the $\mathcal{M}$ configurations of the system.

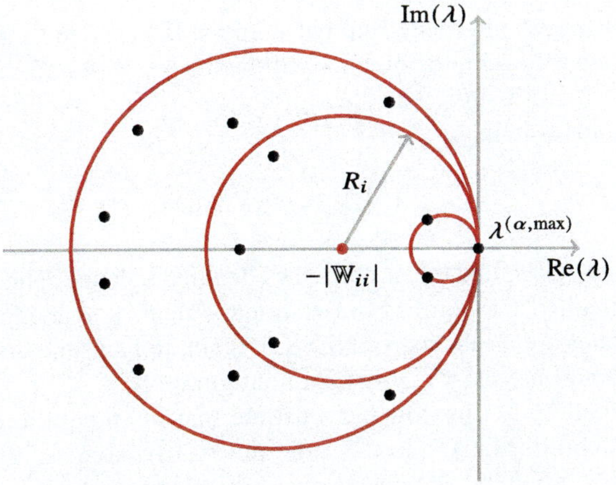


**Fig. 2.1 Illustration of the Gershgorin theorem at work.** Shown are the Gershgorin discs together with the eigenvalues (black filled circles) corresponding to the matrix $\mathbb{W}$ defined in Eq. (2.42). The theorem asserts that all the eigenvalues lie in the union of these Gershgorin discs (including their boundaries), with the complex eigenvalues occurring in conjugate pairs

## 2.3 Stationary State and the Condition of Detailed Balance

Let us come back to Eq. (2.40), valid for a homogeneous Markov process, and ask: Under what condition does the equation ensure that one has a stationary state, in the sense that one has $P(C,t) \to P^{\mathrm{st}}(C)$ as $t \to \infty$? In other words, starting at time $t_0$ with configuration $C_0$ distributed according to the probability $p(C_0)$, the question is if and how Eq. (2.40) ensures that in the limit of long times, one has a time-independent probability to be in any configuration $C$. Note that when the homogeneous Markov process is obtained as a subensemble of a stationary Markov process, we have $P^{\mathrm{st}}(C) = P_1(C)$, see Eq. (2.38). It is evident on examining Eq. (2.40) that the condition is

$$\sum_{C' \neq C} W(C|C')P^{\mathrm{st}}(C') = \sum_{C' \neq C} W(C'|C)P^{\mathrm{st}}(C)\ \forall\, C\,, \tag{2.48}$$

that is, the total number of transitions per unit time into any given configuration $C$ must balance the total number of transitions per unit time out of $C$. The resulting stationary state is generically called a nonequilibrium stationary state (NESS). Note that in order to have a stationary state, one should not have a single $C$ for which the condition (2.48) does not hold. From Eq. (2.47), we see that

$$|P_{\mathrm{st}}\rangle = |\lambda_R^{(\alpha,\max)}\rangle\langle\lambda_L^{(\alpha,\max)}|P(0)\rangle\,, \tag{2.49}$$

that is, the stationary state is given by the left and the right eigenvector of the matrix $\mathbb{W}$ that correspond to the largest eigenvalue; the latter has the value zero, as we have already seen in the last section. We also observe from Eq. (2.47) that relaxation to the stationary state (2.49) is at late times dictated by a timescale given by the inverse of the magnitude of the second largest eigenvalue of $\mathbb{W}$, in other words, the gap in the eigenvalue spectrum between the largest and the second largest eigenvalue. The larger the gap, the quicker will be the relaxation to the stationary state (2.49).

Now, a sufficient (though not necessary) condition that ensures that Eq. (2.48) holds is to have

$$W(C|C')P^{\mathrm{st}}(C') = W(C'|C)P^{\mathrm{st}}(C)\ \forall\ C, C', \tag{2.50}$$

which is called the condition of detailed balance. The corresponding stationary state is called the equilibrium stationary state. It is evident that equilibrium stationary states are a subclass of NESSs. Again, to have an equilibrium stationary state, there should not exist any pair of configurations $C, C'$ for which the condition (2.50) is violated. Let us now demonstrate that in an equilibrium stationary state (in short, in equilibrium), one has time-reversal invariance in the system, in the sense that the probability of occurrence of every sequence of $k$ configurations $C_1, C_2, \ldots, C_k$ equals that of the occurrence of the reverse sequence $C_k, C_{k-1}, \ldots, C_1$. Indeed, the probability of occurrence of the forward sequence is

$$p_{\mathrm{forwrad}} = P^{\mathrm{st}}(C_1)W(C_2|C_1)W(C_3|C_2)\ldots W(C_k|C_{k-1}), \tag{2.51}$$

while the probability of occurrence of the reverse sequence is

$$p_{\mathrm{reverse}} = P^{\mathrm{st}}(C_k)W(C_{k-1}|C_k)W(C_{k-2}|C_{k-1})\ldots W(C_1|C_2). \tag{2.52}$$

When the condition of detailed balance, Eq. (2.50), is satisfied, it is easily verified through direct substitution that

$$p_{\mathrm{forward}} = p_{\mathrm{reverse}}. \tag{2.53}$$

Thus, when in equilibrium, on viewing a movie of time evolution of the configurations, one would not be able to make out whether the underlying dynamics is proceeding forward or backward in time.

Now, given the transition rates $W(C|C')$, how does one check that the condition of detailed balance is satisfied or not? Equation (2.50) suggests that one must know $P^{\mathrm{st}}(C)\ \forall\ C$ to do the check. On the other hand, obtaining $P^{\mathrm{st}}(C)\ \forall\ C$ requires that one first solves Eq. (2.40) to obtain $P(C, t)\ \forall\ C$, and then take the limit $t \to \infty$ to obtain $P^{\mathrm{st}}(C)$, a program whose successful execution does not a priori seem possible for arbitrary transition rates. It turns out that there is a criterion due to Kolmogorov that allows to check whether the condition of detailed balance is satisfied or not solely from a knowledge of the transition rates and without requiring to know the function $P^{\mathrm{st}}(C)$. The criterion says that for a given sequence of $k \geq 2$ configurations $C_1, C_2, \ldots, C_k$ observed at respective times $t_1 < t_2 < \ldots < t_k$, provided we have

$$\begin{aligned} &W(C_2|C_1)W(C_3|C_2)\ldots W(C_k|C_{k-1})W(C_1|C_k) \\ &= W(C_k|C_1)W(C_{k-1}|C_k)\ldots W(C_1|C_2), \end{aligned} \tag{2.54}$$

the condition of detailed balance is satisfied.

Note that the quantity $W(C'|C)P(C,t)$ in Eq. (2.40) gives the probability current at time $t$ from configuration $C$ to $C'$. Hence, the condition of stationarity (2.48) may be interpreted as the condition that in the stationary state, the total probability current into any configuration $C$ equals the total probability current out of $C$. The condition of detailed balance (2.50) further implies that the net probability current between any pair of configurations $C$, $C'$, or, more generally, across a closed loop of configurations $C_1 \to C_2 \to C_3 \ldots C_k \to C_1$, is zero (see condition (2.54). If detailed balance is not obeyed, there has to be at least one closed loop of configurations supporting a net current around the loop in the stationary state.

## 2.4 The Fokker-Planck Equation

The celebrated Fokker-Planck equation appears as an approximation to the master equation (2.40), in the case in which the configurations instead of being discrete are continuous, namely, configurations that are random variables denoted by a continuous variable $X$, and whose specific realizations are denoted by $x$ (we consider here the case of one dimension for simplicity, but the treatment below may be extended to higher dimensions). In this case, one defines the transition probability density $W(x|x')$, which is such that $W(x|x')\mathrm{d}x\mathrm{d}x'$ is the transition probability per unit time from a configuration in the interval $[x', x' + \mathrm{d}x']$ to a configuration in the interval $[x, x + \mathrm{d}x]$. The one-point probability density $P_1(x,t)$ is defined such that $P_1(x,t)\mathrm{d}x$ gives the probability to observe a configuration in the interval $[x, x + \mathrm{d}x]$ at time $t$. The master equation (2.40) then takes the form

$$\frac{\partial}{\partial t}P(x,t) = \int \mathrm{d}x' \left[W(x|x')P(x',t) - W(x'|x)P(x,t)\right]; \quad t \geq t_0, \qquad (2.55)$$

with the initial condition $P(x,t_0) = \delta(x - x_0)$.

To state the assumptions behind the approximation that lead from Eq. (2.55) to the Fokker-Planck equation, we first write the transition rate $W(x|x')$ from $x'$ to $x$, denoting the probability to jump from $x'$ to $x$, in terms of the jump size $r \equiv x - x'$ and the starting point $x'$, as

$$W(x|x') = W(x';r); \quad r \equiv x - x'. \qquad (2.56)$$

The Fokker-Planck equation is obtained under the following two assumptions. (i) The function $W(x';r)$ is a sharply peaked function of $r$, but varies very slowly with $x'$. In other words, there exists a scale for jump length that we denote by $\delta$, which is such that $W(x';r) = 0$ for $|r| > \delta$ and $W(x' + \Delta x; r) = W(x';r)$ for $|\Delta x| < \delta$. In physical terms, what this means is that no jump beyond the scale $\delta$ is allowed, and that the function $W(x';r)$ does not vary as $x'$ varies within the scale $\delta$. (ii) The function

$P(x,t)$ varies slowly with $x$, namely, $P(x+\Delta x,t) \approx P(x,t)$ for $|\Delta x| < \delta$. The master equation (2.55) when rewritten using Eq. (2.56) reads as

$$\frac{\partial}{\partial t}P(x,t) = \int_{-\delta}^{\delta} \mathrm{d}r\; W(x-r;r)P(x-r,t) - P(x,t)\int_{-\delta}^{\delta} \mathrm{d}r\; W(x;-r)\,. \quad (2.57)$$

Under the two stated assumptions, one may then Taylor-expand the product $W(x-r;r)P(x-r,t)$, to get

$$\frac{\partial}{\partial t}P(x,t) = \int \mathrm{d}r\; W(x;r)P(x,t) - \int \mathrm{d}r\; r\frac{\partial}{\partial x}[W(x;r)P(x,t)] + \frac{1}{2}\int \mathrm{d}r\; r^2\frac{\partial^2}{\partial x^2}[W(x;r)P(x,t)] - P(x,t)\int \mathrm{d}r\; W(x;-r)\,, \quad (2.58)$$

where we have retained terms up to second order in $r$, and where the integration over $r$, which runs from $-\delta$ to $\delta$, may be extended over the range $-\infty$ to $\infty$ on using the fact that $W(x;r) = 0$ for $|r| > \delta$. The first and the last term on the right hand side cancel. Then, in terms of jump moments defined as

$$a_\nu(x) \equiv \int \mathrm{d}r\; r^\nu W(x;r);\;\; \nu = 1, 2, \ldots\,, \quad (2.59)$$

we get the desired Fokker-Planck equation

$$\frac{\partial P(x,t)}{\partial t} = -\frac{\partial}{\partial x}[a_1(x)P] + \frac{1}{2}\frac{\partial^2}{\partial x^2}[a_2(x)P]\,. \quad (2.60)$$

On retaining all the terms in the Taylor expansion, one has more generally the Kramers-Moyal expansion:

$$\frac{\partial P(x,t)}{\partial t} = \sum_{\nu=1}^{\infty}\frac{(-1)^\nu}{\nu!}\left(\frac{\partial}{\partial x}\right)^\nu [a_\nu(x)P]\,. \quad (2.61)$$

The quantity $r$ being the jump length, we may relate the quantity $a_\nu(x)$ to the average of the $\nu$-th power of the small jump $x \to x + \Delta x$ taking place in a small time $\Delta t$, as

$$a_\nu(x) = \lim_{\Delta t\to 0}\frac{\langle[\Delta x]^\nu\rangle}{\Delta t}\,. \quad (2.62)$$

Here, the angular brackets $\langle\cdot\rangle$ denote averaging with respect to the stochasticity in the dynamics of the variable $x$, which in our set up is contained in the function $W(x;r)$.

## 2.5 An Example: The Brownian Motion

Having discussed the background material thus far, we now turn to an illustration of the application of the techniques developed by analyzing the paradigmatic example of the Brownian motion. The motion pertains to the random motion of a mesoscopic particle (say, a colloidal particle, which is of the size of a micron) that is immersed in a fluid and is being constantly buffeted by the microscopic fluid molecules that are angstrom-sized. We consider not just one but a collection of many such colloidal "Brownian" particles, which are interacting rather weakly with one another and thus may be treated as executing independent motion. The fluid is contained in a fixed volume and is constituted by a fixed number of fluid molecules. The Brownian particles are macroscopic in number and are yet fewer than the fluid molecules. The fluid is considered to be in thermal equilibrium at temperature $T$, and constitutes a heat bath with which the system of Brownian particles is in contact and is continually exchanging energy so as to be also in equilibrium at the same temperature. The only force acting on the Brownian particles is attributed to that arising from the collision of individual particles with the surrounding fluid molecules.

Let us consider the case of motion in one dimension (generalization to higher dimensions is straightforward), and denote by $M$ and $m$ the mass of the Brownian particles and the fluid molecules, respectively. The velocity of both the Brownian particles and the fluid molecules is a random variable, in the sense that on measuring its value at any given time instant, one may get any number in the range $[-\infty, \infty]$, and then it makes sense to ask for the probability distribution of these random variables. Being in equilibrium, the Brownian particles will have a velocity distribution given by the well-known Maxwell distribution:

$$p_{\text{eq}}(v) = \left(\frac{M}{2\pi k_B T}\right)^{1/2} \exp\left(-\frac{Mv^2}{2k_B T}\right), \tag{2.63}$$

where $k_B$ denotes as usual the Boltzmann constant, and $\int_{-\infty}^{\infty} \mathrm{d}v\, p_{\text{eq}}(v) = 1$. In other words, $p_{\text{eq}}(v)$ gives the probability density to find a Brownian particle with velocity $v$.

In the above backdrop, we may ask: suppose that at a given instant of time $t_0$, a tagged Brownian particle has the velocity $v_0$. How does its velocity $v(t)$ evolve as a function of time for $t > t_0$, while starting from $v(t_0) = v_0$? Note that $v(t \geq t_0)$ is a stochastic process. One may characterize the values of $v(t)$ at time $t \geq t_0$ by a conditional probability distribution $p(v, t|v_0, t_0)$, defined such that $p(v, t|v_0, t_0)\mathrm{d}v\mathrm{d}v_0$ gives the probability to observe the velocity to be in the interval $[v, v + \mathrm{d}v]$ at time $t$, given that it had the value in the interval $[v_0, v_0 + \mathrm{d}v_0]$ at time $t_0$. Based on our discussion earlier in this chapter, one may proceed as follows: consider (i) $v(t)$ as a function of time as a stationary Markov process, for which the one-point probability density is $p_{\text{eq}}(v)$, (ii) treat as a subensemble of this stationary process the stochastic process $v(t \geq t_0)$, with $v(t_0) = v_0$ and the conditional probability $p(v, t|v_0, t_0)$, which by virtue of the stationarity assumption, is a function of only the time difference $t - t_0$:

$$p(v,t|v_0,t_0)=T_{t-t_0}(v|v_0)\,, \tag{2.64}$$

with

$$T_{t-t_0}(v|v_0)\to p_{\rm eq}(v)\text{ as } t\to\infty\,; \tag{2.65}$$

see the discussion in Sect. (2.2.1). The question then is: what sort of time dependence of $v(t\geq t_0)$ would ensure that the property (2.65) is borne out? Equation (2.65) implies that

$$\langle v|v_0\rangle(t)\to\langle v\rangle_{\rm eq}\equiv\int_{-\infty}^{\infty}{\rm d}v\,v\,p_{\rm eq}(v)=0\text{ as } t\to\infty\,, \tag{2.66}$$

$$\langle v^2|v_0\rangle(t)\to\langle v^2\rangle_{\rm eq}\equiv\int_{-\infty}^{\infty}{\rm d}v\,v^2\,p_{\rm eq}(v)=\frac{k_BT}{M}\text{ as } t\to\infty\,, \tag{2.67}$$

$$\langle v^2|v_0\rangle(t)-[\langle v|v_0\rangle(t)]^2\to\frac{k_BT}{M}\text{ as } t\to\infty\,. \tag{2.68}$$

### 2.5.1 Langevin's Theory

To proceed, we write down the Newton's equation of motion for the tagged particle for times $t\geq t_0$, as

$$M\frac{{\rm d}v(t)}{{\rm d}t}=F(t);\;\; v(t_0)=v_0\,, \tag{2.69}$$

where the force $F(t)$ is a random function of time. We may split $F(t)$ into a systematic random force $F_{\rm av}(t)$ that depends on the state of motion of the tagged particle and a truly random force $\xi(t)$ that has zero average and which does not depend on the state of motion of the tagged particle, as

$$F(t)=F_{\rm av}(t)+\xi(t);\;\; \langle\xi\rangle(t)=0\,. \tag{2.70}$$

Here, the averaging, denoted by the angular brackets, may be implemented by considering many independent experiments performed under identical conditions in all of which the tagged Brownian particle has the velocity $v_0$ at time $t_0$, and by averaging at time $t>t_0$ the different values of $F(t)$ realized across these experiments. Note that the result of the average is independent of the choice for $v_0$. Now, Eq. (2.69) implies the dynamics for the average velocity, as

$$M\frac{{\rm d}\langle v\rangle(t)}{{\rm d}t}=\langle F_{\rm av}\rangle(t);\;\; \langle v\rangle(t_0)=v_0\,. \tag{2.71}$$

By virtue of Eq. (2.65), $F_{\text{av}}(t)$ must have the form that one has $\langle v\rangle(t\to\infty)=0$. On physical grounds, we expect the tagged particle to be continually slowing down in time as it tries to make its way through the surrounding fluid molecules that are constantly colliding with it. If the particle has a large velocity at some instant of time, it will encounter more collisions per unit time from the direction opposite to its direction of motion, i.e., from its front than from its rear. Eventually, at long times, the particle comes to a standstill. These observations may be expressed by modeling $F_{\text{av}}(t)$ as a frictional force directed opposite to the motion of the tagged particle, as

$$F_{\text{av}}(t) = -\gamma v(t)\,. \tag{2.72}$$

Here, the friction coefficient $\gamma > 0$ is proportional to the viscosity of the fluid (glycerine, being more viscous, will surely offer greater resistance to the motion than say water that is much less viscous.)[2] Let us note in passing that the form (2.72) constitutes a model for the average force, and how good or bad is the model is to be judged a posteriori by checking its predictions against experimental results. We finally arrive at the so-called Langevin equation for the velocity of the tagged particle, as

$$M\frac{\mathrm{d}v(t)}{\mathrm{d}t} = -\gamma v(t) + \xi(t);\quad v(t_0) = v_0\,. \tag{2.73}$$

Equation (2.73) is an example of what are known as stochastic differential equations, namely, differential equations in which one or more terms is a stochastic process, and whose solution is consequently also a stochastic process.

Equation (2.73) has the formal solution

$$v(t) = v_0\mathrm{e}^{-\gamma t/M} + \frac{1}{M}\int_{t_0}^{t}\mathrm{d}t'\;\mathrm{e}^{-\gamma(t-t')/M}\xi(t')\,, \tag{2.74}$$

implying

$$\langle v\rangle(t) = v_0\mathrm{e}^{-\gamma t/M} \to 0 \text{ as } t\to\infty\,, \tag{2.75}$$

$$\langle v^2\rangle(t) - [\langle v\rangle(t)]^2 = \frac{1}{M^2}\int_{t_0}^{t}\int_{t_0}^{t}\mathrm{d}t'\mathrm{d}t''\;\langle\xi(t')\xi(t'')\rangle\mathrm{e}^{-\gamma/M(2t-t'-t'')}\,. \tag{2.76}$$

While the first equation is consistent with Eq. (2.66), in order that the second equation is consistent with Eq. (2.68), we must have

$$\langle\xi(t)\xi(t')\rangle = 2\gamma k_B T\delta(t-t')\,. \tag{2.77}$$

We may further assume that all higher-order moments of $\xi(t)$ may be expressed in terms of the second moments, as

[2] In three dimensions, e.g., considering the Brownian particle to be a sphere of radius $R$, the celebrated Stoke's law states that $\gamma = 6\pi\eta R$, where $\eta$ is the coefficient of viscosity of the fluid.

$$\langle \xi(t_1)\xi(t_2)\dots\xi(t_{2n+1})\rangle = 0\,, \tag{2.78}$$

$$\langle \xi(t_1)\xi(t_2)\dots\xi(t_{2n})\rangle = \sum \langle \xi(t_{i_1})\xi(t_{i_2})\rangle\langle \xi(t_{i_3})\xi(t_{i_4})\rangle \dots \langle \xi(t_{i_{2n-1}})\xi(t_{i_{2n}})\rangle\,, \tag{2.79}$$

where the sum is over all possible ways of grouping the $2n$ variables $t_1, t_2, \dots, t_{2n}$ into pairs. The above assumption on moments is equivalent to the assumption that $\xi(t)$ is a Gaussian-distributed random variable.[3] Now that we have discussed the properties of all possible moments of $\xi(t)$, we have specified the complete statistical properties of $\xi(t)$. Thus, we arrive at the Langevin description of the motion of the Brownian particle as

$$M\frac{\mathrm{d}v(t)}{\mathrm{d}t} = -\gamma v(t) + \xi(t); \quad v(t_0) = v_0\,, \tag{2.80}$$

with $\xi(t)$ having the properties

$$\langle \xi(t)\rangle = 0, \quad \langle \xi(t)\xi(t')\rangle = 2\gamma k_B T\delta(t-t')\,. \tag{2.81}$$

Computing the power spectrum of $\xi(t)$ by using formula (2.16) yields

$$S(\omega) = 2\gamma k_B T \int_{-\infty}^{\infty} \mathrm{d}\tau\ \delta(\tau)\mathrm{e}^{\mathrm{i}\omega\tau} = 2\gamma k_B T\,, \tag{2.82}$$

implying that $\xi(t)$ is a white noise. Additionally, being Gaussian, it is customary to refer to $\xi(t)$ as a Gaussian, white noise.

In the case in which the Brownian particle is acted upon by a conservative force $F(x) = -V'(x)$, with $V(x)$ being the potential, the corresponding Langevin equation is obtained by adding the force term to the right hand side of Eq. (2.80), to obtain

$$M\frac{\mathrm{d}v(t)}{\mathrm{d}t} = -V'(x) - \gamma v(t) + \xi(t); \quad v(t_0) = v_0\,. \tag{2.83}$$

### 2.5.2 *Einstein's Theory*

While the approach due to Langevin is based on a phenomenological equation of motion for the Brownian particle, Eq. (2.80), Einstein in one of his 1905 annus mirabilis papers argued differently to deduce a form for the evolution equation of the probability distribution of the degrees of freedom of the Brownian particle. We will not attempt to reproduce the arguments here; instead, we will present and solve the evolution equation, as an application of the formalism of the Fokker-Planck equation

[3] One may easily check that all moments of a Gaussian distribution may be expressed in terms of the mean and the variance of the distribution.

discussed in Sect. 2.4. The relevant variable being $v$, the velocity of the Brownian particle, let us rewrite the Fokker-Planck equation (2.60) as

$$\frac{\partial P(v,t)}{\partial t} = -\frac{\partial}{\partial v}[a_1(v)P] + \frac{1}{2}\frac{\partial^2}{\partial v^2}[a_2(v)P], \tag{2.84}$$

subject to the initial condition

$$P(v,t_0) = \delta(v - v_0). \tag{2.85}$$

Next, we aim to obtain the quantities $a_1(v)$ and $a_2(v)$ from the Langevin equation (2.80). The latter when integrated over a small time interval $[t, t+\Delta t]$ writes as

$$\Delta v(t) \equiv v(t+\Delta t) - v(t) = -\frac{\gamma}{M}v(t)\Delta t + \frac{1}{M}\int_t^{t+\Delta t} \mathrm{d}t'\, \xi(t'). \tag{2.86}$$

Using Eq. (2.62), we then obtain

$$a_1(v) = -\frac{\gamma v}{M},\; a_2(v) = \frac{2\gamma k_B T}{M^2}, \tag{2.87}$$

which when substituted in Eq. (2.84) yields the Fokker-Planck equation for the probability distribution of the velocity of the Brownian particle, as

$$\frac{\partial P(v,t)}{\partial t} = \frac{\gamma}{M}\frac{\partial}{\partial v}[vP] + \frac{\gamma k_B T}{M^2}\frac{\partial^2 P(v,t)}{\partial v^2}. \tag{2.88}$$

The solution reads

$$P(v,t) = \sqrt{\frac{M}{2\pi k_B T(1-\mathrm{e}^{-2(\gamma/M)t})}}\exp\left(-\frac{M}{2k_B T}\frac{(v - v_0\mathrm{e}^{-(\gamma/M)t})^2}{1-\mathrm{e}^{-2(\gamma/M)t}}\right). \tag{2.89}$$

It is then readily checked that in the limit $t \to \infty$, one obtains the equilibrium distribution $p_{\mathrm{eq}}(v)$ given in Eq. (2.63).

## 2.6 Generalization of the Langevin Equation

The Langevin description discussed thus far applies to any stochastic process $x(t)$ that evolves in time according to the generic form

$$\frac{\mathrm{d}x}{\mathrm{d}t} = F(x) + \xi(t), \tag{2.90}$$

where $\xi(t)$ is a Gaussian, white noise with the properties given by

$$\langle \xi(t) \rangle = 0, \quad \langle \xi(t)\xi(t') \rangle = \Gamma\delta(t - t'), \tag{2.91}$$

with $\Gamma > 0$ being a constant. Let us now discuss a few relevant generalizations of the above Langevin form of the dynamics. The first obvious generalization applies to an arbitrary number of variables $x_i$, with $i = 1, 2, \ldots, N$, with the dynamics of $x_i$ given by

$$\frac{\mathrm{d}x_i}{\mathrm{d}t} = F_i(x_1, x_2, \ldots, x_N) + \xi_i(t); \quad i = 1, 2, \ldots, N, \tag{2.92}$$

and with each $\xi_i(t)$ being a Gaussian, white noise that may or may not be correlated between the different $x_i$'s:

$$\langle \xi_i(t) \rangle = 0, \quad \langle \xi_i(t)\xi_j(t') \rangle = \Gamma_{ij}\delta(t - t'); \quad i, j = 1, 2, \ldots, N. \tag{2.93}$$

Here, $\Gamma_{ij}$'s are positive constants. A particularly non-trivial generalization is when the coefficient of the noise term in the equation of motion is a function of the stochastic variable itself, that is, the dynamics

$$\frac{\mathrm{d}x}{\mathrm{d}t} = F(x) + G(x)\xi(t), \tag{2.94}$$

where $\xi(t)$ is a Gaussian, white noise with the properties given by

$$\langle \xi(t) \rangle = 0, \quad \langle \xi(t)\xi(t') \rangle = \delta(t - t'), \tag{2.95}$$

Such a noise term, namely, $G(x)\xi(t)$, is called a multiplicative noise, as opposed to the case when $G$ is a constant and then the noise is referred to as an additive noise. In the case of multiplicative noise, extra caution need to be exercised in interpreting the noise term, especially, when discretizing the term in time; two popular discretization schemes are the Itô and the Stratonovich scheme. Appendix B provides a brief primer on these schemes and their implementation.

## 2.7 Phenomenological Derivation of the Langevin Equation

Here, we discuss how the Langevin equation may be arrived at by starting from a Hamiltonian description of a model system that allows an exact analysis. To this end, consider a particle of mass $M$ moving in one dimension in a potential $V(q)$, where $q$ denotes the position coordinate of the particle, with the corresponding conjugate momentum being $p$. The particle is interacting with a certain number $N$ of independent and one-dimensional harmonic oscillators, with $q_j$, $p_j$, $m_j$, and $\omega_j$ denoting respectively the canonical coordinate, the corresponding conjugate momentum, the mass, and the frequency of the $j$-th oscillator; $j = 1, 2, \ldots, N$. The mass of the particle is much larger than that of any of the oscillators: $M \gg m_j \; \forall \; j$. Eventually, we will consider the limit $N \to \infty$, when the system of oscillators will be treated as a heat bath in thermal equilibrium at a certain temperature $T$, while the particle will constitute our system of interest. The Hamiltonian of the combined system is given by

$$H = \frac{p^2}{2M} + V(q) + \sum_{j=1}^{N} \left[ \frac{p_j^2}{2m_j} + \frac{1}{2} m_j \omega_j^2 \left( q_j - \frac{C_j q}{m_j \omega_j^2} \right)^2 \right], \tag{2.96}$$

where $C_j$ denotes the coefficient of linear coupling between the coordinate of the particle and that of the $j$-th oscillator. The Hamilton's equations of motion for the particle and the oscillators are given by

$$\dot{q} = \frac{p}{M}, \quad \dot{p} = -V'(q) + \sum_{j=1}^{N} C_j \left( q_j - \frac{C_j q}{m_j \omega_j^2} \right),$$
$$\dot{q}_j = \frac{p_j}{m_j}, \quad \dot{p}_j = -m_j \omega_j^2 \left( q_j - \frac{C_j q}{m_j \omega_j^2} \right). \tag{2.97}$$

Here, the dot denotes derivative with respect to time, while the prime denotes derivative with respect to $q$. Note that all the above equations are time-reversal invariant. From the above set of equations, we obtain the equation of motion for the coordinate of the $j$-th oscillator as

$$\ddot{q}_j = -\omega_j^2 \left( q_j - \frac{C_j q}{m_j \omega_j^2} \right). \tag{2.98}$$

Let us integrate the above equation forward in time, to get

$$q_j(t) = \cos(\omega_j t) q_j(0) + \sin(\omega_j t) \frac{p_j(0)}{m_j \omega_j} + \frac{C_j}{m_j \omega_j} \int_0^t \mathrm{d}t' \, q(t') \sin(\omega_j (t - t')), \tag{2.99}$$

which on integrating the integral by parts yields

$$q_j(t) - \frac{C_j q(t)}{m_j \omega_j^2} = \cos(\omega_j t)\left(q_j(0) - \frac{C_j q(0)}{m_j \omega_j^2}\right) + \sin(\omega_j t)\frac{p_j(0)}{m_j \omega_j}$$
$$- \frac{C_j}{M m_j \omega_j^2}\int_0^t \mathrm{d}t'\, p(t')\cos(\omega_j(t-t')) \,. \tag{2.100}$$

This when substituted in the second equation in the first line in Eq. (2.97) leads to

$$\dot{p} = -V'(q) - \int_0^t \mathrm{d}t'\, \zeta(t-t')\frac{p(t')}{M} + f(t)\,, \tag{2.101}$$

where we have defined

$$\zeta(t) \equiv \sum_{j=1}^{N} \cos(\omega_j t)\frac{C_j^2}{m_j \omega_j^2}\,,$$
$$f(t) \equiv \sum_{j=1}^{N}\left[C_j \cos(\omega_j t)\left(q_j(0) - \frac{C_j q(0)}{m_j \omega_j^2}\right) + C_j \sin(\omega_j t)\frac{p_j(0)}{m_j \omega_j}\right]. \tag{2.102}$$

Let us now consider an ensemble of initial conditions in which the initial coordinate $q(0)$ and the initial momentum $p(0)$ of the particle have given and fixed values for the different members of the ensemble, while the joint distribution of the set $\{q_j(0), p_j(0)\}$ of the coordinates and the momenta of the oscillators is given by the canonical equilibrium distribution:

$$P(\{q_j(0), p_j(0)\})$$
$$= \frac{\exp\left(-\beta \sum_{j=1}^{N}\left[\frac{p_j^2(0)}{2m_j} + \frac{1}{2}m_j\omega_j^2\left(q_j(0) - \frac{C_j q(0)}{m_j\omega_j^2}\right)^2\right]\right)}{\int\left(\prod_{l=1}^{N} \mathrm{d}q_l(0)\mathrm{d}p_l(0)\right)\exp\left(-\beta \sum_{j=1}^{N}\left[\frac{p_j^2(0)}{2m_j} + \frac{1}{2}m_j\omega_j^2\left(q_j(0) - \frac{C_j q(0)}{m_j\omega_j^2}\right)^2\right]\right)}\,, \tag{2.103}$$

corresponding to the system of oscillators being initially in thermal equilibrium at temperature $T$; here, $\beta = 1/k_B T$. Corresponding to such an ensemble, one finds that the function $f(t)$ in Eq. (2.102) is a random function of time, whose statistical properties may be deduced from that of the quantities $(q_j(0) - C_j q(0)/(m_j\omega_j^2))$ and $p_j(0)$. To this end, one obtains

$$\begin{aligned}
&\left\langle q_j(0) - \frac{C_j q(0)}{m_j \omega_j^2} \right\rangle = 0\,, \quad \langle p_j(0) \rangle = 0\,, \\
&\left\langle \left( q_i(0) - \frac{C_i q(0)}{m_i \omega_i^2} \right) \left( q_j(0) - \frac{C_j q(0)}{m_j \omega_j^2} \right) \right\rangle = \frac{k_B T \delta_{i,j}}{m_j \omega_j^2}\,, \\
&\left\langle \left( q_i(0) - \frac{C_i q(0)}{m_i \omega_i^2} \right) p_j(0) \right\rangle = 0, \quad \langle p_i(0) p_j(0) \rangle = m_j k_B T \delta_{i,j}\,,
\end{aligned} \tag{2.104}$$

where the angular brackets denote averaging with respect to the mentioned ensemble, that is, with respect to the distribution (2.103). Using Eq. (2.102), one then obtains

$$\langle f(t) \rangle = 0, \quad \langle f(t) f(t') \rangle = k_B T \zeta(t - t')\,, \tag{2.105}$$

where $\zeta(t)$ is defined in Eq. (2.102).

We now take the limit $N \to \infty$, in which case one makes the substitution $\sum_{j=1}^{N} \to N \int_0^\infty \mathrm{d}\omega\, g(\omega)$, where the density of states $g(\omega)$ is defined such that $g(\omega)\mathrm{d}\omega$ gives the probability of finding an oscillator with frequency in the range $[\omega, \omega + \mathrm{d}\omega]$. Assuming all the oscillators to have the same mass (i.e., $m_j = m \;\forall\; j$) and all the couplings constants to be equal and to have the scaling form $C_j = C/\sqrt{N}$, one obtains

$$\zeta(t) = C^2 \int_0^\infty \mathrm{d}\omega\, g(\omega) \frac{\cos(\omega t)}{m\omega^2}\,. \tag{2.106}$$

A particular form of $g(\omega)$ that is of interest is given by the so-called Ohmic form:

$$g(\omega) = \begin{cases} \frac{3\omega^2}{\omega_c^3} & \text{if } \omega < \omega_c\,, \\ 0 & \text{if } \omega > \omega_c\,, \end{cases} \tag{2.107}$$

where $\omega_c$ is a high-frequency cut-off. For such a choice of $g(\omega)$, one obtains

$$\zeta(t) = \frac{3C^2}{m\omega_c^3} \frac{\sin(\omega_c t)}{t}\,. \tag{2.108}$$

In the limit $\omega_c \to \infty$, we get

$$\zeta(t) = 2\gamma \delta(t)\,, \tag{2.109}$$

where, provided $C$ scales as $\omega_c^{3/2}$ (i.e., $C \sim \omega_c^{3/2}$, with a suitable prefactor to make $C$ have the right dimension), we have $\gamma$ a constant given by

$$\gamma \equiv \frac{3\pi C^2}{2m\omega_c^3}\,. \tag{2.110}$$

Here, $\delta(t)$ is the Dirac delta function, and we have used $\lim_{\epsilon\to 0}\ [\sin(x/\epsilon)/(\pi x)] = \delta(x)$.

Using Eq. (2.109) in Eq. (2.101) and using $\int_{-\infty}^{a} dx\ \delta(x-a) = 1/2$ lead to

$$\dot{p} = -V'(q) - \gamma\frac{p(t)}{M} + f(t), \tag{2.111}$$

with $f(t)$ being a random function with statistical properties given by

$$\langle f(t)\rangle = 0, \quad \langle f(t)f(t')\rangle = 2\gamma k_B T\delta(t-t'). \tag{2.112}$$

Writing $p = Mv$ in Eq. (2.111), with $v$ being the velocity of the particle, one immediately recognizes Eq. (2.111) as of the same form as the Langevin equation, Eq. (2.83).

## References

1. N.G.V. Kampen, *Stochastic Processes in Physics and Chemistry*, 3rd edn. (Elsevier, Amsterdam, 2007)
2. C.W. Gardiner, *Handbook of Stochastic Methods for Physics, Chemistry and the Natural Sciences* (Springer, Berlin, 1983)
3. H. Risken, *The Fokker-Planck Equation: Methods of Solution and Applications* (Springer, Berlin, 1996)
4. R. Balescu, *Equilibrium and Nonequilibrium Statistical Mechanics* (Wiley, New York, 1975)
5. K. Huang, *Introduction to Statistical Physics*, 2nd edn. (CRC Press, Boca Raton, 2010)
6. V. Balakrishnan, *Elements of Nonequilibrium Statistical Mechanics* (Ane Books India, New Delhi, 2008)
7. K. Jacobs, *Stochastic Processes for Physicists: Understanding Noisy Systems*, 1st edn. (Cambridge University Press, Cambridge, 2010). https://doi.org/10.1017/CBO9780511815980
8. N. Wax, *Selected Papers on Noise and Stochastic Processes* (Dover, New York, 1954)
9. V. Balakrishnan, *Mathematical Physics: Applications and Problems* (Springer, Cham, 2020). https://doi.org/10.1007/978-3-030-39680-0

# Chapter 3
# Kinetic Theory

**Abstract** In this chapter, we lay the foundation of the kinetic theory approach to study of interacting particle systems. We start with defining microstates, macrostates, and the phase space (the $\Gamma$-space) of interacting particle systems involving $N \gg 1$ particles, and then discuss two broad classes of inter-particle interactions: Short and long-range interactions. In a suitably-defined thermodynamic limit, we then introduce the notion of $N$-particle distribution function in the $\Gamma$ space and the single-particle distribution function in the single-particle phase space (the $\mu$-space). For the evolution of the $N$-particle distribution function, we discuss in turn the Liouville equation and the BBGKY hierarchy, and obtain the evolution of the single-particle distribution function by suitably closing the hierarchy. For short-range systems, we obtain the corresponding Boltzmann equation. The derivation of the corresponding equation for long-range systems, which is known as the Vlasov equation, is then discussed in detail. As an illustration of the techniques developed, we then obtain for a model of interacting classical Heisenberg spins the corresponding Vlasov equation, via the aforementioned route involving the BBGKY hierarchy and also via an alternative Klimontovich approach.

## 3.1 System of Interacting Particles

Let us consider a system of $N$ interacting and identical particles of mass $m$ contained in a macroscopic volume $V$ in three dimensions. In this chapter, we will treat the system classically. The total energy of the system is denoted by $E$. The position and the momentum of the $i$th particle, $i = 1, \ldots, N$, are denoted by the three-dimensional vectors $\boldsymbol{q}_i$ and $\boldsymbol{p}_i$, respectively. The components of the position and momentum vectors satisfy the Poisson-bracket identity

$$\{q_{i\alpha}, p_{j\beta}\} = \delta_{i,j}\, \delta_{\alpha,\beta}\,, \tag{3.1}$$

where one has $\alpha, \beta = x, y, z$. The system is characterized by a time-independent Hamiltonian given by



D. Das and S. Gupta, *Facets of Noise*, Fundamental Theories of Physics 214,
https://doi.org/10.1007/978-3-031-45312-0_3

$$H = \sum_{i=1}^{N} \frac{\boldsymbol{p}_i \cdot \boldsymbol{p}_i}{2m} + \sum_{\substack{i,j=1 \\ i<j}}^{N} V\left(|\boldsymbol{q}_i - \boldsymbol{q}_j|\right), \tag{3.2}$$

where $V(q)$ is the inter-particle potential, which due to assumed isotropy of space is taken to be a function solely of inter-particle separation. We also assume the embedding space to be homogeneous. For brevity, denoting $p_i = |\boldsymbol{p}_i|$ and $V_{ij} = V\left(|\boldsymbol{q}_i - \boldsymbol{q}_j|\right)$, Eq. (3.2) is rewritten as

$$H = \sum_{i=1}^{N} \frac{p_i^2}{2m} + \sum_{\substack{i,j=1 \\ i<j}}^{N} V_{ij} \, . \tag{3.3}$$

The corresponding Hamilton's equations of motion are given by

$$\dot{\boldsymbol{q}}_i = \frac{\partial H}{\partial \boldsymbol{p}_i} = \frac{\boldsymbol{p}_i}{m}, \quad \dot{\boldsymbol{p}}_i = -\frac{\partial H}{\partial \boldsymbol{q}_i} = -\sum_{\substack{j=1 \\ j\neq i}}^{N} \frac{\partial V_{ij}}{\partial \boldsymbol{q}_i}, \tag{3.4}$$

where the dot denotes as usual derivative with respect to time. Note that the equations of motion (3.4) are time-reversal invariant: changing $t$ to $-t$ leaves the equations invariant. Using Eq. (3.4), one may easily check from Eq. (3.3) that $\mathrm{d}H/\mathrm{d}t = 0$, which implies that the total energy $E$ is a constant of motion.

## 3.2 Microstates, Macrostates, and Phase Space

A microstate of the system at time $t$ is completely specified by $N$ position vectors $\boldsymbol{q}_i(t)$ and $N$ momentum vectors $\boldsymbol{p}_i(t)$. Counting the components of $\boldsymbol{q}_i$ and $\boldsymbol{p}_i$, one concludes that one needs to specify values of $3N$ position variables and $3N$ momentum variables for a complete specification of the microstate. In contrast to a microstate, specifying values of macroscopic observables (e.g., values of number of particles $N$, volume $V$, total energy $E$) defines a macroscopic state or a macrostate of the system. The set of microstates corresponding to a given macrostate is obtained by solving an equation of the form $H = H(\{\boldsymbol{q}_i, \boldsymbol{p}_i\}) = E$, which evidently shows that one will have a very large number (which increases with increase of $N$, typically exponentially) of microstates consistent with a given macrostate. This collection of microstates, all of which are consistent with a given macrostate, is what is referred to as an ensemble.

Now, we may in principle solve the Hamilton's equations of motion (3.4) for any given initial condition $\{\boldsymbol{q}_i(0), \boldsymbol{p}_i(0)\}\ \forall\ i$ to obtain the microstate of the system at all future times. However, in practice, it is extremely difficult, if not impossible, to solve the set of $6N$ coupled (typically nonlinear) differential equations (3.4) when $N$ is large (typically, $N \sim N_A$, the Avogadro's number). In particular, in the limit $N \to \infty$, we are left with no choice but to study the dynamics of the system statistically.

In order to proceed, one defines the $6N$-dimensional phase space $\Gamma$ spanned by the components of $\boldsymbol{q}_i$ and $\boldsymbol{p}_i\ \forall\, i$. In the following, $\boldsymbol{q}_i$'s and $\boldsymbol{p}_i$'s when written explicitly as functions of time refer to the Lagrangian coordinates of the particles, and otherwise to the Eulerian coordinates of the phase space[1]. In the phase space, the microstate of the system at time $t$ is represented by a single representative point. While performing measurements of macroscopic observables of the system at a given time instant, we would not be able to make out the particular microstate the system is existing in at that instant. Consequently, for a given macrostate, it is reasonable while computing the value of a macroscopic observable $O \equiv O(\{\boldsymbol{q}_i(t), \boldsymbol{p}_i(t)\})$ at any time $t$ to consider its value averaged over all consistent microstates (the ensemble) according to an appropriate probability measure. This averaging procedure is what lies at the heart of the statistical mechanics approach to study of interacting particle systems.

## 3.3 Inter-particle Potential: Short Verses Long-Range Interactions

Long-range interacting systems are characterized by a two-body interparticle potential that decays at large separation $r$ as [1–5]

$$V(r) \sim \frac{J}{r^\alpha};\ 0 \le \alpha \le d\,, \tag{3.5}$$

where $d$ is the dimension of the embedding space, and $J$ is the coupling strength. The case $\alpha = 0$ corresponds to mean-field interactions, whereby every particle is interacting with every other with equal strength (global coupling). The range of allowed values of the decay exponent $\alpha$ implies that the energy per particle, denoted by $\epsilon$, scales super-linearly with the enclosing volume. To see this, consider the example of a particle placed at the center of a hypersphere of radius $R$ in $d$ dimensions, with the other particles homogeneously distributed with a mass density $\rho$. For the interaction potential (3.5), the energy per particle is given by

$$\epsilon = \int_\delta^R \mathrm{d}^d r\, \rho \frac{J}{r^\alpha} = \frac{\rho J \Omega_d}{d-\alpha}\left[R^{d-\alpha} - \delta^{d-\alpha}\right], \tag{3.6}$$

[1] This is in analogy with the convention adopted in studying fluid flows, whereby the Lagrangian characterization of the flow field corresponds to monitoring of fluid motion in terms of motion of an individual fluid 'packet' as it moves through space and time. The Eulerian characterization instead corresponds to monitoring as a function of time the motion at specific locations of the space through which the fluid flows. A parallel may be drawn between the Lagrangian viewpoint and the viewpoint of an observer on a boat that is drifting down a river, while the Eulerian viewpoint corresponds to an observer stationed on the bank of the river and monitoring the river flowing by.

where $\delta \to 0$ is a short distance cut-off that one has to introduce to exclude the contribution to the energy from particles located in a small neighborhood of radius $\delta$, and is a way to regularize the divergence of the potential (3.5) at short distances. In Eq. (3.6), the quantity $\Omega_d$ denotes the angular volume in $d$ dimensions. It then follows from the equation that as $R$ is increased keeping $\rho$ fixed, the energy $\epsilon$ remains finite ($= \rho J \Omega_d/(\alpha - d)\,\delta^{d-\alpha}$) so long as $\alpha > d$, implying linear scaling of the total energy $E$ with the volume $V \sim R^d$, and consequently, the system is extensive. Such systems are called short-range interacting systems. On the other hand, for $0 \leq \alpha < d$, the energy $\epsilon$ scales with the volume as $\epsilon \sim V^{1-\alpha/d}$ (the energy scales logarithmically with $V$ in the marginal case $\alpha = d$), implying a super-linear scaling of the total energy of long-range systems with $V$, as $E \propto V^{2-\alpha/d}$. These systems are thus generically non-extensive.

For long-range systems, on computing the Helmholtz free energy $F \equiv U - TS$, with $U$ being the internal energy, $T$ the intensive temperature and $S$ being the entropy that typically scales linearly with the volume, $S \sim V$, we find due to the super-linear scaling of $U$ with $V$ that the thermodynamic properties are dominated by the energy. Consequently, the equilibrium state of a mechanically isolated long-range system at constant temperature, which is obtained by minimizing $F$, would correspond to the state having the minimum energy, that is, the ground state. One would like to have a theoretical framework allowing for a competition between the energy and the entropy contribution to the free energy so that one may have such phenomena as phase transitions that are known to occur in long-range systems. This may be achieved by scaling the coupling constant as

$$J \to \frac{J}{V^{1-\alpha/d}}, \tag{3.7}$$

thereby making the energy extensive in the volume. This "mathematical trick" (called the Kac's trick [6]) allows to properly study long-range systems within the same framework of equilibrium statistical mechanics as the one that exists for short-range ones.

An important remark is in order. Although Kac's trick allows to obtain an energy that is extensive in the volume, it is not necessarily additive (additivity implies extensivity, but not the converse). This may be illustrated by considering a representative example, the well-studied Curie-Weiss model of magnetism, with the Hamiltonian given by

$$H_{\mathrm{CW}} = -\frac{J}{2N}\sum_{i,j=1}^{N} \sigma_i \sigma_j, \tag{3.8}$$

where $\sigma_i = \pm 1$ are spin variables occupying the sites $i$ of a lattice. The model involves a mean-field system, whereby every spin is interacting with every other with the same strength, and which may be considered as the $\alpha \to 0$ limit of the potential (3.5). Since we have a mean-field system, one does not need to specify the structure of the

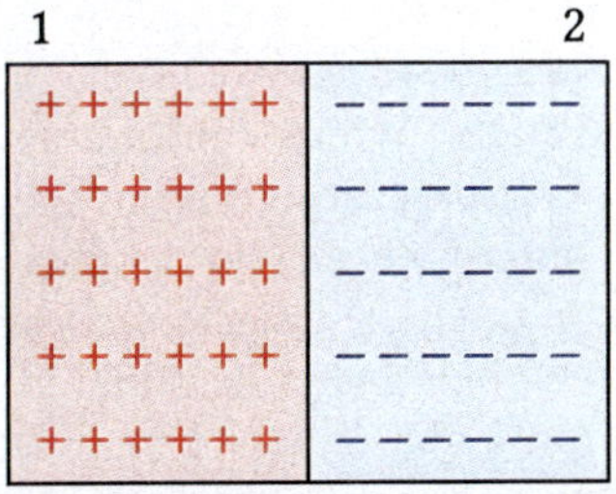


**Fig. 3.1 Non-additivity in the extensive Curie-Weiss model**: Here, we show a zero-magnetization macrostate of the Curie-Weiss system, Eq. (3.8), that is constituted by $N/2$ up-spins in part 1 and $N/2$ down-spins in part 2. An up-spin is denoted by a $+$ sign and a down-spin by a $-$ spin

underlying lattice; it suffices to mention that every site is connected to every other. In the Hamiltonian (3.8), one may notice that the coupling strength has been rescaled by using Kac's trick, so that the energy is extensive in the number of spins given by $N$. Now, let us consider a macrostate with zero total magnetization: $M \equiv \sum_{i=1}^{N} \sigma_i = 0$, which is composed of $N/2$ spin-$(+1)$ sites and $N/2$ spin-$(-1)$ sites, see Fig. 3.1. Since the energy is proportional to the square of magnetization (see Eq. (3.8)), the total energy of the system is $E_{1+2} = 0$. However, the energy of the two parts, namely, $E_1 = E_2 = -JN/4$, does not vanish, and, therefore, we have $E_{1+2} \neq E_1 + E_2$. Thus, the system although extensive is not additive. We have hereby illustrated that extensivity does not guarantee additivity.

## 3.4 Definition of the Thermodynamic Limit

Let us recall for future purposes the notion of thermodynamic limit. In this limit, one has $N \to \infty$, $V \to \infty$ and $E \to \infty$, such that the particle density $n \equiv N/V$ and the energy density $\epsilon \equiv E/V$ are both finite.

## 3.5 Distribution Functions

### *3.5.1 N-particle Distribution Function in Γ Space*

It is evident that an ensemble corresponds to a collection of points in the allowed region of the phase space, with the allowed region defined by the given macrostate of the system. In order to characterize this ensemble of points, let us define for large $N$ an $N$-particle distribution function $f_N(\boldsymbol{q}_1, \boldsymbol{p}_1, \ldots, \boldsymbol{q}_N, \boldsymbol{p}_N, t)$, such that the quantity $f_N(\boldsymbol{q}_1, \boldsymbol{p}_1, \ldots, \boldsymbol{q}_N, \boldsymbol{p}_N, t)\mathrm{d}\boldsymbol{q}_1\mathrm{d}\boldsymbol{p}_1 \ldots \mathrm{d}\boldsymbol{q}_N\mathrm{d}\boldsymbol{p}_N$ is proportional to the number of microstates (consistent with the given macrostate) present in the infinitesimal phase-space volume element $d\boldsymbol{q}_1 d\boldsymbol{p}_1 \ldots d\boldsymbol{q}_N d\boldsymbol{p}_N$ around the point $(\boldsymbol{q}_1, \boldsymbol{p}_1, \ldots, \boldsymbol{q}_N, \boldsymbol{p}_N)$ at time $t$. Each of the representative points constituting the ensemble evolves in the allowed region of the phase space according to the Hamilton's equations of motion (3.4), thereby tracing out a trajectory in the phase space. As the equations of motion (3.4) are first order in time, the trajectory corresponding to a representative point is uniquely determined in terms of its initial location in the phase space. Moreover, the trajectory of a representative point cannot self-intersect, since if it did, there would be two possible future directions at the point of intersection. For notational simplicity, in this chapter, we will use the notation $z_i \equiv (\boldsymbol{q}_i, \boldsymbol{p}_i)$, and denote the $N$-particle distribution function by $f_N(z_1, \ldots, z_N, t)$. Note that, as discussed in Sect. 3.2, here $z_i$'s are the Eulerian coordinates of the $\Gamma$ space.

### 3.5.2 Single-Particle Distribution Function in $\mu$ Space

In contrast to the $\Gamma$ space, one may define a single-particle $\mu$ space that is a six dimensional space spanned by the six axes (the components of $\boldsymbol{q}$ and $\boldsymbol{p}$). In this space is defined a single-particle distribution function $f_1(z, t)$, which is such that $f_1(z, t)dz$ is proportional to the number of particles that are present in the infinitesimal volume element $d\boldsymbol{q}d\boldsymbol{p}$ around the point $(\boldsymbol{q}, \boldsymbol{p})$ at time $t$. Note that the variable $z$ in $f_1(z, t)$ refers to the Eulerian coordinates of the $\mu$ space. We will come back to a more detailed discussion of the single-particle distribution later in the chapter, in Sect. 3.6.2.

## 3.6 Evolution of the $N$-particle Distribution Function

For $N$ not necessarily large, the distribution function $f_N$ has to be replaced with a discrete density function $f_N^{(d)}$ defined as

$$f_N^{(d)}(z_1, \ldots, z_N, t) = \prod_{i=1}^{N} \delta\big(z_i - z_i(t)\big), \tag{3.9}$$

where $z_i(t)$ refers to the $i$th particle. The quantity $f_N^{(d)}$ is nothing but the joint probability at time $t$ to find one particle at $z_1$, another particle at $z_2$, and so on. As the swarm of representative points, denoting microstates corresponding to a given macrostate, evolves in the phase space, the discrete function $f_N^{(d)}(z_1, \ldots, z_N, t)$ also changes its form. The time evolution of $f_N^{(d)}(z_1, \ldots, z_N, t)$ is obtained as

$$\frac{\partial f_N^{(\mathrm{d})}}{\partial t} = \frac{\partial}{\partial t}\left[\delta^3(\boldsymbol{q}_1 - \boldsymbol{q}_1(t))\delta^3(\boldsymbol{p}_1 - \boldsymbol{p}_1(t)) \ldots \delta^3(\boldsymbol{q}_N - \boldsymbol{q}_N(t))\delta^3(\boldsymbol{p}_N - \boldsymbol{p}_N(t))\right]$$
$$= -\dot{\boldsymbol{q}}_1(t) \cdot \frac{\partial \delta^3(\boldsymbol{q}_1 - \boldsymbol{q}_1(t))}{\partial \boldsymbol{q}_1}\delta^3(\boldsymbol{p}_1 - \boldsymbol{p}_1(t)) \ldots \delta^3(\boldsymbol{q}_N - \boldsymbol{q}_N(t))\delta^3(\boldsymbol{p}_N - \boldsymbol{p}_N(t))$$
$$- \dot{\boldsymbol{p}}_1(t) \cdot \delta^3(\boldsymbol{q}_1 - \boldsymbol{q}_1(t))\frac{\partial \delta^3(\boldsymbol{p}_1 - \boldsymbol{p}_1(t))}{\partial \boldsymbol{p}_1} \ldots \delta^3(\boldsymbol{q}_N - \boldsymbol{q}_N(t))\delta^3(\boldsymbol{p}_N - \boldsymbol{p}_N(t))$$
$$- \ldots, \tag{3.10}$$

where we have used $\partial\delta(x-a)/\partial x = -\,\partial\delta(x-a)/\partial a$. Note that a single dot $(\cdot)$ occurring between the factors in different terms on the right hand side of Eq. (3.10) refers to the vector dot product. Substituting the Hamilton's equations (3.4) in Eq. (3.10), we get

$$\frac{\partial f_N^{(\mathrm{d})}}{\partial t}$$
$$= -\frac{\boldsymbol{p}_1(t)}{m} \cdot \frac{\partial \delta^3(\boldsymbol{q}_1 - \boldsymbol{q}_1(t))}{\partial \boldsymbol{q}_1}\delta^3(\boldsymbol{p}_1 - \boldsymbol{p}_1(t)) \ldots \delta^3(\boldsymbol{q}_N - \boldsymbol{q}_N(t))\delta^3(\boldsymbol{p}_N - \boldsymbol{p}_N(t))$$
$$+ \left(\sum_{\substack{j=1 \\ j\neq 1}}^{N} \frac{\partial V_{1j}}{\partial \boldsymbol{q}_1(t)}\right) \cdot \delta^3(\boldsymbol{q}_1 - \boldsymbol{q}_1(t))\frac{\partial \delta^3(\boldsymbol{p}_1 - \boldsymbol{p}_1(t))}{\partial \boldsymbol{p}_1} \times \ldots$$
$$\times\, \delta^3(\boldsymbol{q}_N - \boldsymbol{q}_N(t))\delta^3(\boldsymbol{p}_N - \boldsymbol{p}_N(t)) - \ldots. \tag{3.11}$$

Now, using the identity $f(x)\delta(x-a) = f(a)\delta(x-a)$ in Eq. (3.11), one obtains

$$\frac{\partial f_N^{(\mathrm{d})}}{\partial t}$$
$$= -\frac{\boldsymbol{p}_1}{m} \cdot \frac{\partial \delta^3(\boldsymbol{q}_1 - \boldsymbol{q}_1(t))}{\partial \boldsymbol{q}_1}\delta^3(\boldsymbol{p}_1 - \boldsymbol{p}_1(t)) \ldots \delta^3(\boldsymbol{q}_N - \boldsymbol{q}_N(t))\delta^3(\boldsymbol{p}_N - \boldsymbol{p}_N(t))$$
$$+ \left(\sum_{\substack{j=1 \\ j\neq 1}}^{N} \frac{\partial V_{1j}}{\partial \boldsymbol{q}_1}\right) \cdot \delta^3(\boldsymbol{q}_1 - \boldsymbol{q}_1(t))\frac{\partial \delta^3(\boldsymbol{p}_1 - \boldsymbol{p}_1(t))}{\partial \boldsymbol{p}_1} \times \ldots$$
$$\times\, \delta^3(\boldsymbol{q}_N - \boldsymbol{q}_N(t))\delta^3(\boldsymbol{p}_N - \boldsymbol{p}_N(t)) - \ldots, \tag{3.12}$$

which using the defining equation (3.9) of $f_N^{(\mathrm{d})}$ may be written as

$$\frac{\partial f_N^{(\mathrm{d})}}{\partial t} = -\frac{\boldsymbol{p}_1}{m} \cdot \frac{\partial f_N^{(\mathrm{d})}}{\partial \boldsymbol{q}_1} + \left(\sum_{\substack{j=1 \\ j\neq 1}}^{N} \frac{\partial V_{1j}}{\partial \boldsymbol{q}_1}\right) \cdot \frac{\partial f_N^{(\mathrm{d})}}{\partial \boldsymbol{p}_1} - \ldots. \tag{3.13}$$

Considering all the terms on the right hand side of Eq. (3.13), we finally obtain

$$\frac{\partial f_N^{(\mathrm{d})}}{\partial t} = -\sum_{i=1}^{N} \frac{\boldsymbol{p}_i}{m} \cdot \frac{\partial f_N^{(\mathrm{d})}}{\partial \boldsymbol{q}_i} + \sum_{i=1}^{N} \sum_{\substack{j=1 \\ j \neq i}}^{N} \left( \frac{\partial V_{ij}}{\partial \boldsymbol{q}_i} \right) \cdot \frac{\partial f_N^{(\mathrm{d})}}{\partial \boldsymbol{p}_i} . \tag{3.14}$$

Note that the derivation of Eq. (3.14) is exact and holds for any number of particles.

### 3.6.1 *Liouville Equation*

For sufficiently large $N$, one has $f_N^{(\mathrm{d})} \to f_N(z_1, \ldots, z_N, t)$ and from Eq. (3.14), one obtains the dynamical equation for the evolution of $f_N$, namely,

$$\frac{\partial f_N}{\partial t} = -\sum_{i=1}^{N} \frac{\boldsymbol{p}_i}{m} \cdot \nabla_{\boldsymbol{q}_i} f_N + \sum_{i=1}^{N} \sum_{\substack{j=1 \\ j \neq i}}^{N} (\nabla_{\boldsymbol{q}_i} V_{ij}) \cdot (\nabla_{\boldsymbol{p}_i} f_N) , \tag{3.15}$$

where we have used the notation $\nabla_{\boldsymbol{q}_i} \equiv \partial/\partial \boldsymbol{q}_i$ and $\nabla_{\boldsymbol{p}_i} \equiv \partial/\partial \boldsymbol{p}_i$. Equation (3.15) is known as the Liouville equation. The total number of microstates consistent with a given macrostate is constant in time. Then, one may redefine $f_N$ by dividing it by this constant, so that $f_N$ is normalized to unity at all times:

$$\int \mathrm{d}z_1 \ldots \mathrm{d}z_N \; f_N(z_1, \ldots, z_N, t) = 1 . \tag{3.16}$$

Once normalized, $f_N(z_1, \ldots, z_N)$ becomes the probability density to find a microstate at time $t$ around the phase-space point $(z_1, \ldots, z_N)$. Note that the function $f_N$ is defined over the region of the phase space as allowed by the given macrostate of the system. However, in the thermodynamic limit, when the allowed region extends effectively to infinity, we require $f_N$ to vanish at the boundaries of the phase space. This would ensure that the integral on the left hand side of Eq. (3.16) is finite, and consequently, the equality in Eq. (3.16) holds. From Eq. (3.3), it is seen that the Hamiltonian $H$ is invariant with respect to exchange of any two particles. The Liouville dynamics (3.15) ensures that $f_N$ also respects this symmetry.

### 3.6.2 *BBGKY Hierarchy*

Equilibrium statistical mechanics deals with the stationary solution of the Liouville equation, i.e., the equilibrium state. However, in order to understand how a system relaxes to equilibrium while starting from generic initial conditions, one needs to know the full $N$-particle distribution function as a function of time, a task extremely difficult to accomplish. To proceed, we start with a reduced $s$-particle distribution function defined as

$$f_s(z_1, \dots, z_s, t) \equiv \frac{N!}{(N-s)!} \int \mathrm{d}z_{s+1} \dots \mathrm{d}z_N \, f_N(z_1, \dots, z_N, t) ; \quad s \in [1, (N-1)]. \tag{3.17}$$

Note that $f_1(z_1, t)$ is the single-particle distribution function defined in Sect. 3.5.2. The prefactor on the right hand side of Eq. (3.17) takes care of the fact that because of the particle-permutation symmetry of $f_N$ mentioned above, it is immaterial as to which of the $(N-s)$ particles we integrate over on the right hand side. For example, let us consider $N = 3$ and $s = 2$. In this case, let us obtain the two-particle distribution function $f_2$. We recall that $f_2(z_1, z_2, t)$ is the probability distribution to observe any two particles at positions around $z_1$ and $z_2$ at time $t$, whatever be the position of the rest of the particles (here, only one particle has been left out, as we have $N = 3$). Then, one writes

$$\begin{aligned} f_2(z_1, z_2, t) = & \int \mathrm{d}z_1'\mathrm{d}z_2'\mathrm{d}z_3' \, \delta(z_1' - z_1) \, \delta(z_2' - z_2) \, f_3(z_1', z_2', z_3', t) \\ & + \int \mathrm{d}z_1'\mathrm{d}z_2'\mathrm{d}z_3' \, \delta(z_1' - z_1) \, \delta(z_3' - z_2) \, f_3(z_1', z_2', z_3', t) \\ & + \int \mathrm{d}z_1'\mathrm{d}z_2'\mathrm{d}z_3' \, \delta(z_2' - z_1) \, \delta(z_1' - z_2) \, f_3(z_1', z_2', z_3', t) \\ & + \int \mathrm{d}z_1'\mathrm{d}z_2'\mathrm{d}z_3' \, \delta(z_2' - z_1) \, \delta(z_3' - z_2) \, f_3(z_1', z_2', z_3', t) \\ & + \int \mathrm{d}z_1'\mathrm{d}z_2'\mathrm{d}z_3' \, \delta(z_3' - z_1) \, \delta(z_1' - z_2) \, f_3(z_1', z_2', z_3', t) \\ & + \int \mathrm{d}z_1'\mathrm{d}z_2'\mathrm{d}z_3' \, \delta(z_3' - z_1) \, \delta(z_2' - z_2) \, f_3(z_1', z_2', z_3', t) . \end{aligned} \tag{3.18}$$

The contributions of all the terms on the right hand side of Eq. (3.18) are the same. In order to prove this fact, here, we show how to recover the first term from say the fourth term. The steps involved in this proof may be applied to the other terms on the right hand side of Eq. (3.18) to recover the first term. We write the fourth term on the right hand side of Eq. (3.18) as

$$\begin{aligned} & \int \mathrm{d}z_1'\mathrm{d}z_2'\mathrm{d}z_3' \, \delta(z_2' - z_1) \, \delta(z_3' - z_2) \, f_3(z_1', z_2', z_3', t) \\ & = \int \mathrm{d}z_2'\mathrm{d}z_1'\mathrm{d}z_3' \, \delta(z_1' - z_1) \, \delta(z_3' - z_2) \, f_3(z_2', z_1', z_3', t) , \end{aligned} \tag{3.19}$$

where we have interchanged the dummy integration variables $z_1'$ and $z_2'$. Using the permutation symmetry $f_3(z_2', z_1', z_3', t) = f_3(z_1', z_2', z_3', t)$, the above equation is reduced to

$$\int dz_1' dz_2' dz_3' \, \delta(z_2' - z_1) \, \delta(z_3' - z_2) \, f_3(z_1', z_2', z_3', t)$$
$$= \int dz_2' dz_1' dz_3' \, \delta(z_1' - z_1) \, \delta(z_3' - z_2) \, f_3(z_1', z_2', z_3', t) \,. \tag{3.20}$$

Now, interchanging the integration variables $z_2'$ and $z_3'$ on the right hand side of the last equation, then using $f_3(z_1', z_3', z_2', t) = f_3(z_1', z_2', z_3', t)$, one recovers the first term on the right hand side of Eq. (3.18). On following similar mathematical steps for all terms other than the first term on the right hand side of Eq. (3.18), one obtains

$$f_2(z_1, z_2, t) = 6 \int dz_1' dz_2' dz_3' \, \delta(z_1' - z_1) \, \delta(z_2' - z_2) \, f_3(z_1', z_2', z_3', t)$$
$$= 6 \int dz_3' \, f_3(z_1, z_2, z_3', t) \,. \tag{3.21}$$

The normalization of $f_s$ is obtained by using Eqs. (3.17) and (3.16), and is given by

$$\int dz_1 \ldots dz_s \, f_s(z_1, \ldots, z_s, t) = \frac{N!}{(N-s)!} = (N-s+1)(N-s+2)\ldots N \,. \tag{3.22}$$

In the thermodynamic limit, it is then evident that the normalization of $f_s$ scales with $N$ as $N^s$.

The time evolution of $f_s$ is obtained from Eq. (3.17) with the help of the Liouville equation (3.15), as

$$\frac{\partial f_s}{\partial t} = \frac{N!}{(N-s)!} \int dz_{s+1} \ldots dz_N \left[ -\sum_{i=1}^{N} \frac{\boldsymbol{p}_i}{m} \cdot \nabla_{\boldsymbol{q}_i} f_N + \sum_{i=1}^{N} \sum_{\substack{j=1 \\ j \neq i}}^{N} (\nabla_{\boldsymbol{q}_i} V_{ij}) \cdot \nabla_{\boldsymbol{p}_i} f_N \right] . \tag{3.23}$$

We now proceed to evaluating the right hand side of Eq. (3.23) by rewriting it as

$$\frac{\partial f_s}{\partial t} = \frac{N!}{(N-s)!} \int dz_{s+1} \ldots dz_N \left[ -\sum_{i=1}^{s} \frac{\boldsymbol{p}_i}{m} \cdot \nabla_{\boldsymbol{q}_i} f_N - \sum_{i=s+1}^{N} \frac{\boldsymbol{p}_i}{m} \cdot \nabla_{\boldsymbol{q}_i} f_N \right.$$
$$\left. + \sum_{i=1}^{s} \sum_{\substack{j=1 \\ j \neq i}}^{N} (\nabla_{\boldsymbol{q}_i} V_{ij}) \cdot \nabla_{\boldsymbol{p}_i} f_N + \sum_{i=s+1}^{N} \sum_{\substack{j=1 \\ j \neq i}}^{N} (\nabla_{\boldsymbol{q}_i} V_{ij}) \cdot \nabla_{\boldsymbol{p}_i} f_N \right] . \tag{3.24}$$

Let us be careful and compute the right hand side of the above equation term by term. The first integral on the right hand side of Eq. (3.24) yields

$$I_1 \equiv -\frac{N!}{(N-s)!}\sum_{i=1}^{s}\frac{\boldsymbol{p}_i}{m}\cdot\nabla_{\boldsymbol{q}_i}\left(\int \mathrm{d}z_{s+1}\ldots\mathrm{d}z_N\, f_N\right) = -\sum_{i=1}^{s}\frac{\boldsymbol{p}_i}{m}\cdot\nabla_{\boldsymbol{q}_i} f_s\,, \tag{3.25}$$

where we have used the definition of $f_s$ given in Eq. (3.17).

Now, the second integral on the right hand side of Eq. (3.24) is given by

$$I_2 \equiv -\frac{N!}{(N-s)!}\sum_{i=s+1}^{N}\int \mathrm{d}z_{s+1}\ldots\mathrm{d}z_N\,\frac{\boldsymbol{p}_i}{m}\cdot\nabla_{\boldsymbol{q}_i} f_N\,, \tag{3.26}$$

where we have a gradient term in $\boldsymbol{q}_i$ with $\boldsymbol{q}_i$-independent coefficient. This type of integrals evaluates $f_N$ at the boundary of the phase space. We have to deal with such integrals very often in the following. Therefore, let us evaluate one such integral and see how it works. We take the first term from the integrand on the right hand side of Eq. (3.26), and write

$$\begin{aligned} I_2^{(1)} &= \int \mathrm{d}\boldsymbol{q}_{s+1}\mathrm{d}\boldsymbol{p}_{s+1}\ldots\mathrm{d}z_N\,\frac{\boldsymbol{p}_{s+1}}{m}\cdot\nabla_{\boldsymbol{q}_{s+1}} f_N(z_1,\ldots,\boldsymbol{q}_{s+1},\boldsymbol{p}_{s+1},\ldots,z_N,t) \\ &= \int \mathrm{d}\boldsymbol{p}_{s+1}\ldots\mathrm{d}z_N\,\frac{\boldsymbol{p}_{s+1}}{m}\int_{-\infty}^{\infty}\mathrm{d}\boldsymbol{q}_{s+1}\,\nabla_{\boldsymbol{q}_{s+1}} f_N(z_1,\ldots,\boldsymbol{q}_{s+1},\boldsymbol{p}_{s+1},\ldots,z_N,t)\,, \end{aligned} \tag{3.27}$$

where the integrand is a total derivative in $\boldsymbol{q}_{s+1}$. Then, performing the integral over $\boldsymbol{q}_{s+1}$, one obtains from the last equation that

$$I_2^{(1)} = \int \mathrm{d}\boldsymbol{p}_{s+1}\ldots\mathrm{d}z_N\,\frac{\boldsymbol{p}_{s+1}}{m}\Big[f_N(z_1,\ldots,\boldsymbol{q}_{s+1},\boldsymbol{p}_{s+1},\ldots,z_N,t)\Big]_{\text{boundaries of }\boldsymbol{q}_{s+1}} = 0\,, \tag{3.28}$$

by virtue of $f_N$ vanishing on the boundaries of the phase space. Similarly, one may show that all the integrals in Eq. (3.26) evaluate to zero. We therefore conclude that $I_2 = 0$.

The third integral on the right hand side of Eq. (3.24) may be reduced to

$$
\begin{aligned}
I_3 &\equiv \frac{N!}{(N-s)!}\int \mathrm{d}z_{s+1}\ldots \mathrm{d}z_N \sum_{i=1}^{s}\Bigg[\sum_{\substack{j=1\\ j\neq i}}^{s}(\nabla_{q_i}V_{ij})\cdot\nabla_{p_i}f_N + \sum_{j=s+1}^{N}(\nabla_{q_i}V_{ij})\cdot\nabla_{p_i}f_N\Bigg]\\
&= \frac{N!}{(N-s)!}\sum_{i=1}^{s}\sum_{\substack{j=1\\ j\neq i}}^{s}(\nabla_{q_i}V_{ij})\cdot\nabla_{p_i}\left(\int \mathrm{d}z_{s+1}\ldots \mathrm{d}z_N\, f_N\right)\\
&\quad + \frac{N!}{(N-s)!}\sum_{i=1}^{s}\sum_{j=s+1}^{N}\int \mathrm{d}z_{s+1}\ldots \mathrm{d}z_N\,(\nabla_{q_i}V_{ij})\cdot\nabla_{p_i}f_N\\
&= \sum_{i=1}^{s}\sum_{\substack{j=1\\ j\neq i}}^{s}(\nabla_{q_i}V_{ij})\cdot\nabla_{p_i}f_s + \frac{N!}{(N-s)!}\sum_{i=1}^{s}\nabla_{p_i}\cdot\sum_{j=s+1}^{N}\int \mathrm{d}z_{s+1}\ldots \mathrm{d}z_N(\nabla_{q_i}V_{ij})f_N\,.
\end{aligned}
\tag{3.29}
$$

As $f_N$ is symmetric under interchange of any two particles, all the $(N-s)$ integrals in the sum occurring in the second term on the right hand side of Eq. (3.29) are equal to one another. Then, one has

$$
\begin{aligned}
I_3 &= \sum_{i=1}^{s}\sum_{\substack{j=1\\ j\neq i}}^{s}(\nabla_{q_i}V_{ij})\cdot\nabla_{p_i}f_s + \frac{N!}{(N-s-1)!}\sum_{i=1}^{s}\nabla_{p_i}\cdot\int \mathrm{d}z_{s+1}\ldots \mathrm{d}z_N\,(\nabla_{q_i}V_{i,s+1})f_N\\
&= \sum_{i=1}^{s}\sum_{\substack{j=1\\ j\neq i}}^{s}(\nabla_{q_i}V_{ij})\cdot\nabla_{p_i}f_s + \sum_{i=1}^{s}\int \mathrm{d}z_{s+1}\,(\nabla_{q_i}V_{i,s+1})\cdot\nabla_{p_i}f_{s+1}\,.
\end{aligned}
\tag{3.30}
$$

Finally, we rewrite the last integral on the right hand side of Eq. (3.24) as

$$
I_4 \equiv \frac{N!}{(N-s)!}\sum_{i=s+1}^{N}\sum_{\substack{j=1\\ j\neq i}}^{s}\int \mathrm{d}z_{s+1}\ldots \mathrm{d}z_N\,(\nabla_{q_i}V_{ij})\cdot\nabla_{p_i}f_N\,, \tag{3.31}
$$

and note that there is a gradient term in $\boldsymbol{p}_i$ with $\boldsymbol{p}_i$-independent coefficient. Therefore, the integral evaluates $f_N$ at the boundary of phase space, where it is assumed to vanish (similar to what we have encountered in Eq. (3.27)). Then, we have $I_4 = 0$.

Now, using the above mentioned results and substituting Eq. (3.25) and (3.30) in Eq. (3.24), we obtain

$$
\frac{\partial f_s}{\partial t} = -\sum_{i=1}^{s}\frac{\boldsymbol{p}_i}{m}\cdot\nabla_{q_i}f_s + \sum_{i=1}^{s}\sum_{\substack{j=1\\ j\neq i}}^{s}(\nabla_{q_i}V_{ij})\cdot\nabla_{p_i}f_s + \sum_{i=1}^{s}\int \mathrm{d}z_{s+1}(\nabla_{q_i}V_{i,s+1})\cdot\nabla_{p_i}f_{s+1}\,,
\tag{3.32}
$$

for $s = 1, \ldots, (N-1)$. It may be noted that for $s = 1$, the second term on the right hand side of Eq. (3.32) is absent. From Eq. (3.32), it is seen that the time evolution of the $s$-particle distribution function $f_s$ is coupled to the $(s+1)$-particle distribution function because of the last term on the right hand side of Eq. (3.32), which is called the 'collision integral'. The collision integral actually represents interaction between each of the particles in the group of $s$ particles defining $f_s$ with one particle outside the group. Because of this collision integral, there exists no closed equation for any of the $f_s$'s in the set of $(N-1)$ integro-differential equations (3.32). This $(N-1)$ set of equations (3.32) along with the Liouville equation (3.15), which is a closed equation for $f_N$, constitute the BBGKY hierarchy.[2]

## 3.7 Evolution of the Single-Particle Distribution Function

For all practical purposes, one needs to truncate the BBGKY hierarchy at small values of $s$ with some suitable approximation depending on the underlying dynamics. Although such a treatment would not yield results as complete as the knowledge of $f_N$, we would be able to compute the expectation values of the usual observables. For example, let us consider an observable given by the sum of identical single-particle functions $a_1(z_i)$:

$$O_1(z_1, \ldots, z_N) \equiv \sum_{i=1}^{N} a_1(z_i) . \tag{3.33}$$

Note that $O_1(z_1, \ldots, z_N)$ defines a macroscopic observable, since it concerns the whole collection of $N$ particles and not just one or a few particles. The value of $O_1(z_1, \ldots, z_N)$ is different for different microstates, and its average with respect to different microstates, or, its expectation value, is obtained as

$$\begin{aligned}
\langle O_1 \rangle(t) &= \int \mathrm{d}z_1 \ldots \mathrm{d}z_N \sum_{i=1}^{N} a_1(z_i) f_N(z_1, \ldots, z_N, t) \\
&= \sum_{i=1}^{N} \int \mathrm{d}z_i \, a_1(z_i) \int \mathrm{d}z_1 \ldots \mathrm{d}z_{i-1} \mathrm{d}z_{i+1} \ldots \mathrm{d}z_N \; f_N(z_1, \ldots, z_N, t) \\
&= \sum_{i=1}^{N} \int \mathrm{d}z_i \, a_1(z_i) \, \frac{(N-1)!}{N!} \; f_1(z_i, t) \\
&= \int \mathrm{d}z_1 \, a_1(z_1) \, f_1(z_1, t) .
\end{aligned} \tag{3.34}$$

[2] The hierarchy was proposed independently by Nikolay **B**ogolyubov, Max **B**orn, Herbert S. **G**reen, John Gamble **K**irkwood, and Jacques **Y**von. Therefore, it is named as BBGKY hierarchy.

Similarly, for an observable which is a sum of two-particle functions given by

$$O_2(z_1, \ldots, z_N) \equiv \sum_{i=i}^{N} \sum_{\substack{j=1 \\ j \neq i}}^{N} a_2(z_i, z_j), \tag{3.35}$$

one obtains the expectation value as

$$\begin{aligned}
\langle O_2 \rangle(t) &= \int \mathrm{d}z_1 \ldots \mathrm{d}z_N \sum_{i=1}^{N} \sum_{\substack{j=1 \\ j \neq i}}^{N} a_2(z_i, z_j)\, f_N(z_1, \ldots, z_N, t) \\
&= \sum_{i=1}^{N} \sum_{\substack{j=1 \\ j \neq i}}^{N} \int \mathrm{d}z_i \mathrm{d}z_j\, a_2(z_i, z_j)\, \frac{(N-2)!}{N!} f_2(z_i, z_j, t) \\
&= \int \mathrm{d}z_1 \mathrm{d}z_2\, a_2(z_1, z_2)\, f_2(z_1, z_2, t)\,.
\end{aligned} \tag{3.36}$$

The first two equations from the BBGKY hierarchy (3.32) reads

$$\frac{\partial}{\partial t} f_1(z_1, t) = -\left[\frac{\boldsymbol{p}_1}{m} \cdot \nabla_{\boldsymbol{q}_1}\right] f_1(z_1, t) + \int \mathrm{d}z_2 \left[\left(\nabla_{\boldsymbol{q}_1} V_{12}\right) \cdot \nabla_{\boldsymbol{p}_1}\right] f_2(z_1, z_2, t)\,, \tag{3.37}$$

and

$$\begin{aligned}
\frac{\partial}{\partial t} f_2(z_1, z_2, t) = &- \left[\frac{\boldsymbol{p}_1}{m} \cdot \nabla_{\boldsymbol{q}_1} + \frac{\boldsymbol{p}_2}{m} \cdot \nabla_{\boldsymbol{q}_2}\right] f_2(z_1, z_2, t) \\
&+ \left[\left(\nabla_{\boldsymbol{q}_1} V_{12}\right) \cdot \left(\nabla_{\boldsymbol{p}_1} - \nabla_{\boldsymbol{p}_2}\right)\right] f_2(z_1, z_2, t) \\
&+ \int \mathrm{d}z_3 \left[\left(\nabla_{\boldsymbol{q}_1} V_{13}\right) \cdot \nabla_{\boldsymbol{p}_1} + \left(\nabla_{\boldsymbol{q}_2} V_{23}\right) \cdot \nabla_{\boldsymbol{p}_2}\right] f_3(z_1, z_2, z_3, t)\,.
\end{aligned} \tag{3.38}$$

In writing Eq. (3.38) we have used the fact that the interaction forces between two particles are of same magnitude but with opposite direction, i.e., $\nabla_{\boldsymbol{q}_1} V_{12} = -\nabla_{\boldsymbol{q}_2} V_{21}$.

Now, the $s$-particle distribution function $f_s(z_1, \ldots, z_s, t)$ may be written in terms of product of single-particle distributions $f_1$ and correlation functions $g_{s'}(z_1, \ldots, z_{s'}, t)$; $s' \leq s$. The arguments are as follows. If the particles are sufficiently far from each other that they are uncorrelated, then the probability to find a particle at $z_i$ is independent of the probability to find another particle at $z_j$ for all pairs $(i, j)$. Consequently, one may approximate $f_s$ as

$$f_s(z_1, \ldots, z_s) \approx \prod_{i=1}^{s} f_1(z_i, t)\,. \tag{3.39}$$

However, the particles are correlated when the inter-particle separation is small, and consequently, Eq. (3.39) is then no longer valid. Now, considering the fact that there should be a mathematical interpolation between the uncorrelated form (3.39) and a correlated form, one introduces suitable correlation functions $g_s$'s. For instance, $g_2$ and $g_3$ are defined, such that

$$f_2(z_1, z_2, t) = f_1(z_1, t) f_1(z_2, t) + g_2(z_1, z_2, t)\,, \tag{3.40}$$

and

$$\begin{aligned} f_3(z_1, z_2, z_3, t) =& f_1(z_1, t) f_1(z_2, t) f_1(z_3, t) + f_1(z_1, t) g_2(z_2, z_3, t) \\ &+ f_1(z_2, t) g_2(z_1, z_3, t) + f_1(z_3, t) g_2(z_1, z_2, t) + g_3(z_1, z_2, z_3, t)\,. \end{aligned} \tag{3.41}$$

Integrating both sides of Eqs. (3.40) and (3.41) with the help of Eq. (3.17), we get

$$N(N-1) = N^2 + \int \mathrm{d}z_1 \mathrm{d}z_2\, g_2(z_1, z_2, t)\,, \tag{3.42}$$

and

$$N(N-1)(N-2) = N^3 + 3N \int \mathrm{d}z_1 \mathrm{d}z_2\, g_2(z_1, z_2, t) + \int \mathrm{d}z_1 \mathrm{d}z_2 \mathrm{d}z_3\, g_3(z_1, z_2, z_3, t)\,, \tag{3.43}$$

respectively. For large but finite $N$, it is seen from Eq. (3.42) that the normalization of $g_2$ scales as $N$. Similarly, from Eq. (3.43), one may see with the help of Eq. (3.42) that the normalization of $g_3$ also scales as $N$.

Substituting the form (3.40) in Eq. (3.37), we obtain

$$\begin{aligned} \frac{\partial}{\partial t} f_1(z_1, t) =& -\frac{\boldsymbol{p}_1}{m} \cdot \nabla_{\boldsymbol{q}_1} f_1(z_1, t) \\ &+ \int \mathrm{d}z_2 \left(\nabla_{\boldsymbol{q}_1} V_{12}\right) \cdot \nabla_{\boldsymbol{p}_1} \left[f_1(z_1, t)\, f_1(z_2, t) + g_2(z_1, z_2, t)\right]\,. \end{aligned} \tag{3.44}$$

Similarly, substituting Eqs. (3.40) and (3.41) in Eq. (3.38), and then using Eq. (3.37), we have

$$
\begin{aligned}
\frac{\partial}{\partial t} g_2(z_1, z_2, t) = &-\frac{p_1}{m} \cdot \nabla_{q_1} g_2(z_1, z_2, t) - \frac{p_2}{m} \cdot \nabla_{q_2} g_2(z_1, z_2, t) \\
&+ \left(\nabla_{q_1} V_{12}\right) \cdot \left(\nabla_{p_1} - \nabla_{p_2}\right) [f_1(z_1, t) f_1(z_2, t) + g_2(z_1, z_2, t)] \\
&+ \int \mathrm{d}z_3 \left(\nabla_{q_1} V_{13}\right) \cdot \nabla_{p_1} [f_1(z_1, t) g_2(z_2, z_3, t) + f_1(z_3, t) g_2(z_1, z_2, t)] \\
&+ \int \mathrm{d}z_3 \left(\nabla_{q_2} V_{23}\right) \cdot \nabla_{p_2} [f_1(z_2, t) g_2(z_1, z_3, t) + f_1(z_3, t) g_2(z_1, z_2, t)] \\
&+ \int \mathrm{d}z_3 \left[\left(\nabla_{q_1} V_{13}\right) \cdot \nabla_{p_1} + \left(\nabla_{q_2} V_{23}\right) \cdot \nabla_{p_2}\right] g_3(z_1, z_2, z_3, t) . \qquad (3.45)
\end{aligned}
$$

We rewrite Eq. (3.44) as

$$
\begin{aligned}
&\frac{\partial}{\partial t} f_1(z_1, t) + \frac{p_1}{m} \cdot \nabla_{q_1} f_1(z_1, t) \\
&= \nabla_{q_1} \underbrace{\left[\int \mathrm{d}z_2 \, V_{12} f_1(z_2, t)\right]}_{\text{average of two-body interaction potential}} \cdot \nabla_{p_1} f_1(z_1, t) + \underbrace{\int \mathrm{d}z_2 \left(\nabla_{q_1} V_{12}\right) \cdot \nabla_{p_1} g_2(z_1, z_2, t)}_{\text{depends on two-particle correlations}} . \qquad (3.46)
\end{aligned}
$$

Let us denote the averaged two-particle interaction potential by

$$
V[f_1](z_1, t) \equiv \int \mathrm{d}z_2 \, V_{12} \, f_1(z_2, t) . \qquad (3.47)
$$

Note that the averaged potential, which may also be interpreted as the mean-field potential, is a functional of the single-particle distribution function. Then, one may reduce Eq. (3.46) to

$$
\begin{aligned}
\frac{\partial}{\partial t} f_1(z_1, t) + \frac{p_1}{m} \cdot \nabla_{q_1} f_1(z_1, t) = &\left(\nabla_{q_1} V[f_1](z_1, t)\right) \cdot \nabla_{p_1} f_1(z_1, t) \\
&+ \int \mathrm{d}z_2 \left(\nabla_{q_1} V_{12}\right) \cdot \nabla_{p_1} g_2(z_1, z_2, t) . \qquad (3.48)
\end{aligned}
$$

The purpose of the kinetic theory is to obtain a closed equation for the single-particle distribution function $f_1(z_1, t)$, and to describe the statistical dynamics of a many-body system through it. The justification for the latter comes from the fact that in statistical mechanics, we mostly deal with observables of the form defined in Eq. (3.33), whose expectation value is determined solely by $f_1(z_1, t)$. From Eq. (3.48), it may be seen that the time evolution of $f_1$ effectively depends on two terms: the first term on the right hand side, corresponding to a mean-field potential, and the second term on the right hand side, corresponding to the two-particle correlations. The relative strength of these two terms changes according to the nature of interactions (short-range or long-range) present in the system. Let us recall that if the inter-particle potential decays slowly as a function of the separation $q$ between the particles, specifically, as $V(q) = 1/q^{\alpha};\ 0 \leq \alpha \leq d$, with $d$ being the embedding spatial dimension of the system, it is regarded as long-range interactions, otherwise the interaction is short-ranged. We now aim to write down a closed-form equation for the single-particle distribution function $f_1(z_1, t)$.

### 3.7.1 Boltzmann Equation

For short-ranged inter-particle potentials, one may derive an evolution equation for $f_1(z_1, t)$. This equation is known as the Boltzmann equation, which we now derive using the first two equations from the BBGKY hierarchy given in Eqs. (3.37) and (3.38). In the following, we state the assumptions related to the derivation.

- The interactions between the particles are short-ranged, i.e., two particles interact with each other only if one of them comes within a small sphere of influence (with radius $r_0 \sim$ molecular diameter) due to the other. Therefore, the spatial parts of the collision integrals in Eqs. (3.37) and (3.38) extend only over a volume of radius $r_0$, and they produce terms of order $r_0^3$.
- The system is assumed to be dilute, i.e., the average inter-particle distance is large compared to $r_0$. Therefore, the average particle density $n$ is a small parameter. From the normalization condition given in Eq. (3.22), we see that $f_s \propto n f_{s-1}$.
- The dimensions of the bracketed terms (terms inside $[\cdot]$) on the right hand side of Eqs. (3.37) and (3.38) are (time)$^{-1}$. The terms involving the gradient of the inter-particle potentials are associated with a timescale $t_c$, such that

$$\nabla_{q_i} V_{ij} \cdot \nabla_{p_i} \sim \frac{1}{t_c}. \tag{3.49}$$

  From Eq. (3.49), we see that $t_c$ is related to the time during which one particle experiences the inter-particle potential due to another particle, and is therefore the time that it takes to pass through the sphere of influence of the other particle. In other words, $t_c$ is nothing but the timescale of collision between two particles. Since we are considering only short-range potentials, $t_c$ is assumed to be very small.
- There is another timescale $t_1$ associated with the term involving the spatial gradient of $f_1$ in Eq. (3.37). It is the time taken by a particle to traverse a distance over which $f_1$ varies significantly. In Fig. 3.2, we have shown a schematic diagram to illustrate the physical meaning of the timescale $t_1$ in one dimension. The blue curve in the figure shows the distribution function $f_1(q_1, p_1, t)$ at time $t$. Let us assume that a particle is at position $q_a$ at time $t$. The characteristic length scale over which $f_1(q_1, p_1, t)$ changes significantly is $l_1$. In time $t_1$, the particle moves from $q_a$ to $q_b$. Accordingly, the distribution function $f_1(q_1, p_1, t)$ changes its form to another, shown as the red curve in the figure. Denoting the characteristic velocity

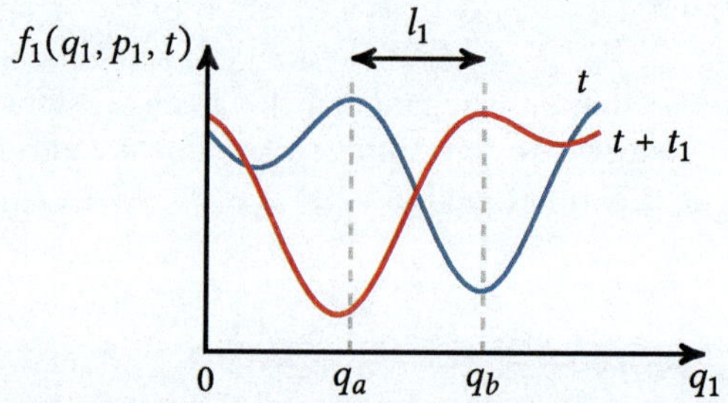


**Fig. 3.2 Schematic diagram for $f_1(q_1, p_1, t)$ in one dimension to illustrate the physical meaning of the timescale $t_1$**. The blue curve shows the the distribution function $f_1(q_1, p_1, t)$ at time $t$. The characteristic length scale over which $f_1(q_1, p_1, t)$ changes significantly is denoted by $l_1$. Suppose a particle is at position $q_a$ at time $t$. As the particle moves with time, the form of $f_1$ also changes. The typical time that a particle takes to traverse the characteristic length $l_1$ is given by $t_1$. The red curve shows the form of $f_1$ at time $t + t_1$

of a particle by $\boldsymbol{v}_{\text{char}}$, the time taken by the particle to traverse the characteristic length $l_1$ is given by $t_1 = l_1/|\boldsymbol{v}_{\text{char}}|$. This timescale is much larger than the collision time: $t_1 \gg t_c$. This same timescale $t_1$ is associated with the terms involving the spatial gradient of $f_2$ in Eq. (3.38). Note that $t_1 \gg t_c$ implies that we have $l_1 \gg r_0$.

Considering all the points mentioned above, we rewrite Eqs. (3.37) and (3.38) with the order of magnitude of each term as

$$\frac{\partial}{\partial t} f_1(z_1, t) = \underbrace{-\frac{\boldsymbol{p}_1}{m} \cdot \nabla_{\boldsymbol{q}_1} f_1(z_1, t)}_{\mathcal{O}(f_1/t_1)} + \underbrace{\int dz_2 \left(\nabla_{\boldsymbol{q}_1} V_{12}\right) \cdot \nabla_{\boldsymbol{p}_1} f_2(z_1, z_2, t)}_{\mathcal{O}(n r_0^3 f_1/t_c)}, \tag{3.50}$$

and

$$\begin{aligned} \frac{\partial}{\partial t} f_2(z_1, z_2, t) = & \underbrace{-\frac{\boldsymbol{p}_1}{m} \cdot \nabla_{\boldsymbol{q}_1} f_2(z_1, z_2, t) - \frac{\boldsymbol{p}_2}{m} \cdot \nabla_{\boldsymbol{q}_2} f_2(z_1, z_2, t)}_{\mathcal{O}(f_2/t_1)} \\ & + \underbrace{\left(\nabla_{\boldsymbol{q}_1} V_{12}\right) \cdot \left(\nabla_{\boldsymbol{p}_1} - \nabla_{\boldsymbol{p}_2}\right) f_2(z_1, z_2, t)}_{\mathcal{O}(f_2/t_c)} \\ & + \underbrace{\int dz_3 \left[\left(\nabla_{\boldsymbol{q}_1} V_{13}\right) \cdot \nabla_{\boldsymbol{p}_1} + \left(\nabla_{\boldsymbol{q}_2} V_{23}\right) \cdot \nabla_{\boldsymbol{p}_2}\right] f_3(z_1, z_2, z_3, t)}_{\mathcal{O}(n r_0^3 f_2/t_c)} . \end{aligned} \tag{3.51}$$

Now, noting that $nr_0^3 f_2/t_c \ll f_2/t_c$, we neglect the collision integral on the right hand side of Eq. (3.51). Thus, Eqs. (3.50) and (3.51) reduce to

$$\frac{\partial}{\partial t} f_1(z_1,t) = -\frac{\boldsymbol{p}_1}{m} \cdot \nabla_{\boldsymbol{q}_1} f_1(z_1,t) + \int \mathrm{d}z_2 \left(\nabla_{\boldsymbol{q}_1} V_{12}\right) \cdot \nabla_{\boldsymbol{p}_1} f_2(z_1,z_2,t)\,, \tag{3.52}$$

and

$$\begin{aligned}\frac{\partial}{\partial t} f_2(z_1,z_2,t) = &-\frac{\boldsymbol{p}_1}{m} \cdot \nabla_{\boldsymbol{q}_1} f_2(z_1,z_2,t) - \frac{\boldsymbol{p}_2}{m} \cdot \nabla_{\boldsymbol{q}_2} f_2(z_1,z_2,t) \\ &+ \left(\nabla_{\boldsymbol{q}_1} V_{12}\right) \cdot \left(\nabla_{\boldsymbol{p}_1} - \nabla_{\boldsymbol{p}_2}\right) f_2(z_1,z_2,t)\,. \end{aligned} \tag{3.53}$$

Note that Eqs. (3.52) and (3.53) form a closed set.

The two-particle distribution function $f_2(z_1, z_2, t)$ varies at a much faster rate compared to the single-particle distribution function $f_1(z_1, t)$. The reason for this is that the last term on the right hand side of Eq. (3.53) is of the order of $f_2/t_c$, whose variation in time is the fastest among all the timescales in Eqs. (3.52) and (3.53). Therefore, by the time over which $f_1$ varies significantly, $f_2$ would have reached its stationary value. So, as far as the time evolution of $f_1$ is concerned, we may put $\partial f_2/\partial t = 0$. Then, from Eq. (3.53), we get

$$\begin{aligned}\left(\nabla_{\boldsymbol{q}_1} V_{12}\right) \cdot \nabla_{\boldsymbol{p}_1} f_2(z_1,z_2,t) = &\frac{\boldsymbol{p}_1}{m} \cdot \nabla_{\boldsymbol{q}_1} f_2(z_1,z_2,t) + \frac{\boldsymbol{p}_2}{m} \cdot \nabla_{\boldsymbol{q}_2} f_2(z_1,z_2,t) \\ &+ \left(\nabla_{\boldsymbol{q}_1} V_{12}\right) \cdot \nabla_{\boldsymbol{p}_2} f_2(z_1,z_2,t)\,, \end{aligned} \tag{3.54}$$

which when substituted in the collision integral on the right hand side of Eq. (3.52) yields

$$\begin{aligned}\frac{\partial}{\partial t} f_1(z_1,t) = &-\frac{\boldsymbol{p}_1}{m} \cdot \nabla_{\boldsymbol{q}_1} f_1(z_1,t) + \int \mathrm{d}z_2 \left(\frac{\boldsymbol{p}_1}{m} \cdot \nabla_{\boldsymbol{q}_1} + \frac{\boldsymbol{p}_2}{m} \cdot \nabla_{\boldsymbol{q}_2}\right) f_2(z_1,z_2,t) \\ &+ \int \mathrm{d}z_2 \left(\nabla_{\boldsymbol{q}_1} V_{12}\right) \cdot \nabla_{\boldsymbol{p}_2} f_2(z_1,z_2,t)\,. \end{aligned} \tag{3.55}$$

In the last integral on the right hand side of Eq. (3.55), we have a gradient term in $\boldsymbol{p}_2$ with a $\boldsymbol{p}_2$-independent coefficient. Therefore, the integral evaluates $f_2$ at the boundaries of phase space, where, as usual, it is assumed to vanish. Then, Eq. (3.55) reduces to

$$\frac{\partial}{\partial t} f_1(z_1,t) = -\frac{\boldsymbol{p}_1}{m} \cdot \nabla_{\boldsymbol{q}_1} f_1(z_1,t) + \int \mathrm{d}z_2 \left(\frac{\boldsymbol{p}_1}{m} \cdot \nabla_{\boldsymbol{q}_1} + \frac{\boldsymbol{p}_2}{m} \cdot \nabla_{\boldsymbol{q}_2}\right) f_2(z_1,z_2,t)\,. \tag{3.56}$$

Let us now introduce the center-of-mass coordinates[3] $\boldsymbol{Q}$ and $\boldsymbol{q}$:

$$\boldsymbol{Q} \equiv \frac{1}{2}(\boldsymbol{q}_1 + \boldsymbol{q}_2), \quad \boldsymbol{q} \equiv \boldsymbol{q}_1 - \boldsymbol{q}_2, \tag{3.57}$$

so that

$$\boldsymbol{q}_1 = \boldsymbol{Q} + \frac{\boldsymbol{q}}{2}, \quad \boldsymbol{q}_2 = \boldsymbol{Q} - \frac{\boldsymbol{q}}{2}. \tag{3.58}$$

From Eq. (3.57), one obtains the relationship between the gradient operators in $\boldsymbol{q}_1, \boldsymbol{q}_2, \boldsymbol{Q}$, and $\boldsymbol{q}$ as

$$\nabla_{\boldsymbol{q}_1}(\cdot) = \frac{1}{2}\nabla_{\boldsymbol{Q}}(\cdot) + \nabla_{\boldsymbol{q}}(\cdot), \quad \nabla_{\boldsymbol{q}_2}(\cdot) = \frac{1}{2}\nabla_{\boldsymbol{Q}}(\cdot) - \nabla_{\boldsymbol{q}}(\cdot). \tag{3.59}$$

Using Eq. (3.59), the integrand in the collision integral in Eq. (3.56) may be written as

$$\left(\frac{\boldsymbol{p}_1}{m} \cdot \nabla_{\boldsymbol{q}_1} + \frac{\boldsymbol{p}_2}{m} \cdot \nabla_{\boldsymbol{q}_2}\right) f_2(z_1, z_2, t) = \frac{\boldsymbol{p}_1 - \boldsymbol{p}_2}{m} \cdot \nabla_{\boldsymbol{q}} f_2(z_1, z_2, t) + \frac{\boldsymbol{p}_1 + \boldsymbol{p}_2}{2m} \cdot \nabla_{\boldsymbol{Q}} f_2(z_1, z_2, t), \tag{3.60}$$

where the right hand side is to be evaluated by using Eq. (3.59) and noting that $\boldsymbol{q}_1$ and $\boldsymbol{q}_2$ occurring on the right hand side (recall that $z_i \equiv (\boldsymbol{q}_i, \boldsymbol{p}_i)$) are functions of both $\boldsymbol{q}$ and $\boldsymbol{Q}$. In a homogeneous system, all functions depend on positions only through the relative distance $\boldsymbol{q}$, and so the term $\nabla_{\boldsymbol{Q}} f_2$ is effectively zero. Neglecting this term in Eq. (3.60), and then using it in Eq. (3.56), one obtains

$$\frac{\partial}{\partial t} f_1(z_1, t) = -\frac{\boldsymbol{p}_1}{m} \cdot \nabla_{\boldsymbol{q}_1} f_1(z_1, t) + \frac{1}{m}\int \mathrm{d}z_2\, (\boldsymbol{p}_1 - \boldsymbol{p}_2) \cdot \nabla_{\boldsymbol{q}} f_2(z_1, z_2, t), \tag{3.61}$$

where, on the basis of the assumption that $\nabla_{\boldsymbol{Q}} f_2$ is zero, we have $\boldsymbol{q}_1$ and $\boldsymbol{q}_2$ in the second term on the right hand side effectively functions of $\boldsymbol{q}$ alone.

The collision integral on the right hand side of Eq. (3.61) has to be evaluated for a given $z_1$, i.e., for given $\boldsymbol{q}_1, \boldsymbol{p}_1$. Note that for given $\boldsymbol{q}_1$, we have $\mathrm{d}\boldsymbol{q}_2 = -\mathrm{d}\boldsymbol{q}$, see Eq. (3.57). Then, the spatial part of the collision integral on the right hand side of Eq. (3.61) may be written in the center-of-mass coordinate system as

[3] For a system of $N$ particles, each with mass $m_i$ and position vector $\boldsymbol{q}_i$; $1 \le i \le N$, the position vector of the center of mass is given by $\boldsymbol{Q} = (\sum_{i=1}^{N} m_i \boldsymbol{q}_i)/(\sum_{i=1}^{N} m_i)$ . Here, we have two particles with the same mass $m$, and hence, $\boldsymbol{Q} = (\boldsymbol{q}_1 + \boldsymbol{q}_2)/2$.

$$\int d\boldsymbol{q}_2\,(\boldsymbol{p}_1-\boldsymbol{p}_2)\cdot\nabla_{\boldsymbol{q}} f_2(z_1,z_2,t)=\int d\boldsymbol{q}\,(\boldsymbol{p}_2-\boldsymbol{p}_1)\cdot\nabla_{\boldsymbol{q}} f_2\big(\boldsymbol{q}_1(\boldsymbol{q}),\boldsymbol{q}_2(\boldsymbol{q}),\boldsymbol{p}_1,\boldsymbol{p}_2,t\big)\,, \tag{3.62}$$

where we have shown explicitly the functional dependence of $\boldsymbol{q}_1$ and $\boldsymbol{q}_2$. Now, we decompose the vector $\boldsymbol{q}$ parallel and perpendicular to the vector $(\boldsymbol{p}_2-\boldsymbol{p}_1)$, as

$$\boldsymbol{q}=q_{\parallel}\frac{(\boldsymbol{p}_2-\boldsymbol{p}_1)}{|\boldsymbol{p}_1-\boldsymbol{p}_2|}+\boldsymbol{q}_{\perp}\,, \tag{3.63}$$

so that

$$(\boldsymbol{p}_2-\boldsymbol{p}_1)\cdot\boldsymbol{q}=|\boldsymbol{p}_1-\boldsymbol{p}_2|\,q_{\parallel}\,, \tag{3.64}$$

and hence

$$(\boldsymbol{p}_2-\boldsymbol{p}_1)\cdot\nabla_{\boldsymbol{q}}(\cdot)=|\boldsymbol{p}_1-\boldsymbol{p}_2|\,\frac{\partial}{\partial q_{\parallel}}(\cdot)\,. \tag{3.65}$$

Using Eqs. (3.63) and (3.65), one obtains from Eq. (3.62) that

$$\int d\boldsymbol{q}_2\,(\boldsymbol{p}_1-\boldsymbol{p}_2)\cdot\nabla_{\boldsymbol{q}} f_2(z_1,z_2,t)$$
$$=\int d\boldsymbol{q}_{\perp}|\boldsymbol{p}_1-\boldsymbol{p}_2|\int dq_{\parallel}\frac{\partial}{\partial q_{\parallel}}f_2\big(\boldsymbol{q}_1(q_{\parallel},\boldsymbol{q}_{\perp}),\boldsymbol{q}_2(q_{\parallel},\boldsymbol{q}_{\perp}),\boldsymbol{p}_1,\boldsymbol{p}_2,t\big)\,. \tag{3.66}$$

We parametrize the $\boldsymbol{q}_{\perp}$-plane by the polar coordinates $(b,\phi)$ with $b=|\boldsymbol{q}_{\perp}|$, see Fig. 3.3 for a schematic diagram of a collision. The parameter $b$ is commonly known as the impact parameter. Hence, we have

$$d\boldsymbol{q}_{\perp}=b\,db\,d\phi\,, \tag{3.67}$$

which when substituted in Eq. (3.66) yields

$$\int d\boldsymbol{q}_2\,(\boldsymbol{p}_1-\boldsymbol{p}_2)\cdot\nabla_{\boldsymbol{q}} f_2(z_1,z_2,t)$$
$$=\int db\,d\phi\,b|\boldsymbol{p}_1-\boldsymbol{p}_2|\int dq_{\parallel}\,\frac{\partial}{\partial q_{\parallel}}f_2\big(\boldsymbol{q}_1(q_{\parallel},b,\phi),\boldsymbol{q}_2(q_{\parallel},b,\phi),\boldsymbol{p}_1,\boldsymbol{p}_2,t\big)\,. \tag{3.68}$$

We recall that the limits of the spatial part in the collision integral extends only to the sphere of influence of a particle. As seen from the figure, the limits of the integral over $q_{\parallel}$ will extend from $q_{a\parallel}$ to $q_{b\parallel}$. Then, from Eq. (3.68), one obtains

$$
\begin{aligned}
\int \mathrm{d}\boldsymbol{q}_2\, (\boldsymbol{p}_1 - \boldsymbol{p}_2) \cdot \nabla_q f_2(z_1, z_2, t) \\
&= \int \mathrm{d}b\, \mathrm{d}\phi\, b|\boldsymbol{p}_1 - \boldsymbol{p}_2| \int_{q_{a\parallel}}^{q_{b\parallel}} \mathrm{d}q_\parallel \frac{\partial}{\partial q_\parallel} f_2\big(\boldsymbol{q}_1(q_\parallel, b, \phi), \boldsymbol{q}_2(q_\parallel, b, \phi), \boldsymbol{p}_1, \boldsymbol{p}_2, t\big) \\
&= \int \mathrm{d}b\, \mathrm{d}\phi\, b|\boldsymbol{p}_1 - \boldsymbol{p}_2| \Big[ f_2\big(\boldsymbol{q}_1(q_{b\parallel}, b, \phi), \boldsymbol{q}_2(q_{b\parallel}, b, \phi), \boldsymbol{p}_1, \boldsymbol{p}_2, t\big) \\
&\qquad - f_2\big(\boldsymbol{q}_1(q_{a\parallel}, b, \phi), \boldsymbol{q}_2(q_{a\parallel}, b, \phi), \boldsymbol{p}_1, \boldsymbol{p}_2, t\big) \Big]. \qquad (3.69)
\end{aligned}
$$

Now, we assume that just before as well as just after the collision, i.e., when one particle is outside the sphere of influence due to the other, they are uncorrelated. This assumption is famously known as molecular chaos hypothesis. Due to this hypothesis, one may express the two-particle distribution function $f_2$ as a product of two single-particle distribution functions. Let us recall that we have $r_0 \ll l_1$, the distance over which $f_1$ changes significantly. Since we have $q_{b\parallel} - q_{a\parallel} < 2r_0$, see

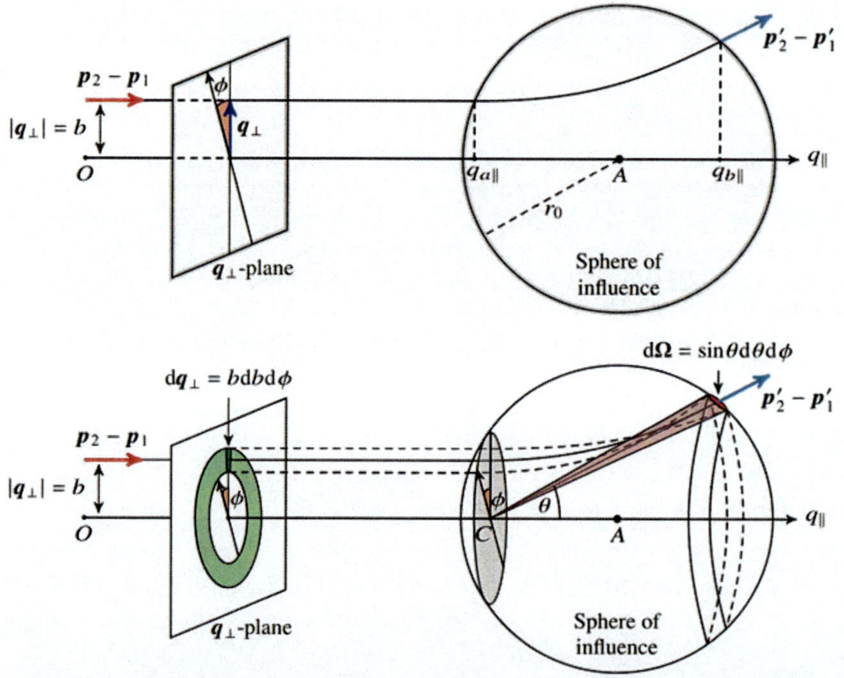


**Fig. 3.3 Schematic diagram of a collision, to evaluate the right hand side of Eq.** (3.66). The straight line $\boldsymbol{OA}$ denotes the collision axis. The $\boldsymbol{q}_\perp$-plane is parametrized by the polar coordinates $(b, \phi)$. Then, an elemental area on the $\boldsymbol{q}_\perp$-plane is given by $\mathrm{d}\boldsymbol{q}_\perp = b\, \mathrm{d}b\, \mathrm{d}\phi$. Parametrizing the solid angle $\boldsymbol{\Omega}$ by the spherical-polar angles $(\theta, \phi)$, one has the elemental solid angle formed by the vector $\boldsymbol{p}_2' - \boldsymbol{p}_1'$ at point $C$ as $\mathrm{d}\boldsymbol{\Omega} = \sin\theta \mathrm{d}\theta \mathrm{d}\phi$

Fig. 3.3, we may then evaluate the single-particle distribution functions at the same point $\boldsymbol{q}_1$. Thus, one may write

$$f_2\big(\boldsymbol{q}_1(q_{a\|}, b, \phi), \boldsymbol{q}_2(q_{a\|}, b, \phi), \boldsymbol{p}_1, \boldsymbol{p}_2, t\big) = f_1(\boldsymbol{q}_1, \boldsymbol{p}_1, t) f_1(\boldsymbol{q}_1, \boldsymbol{p}_2, t), \quad (3.70)$$

$$f_2\big(\boldsymbol{q}_1(q_{b\|}, b, \phi), \boldsymbol{q}_2(q_{b\|}, b, \phi), \boldsymbol{p}_1, \boldsymbol{p}_2, t\big) = f_1(\boldsymbol{q}_1, \boldsymbol{p}'_1, t) f_1(\boldsymbol{q}_1, \boldsymbol{p}'_2, t), \quad (3.71)$$

where we have suppressed the arguments of $\boldsymbol{q}_1$ and used the final momenta $\boldsymbol{p}'_1$ and $\boldsymbol{p}'_2$, which depend on $\boldsymbol{p}_1$, $\boldsymbol{p}_2$, $b$, and $\phi$. To see this first note that energy and momentum conservation imply respectively that

$$(p_1)^2 + (p_2)^2 = (p'_1)^2 + (p'_2)^2, \quad (3.72)$$

$$\boldsymbol{p}_1 + \boldsymbol{p}_2 = \boldsymbol{p}'_1 + \boldsymbol{p}'_2, \quad (3.73)$$

implying that $\boldsymbol{p}_1 \cdot \boldsymbol{p}_2 = \boldsymbol{p}'_1 \cdot \boldsymbol{p}'_2$, and hence that

$$|\boldsymbol{p}_2 - \boldsymbol{p}_1| = |\boldsymbol{p}'_2 - \boldsymbol{p}'_1|. \quad (3.74)$$

Note that in expressing energy conservation in Eq. (3.72), we have written down only the kinetic-energy contribution. This is because in the strategy adopted for our current analysis, collisions are assumed to take place at the same spatial location (see the remark preceding Eq. (3.70)), resulting in no change in the potential energy of the two colliding particles before and after the collision. Equation (3.74) suggests that the relative momentum after collision preserves its magnitude and only rotates to take up a direction indicated by the spherical-polar angles $\phi$ and $\theta$ in Fig.3.3. Now, conservation of angular momentum implies that the post-collision relative momentum $\boldsymbol{p}'_2 - \boldsymbol{p}'_1$ lies on a plane perpendicular to the $\boldsymbol{q}_\perp$-plane and containing the momentum $\boldsymbol{p}_2 - \boldsymbol{p}_1$ and the collision axis $\boldsymbol{OA}$ (see Fig. 3.3). In other words, the azimuthal angle $\phi$ corresponding to the relative momentum does not change after collision. However, the angle $\theta$ between $\boldsymbol{p}_2 - \boldsymbol{p}_1$ and $\boldsymbol{p}'_2 - \boldsymbol{p}'_1$, known as the scattering angle, has a one-to-one correspondence with the impact parameter $b$ with $\theta(b)$ being a decreasing function of $b$. The latter feature follows from the fact that larger is $b$, farther is the particle from the scattering center and hence smaller will be the amount of scattering, as measured by the angle $\theta$, that it suffers. As a result, one can always transform $\boldsymbol{q}_\perp$(parametrized by $b$ and $\phi$) to a solid angle $\boldsymbol{\Omega}$ (parametrized by $\theta$ and $\phi$). The Jacobian for this transformation $|\mathrm{d}\boldsymbol{q}_\perp/\mathrm{d}\boldsymbol{\Omega}| = (b/\sin\theta)|\mathrm{d}b/\mathrm{d}\theta|$ has the dimension of area, and is referred to as the differential cross-section, see Fig. 3.3. Thus, the final momenta $\boldsymbol{p}'_1$ and $\boldsymbol{p}'_2$ are uniquely determined by the vectors $\boldsymbol{p}_1$, $\boldsymbol{p}_2$, and $\boldsymbol{q}_\perp$.

Using Eqs. (3.70) and (3.71) for the two-particle distribution functions in Eq. (3.69), one obtains

$$\int \mathrm{d}\boldsymbol{q}_2\, (\boldsymbol{p}_1 - \boldsymbol{p}_2) \cdot \nabla_q f_2(z_1, z_2, t)$$
$$= \int \mathrm{d}b\mathrm{d}\phi\, b|\boldsymbol{p}_1 - \boldsymbol{p}_2| \Big[ f_1(\boldsymbol{q}_1, \boldsymbol{p}_1', t) f_1(\boldsymbol{q}_1, \boldsymbol{p}_2', t) - f_1(\boldsymbol{q}_1, \boldsymbol{p}_1, t) f_1(\boldsymbol{q}_1, \boldsymbol{p}_2, t) \Big]. \tag{3.75}$$

Substituting Eq. (3.75) in Eq. (3.61), we finally obtain the Boltzmann equation:

$$\frac{\partial}{\partial t} f_1(z_1, t) = -\frac{\boldsymbol{p}_1}{m} \cdot \nabla_{\boldsymbol{q}_1} f_1(z_1, t) + \int \mathrm{d}b\, \mathrm{d}\phi\, \mathrm{d}\boldsymbol{p}_2\, \frac{b|\boldsymbol{p}_1 - \boldsymbol{p}_2|}{m}$$
$$\times \left[ f_1(\boldsymbol{q}_1, \boldsymbol{p}_1', t) f_1(\boldsymbol{q}_1, \boldsymbol{p}_2', t) - f_1(\boldsymbol{q}_1, \boldsymbol{p}_1, t) f_1(\boldsymbol{q}_1, \boldsymbol{p}_2, t) \right]. \tag{3.76}$$

### 3.7.2 The Boltzmann *H*-theorem

It can be shown rigorously [3] as the content of the so-called $H$-theorem that the evolution determined by the Boltzmann equation (3.76) would drive the system over the timescale $t_c$ (recall that $t_c$ is the timescale of collision between two particles) towards Boltzmann-Gibbs equilibrium. The evolution towards equilibrium is essentially dictated by the second term of $\mathcal{O}(1/t_c)$ on the right hand side of Eq. (3.76), which as has been discussed previously arises from the dynamics of collision between two particles. The second term is usually referred to as the collision integral term of the Boltzmann equation. In equilibrium, the single-particle distribution has the well-known form

$$f_1(\boldsymbol{q}_1, \boldsymbol{p}_1) = n \left( \frac{\beta}{2\pi m} \right)^{3/2} \mathrm{e}^{-\beta p_1^2/(2m)}. \tag{3.77}$$

Here, $\beta$ is a parameter related to the energy density $\epsilon$ as $\beta = 3/(2\epsilon)$, while we recall that $n$ is the particle density.

In order to discuss the $H$-theorem, we follow here the discussion presented in Ref. [3]. Consider the so-called $H$ functional

$$H(t) \equiv \int \mathrm{d}\boldsymbol{q}_1 \mathrm{d}\boldsymbol{p}_1\, f_1(\boldsymbol{q}_1, \boldsymbol{p}_1, t) \log[f_1(\boldsymbol{q}_1, \boldsymbol{p}_1, t)]. \tag{3.78}$$

Taking derivative with respect to time, one gets from Eq. (3.78) that

$$\frac{\mathrm{d}H(t)}{\mathrm{d}t} = \int \mathrm{d}\boldsymbol{q}_1 \mathrm{d}\boldsymbol{p}_1\, (1 + \log[f_1(\boldsymbol{q}_1, \boldsymbol{p}_1, t)]) \frac{\partial}{\partial t} f_1(\boldsymbol{q}_1, \boldsymbol{p}_1, t). \tag{3.79}$$

We may now use Eq. (3.76) to evaluate the right hand side of Eq. (3.79). One gets

$$
\begin{aligned}
\frac{\mathrm{d}H(t)}{\mathrm{d}t} &= -\int \mathrm{d}\boldsymbol{q}_1 \mathrm{d}\boldsymbol{p}_1 \left(1 + \log[f_1(\boldsymbol{q}_1, \boldsymbol{p}_1, t)]\right) \frac{\boldsymbol{p}_1}{m} \cdot \nabla_{\boldsymbol{q}_1} f_1(\boldsymbol{q}_1, \boldsymbol{p}_1, t) \\
&+ \frac{1}{m} \int \mathrm{d}\boldsymbol{q}_1 \mathrm{d}\boldsymbol{p}_1 \mathrm{d}\boldsymbol{p}_2 \mathrm{d}b\, \mathrm{d}\phi\, b |\boldsymbol{p}_1 - \boldsymbol{p}_2| \left(1 + \log[f_1(\boldsymbol{q}_1, \boldsymbol{p}_1, t)]\right) \\
&\times \left[f_1(\boldsymbol{q}_1, \boldsymbol{p}_1', t) f_1(\boldsymbol{q}_1, \boldsymbol{p}_2', t) - f_1(\boldsymbol{q}_1, \boldsymbol{p}_1, t) f_1(\boldsymbol{q}_1, \boldsymbol{p}_2, t)\right].
\end{aligned}
\tag{3.80}
$$

Let us consider the first integral on the right hand side of Eq. (3.80). One has

$$
\begin{aligned}
&\int \mathrm{d}\boldsymbol{q}_1 \mathrm{d}\boldsymbol{p}_1 \left(1 + \log[f_1(\boldsymbol{q}_1, \boldsymbol{p}_1, t)]\right) \frac{\boldsymbol{p}_1}{m} \cdot \nabla_{\boldsymbol{q}_1} f_1(\boldsymbol{q}_1, \boldsymbol{p}_1, t) \\
&= \int \mathrm{d}\boldsymbol{q}_1 \mathrm{d}\boldsymbol{p}_1 \frac{\boldsymbol{p}_1}{m} \cdot \nabla_{\boldsymbol{q}_1} [f_1(\boldsymbol{q}_1, \boldsymbol{p}_1, t) \log(f_1(\boldsymbol{q}_1, \boldsymbol{p}_1, t)) \\
&= 0,
\end{aligned}
\tag{3.81}
$$

where we have used the fact that the integral on the right hand of the second last equation can be converted into one that evaluates $f_1$ on the boundaries of $\boldsymbol{q}_1$ where it is assumed to vanish. We thus have

$$
\begin{aligned}
\frac{\mathrm{d}H(t)}{\mathrm{d}t} &= \frac{1}{m} \int \mathrm{d}\boldsymbol{q}_1 \mathrm{d}\boldsymbol{p}_1 \mathrm{d}\boldsymbol{p}_2 \mathrm{d}b\, \mathrm{d}\phi\, b |\boldsymbol{p}_1 - \boldsymbol{p}_2| \left(1 + \log[f_1(\boldsymbol{q}_1, \boldsymbol{p}_1, t)]\right) \\
&\times \left[f_1(\boldsymbol{q}_1, \boldsymbol{p}_1', t) f_1(\boldsymbol{q}_1, \boldsymbol{p}_2', t) - f_1(\boldsymbol{q}_1, \boldsymbol{p}_1, t) f_1(\boldsymbol{q}_1, \boldsymbol{p}_2, t)\right].
\end{aligned}
\tag{3.82}
$$

We now proceed to evaluate the right hand side of Eq. (3.82). Interchanging the dummy variables $\boldsymbol{p}_1$ and $\boldsymbol{p}_2$, the right hand side of Eq. (3.82) may be written as

$$
\begin{aligned}
I &= \frac{1}{m} \int \mathrm{d}\boldsymbol{q}_1 \mathrm{d}\boldsymbol{p}_1 \mathrm{d}\boldsymbol{p}_2 \mathrm{d}b\, \mathrm{d}\phi\, b |\boldsymbol{p}_1 - \boldsymbol{p}_2| \left(1 + \log[f_1(\boldsymbol{q}_1, \boldsymbol{p}_1, t)]\right) \\
&\times \left[f_1(\boldsymbol{q}_1, \boldsymbol{p}_1', t) f_1(\boldsymbol{q}_1, \boldsymbol{p}_2', t) - f_1(\boldsymbol{q}_1, \boldsymbol{p}_1, t) f_1(\boldsymbol{q}_1, \boldsymbol{p}_2, t)\right] \\
&= \frac{1}{m} \int \mathrm{d}\boldsymbol{q}_1 \mathrm{d}\boldsymbol{p}_1 \mathrm{d}\boldsymbol{p}_2 \mathrm{d}b\, \mathrm{d}\phi\, b |\boldsymbol{p}_2 - \boldsymbol{p}_1| \left(1 + \log[f_1(\boldsymbol{q}_1, \boldsymbol{p}_2, t)]\right) \\
&\times \left[f_1(\boldsymbol{q}_1, \boldsymbol{p}_1', t) f_1(\boldsymbol{q}_1, \boldsymbol{p}_2', t) - f_1(\boldsymbol{q}_1, \boldsymbol{p}_1, t) f_1(\boldsymbol{q}_1, \boldsymbol{p}_2, t)\right] \\
&= \frac{1}{2m} \int \mathrm{d}\boldsymbol{q}_1 \mathrm{d}\boldsymbol{p}_1 \mathrm{d}\boldsymbol{p}_2 \mathrm{d}b\, \mathrm{d}\phi\, b |\boldsymbol{p}_2 - \boldsymbol{p}_1| \left(2 + \log[f_1(\boldsymbol{q}_1, \boldsymbol{p}_1, t) f_1(\boldsymbol{q}_1, \boldsymbol{p}_2, t)]\right) \\
&\times \left[f_1(\boldsymbol{q}_1, \boldsymbol{p}_1', t) f_1(\boldsymbol{q}_1, \boldsymbol{p}_2', t) - f_1(\boldsymbol{q}_1, \boldsymbol{p}_1, t) f_1(\boldsymbol{q}_1, \boldsymbol{p}_2, t)\right].
\end{aligned}
\tag{3.83}
$$

Interchange now the variables of integration from those describing quantities before the collision, $(\boldsymbol{p}_1, \boldsymbol{p}_2, b, \phi)$, to those describing quantities after the collision, $(\boldsymbol{p}_1', \boldsymbol{p}_2', b', \phi')$ and then use the fact that for every collision, there is an inverse one that is obtained by reversing the momenta of the products. Consequently, the Jacobian of the transformation $(\boldsymbol{p}_1, \boldsymbol{p}_2, b, \phi) \to (\boldsymbol{p}_1', \boldsymbol{p}_2', b', \phi')$ is unity, and we have

$$I = \frac{1}{2m}\int \mathrm{d}\boldsymbol{q}_1 \mathrm{d}\boldsymbol{p}_1' \mathrm{d}\boldsymbol{p}_2' \mathrm{d}b' \,\mathrm{d}\phi' \, b' |\boldsymbol{p}_2 - \boldsymbol{p}_1| \left(2 + \log[f_1(\boldsymbol{q}_1, \boldsymbol{p}_1, t) f_1(\boldsymbol{q}_1, \boldsymbol{p}_2, t)]\right)$$
$$\times \left[f_1(\boldsymbol{q}_1, \boldsymbol{p}_1', t) f_1(\boldsymbol{q}_1, \boldsymbol{p}_2', t) - f_1(\boldsymbol{q}_1, \boldsymbol{p}_1, t) f_1(\boldsymbol{q}_1, \boldsymbol{p}_2, t)\right], \tag{3.84}$$

where $(\boldsymbol{p}_1, \boldsymbol{p}_2)$ are to be regarded as functions of $(\boldsymbol{p}_1', \boldsymbol{p}_2', b', \phi')$. Using Eq. (3.74), we get

$$I = \frac{1}{2m}\int \mathrm{d}\boldsymbol{q}_1 \mathrm{d}\boldsymbol{p}_1' \mathrm{d}\boldsymbol{p}_2' \mathrm{d}b' \,\mathrm{d}\phi' \, b' |\boldsymbol{p}_2' - \boldsymbol{p}_1'| \left(2 + \log[f_1(\boldsymbol{q}_1, \boldsymbol{p}_1, t) f_1(\boldsymbol{q}_1, \boldsymbol{p}_2, t)]\right)$$
$$\times \left[f_1(\boldsymbol{q}_1, \boldsymbol{p}_1', t) f_1(\boldsymbol{q}_1, \boldsymbol{p}_2', t) - f_1(\boldsymbol{q}_1, \boldsymbol{p}_1, t) f_1(\boldsymbol{q}_1, \boldsymbol{p}_2, t)\right]. \tag{3.85}$$

Next, we relabel the integration variables so as to get rid of the primes, to get

$$I = \frac{1}{2m}\int \mathrm{d}\boldsymbol{q}_1 \mathrm{d}\boldsymbol{p}_1 \mathrm{d}\boldsymbol{p}_2 \mathrm{d}b \,\mathrm{d}\phi \, b |\boldsymbol{p}_2 - \boldsymbol{p}_1| \left(2 + \log[f_1(\boldsymbol{q}_1, \boldsymbol{p}_1', t) f_1(\boldsymbol{q}_1, \boldsymbol{p}_2', t)]\right)$$
$$\times \left[f_1(\boldsymbol{q}_1, \boldsymbol{p}_1, t) f_1(\boldsymbol{q}_1, \boldsymbol{p}_2, t) - f_1(\boldsymbol{q}_1, \boldsymbol{p}_1', t) f_1(\boldsymbol{q}_1, \boldsymbol{p}_2', t)\right]. \tag{3.86}$$

Adding Eq. (3.86) with Eq. (3.83), we get

$$I = \frac{1}{4m}\int \mathrm{d}\boldsymbol{q}_1 \mathrm{d}\boldsymbol{p}_1 \mathrm{d}\boldsymbol{p}_2 \mathrm{d}b \,\mathrm{d}\phi \, b |\boldsymbol{p}_2 - \boldsymbol{p}_1|$$
$$\times \left(\log[f_1(\boldsymbol{q}_1, \boldsymbol{p}_1, t) f_1(\boldsymbol{q}_1, \boldsymbol{p}_2, t)] - \log[f_1(\boldsymbol{q}_1, \boldsymbol{p}_1', t) f_1(\boldsymbol{q}_1, \boldsymbol{p}_2', t)]\right)$$
$$\times \left[f_1(\boldsymbol{q}_1, \boldsymbol{p}_1', t) f_1(\boldsymbol{q}_1, \boldsymbol{p}_2', t) - f_1(\boldsymbol{q}_1, \boldsymbol{p}_1, t) f_1(\boldsymbol{q}_1, \boldsymbol{p}_2, t)\right]. \tag{3.87}$$

Now, logarithm is a monotonically increasing function, implying thereby that for $x \neq y$, we have $(x - y)[\log y - \log x] < 0$. Hence, the right hand side of the above equation is never positive. We thus conclude that under evolution according to the Boltzmann equation (3.76), we have

$$\frac{\mathrm{d}H(t)}{\mathrm{d}t} < 0\,. \tag{3.88}$$

On the basis of the above, we conclude that $f_1(\boldsymbol{q}_1, \boldsymbol{p}_1, t)$ will relax in a time of the order of the collision time scale $t_c$ to a function that would make $\mathrm{d}H(t)/\mathrm{d}t = 0$. In such a state, we must have $f_1(\boldsymbol{q}_1, \boldsymbol{p}_1, t)$ of such a form that

$$f_1(\boldsymbol{q}_1, \boldsymbol{p}_1', t) f_1(\boldsymbol{q}_1, \boldsymbol{p}_2', t) = f_1(\boldsymbol{q}_1, \boldsymbol{p}_1, t) f_1(\boldsymbol{q}_1, \boldsymbol{p}_2, t)\,. \tag{3.89}$$

We may propose $\log[f_1(\boldsymbol{q}_1, \boldsymbol{p}_1, t)]$ to be a function of the two conserved quantities in binary collisions, namely, energy and momentum, and write

$$\log[f_1(\boldsymbol{q}_1, \boldsymbol{p}_1, t)] = -c_1(\boldsymbol{q}_1, t)\frac{p_1^2}{2m} + \boldsymbol{c}_2(\boldsymbol{q}_1, t) \cdot \boldsymbol{p}_1 + c_3(\boldsymbol{q}_1, t)\,, \tag{3.90}$$

where $c_1(\boldsymbol{q}_1, t)$ and $c_3(\boldsymbol{q}_1, t)$ are arbitrary scalar functions of $\boldsymbol{q}_1$, while $\boldsymbol{c}_2(\boldsymbol{q}_1, t)$ is an arbitrary vector function of $\boldsymbol{q}_1$. One may easily check using Eqs. (3.72) and (3.73) that the functional form of $f_1(\boldsymbol{q}_1, \boldsymbol{p}_1, t)$ specified by Eq. (3.90) satisfies Eq. (3.89).

Putting

$$\beta \equiv c_1, \quad \boldsymbol{p}_0 \equiv \frac{m}{c_1}\boldsymbol{c}_2, \quad n \equiv \left(\frac{2\pi m}{c_1}\right)^{3/2} \mathrm{e}^{mc_2^2/(2c_1)+c_3}, \tag{3.91}$$

we get

$$f_1(\boldsymbol{q}_1, \boldsymbol{p}_1, t) = n(\boldsymbol{q}_1, t)\left(\frac{\beta(\boldsymbol{q}_1, t)}{2\pi m}\right)^{3/2} \times \exp\left[-\beta(\boldsymbol{q}_1, t)\frac{(\boldsymbol{p}_1 - \boldsymbol{p}_0(\boldsymbol{q}_1, t)) \cdot (\boldsymbol{p}_1 - \boldsymbol{p}_0(\boldsymbol{q}_1, t))}{2m}\right], \tag{3.92}$$

in which the density $n$ and the inverse temperature $\beta$ are arbitrary functions of space and time, and moreover, for every $\boldsymbol{q}_1$, the velocity distribution is a Gaussian centered at an arbitrary average velocity $\boldsymbol{p}_0(\boldsymbol{q}_1, t)/m$. The collision integral term of the Boltzmann equation (3.76) may be checked to vanish for distribution functions of the form (3.92). However, the streaming term, i.e., the first term on the right hand side of the Boltzmann equation (3.76), does not make the distribution (3.92) a stationary solution of the Boltzmann equation. Indeed, if at a particular time instant, $f_1$ has the form (3.92), it will be changed in time under the action of the streaming term. Nevertheless, the collision integral acting on a much shorter time scale $t_c$ will soon restore the distribution to the form (3.92). Consequently, $f_1$ will reach the form (3.92) with constant $n$ and $\beta$ and with $\boldsymbol{p}_0 = \boldsymbol{0}$ for a system that is globally at rest. Thus, the stationary solution of the Boltzmann equation (3.76) is the equilibrium distribution given in Eq. (3.77). Note that if at time $t = 0$, we have an $f_1$ that is uniform in space, it remains so under evolution due to the Boltzmann equation.

### 3.7.3 *Vlasov Equation*

For a long-range interacting system, invoking the Kac's trick, we have $J \to J/N$, implying $V \to V/N$ (where $V$ is the potential). On using Eq. (3.48), and noting that $f_1 \sim N$ (so that $V[f_1](z_1, t) \sim \mathcal{O}(1)$, see Eq. (3.47)) and that $g_2 \sim N$, one obtains, neglecting corrections of $\mathcal{O}(1/N)$ arsing from the term involving $g_2$, the so-called Vlasov equation for the time evolution of the single-particle distribution function $f_1(z_1, t)$:

$$\frac{\partial}{\partial t} f_1\,(z_1, t) + \frac{\boldsymbol{p}_1}{m} \cdot \nabla_{\boldsymbol{q}_1} f_1\,(z_1, t) = \left(\nabla_{\boldsymbol{q}_1} V[f_1](z_1, t)\right) \cdot \nabla_{\boldsymbol{p}_1} f_1(z_1, t)\,. \tag{3.93}$$

Let us make some remarks comparing the Vlasov equation, valid for long-range systems, with the corresponding equation (3.50) valid for short-range systems. For convenience, the latter equation is rewritten below in a convenient form:

$$\frac{\partial}{\partial t} f_1(z_1,t) + \frac{\boldsymbol{p}_1}{m} \cdot \nabla_{q_1} f_1(z_1,t) = \int \mathrm{d}z_2 \left(\nabla_{q_1} V_{12}\right) \cdot \nabla_{p_1} f_2(z_1,z_2,t)\,, \qquad (3.94)$$

We see that to leading order in $1/N$, the time evolution of the single-particle distribution is governed in the case of long-range systems by the averaged two-particle interaction potential or the mean-field potential $V[f_1](z_1,t)$ (besides the trivial "streaming term" on the left hand side of both Eqs. (3.93) and (3.94) that is present even when the particles are noninteracting), with the two-particle correlation term $g_2$ providing the leading higher-order correction. In contrast, it is evident from Eq. (3.94) that for a short-range system, it is the two-particle correlation $g_2$ contained in $f_2$ (see Eq. (3.40)) that provides the leading time-evolution of the single-particle distribution.

We now discuss about some relevant features of the Vlasov equation (3.93). The equation is time-reversal invariant. Next, one can associate with the equation a Hamilton dynamics corresponding to a single-particle (mean-field) Hamiltonian

$$\mathcal{H}[f_1](p_1,q_1,t) = \frac{p_1^2}{2m} + V[f_1](q_1,t)\,. \qquad (3.95)$$

so that one can rewrite the Vlasov equation in terms of a Poisson bracket:

$$\frac{\partial f_1}{\partial t} + \{f_1,\mathcal{H}\} = 0\,. \qquad (3.96)$$

This implies that the Vlasov equation has Vlasov-stationary solutions ($\partial f_1/\partial t = 0$) given by arbitrary normalizable functions of the single-particle Hamiltonian, as $f_1(z_1) = F(\mathcal{H})$. A particular class of stationary solutions involves those that are homogeneous in the position coordinate ($\partial f_1/\partial q_1 = 0$), i.e., $f_1(z_1) \propto h(p_1)$, with $h(p_1)$ being a normalizable function of the momentum. In this case, one has from Eq. (3.47) that $V[f_1](q_1,t) \propto \int \mathrm{d}q_2 V(|q_1 - q_2|)$, which for the integration limit from $-\infty$ to $+\infty$, or from 0 to $2\pi$ for periodic $V$, vanishes. As a result, one obtains $\{f_1,\mathcal{H}\} = 0$ and $\partial f_1/\partial t = 0$. Noting the extent of arbitrariness in choosing the functions $F$ and $h$, it may be concluded that the Vlasov equation allows for an infinite number of stationary solutions, which define the "Vlasov equilibrium" state of the system. In addition, the equation admits an infinite number of conserved quantities, the Casimirs $C[f_1] = \int \mathrm{d}z_1\, c(f_1(z_1))$, where $c$ is an arbitrary function. In particular, the single-particle entropy $S[f_1] \equiv -\int \mathrm{d}z_1\, f_1 \ln f_1$ is a conserved quantity, namely, its value does not change in time.

The leading correction to the Vlasov equation is obtained from Eqs. (3.48) and (3.94)

$$\frac{\partial}{\partial t} f_1(z_1, t) + \frac{\mathbf{p}_1}{m} \cdot \nabla_{\mathbf{q}_1} f_1(z_1, t) - \left(\nabla_{\mathbf{q}_1} V[f_1](z_1, t)\right) \cdot \nabla_{\mathbf{p}_1} f_1(z_1, t)$$
$$= \int \mathrm{d}z_2 \left(\nabla_{\mathbf{q}_1} V_{12}\right) \cdot \nabla_{\mathbf{p}_1} g_2(z_1, z_2, t). \tag{3.97}$$

Noting that $f_1 \sim N$ and $g_2 \sim N$, we obtain from the above equation a time evolution of Vlasov-stationary solutions on timescales of $\mathcal{O}(N)$.

Let us now turn to the second equation (3.45) of the BBGKY hierarchy, which, using the Kac's scaling $V \to V/N$, gives us to leading order in $1/N$ the result

$$\begin{aligned}
\frac{\partial}{\partial t} g_2(z_1, z_2, t) &= -\frac{\mathbf{p}_1}{m} \cdot \nabla_{\mathbf{q}_1} g_2(z_1, z_2, t) - \frac{\mathbf{p}_2}{m} \cdot \nabla_{\mathbf{q}_2} g_2(z_1, z_2, t) \\
&+ \left(\nabla_{\mathbf{q}_1} V_{12}\right) \cdot \left(\nabla_{\mathbf{p}_1} - \nabla_{\mathbf{p}_2}\right) f_1(z_1, t) f_1(z_2, t) \\
&+ \int \mathrm{d}z_3 \left(\nabla_{\mathbf{q}_1} V_{13}\right) \cdot \nabla_{\mathbf{p}_1} \left[f_1(z_1, t) g_2(z_2, z_3, t) + f_1(z_3, t) g_2(z_1, z_2, t)\right] \\
&+ \int \mathrm{d}z_3 \left(\nabla_{\mathbf{q}_2} V_{23}\right) \cdot \nabla_{\mathbf{p}_2} \left[f_1(z_2, t) g_2(z_1, z_3, t) + f_1(z_3, t) g_2(z_1, z_2, t)\right].
\end{aligned} \tag{3.98}$$

On the basis of the above discussion, we may now summarize the general scenario of relaxation in long-range systems; the relaxation process is shown schematically in Fig. 3.4. In a first stage, known as the stage of *violent relaxation*, the system goes from a generic initial condition towards a Vlasov-stable stationary state on a fast timescale that is independent of the number of particles. In a second stage, known as the stage of *collisional relaxation*, finite-$N$ effects come into play that drive the system through a sequence of Vlasov-stable states towards a Boltzmann-Gibbs equilibrium state on a timescale that is strongly dependent on $N$. The latter scale is often a power law $\sim N^\alpha$ with $\alpha > 0$; A typical example is the Chandrasekhar relaxation timescale for stellar systems, which is proportional to $N/\ln N$. The Vlasov-stable stationary states have been named the quasistationary states (QSSs), because such states emerge as the true stationary states on taking the $N \to \infty$ limit first, followed by the limit $t \to \infty$.

## 3.8 An Example: Interacting Classical Heisenberg Spins

We consider $N$ globally-coupled classical Heisenberg spins of unit length denoted by $\mathbf{S}_i = (S_{ix}, S_{iy}, S_{iz})$; $i = 1, 2, \ldots, N$. The spins are interacting with each other via a mean-field coupling, which as already discussed is the extreme limit of long-range interactions. Expressed in terms of spherical polar angles $\theta_i \in [0, \pi]$ and $\phi_i \in [0, 2\pi)$, one has $S_{ix} = \sin\theta_i \cos\phi_i$, $S_{iy} = \sin\theta_i \sin\phi_i$, $S_{iz} = \cos\theta_i$. The Hamiltonian of the system is given by

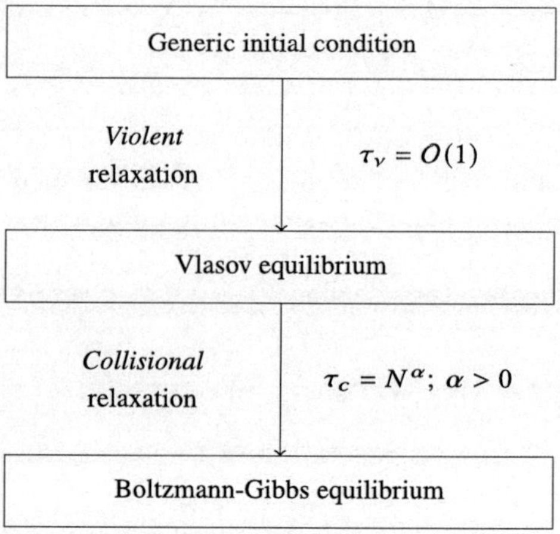


**Fig. 3.4 Schematic description of dynamical evolution of generic long-range interacting systems. Here, $\tau_v$ and $\tau_c$ denote respectively the *violent relaxation* and the *collisional relaxation* timescale**

$$H = -\frac{J}{2N}\sum_{i,j=1}^{N} \mathbf{S}_i \cdot \mathbf{S}_j + D\sum_{i=1}^{N} S_{iz}^{2n} , \tag{3.99}$$

where $n > 0$ is an integer. The first term with $J > 0$ on the right hand side models a ferromagnetic mean-field interaction between the spins. On the other hand, the second term with $D > 0$ on the right hand side accounts for local anisotropy: for example, $n = 1$ (respectively, $n = 2$) models quadratic (respectively, quartic) anisotropy, and will be referred to below as the quadratic (respectively, the quartic) model. Single-spin Hamiltonian of Heisenberg spins and involving quadratic and quartic terms has been considered previously in the literature, see, e.g., Ref. [7].

The coupling constant $J$ in Eq. (3.99) has been scaled down by the system size $N$ in order to make the energy extensive, in accordance with the Kac's prescription [6]. The system (3.99) however is intrinsically non-additive. This is because, as already discussed earlier in the chapter, extensivity does not guarantee additivity, although the converse is true. As a result, the system cannot be trivially subdivided into independent macroscopic parts. In the following, we set $J$ to unity without loss of generality. The anisotropy term in Eq. (3.99) lowers energy by having the magnetization vector

$$\mathbf{m} \equiv \frac{\sum_{i=1}^{N} \mathbf{S}_i}{N} = (m_x, m_y, m_z) \tag{3.100}$$

pointing in the $xy$ plane. However, no preferred direction of orientation of spins exists for $n = 0$, and the model becomes isotropic in the spin space. The length of the magnetization vector is given by

$$m \equiv \sqrt{m_x^2 + m_y^2 + m_z^2} . \tag{3.101}$$

Noting that the canonical variables for a classical Heisenberg spin are $\phi_i$ and

$$p_i \equiv \cos\theta_i\,, \tag{3.102}$$

the Poisson bracket $\{A, B\}$ for two functions of the spins are given by [8]

$$\{A, B\} \equiv \sum_{i=1}^{N} \left( \frac{\partial A}{\partial \phi_i}\frac{\partial B}{\partial p_i} - \frac{\partial A}{\partial p_i}\frac{\partial B}{\partial \phi_i} \right). \tag{3.103}$$

Then, one has

$$\{\phi_i, p_j\} = \delta_{i,j}\,, \tag{3.104}$$

while the individual spin components may be written as

$$S_{ix} = \sqrt{1-p_i^2}\cos\phi_i\,, \quad S_{iy} = \sqrt{1-p_i^2}\sin\phi_i\,, \quad S_{iz} = p_i\,. \tag{3.105}$$

Using Eq. (3.105), the $N$-spin Hamiltonian (3.99) may be written in terms of the canonical coordinates $\{\phi_i, p_i\}$ as

$$H = -\frac{1}{2N}\sum_{i=1}^{N}\sum_{j=1}^{N}\left[p_i p_j + \sqrt{1-p_i^2}\sqrt{1-p_j^2}\cos(\phi_i-\phi_j)\right] + D\sum_{i=1}^{N} p_i^{2n}\,. \tag{3.106}$$

The Hamilton's equations of motion obtained from Eq. (3.106) are given by

$$\dot{\phi}_i = \frac{\partial H}{\partial p_i} = 2nDp_i^{2n-1} - \frac{1}{N}\sum_{k=1}^{N}\left[p_k - \cos(\phi_i-\phi_k)\frac{p_i\sqrt{1-p_k^2}}{\sqrt{1-p_i^2}}\right], \tag{3.107}$$

$$\dot{p}_i = -\frac{\partial H}{\partial \phi_i} = -\frac{1}{N}\sum_{k=1}^{N}\sqrt{1-p_i^2}\sqrt{1-p_k^2}\sin(\phi_i-\phi_k)\,. \tag{3.108}$$

### 3.8.1 Vlasov Equation for the System

We now proceed to derive the Vlasov equation for the Heisenberg spin model (3.99). Comparing the form of the $N$-spin Hamiltonian of the model written in terms of the canonical coordinates in Eq. (3.106) with the $N$-particle Hamiltonian (3.3), one observes the following: in the latter, the canonical variables $p_i$ and $q_i$ appear separately in the two terms of the Hamiltonian, while one of the terms in the former contains both the canonical variables $p_i$ and $\phi_i$. Consequently, the Vlasov equation (3.93) obtained for the $N$-particle Hamiltonian (3.3) cannot be straightforwardly adapted to the $N$-spin Hamiltonian (3.106), and we need to derive afresh the Vlasov equation in the present case. To this end, we start with a discrete $N$-spin density function similar to the one defined in Eq. (3.9) with $z_i \equiv (\phi_i,\ p_i)$:

$$f_N^{(\mathrm{d})} = \prod_{i=1}^{N} \delta\big(\phi_i - \phi_i(t)\big)\, \delta\big(p_i - p_i(t)\big)\,. \tag{3.109}$$

Proceeding as was done following Eq. (3.9) in Sect. 3.6, and using the Hamilton's equations (3.107) and (3.108), one obtains in the limit $N \to \infty$ the Liouville equation for the time evolution of the corresponding $N$-spin distribution function $f_N$ as

$$\begin{aligned}\frac{\partial f_N}{\partial t} =& \frac{1}{N}\sum_{i=1}^{N}\sum_{k=1}^{N}\sqrt{1-p_i^2}\sqrt{1-p_k^2}\,\sin(\phi_i-\phi_k)\frac{\partial f_N}{\partial p_i} - \sum_{i=1}^{N} 2nDp_i^{2n-1}\frac{\partial f_N}{\partial \phi_i}\\ &+\frac{1}{N}\sum_{i=1}^{N}\sum_{k=1}^{N} p_k\frac{\partial f_N}{\partial \phi_i} - \frac{1}{N}\sum_{i=1}^{N}\sum_{k=1}^{N}\cos(\phi_i-\phi_k)\,\frac{p_i\sqrt{1-p_k^2}}{\sqrt{1-p_i^2}}\frac{\partial f_N}{\partial \phi_i}\,.\end{aligned} \tag{3.110}$$

Note that $f_N$ satisfies the periodicity condition: $f_N(\{\phi, p_i\}, t) = f_N(\{\phi_i + 2\pi, p_i\}, t)$ for any number of $i$'s in the set $\{\phi_i, p_i\}$ and for all $t$. The time evolution of the $s$-spin distribution function $f_s(z_1, \ldots, z_s)$ is obtained from Eq. (3.17) with the help of the Liouville equation (3.110) as

$$\begin{aligned}\frac{\partial f_s}{\partial t} =& \frac{N!}{(N-s)!}\int \mathrm{d}z_{s+1}\ldots\mathrm{d}z_N \Bigg[\frac{1}{N}\sum_{i=1}^{N}\sum_{k=1}^{N}\sqrt{1-p_i^2}\sqrt{1-p_k^2}\,\sin(\phi_i-\phi_k)\frac{\partial f_N}{\partial p_i}\\ &-\sum_{i=1}^{N} 2nDp_i^{2n-1}\frac{\partial f_N}{\partial \phi_i} + \frac{1}{N}\sum_{i=1}^{N}\sum_{k=1}^{N} p_k\frac{\partial f_N}{\partial \phi_i}\\ &-\frac{1}{N}\sum_{i=1}^{N}\sum_{k=1}^{N}\cos(\phi_i-\phi_k)\,\frac{p_i\sqrt{1-p_k^2}}{\sqrt{1-p_i^2}}\frac{\partial f_N}{\partial \phi_i}\Bigg]\,.\end{aligned} \tag{3.111}$$

Note that here, we have

$$\int \mathrm{d}z_i \equiv \int_0^{2\pi} \mathrm{d}\phi_i \int_{-1}^{1} \mathrm{d}p_i\,. \tag{3.112}$$

Let us be very careful in evaluating the integrals on the right hand side of Eq. (3.111). We go term by term. Let us consider the first integral on the right hand side of Eq. (3.111). We denote

$$
\begin{aligned}
I_1 &\equiv \frac{N!}{(N-s)!}\int \mathrm{d}z_{s+1}\ldots \mathrm{d}z_N\, \frac{1}{N}\sum_{i=1}^{N}\sum_{k=1}^{N}\sqrt{1-p_i^2}\sqrt{1-p_k^2}\,\sin(\phi_i-\phi_k)\frac{\partial f_N}{\partial p_i} \\
&= \frac{N!}{N(N-s)!}\int \mathrm{d}z_{s+1}\ldots \mathrm{d}z_N \left[\sum_{i=1}^{s}\sum_{k=1}^{s}+\sum_{i=1}^{s}\sum_{k=s+1}^{N}+\sum_{i=s+1}^{N}\sum_{k=1}^{s}+\sum_{i=s+1}^{N}\sum_{k=s+1}^{N}\right] \\
&\times \sqrt{1-p_i^2}\sqrt{1-p_k^2}\,\sin(\phi_i-\phi_k)\frac{\partial f_N}{\partial p_i}\,.
\end{aligned}
\tag{3.113}
$$

Using Eq. (3.17), we reduce Eq. (3.113) to

$$
\begin{aligned}
I_1 =& \frac{1}{N}\sum_{i=1}^{s}\sum_{k=1}^{s}\sqrt{1-p_i^2}\sqrt{1-p_k^2}\,\sin(\phi_i-\phi_k)\frac{\partial f_s}{\partial p_i} \\
&+\frac{N!}{N(N-s)!}\sum_{i=1}^{s}\sqrt{1-p_i^2}\int \mathrm{d}z_{s+1}\ldots \mathrm{d}z_N \sum_{k=s+1}^{N}\sin(\phi_i-\phi_k)\sqrt{1-p_k^2}\,\frac{\partial f_N}{\partial p_i} \\
&+\frac{N!}{N(N-s)!}\sum_{i=1}^{s}\sqrt{1-p_i^2}\int \mathrm{d}z_{s+1}\ldots \mathrm{d}z_N \sum_{k=s+1}^{N}\sin(\phi_k-\phi_i)\sqrt{1-p_k^2}\,\frac{\partial f_N}{\partial p_k} \\
&+\frac{N!}{N(N-s)!}\int \mathrm{d}z_{s+1}\ldots \mathrm{d}z_N \sum_{i=s+1}^{N}\sum_{k=s+1}^{N}\sin(\phi_i-\phi_k)\sqrt{1-p_i^2}\sqrt{1-p_k^2}\,\frac{\partial f_N}{\partial p_i},
\end{aligned}
\tag{3.114}
$$

where in obtaining the third term on the right hand side, we have interchanged the dummy indices $i$ and $k$. Then, the third term on the right hand side of Eq. (3.114) is written as

$$
\begin{aligned}
I_1^{(3)} &\equiv \frac{N!}{N(N-s)!}\sum_{i=1}^{s}\sqrt{1-p_i^2}\sum_{k=s+1}^{N}\int \mathrm{d}z_{s+1}\ldots \mathrm{d}z_N \\
&\times \left[\frac{\partial}{\partial p_k}\left(\sin(\phi_k-\phi_i)\sqrt{1-p_k^2}\,f_N\right)+\sin(\phi_k-\phi_i)\frac{p_k}{\sqrt{1-p_k^2}}f_N\right],
\end{aligned}
\tag{3.115}
$$

the first term (having total derivative in $p_k$) of which vanishes after performing the integral over $p_k$, since the factor $\sqrt{1-p_k^2}$ vanishes at the integration limits of $p_k$ given by $p_k=\pm 1$. Consequently, one obtains

$$
I_1^{(3)} = \frac{N!}{N(N-s)!}\sum_{i=1}^{s}\sqrt{1-p_i^2}\sum_{k=s+1}^{N}\int \mathrm{d}z_{s+1}\ldots \mathrm{d}z_N\,\sin(\phi_k-\phi_i)\frac{p_k}{\sqrt{1-p_k^2}}f_N\,.
\tag{3.116}
$$

Similarly, expressing it as a total derivative term in $p_i$ and performing the corresponding integral, one may reduce the last integral on the right hand side of Eq. (3.114) to

$$I_1^{(4)} \equiv \frac{N!}{N(N-s)!} \sum_{i=s+1}^{N} \sum_{k=s+1}^{N} \int \mathrm{d}z_{s+1} \ldots \mathrm{d}z_N \, \sin(\phi_i - \phi_k) \, \frac{p_i\sqrt{1-p_k^2}}{\sqrt{1-p_i^2}} \, f_N \,. \tag{3.117}$$

Substituting Eqs. (3.116) and (3.117) in Eq. (3.114), we obtain

$$\begin{aligned} I_1 =& \frac{1}{N} \sum_{i=1}^{s} \sum_{k=1}^{s} \sqrt{1-p_i^2}\sqrt{1-p_k^2} \, \sin(\phi_i - \phi_k) \frac{\partial f_s}{\partial p_i} \\ &+ \frac{N!}{N(N-s)!} \sum_{i=1}^{s} \sqrt{1-p_i^2} \sum_{k=s+1}^{N} \int \mathrm{d}z_{s+1} \ldots \mathrm{d}z_N \, \sin(\phi_i - \phi_k) \sqrt{1-p_k^2} \, \frac{\partial f_N}{\partial p_i} \\ &+ \frac{N!}{N(N-s)!} \sum_{i=1}^{s} \sqrt{1-p_i^2} \sum_{k=s+1}^{N} \int \mathrm{d}z_{s+1} \ldots \mathrm{d}z_N \, \sin(\phi_k - \phi_i) \frac{p_k}{\sqrt{1-p_k^2}} f_N \\ &+ \frac{N!}{N(N-s)!} \sum_{i=s+1}^{N} \sum_{k=s+1}^{N} \int \mathrm{d}z_{s+1} \ldots \mathrm{d}z_N \, \sin(\phi_i - \phi_k) \, \frac{p_i\sqrt{1-p_k^2}}{\sqrt{1-p_i^2}} \, f_N \,. \end{aligned} \tag{3.118}$$

The second integral on the right hand side of Eq. (3.111) reads

$$\begin{aligned} I_2 \equiv & -\frac{N!}{(N-s)!} \int \mathrm{d}z_{s+1} \ldots \mathrm{d}z_N \sum_{i=1}^{N} 2nDp_i^{2n-1} \frac{\partial f_N}{\partial \phi_i} \\ =& -2nD \frac{N!}{(N-s)!} \int \mathrm{d}z_{s+1} \ldots \mathrm{d}z_N \left[ \sum_{i=1}^{s} + \sum_{i=s+1}^{N} \right] p_i^{2n-1} \frac{\partial f_N}{\partial \phi_i} \\ =& -2nD \sum_{i=1}^{s} p_i^{2n-1} \frac{\partial f_s}{\partial \phi_i} - 2nD \frac{N!}{(N-s)!} \sum_{i=s+1}^{N} \int \mathrm{d}z_{s+1} \ldots \mathrm{d}z_N \frac{\partial}{\partial \phi_i} \left( p_i^{2n-1} f_N \right) , \end{aligned} \tag{3.119}$$

where in the last step, we have used Eq. (3.17) in obtaining the first term on the right hand side. The last term on the right hand side of Eq. (3.119) is a total derivative term in $\phi_i$. Now, one performing the integral over $\phi_i$, one finds that the contribution of this term is zero, as $f_N(z_1, \ldots, \phi_i = 0, \ldots, z_N) = f_N(z_1, \ldots, \phi_i = 2\pi, \ldots, z_N) \ \forall \ i$. Consequently, Eq. (3.119) reduces to

$$I_2 = -2nD \sum_{i=1}^{s} p_i^{2n-1} \frac{\partial f_s}{\partial \phi_i}. \tag{3.120}$$

From the third integral on the right hand side of Eq. (3.111), one obtains

$$\begin{aligned}
I_3 &\equiv \frac{N!}{N(N-s)!} \int \mathrm{d}z_{s+1} \dots \mathrm{d}z_N \sum_{i=1}^{N} \sum_{k=1}^{N} p_k \frac{\partial f_N}{\partial \phi_i} \\
&= \frac{N!}{N(N-s)!} \int \mathrm{d}z_{s+1} \dots \mathrm{d}z_N \\
&\times \left[ \sum_{i=1}^{s} \sum_{k=1}^{s} + \sum_{i=1}^{s} \sum_{k=s+1}^{N} + \sum_{i=s+1}^{N} \sum_{k=1}^{s} + \sum_{i=s+1}^{N} \sum_{k=s+1}^{N} \right] p_k \frac{\partial f_N}{\partial \phi_i} \\
&= \frac{1}{N} \sum_{i=1}^{s} \sum_{k=1}^{s} p_k \frac{\partial f_s}{\partial \phi_i} + \frac{N!}{N(N-s)!} \sum_{i=1}^{s} \sum_{k=s+1}^{N} \int \mathrm{d}z_{s+1} \dots \mathrm{d}z_N \, p_k \frac{\partial f_N}{\partial \phi_i} \\
&+ \frac{N!}{N(N-s)!} \sum_{k=1}^{s} p_k \sum_{i=s+1}^{N} \int \mathrm{d}z_{s+1} \dots \mathrm{d}z_N \frac{\partial f_N}{\partial \phi_i} \\
&+ \frac{N!}{N(N-s)!} \sum_{i=s+1}^{N} \sum_{k=s+1}^{N} \int \mathrm{d}z_{s+1} \dots \mathrm{d}z_N \frac{\partial}{\partial \phi_i} \Big( p_k f_N \Big),
\end{aligned} \tag{3.121}$$

where in the last step, we have used Eq. (3.17) in obtaining the first term on the right hand side. The last two terms on the right hand side of Eq. (3.121) having total derivative in $\phi_i$ vanishes after the integral over $\phi_i$ is performed due to $f_N$ being $2\pi$-periodic in $\phi_i$. Then, we get

$$I_3 = \frac{1}{N} \sum_{i=1}^{s} \sum_{k=1}^{s} p_k \frac{\partial f_s}{\partial \phi_i} + \frac{N!}{N(N-s)!} \sum_{i=1}^{s} \sum_{k=s+1}^{N} \int \mathrm{d}z_{s+1} \dots \mathrm{d}z_N \, p_k \frac{\partial f_N}{\partial \phi_i}. \tag{3.122}$$

Now, we evaluate the last integral on the right hand side of Eq. (3.111). To this end, we denote

$$\begin{aligned}
I_4 &\equiv -\frac{N!}{N(N-s)!} \int \mathrm{d}z_{s+1} \dots \mathrm{d}z_N \sum_{i=1}^{N} \sum_{k=1}^{N} \cos(\phi_i - \phi_k) \frac{p_i \sqrt{1-p_k^2}}{\sqrt{1-p_i^2}} \frac{\partial f_N}{\partial \phi_i} \\
&= -\frac{N!}{N(N-s)!} \int \mathrm{d}z_{s+1} \dots \mathrm{d}z_N \left[ \sum_{i=1}^{s} \sum_{k=1}^{s} + \sum_{i=1}^{s} \sum_{k=s+1}^{N} + \sum_{i=s+1}^{N} \sum_{k=1}^{s} + \sum_{i=s+1}^{N} \sum_{k=s+1}^{N} \right] \\
&\quad \times \cos(\phi_i - \phi_k) \frac{p_i \sqrt{1-p_k^2}}{\sqrt{1-p_i^2}} \frac{\partial f_N}{\partial \phi_i}.
\end{aligned} \tag{3.123}$$

Using Eq. (3.17), we reduce Eq. (3.123) to

$$
\begin{aligned}
I_4 =& -\frac{1}{N}\sum_{i=1}^{s}\sum_{k=1}^{s}\cos(\phi_i-\phi_k)\,\frac{p_i\sqrt{1-p_k^2}}{\sqrt{1-p_i^2}}\frac{\partial f_s}{\partial \phi_i} \\
&-\frac{N!}{N(N-s)!}\sum_{i=1}^{s}\frac{p_i}{\sqrt{1-p_i^2}}\int \mathrm{d}z_{s+1}\ldots\mathrm{d}z_N\sum_{k=s+1}^{N}\cos(\phi_i-\phi_k)\sqrt{1-p_k^2}\,\frac{\partial f_N}{\partial \phi_i} \\
&-\frac{N!}{N(N-s)!}\sum_{i=1}^{s}\sqrt{1-p_i^2}\int \mathrm{d}z_{s+1}\ldots\mathrm{d}z_N\sum_{k=s+1}^{N}\cos(\phi_k-\phi_i)\frac{p_k}{\sqrt{1-p_k^2}}\frac{\partial f_N}{\partial \phi_k} \\
&-\frac{N!}{N(N-s)!}\int \mathrm{d}z_{s+1}\ldots\mathrm{d}z_N\sum_{i=s+1}^{N}\sum_{k=s+1}^{N}\cos(\phi_i-\phi_k)\frac{p_i\sqrt{1-p_k^2}}{\sqrt{1-p_i^2}}\frac{\partial f_N}{\partial \phi_i},
\end{aligned}
\tag{3.124}
$$

where in obtaining the third term on the right hand side, we have interchanged the dummy indices $i$ and $k$. Then, the third term on the right hand side of Eq. (3.124) is written as

$$
\begin{aligned}
I_4^{(3)} \equiv& -\frac{N!}{N(N-s)!}\sum_{i=1}^{s}\sqrt{1-p_i^2}\sum_{k=s+1}^{N}\int \mathrm{d}z_{s+1}\ldots\mathrm{d}z_N \\
&\times\left[\frac{\partial}{\partial \phi_k}\left(\cos(\phi_k-\phi_i)\frac{p_k}{\sqrt{1-p_k^2}}f_N\right)+\sin(\phi_k-\phi_i)\frac{p_k}{\sqrt{1-p_k^2}}f_N\right].
\end{aligned}
\tag{3.125}
$$

The first term of Eq. (3.125) vanishes after performing the integral over $\phi_k$ and using the condition $f_N(Z_1,\ldots,\phi_k=0,\ldots,z_N)=f_N(Z_1,\ldots,\phi_k=2\pi,\ldots,z_N)\ \forall\ k$. Then, we obtain

$$
I_4^{(3)} = -\frac{N!}{N(N-s)!}\sum_{i=1}^{s}\sqrt{1-p_i^2}\sum_{k=s+1}^{N}\int \mathrm{d}z_{s+1}\ldots\mathrm{d}z_N\ \sin(\phi_k-\phi_i)\frac{p_k}{\sqrt{1-p_k^2}}f_N\,.
\tag{3.126}
$$

Similarly, expressing it as a total derivative term in $\phi_i$ and performing the corresponding integral, one may reduce the last integral on the right hand side of Eq. (3.124) to

$$
I_4^{(4)} = -\frac{N!}{N(N-s)!}\sum_{i=s+1}^{N}\sum_{k=s+1}^{N}\int \mathrm{d}z_{s+1}\ldots\mathrm{d}z_N\ \sin(\phi_i-\phi_k)\frac{p_i\sqrt{1-p_k^2}}{\sqrt{1-p_i^2}}f_N\,.
\tag{3.127}
$$

Substituting Eqs. (3.126) and (3.127) in Eq. (3.124), one obtains

$$
\begin{aligned}
I_4 = &-\frac{1}{N}\sum_{i=1}^{s}\sum_{k=1}^{s}\cos(\phi_i-\phi_k)\,\frac{p_i\sqrt{1-p_k^2}}{\sqrt{1-p_i^2}}\frac{\partial f_s}{\partial \phi_i} \\
&-\frac{N!}{N(N-s)!}\sum_{i=1}^{s}\frac{p_i}{\sqrt{1-p_i^2}}\sum_{k=s+1}^{N}\int \mathrm{d}z_{s+1}\ldots \mathrm{d}z_N\,\cos(\phi_i-\phi_k)\sqrt{1-p_k^2}\,\frac{\partial f_N}{\partial \phi_i} \\
&-\frac{N!}{N(N-s)!}\sum_{i=1}^{s}\sqrt{1-p_i^2}\sum_{k=s+1}^{N}\int \mathrm{d}z_{s+1}\ldots \mathrm{d}z_N\,\sin(\phi_k-\phi_i)\frac{p_k}{\sqrt{1-p_k^2}}f_N \\
&-\frac{N!}{N(N-s)!}\sum_{i=s+1}^{N}\sum_{k=s+1}^{N}\int \mathrm{d}z_{s+1}\ldots \mathrm{d}z_N\,\sin(\phi_i-\phi_k)\frac{p_i\sqrt{1-p_k^2}}{\sqrt{1-p_i^2}}f_N\,.
\end{aligned}
\tag{3.128}
$$

Finally, substituting Eqs. (3.118), (3.120), (3.122), and (3.128) in Eq. (3.111), we obtain the time evolution of the $s$-spin distribution function $f_s$ as

$$
\begin{aligned}
\frac{\partial f_s}{\partial t} = &\frac{1}{N}\sum_{i=1}^{s}\sum_{k=1}^{s}\sqrt{1-p_i^2}\sqrt{1-p_k^2}\,\sin(\phi_i-\phi_k)\frac{\partial f_s}{\partial p_i} \\
&-\frac{1}{N}\sum_{i=1}^{s}\sum_{k=1}^{s}\cos(\phi_i-\phi_k)\,\frac{p_i\sqrt{1-p_k^2}}{\sqrt{1-p_i^2}}\frac{\partial f_s}{\partial \phi_i} \\
&-2nD\sum_{i=1}^{s}p_i^{2n-1}\frac{\partial f_s}{\partial \phi_i}+\frac{1}{N}\sum_{i=1}^{s}\sum_{k=1}^{s}p_k\frac{\partial f_s}{\partial \phi_i} \\
&+\frac{N!}{N(N-s)!}\sum_{i=1}^{s}\sum_{k=s+1}^{N}\int \mathrm{d}z_{s+1}\ldots \mathrm{d}z_N\,p_k\frac{\partial f_N}{\partial \phi_i} \\
&+\frac{N!}{N(N-s)!}\sum_{i=1}^{s}\sqrt{1-p_i^2}\sum_{k=s+1}^{N}\int \mathrm{d}z_{s+1}\ldots \mathrm{d}z_N\,\sin(\phi_i-\phi_k)\sqrt{1-p_k^2}\,\frac{\partial f_N}{\partial p_i} \\
&-\frac{N!}{N(N-s)!}\sum_{i=1}^{s}\frac{p_i}{\sqrt{1-p_i^2}}\sum_{k=s+1}^{N}\int \mathrm{d}z_{s+1}\ldots \mathrm{d}z_N\,\cos(\phi_i-\phi_k)\sqrt{1-p_k^2}\,\frac{\partial f_N}{\partial \phi_i}\,.
\end{aligned}
\tag{3.129}
$$

Let us now concentrate on the last three terms having integrals on the right hand side of Eq. (3.129). Denoting the fifth term on the right hand side of Eq. (3.129) as $I^{(5)}$, we have

$$I^{(5)} \equiv \frac{N!}{N(N-s)!} \sum_{i=1}^{s} \frac{\partial}{\partial \phi_i} \sum_{k=s+1}^{N} \int \mathrm{d}z_{s+1} \ldots \mathrm{d}z_N \, p_k f_N(z_1, \ldots, z_N, t) \,. \quad (3.130)$$

As $f_N$ is symmetric with respect to exchange of any two $z$'s, the integral on the right hand side of Eq. (3.130) may be written as $(N-s)$ similar integrals. Then, we obtain

$$\begin{aligned} I^{(5)} &= \frac{N!}{N(N-s)!} \sum_{i=1}^{s} \frac{\partial}{\partial \phi_i} (N-s) \int \mathrm{d}z_{s+1} \ldots \mathrm{d}z_N \, p_{s+1} \, f_N(z_1, \ldots, z_N, t) \\ &= \frac{1}{N} \sum_{i=1}^{s} \frac{\partial}{\partial \phi_i} \int \mathrm{d}z_{s+1} \, p_{s+1} \frac{N!}{(N-s-1)!} \int \mathrm{d}z_{s+2} \ldots \mathrm{d}z_N \, f_N(z_1, \ldots, z_N, t) \,, \end{aligned} \quad (3.131)$$

which with the help of Eq. (3.17) is reduced to

$$I^{(5)} = \frac{1}{N} \sum_{i=1}^{s} \int \mathrm{d}z_{s+1} \, p_{s+1} \frac{\partial f_{s+1}}{\partial \phi_i} \,. \quad (3.132)$$

The sixth term on the right hand side of Eq. (3.129) is written as

$$I^{(6)} \equiv \frac{N!}{N(N-s)!} \sum_{i=1}^{s} \sqrt{1-p_i^2} \sum_{k=s+1}^{N} \int \mathrm{d}z_{s+1} \ldots \mathrm{d}z_N \, \sin(\phi_i - \phi_k) \sqrt{1-p_k^2} \frac{\partial f_N}{\partial p_i} \,. \quad (3.133)$$

Expanding the sine term, we obtain from Eq. (3.133) that

$$\begin{aligned} I^{(6)} &= \frac{N!}{N(N-s)!} \sum_{i=1}^{s} \sqrt{1-p_i^2} \, \sin\phi_i \frac{\partial}{\partial p_i} \sum_{k=s+1}^{N} \int \mathrm{d}z_{s+1} \ldots \mathrm{d}z_N \cos\phi_k \sqrt{1-p_k^2} \, f_N \\ &- \frac{N!}{N(N-s)!} \sum_{i=1}^{s} \sqrt{1-p_i^2} \cos\phi_i \frac{\partial}{\partial p_i} \sum_{k=s+1}^{N} \int \mathrm{d}z_{s+1} \ldots \mathrm{d}z_N \sin\phi_k \sqrt{1-p_k^2} \, f_N \\ &= \frac{N!}{N(N-s)!} \sum_{i=1}^{s} \sqrt{1-p_i^2} \, \sin\phi_i \frac{\partial}{\partial p_i} \, (N-s) \\ &\times \int \mathrm{d}z_{s+1} \cos\phi_{s+1} \sqrt{1-p_{s+1}^2} \int \mathrm{d}z_{s+2} \ldots \mathrm{d}z_N \, f_N \\ &- \frac{N!}{N(N-s)!} \sum_{i=1}^{s} \sqrt{1-p_i^2} \cos\phi_i \frac{\partial}{\partial p_i} \, (N-s) \\ &\times \int \mathrm{d}z_{s+1} \sin\phi_{s+1} \sqrt{1-p_{s+1}^2} \int \mathrm{d}z_{s+2} \ldots \mathrm{d}z_N \, f_N \\ &= \frac{1}{N} \sum_{i=1}^{s} \sqrt{1-p_i^2} \int \mathrm{d}z_{s+1} \, \sin(\phi_i - \phi_{s+1}) \sqrt{1-p_{s+1}^2} \frac{\partial f_{s+1}}{\partial p_i} \,. \end{aligned} \quad (3.134)$$

In obtaining the second equality, we have used the symmetry property of $f_N$. Similarly, the last term on the right hand side of Eq. (3.129) may be reduced to

$$I^{(7)} \equiv -\frac{1}{N}\sum_{i=1}^{s}\frac{p_i}{\sqrt{1-p_i^2}}\int \mathrm{d}z_{s+1}\ \cos(\phi_i-\phi_{s+1})\sqrt{1-p_{s+1}^2}\,\frac{\partial f_{s+1}}{\partial \phi_i}. \tag{3.135}$$

Then, substituting Eqs. (3.132), (3.134), and (3.135) in Eq. (3.129), one finally obtains the time evolution of the $s$-spin distribution function $f_s$ as

$$\begin{aligned}
\frac{\partial f_s}{\partial t} =& \frac{1}{N}\sum_{i=1}^{s}\sum_{k=1}^{s}\sqrt{1-p_i^2}\sqrt{1-p_k^2}\ \sin(\phi_i-\phi_k)\frac{\partial f_s}{\partial p_i} \\
&-\frac{1}{N}\sum_{i=1}^{s}\sum_{k=1}^{s}\cos(\phi_i-\phi_k)\ \frac{p_i\sqrt{1-p_k^2}}{\sqrt{1-p_i^2}}\frac{\partial f_s}{\partial \phi_i} \\
&-2nD\sum_{i=1}^{s}p_i^{2n-1}\frac{\partial f_s}{\partial \phi_i}+\frac{1}{N}\sum_{i=1}^{s}\sum_{k=1}^{s}p_k\frac{\partial f_s}{\partial \phi_i}+\frac{1}{N}\sum_{i=1}^{s}\int \mathrm{d}z_{s+1}\ p_{s+1}\ \frac{\partial f_{s+1}}{\partial \phi_i} \\
&+\frac{1}{N}\sum_{i=1}^{s}\sqrt{1-p_i^2}\int \mathrm{d}z_{s+1}\ \sin(\phi_i-\phi_{s+1})\sqrt{1-p_{s+1}^2}\,\frac{\partial f_{s+1}}{\partial p_i} \\
&-\frac{1}{N}\sum_{i=1}^{s}\frac{p_i}{\sqrt{1-p_i^2}}\int \mathrm{d}z_{s+1}\ \cos(\phi_i-\phi_{s+1})\sqrt{1-p_{s+1}^2}\,\frac{\partial f_{s+1}}{\partial \phi_i},
\end{aligned} \tag{3.136}$$

for $s = 1, \ldots, (N-1)$. Equation (3.136) along with the Liouville equation (3.110) form the BBGKY hierarchy for the Heisenberg spin model (3.99).

Now, one may straightforwardly write down the equation of motion for the single-spin distribution function $f_1$: with $s = 1$, we obtain from Eq. (3.136) that

$$\begin{aligned}
\frac{\partial f_1}{\partial t} =& -2nDp_1^{2n-1}\frac{\partial f_1}{\partial \phi_1}+\frac{1}{N}\int \mathrm{d}z_2\ p_2\ \frac{\partial f_2}{\partial \phi_1} \\
&+\frac{1}{N}\sqrt{1-p_1^2}\int \mathrm{d}z_2\ \sin(\phi_1-\phi_2)\sqrt{1-p_2^2}\,\frac{\partial f_2}{\partial p_1} \\
&-\frac{1}{N}\frac{p_1}{\sqrt{1-p_1^2}}\int \mathrm{d}z_2\ \cos(\phi_1-\phi_2)\sqrt{1-p_2^2}\,\frac{\partial f_2}{\partial \phi_1}.
\end{aligned} \tag{3.137}$$

Substituting in Eq. (3.137) an expansion similar to Eq. (3.40), one obtains

$$\frac{\partial f_1(z_1,t)}{\partial t} = -2nDp_1^{2n-1}\frac{\partial f_1(z_1,t)}{\partial \phi_1}$$
$$+\frac{1}{N}\int \mathrm{d}z_2\ p_2\ \frac{\partial}{\partial \phi_1}\Big[f_1(z_1,t)f_1(z_2,t)+g_2(z_1,z_2,t)\Big]$$
$$+\frac{1}{N}\sqrt{1-p_1^2}\int \mathrm{d}z_2\ \sin(\phi_1-\phi_2)\sqrt{1-p_2^2}\ \frac{\partial}{\partial p_1}\Big[f_1(z_1,t)f_1(z_2,t)+g_2(z_1,z_2,t)\Big]$$
$$-\frac{1}{N}\frac{p_1}{\sqrt{1-p_1^2}}\int \mathrm{d}z_2 \cos(\phi_1-\phi_2)\sqrt{1-p_2^2}\ \frac{\partial}{\partial \phi_1}\Big[f_1(z_1,t)f_1(z_2,t)+g_2(z_1,z_2,t)\Big].$$
$$(3.138)$$

Expanding the sine and cosine terms in the last equation, we get

$$\frac{\partial f_1(z_1,t)}{\partial t} = -2nDp_1^{2n-1}\frac{\partial f_1(z_1,t)}{\partial \phi_1} + \frac{1}{N}\frac{\partial f_1(z_1,t)}{\partial \phi_1}\int \mathrm{d}z_2\ p_2 f_1(z_2,t)$$
$$+\frac{1}{N}\sqrt{1-p_1^2}\sin\phi_1\frac{\partial f_1(z_1,t)}{\partial p_1}\int \mathrm{d}z_2\ \cos\phi_2\sqrt{1-p_2^2}\ f_1(z_2,t)$$
$$-\frac{1}{N}\sqrt{1-p_1^2}\cos\phi_1\frac{\partial f_1(z_1,t)}{\partial p_1}\int \mathrm{d}z_2\ \sin\phi_2\sqrt{1-p_2^2}\ f_1(z_2,t)$$
$$-\frac{1}{N}\frac{p_1}{\sqrt{1-p_1^2}}\cos\phi_1\frac{\partial f_1(z_1,t)}{\partial \phi_1}\int \mathrm{d}z_2\ \cos\phi_2\sqrt{1-p_2^2}\ f_1(z_2,t)$$
$$-\frac{1}{N}\frac{p_1}{\sqrt{1-p_1^2}}\sin\phi_1\frac{\partial f_1(z_1,t)}{\partial \phi_1}\int \mathrm{d}z_2\ \sin\phi_2\sqrt{1-p_2^2}\ f_1(z_2,t)$$
$$+\frac{1}{N}\int \mathrm{d}z_2\ p_2\ \frac{\partial g_2(z_1,z_2,t)}{\partial \phi_1}$$
$$+\frac{1}{N}\sqrt{1-p_1^2}\int \mathrm{d}z_2\ \sin(\phi_1-\phi_2)\sqrt{1-p_2^2}\ \frac{\partial g_2(z_1,z_2,t)}{\partial p_1}$$
$$-\frac{1}{N}\frac{p_1}{\sqrt{1-p_1^2}}\int \mathrm{d}z_2\ \cos(\phi_1-\phi_2)\sqrt{1-p_2^2}\ \frac{\partial g_2(z_1,z_2,t)}{\partial \phi_1}. \qquad (3.139)$$

Now, we define renormalized $s$-spin distributions $\bar{f}_s$, such that

$$\bar{f}_s(z_1,\ldots,z_s,t) \equiv \frac{1}{N^s} f_s(z_1,\ldots,z_s,t). \qquad (3.140)$$

It then follows from Eq. (3.22) that

$$\int \mathrm{d}z_1\ \bar{f}_1(z_1,t) = 1, \qquad (3.141)$$
$$\int \mathrm{d}z_1\mathrm{d}z_2\ \bar{f}_2(z_1,z_2,t) = 1-\frac{1}{N}. \qquad (3.142)$$

In this new definition, the two-spin correlation function is written as

$$\begin{aligned}\bar{g}_2(z_1, z_2, t) &= \bar{f}_2(z_1, z_2, t) - \bar{f}_1(z_1, t)\bar{f}_1(z_2, t) \\ &= \frac{1}{N^2}\left[f_2(z_1, z_2, t) - f_1(z_1, t)f_1(z_2, t)\right] \\ &= \frac{1}{N^2}\, g_2(z_1, z_2, t)\,,\end{aligned} \tag{3.143}$$

where we have used Eqs. (3.140) and (3.40). From Eq. (3.42), we have already seen that $g_2(z_1, z_2, t)$ is of $\mathcal{O}(N)$, so that from Eq. (3.143), we get

$$\bar{g}_2(z_1, z_2, t) = \mathcal{O}\left(\frac{1}{N}\right). \tag{3.144}$$

Multiplying both sides of Eq. (3.139) by $1/N$, we rewrite the equation in terms of the newly defined distribution functions as

$$\begin{aligned}\frac{\partial \bar{f}_1(z_1, t)}{\partial t} &= -2nDp_1^{2n-1}\frac{\partial \bar{f}_1(z_1, t)}{\partial \phi_1} + \frac{\partial \bar{f}_1(z_1, t)}{\partial \phi_1}\int \mathrm{d}z_2\; p_2 \bar{f}_1(z_2, t) \\ &+ \sqrt{1-p_1^2}\sin\phi_1 \frac{\partial \bar{f}_1(z_1, t)}{\partial p_1}\int \mathrm{d}z_2\; \cos\phi_2\sqrt{1-p_2^2}\;\bar{f}_1(z_2, t) \\ &- \sqrt{1-p_1^2}\cos\phi_1 \frac{\partial \bar{f}_1(z_1, t)}{\partial p_1}\int \mathrm{d}z_2\; \sin\phi_2\sqrt{1-p_2^2}\;\bar{f}_1(z_2, t) \\ &- \frac{p_1}{\sqrt{1-p_1^2}}\cos\phi_1 \frac{\partial \bar{f}_1(z_1, t)}{\partial \phi_1}\int \mathrm{d}z_2\; \cos\phi_2\sqrt{1-p_2^2}\;\bar{f}_1(z_2, t) \\ &- \frac{p_1}{\sqrt{1-p_1^2}}\sin\phi_1 \frac{\partial \bar{f}_1(z_1, t)}{\partial \phi_1}\int \mathrm{d}z_2\; \sin\phi_2\sqrt{1-p_2^2}\;\bar{f}_1(z_2, t) \\ &+ \int \mathrm{d}z_2\; p_2\,\frac{\partial \bar{g}_2(z_1, z_2, t)}{\partial \phi_1} + \sqrt{1-p_1^2}\int \mathrm{d}z_2\; \sin(\phi_1-\phi_2)\sqrt{1-p_2^2}\;\frac{\partial \bar{g}_2(z_1, z_2, t)}{\partial p_1} \\ &- \frac{p_1}{\sqrt{1-p_1^2}}\int \mathrm{d}z_2\; \cos(\phi_1-\phi_2)\sqrt{1-p_2^2}\;\frac{\partial \bar{g}_2(z_1, z_2, t)}{\partial \phi_1}\,.\end{aligned} \tag{3.145}$$

Equation (3.141) suggests that $\bar{f}_1 = \mathcal{O}(1)$, while we already have $\bar{g}_2 = \mathcal{O}(1/N)$. Therefore, for times $t \ll N$, we may neglect the terms involving $\bar{g}_2$ on the right hand side of Eq. (3.145). Then, Eq. (3.145) becomes a closed equation involving terms having only $\bar{f}_1$. For brevity, we drop the subscript from $\bar{f}_1$, and rewrite the evolution equation for the single-spin distribution function in terms of its argument $z = (\phi, p)$ as

$$\begin{aligned}&\frac{\partial \bar{f}(\phi, p, t)}{\partial t} \\ &= -2nDp^{2n-1}\frac{\partial \bar{f}(\phi, p, t)}{\partial \phi} + \frac{\partial \bar{f}(\phi, p, t)}{\partial \phi}\int_0^{2\pi}\mathrm{d}\phi_2\int_{-1}^{1}\mathrm{d}p_2\;\; p_2\;\bar{f}(\phi_2, p_2, t)\end{aligned}$$

$$+\sqrt{1-p^2}\,\sin\phi\,\frac{\partial \bar{f}(\phi,p,t)}{\partial p}\int_0^{2\pi}\mathrm{d}\phi_2\int_{-1}^{1}\mathrm{d}p_2\,\cos\phi_2\sqrt{1-p_2^2}\,\bar{f}(\phi_2,p_2,t)$$
$$-\sqrt{1-p^2}\,\cos\phi\,\frac{\partial \bar{f}(\phi,p,t)}{\partial p}\int_0^{2\pi}\mathrm{d}\phi_2\int_{-1}^{1}\mathrm{d}p_2\,\sin\phi_2\sqrt{1-p_2^2}\,\bar{f}(\phi_2,p_2,t)$$
$$-\frac{p}{\sqrt{1-p^2}}\cos\phi\,\frac{\partial \bar{f}(\phi,p,t)}{\partial \phi}\int_0^{2\pi}\mathrm{d}\phi_2\int_{-1}^{1}\mathrm{d}p_2\,\cos\phi_2\sqrt{1-p_2^2}\,\bar{f}(\phi_2,p_2,t)$$
$$-\frac{p}{\sqrt{1-p^2}}\sin\phi\,\frac{\partial \bar{f}(\phi,p,t)}{\partial \phi}\int_0^{2\pi}\mathrm{d}\phi_2\int_{-1}^{1}\mathrm{d}p_2\,\sin\phi_2\sqrt{1-p_2^2}\,\bar{f}(\phi_2,p_2,t)\,, \tag{3.146}$$

which is the Vlasov equation for the Heisenberg spin system (3.99).

The magnetization components are defined as

$$m_x \equiv \int_0^{2\pi}\mathrm{d}\phi\int_{-1}^{1}\mathrm{d}p\,\sqrt{1-p^2}\cos\phi\;\bar{f}(\phi,p,t) = m_x[\bar{f}]\,, \tag{3.147}$$
$$m_y \equiv \int_0^{2\pi}\mathrm{d}\phi\int_{-1}^{1}\mathrm{d}p\,\sqrt{1-p^2}\sin\phi\;\bar{f}(\phi,p,t) = m_y[\bar{f}]\,, \tag{3.148}$$
$$m_z \equiv \int_0^{2\pi}\mathrm{d}\phi\int_{-1}^{1}\mathrm{d}p\;p\,\bar{f}(\phi,p,t) = m_z[\bar{f}]\,, \tag{3.149}$$

which evidently show that the magnetization $\mathbf{m}$ is a functional of the single-spin distribution $\bar{f}(\phi, p, t)$. Indeed, Eqs. (3.147)–(3.149) represent the $N \to \infty$ limit of Eq. (3.100). This may be seen as follows. Suppose out of $N$ spins, $N_\alpha$ spins have a spin value $\mathbf{S}^{(\alpha)}$. Then, from Eq. (3.100), one may write

$$\mathbf{m} = \frac{1}{N}\sum_{\substack{\text{all possible}\\ \text{values of }\alpha}} N_\alpha\,\mathbf{S}^{(\alpha)} = \sum_{\substack{\text{all possible}\\ \text{values of }\alpha}} \frac{N_\alpha}{N}\,\mathbf{S}^{(\alpha)} \equiv \sum_{\substack{\text{all possible}\\ \text{values of }\alpha}} \bar{p}_\alpha\,\mathbf{S}^{(\alpha)}\,, \tag{3.150}$$

where $\bar{p}_\alpha$ is the probability of observing a spin value $\mathbf{S}^{(\alpha)}$. Therefore, in the limit $N \to \infty$, one may write

$$\boldsymbol{m} \to \boldsymbol{m}[f] \equiv \lim_{N\to\infty}\sum_{\substack{\text{all possible}\\ \text{values of }\alpha}} \bar{p}_\alpha\,\mathbf{S}^{(\alpha)} = \int_0^{2\pi}\mathrm{d}\phi\int_{-1}^{1}\mathrm{d}p\,\mathbf{S}(\phi,p)\,\bar{f}(\phi,p,t)\,, \tag{3.151}$$

which in terms of its components yields Eqs. (3.147)–(3.149).

In terms of the magnetization components, the Vlasov equation (3.146) for the system (3.99) may be rewritten in a simplified version as

$$\frac{\partial \bar{f}(\phi, p, t)}{\partial t} = -\Big[2nDp^{2n-1} - m_z + \frac{p}{\sqrt{1-p^2}}\Big(m_x \cos\phi + m_y \sin\phi\Big)\Big]\frac{\partial \bar{f}(\phi, p, t)}{\partial \phi}$$
$$+ \sqrt{1-p^2}\,\Big(m_x \sin\phi - m_y \cos\phi\Big)\frac{\partial \bar{f}(\phi, p, t)}{\partial p}\,. \tag{3.152}$$

We remark that in obtaining the Vlasov equation for the Heisenberg spin model, we have neglected the contribution of the two-spin correlation $\bar{g}_2$. The latter fact implies that the equation is valid for times $t \ll N$ or, in other words, in the limit $N \to \infty$ with $t$ finite.

### 3.8.2 The Klimontovich Approach

In this section, we derive the Vlasov equation (3.152) corresponding to the long-range interacting Heisenberg spin model (3.99) by using a different approach pioneered by Klimontovich [3, 9]. An advantage is that we will be able to provide a physical basis for the various size-dependent timescales of relaxation (to be discussed in detail in Chap. 8) for our model system (3.99). We first rewrite the Hamilton's equations of motion (3.107) and (3.108) in terms of the magnetization components given by Eq. (3.100), as

$$\dot{\phi}_i = 2nDp_i^{2n-1} + \frac{p_i}{\sqrt{1-p_i^2}}\big(m_x \cos\phi_i + m_y \sin\phi_i\big) - m_z\,, \tag{3.153}$$

$$\dot{p}_i = -m_x\sqrt{1-p_i^2}\,\sin\phi_i + m_y\sqrt{1-p_i^2}\,\cos\phi_i\,. \tag{3.154}$$

In this approach, the state of the $N$-spin system is described by the discrete single-spin time-dependent distribution function

$$\bar{f}^{(\mathrm{d})}(\phi, p, t) = \frac{1}{N}\sum_{i=1}^{N} \delta\big(\phi - \phi_i(t)\big)\,\delta\big(p - p_i(t)\big)\,, \tag{3.155}$$

which is defined such that $\bar{f}^{(\mathrm{d})}(\phi, p, t)\mathrm{d}\phi\,\mathrm{d}p$ counts the fraction of the total number of spins with canonical coordinates in $[\phi, \phi + \mathrm{d}\phi]$ and $[p, p + \mathrm{d}p]$. Here, $(\phi, p)$ are the Eulerian coordinates of the two-dimensional phase space, while $(\phi_i, p_i)$ are the Lagrangian coordinates of the spins, see Sect. 3.2. Note that $\bar{f}^{(\mathrm{d})}$ satisfies $\bar{f}^{(\mathrm{d})}(\phi, p, t) = \bar{f}^{(\mathrm{d})}(\phi + 2\pi, p, t)$ and the normalization

$$\int_0^{2\pi} \mathrm{d}\phi \int_{-1}^{1} \mathrm{d}p\; \bar{f}^{(\mathrm{d})}(\phi, p, t) = 1\,. \tag{3.156}$$

We remark that for the Heisenberg spin model (3.99) the single-spin distribution function $\bar{f}^{(\mathrm{d})}$ is defined in a two-dimensional $\mu$-space, while $f_N^{(\mathrm{d})}$ is defined in the

$2N$-dimensional $\Gamma$-space (more generally, for a system in three dimensions with $3N$ generalized coordinates, the dimension of the $\mu$-space is 6, and the dimension of the $\Gamma$-space is $6N$). A microstate of the system is denoted by $N$ representative points in the former and by a single representative point in the latter.

In terms of the function $\bar{f}^{(\mathrm{d})}$, we may express the magnetization component $m_x$ as follows:

$$\begin{aligned}
m_x(t) &= \frac{1}{N}\sum_{i=1}^{N}\sqrt{1-p_i^2(t)}\,\cos\phi_i(t) \\
&= \frac{1}{N}\sum_{i=1}^{N}\int_0^{2\pi}\mathrm{d}\phi\int_{-1}^{1}\mathrm{d}p\sqrt{1-p^2}\,\cos\phi\,\delta(\phi-\phi_i(t))\,\delta(p-p_i(t)) \\
&= \int_0^{2\pi}\mathrm{d}\phi\int_{-1}^{1}\mathrm{d}p\sqrt{1-p^2}\,\cos\phi\,\bar{f}^{(\mathrm{d})}(\phi,p,t)\,.
\end{aligned} \tag{3.157}$$

We thus see that $m_x$ is a functional of $\bar{f}^{(\mathrm{d})}(\phi, p, t)$, and so are $m_y$ and $m_z$. We then have

$$\begin{aligned}
&(m_x, m_y, m_z)[\bar{f}^{(\mathrm{d})}] \\
&= \int_0^{2\pi}\mathrm{d}\phi\int_{-1}^{1}\mathrm{d}p\,\left(\sqrt{1-p^2}\cos\phi,\,\sqrt{1-p^2}\sin\phi,\,p\right)\bar{f}^{(\mathrm{d})}(\phi,p,t)\,.
\end{aligned} \tag{3.158}$$

Differentiating $\bar{f}^{(\mathrm{d})}$ with respect to time, we obtain from Eq. (3.155) that

$$\begin{aligned}
&\frac{\partial\bar{f}^{(\mathrm{d})}(\phi,p,t)}{\partial t} \\
&= \frac{1}{N}\sum_{i=1}^{N}\left[\delta(\phi-\phi_i(t))\,\frac{\partial}{\partial t}\delta(p-p_i(t)) + \delta(p-p_i(t))\,\frac{\partial}{\partial t}\delta(\phi-\phi_i(t))\right] \\
&= \frac{1}{N}\sum_{i=1}^{N}\left[\delta(\phi-\phi_i(t))\,\dot{p}_i\,\frac{\partial}{\partial p_i}\delta(p-p_i(t)) + \delta(p-p_i(t))\,\dot{\phi}_i\,\frac{\partial}{\partial \phi_i}\delta(\phi-\phi_i(t))\right].
\end{aligned} \tag{3.159}$$

Substituting Eqs. (3.153) and (3.154) in Eq. (3.159), and using $\partial\delta(x-a)/\partial x = -\partial\delta(x-a)/\partial a$, and $f(x)\delta(x-a) = f(a)\delta(x-a)$, one obtains the Klimontovich equation for the time evolution of $\bar{f}^{(\mathrm{d})}$ as

$$\frac{\partial\bar{f}^{(\mathrm{d})}(\phi,p,t)}{\partial t} = -g_p\frac{\partial}{\partial p}\bar{f}^{(\mathrm{d})}(\phi,p,t) - g_\phi\frac{\partial}{\partial\phi}\bar{f}^{(\mathrm{d})}(\phi,p,t)\,, \tag{3.160}$$

where we have defined

$$g_p \equiv g_p[\bar{f}^{(d)}](\phi, p) = \sqrt{1-p^2}m_y[\bar{f}^{(d)}]\cos\phi - \sqrt{1-p^2}m_x[\bar{f}^{(d)}]\sin\phi\,, \tag{3.161}$$

$$g_\phi \equiv g_\phi[\bar{f}^{(d)}](\phi, p) = m_x[\bar{f}^{(d)}]\frac{p}{\sqrt{1-p^2}}\cos\phi + m_y[\bar{f}^{(d)}]\frac{p}{\sqrt{1-p^2}}\sin\phi - m_z[\bar{f}^{(d)}] + 2nDp^{2n-1}\,. \tag{3.162}$$

We remark that Eq. (3.160) holds for any number of spins, and its derivation does not involve any approximation whatsoever. Note that the distribution $\bar{f}^{(d)}(\phi, p, t)$ contains information about the trajectory $(\phi(t), p(t))$ of each and every particle. Solving the Klimontovich equation is equivalent to solving the Hamilton's equations of motion for each particle. Indeed, either we solve the equations of motion for a given initial condition $\{\phi_i(0), p_i(0)\}$ to obtain $\{\phi_i(t), p_i(t)\}$ and hence, $\bar{f}^{(d)}(\phi, p, t)$ at time $t$, or, construct corresponding to the condition $\{\phi_i(0), p_i(0)\}$ the initial distribution $\bar{f}^{(d)}(\phi, p, 0)$ and solve the Klimontovich equation to obtain $\bar{f}^{(d)}(\phi, p, t)$. Both these courses of action are equally daunting, especially for large $N$.

To proceed, we now define an averaged single-spin distribution function $\bar{f}(\phi, p, t)$, corresponding to averaging $\bar{f}^{(d)}(\phi, p, t)$ over an ensemble of initial conditions $\{\phi_i(0), p_i(0)\}$ consistent with the same macroscopic initial state. The function $\bar{f}^{(d)}(\phi, p, t)$ being composed of a series of delta functions is a rather singular function. The aforementioned averaging then generates, as the corresponding smoothened version, the continuous function $\bar{f}(\phi, p, t)$, which thus becomes a well-defined one-spin distribution function. It is reasonable to expect that the smoothening would enhance with increase of $N$. Denoting by angular brackets the averaging with respect to the ensemble of initial conditions $\{\phi_i(0), p_i(0)\}$, we may represent this process of smoothening as

$$\bar{f}^{(d)}(\phi, p, t) = \bar{f}(\phi, p, t) + \delta\bar{f}(\phi, p, t)\,, \quad \langle \bar{f}^{(d)}(\phi, p, t)\rangle = \bar{f}(\phi, p, t)\,. \tag{3.163}$$

Here, while the function $\bar{f}^{(d)}$ and hence $\delta\bar{f}$ depend on the details of the initial conditions $\{\phi_i(0), p_i(0)\}$, the averaged distribution $\bar{f}$ depends instead on the whole ensemble of initial conditions and not on any individual initial condition. We remark that since the mentioned averaging is performed at time $t = 0$, Eq. (3.163) is strictly valid at $t = 0$, but deterministic evolution of the individual microstates according to the Hamilton's equations ensures that the equation remains valid for any $t > 0$.

The aforementioned process of smoothening is illustrated schematically in Fig. 3.5. Panels (a)–(c) show three different consistent microstates at $t = 0$ and for a system of 20 particles. The red square, shown to be highly exaggerated in size, is supposed to represent an infinitesimal volume element centered at the point A = $(\phi, p)$ in the two-dimensional $\mu$-space, and the points contained in the box is a measure of $\bar{f}^{(d)}(\phi, p, t = 0)$. While $\bar{f}^{(d)}$ shows strong fluctuations across individual

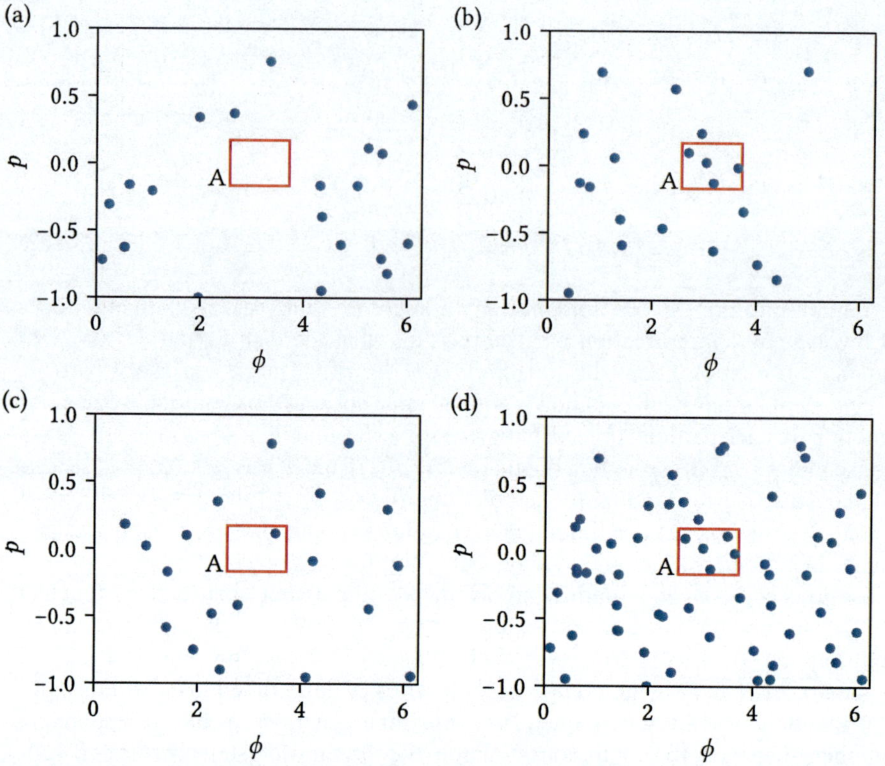


**Fig. 3.5 Schematic diagram in the $\mu$-space illustrating the averaging process of smoothening to obtain $\bar{f}(\phi, p, t)$ from** $\bar{f}^{(d)}(\phi, p, t)$. Panels **a–c** show different microstates for a system with 20 particles at time $t = 0$. Here, the red square is a highly exaggerated volume element around the point A=$(\phi, p)$. It is seen from these three panels that the $\bar{f}^{(d)}(\phi, p, t)$ is a spiky function. Panel **d** is obtained by superimposing the plots in panels **a–c**. In other words, it depicts the average behavior of $\bar{f}^{(d)}(\phi, p, t)$ over these three microstates. We see that $\bar{f}^{(d)}(\phi, p, t)$ becomes a smoother function of $(\phi, p)$. In this way, by averaging over all possible microstates consistent with a given macrostate, one obtains $\bar{f}(\phi, p, t)$, see text

microstates, panel (d) shows that on averaging over the different microstates (illustrated here by implementing a superposition of the pictures in panels (a)–(c)), one obtains a well-defined $\bar{f}(\phi, p, t = 0)$ for A.

In order to obtain the scaling with $N$ of $\delta \bar{f}(\phi, p, t)$ in Eq. (3.163), let us recall some basic elements of probability theory [10]. Let $\{X_i;\ i = 1, 2, \ldots, N\}$ be a set of $N$ independent and identically-distributed random variables sampled from a common distribution $p(X)$ with mean given by $\mu$ and a finite variance given by $\sigma^2$. The law of large numbers combined with the central limit theorem in its simplest form states that for large $N$, the sum $Y \equiv (1/N) \sum_{i=1}^{N} X_i$, itself a random variable, is distributed according to a Gaussian distribution centered at $\mu$ and with variance given by $\sigma^2/N$, for fluctuations $Y - \mu$ of order $\sigma/\sqrt{N}$. An implication is that one may for large $N$

express $Y$ as

$$Y = \mu + \frac{1}{\sqrt{N}} \Delta Y\,, \tag{3.164}$$

where $\Delta Y = \mathcal{O}(N^0)$ is a random variable with zero mean and variance equal to $\sigma^2$. In our case, consider that $\bar{f}^{(\mathrm{d})}$ defined in Eq. (3.155) is a sum of $N$ random variables given by the product $\delta(\phi - \phi_i)\delta(p - p_i)$ (as seen in Fig. 3.5, the product is indeed a random variable that for a given $(\phi, p)$ and at a given time fluctuates from one microstate to another, see panels (a) – (c)). The quantity $\bar{f}^{(\mathrm{d})}$ when averaged over many (ideally infinite) microstates at a given instant $t$ and for a given $(\phi, p)$ would yield the smoothed distribution $\bar{f}$, while in the absence of any further knowledge, we may write for large $N$ and by drawing an analogy with Eq. (3.164) that

$$\bar{f}^{(\mathrm{d})}(\phi, p, t) = \bar{f}(\phi, p, t) + \frac{1}{\sqrt{N}} \delta \bar{f}^{(1)}(\phi, p, t)\,. \tag{3.165}$$

Here, the fluctuation $\delta \bar{f}^{(1)}(\phi, p, t)$ is a random variable satisfying $\delta \bar{f}^{(1)}(\phi, p, t) = \mathcal{O}(N^0)$ and which has zero average. At least at the initial instant $t = 0$, it would be reasonable to expect that the different consistent microstates represent independent samplings of the same parent distribution, so that $\delta(\phi - \phi_i)\delta(p - p_i)$ for different $i$'s and at a given $(\phi, p)$ are independent and identically distributed. More generally, one may expand $\bar{f}^{(\mathrm{d})}$ in powers of $1/\sqrt{N}$ to account for fluctuations at all orders, as

$$\bar{f}^{(\mathrm{d})}(\phi, p, t) = \bar{f}(\phi, p, t) + \frac{1}{\sqrt{N}} \delta \bar{f}^{(1)}(\phi, p, t) + \frac{1}{N} \delta \bar{f}^{(2)}(\phi, p, t) + \cdots\,, \tag{3.166}$$

where all the $\delta \bar{f}^{(\nu)}$'s, with $\nu \geq 1$, and at a given $(\phi, p)$ are random variables with zero mean and of order $N^0$.

Substituting Eq. (3.166) in Eq. (3.160), we get

$$\begin{aligned}
&\frac{\partial \bar{f}}{\partial t} + g_p^{(0)} \frac{\partial \bar{f}}{\partial u} + g_\phi^{(0)} \frac{\partial \bar{f}}{\partial \phi} \\
&\quad + \frac{1}{\sqrt{N}} \left[ \frac{\partial \delta \bar{f}^{(1)}}{\partial t} + \delta g_p^{(1)} \frac{\partial \bar{f}}{\partial p} + g_p^{(0)} \frac{\partial \delta \bar{f}^{(1)}}{\partial p} + \delta g_\phi^{(1)} \frac{\partial \bar{f}}{\partial \phi} + g_\phi^{(0)} \frac{\partial \delta \bar{f}^{(1)}}{\partial \phi} \right] \\
&\quad + \frac{1}{N} \left[ \frac{\partial \delta \bar{f}^{(2)}}{\partial t} + g_p^{(0)} \frac{\partial \delta \bar{f}^{(2)}}{\partial p} + g_\phi^{(0)} \frac{\partial \delta \bar{f}^{(2)}}{\partial \phi} + \delta g_p^{(1)} \frac{\partial \delta \bar{f}^{(1)}}{\partial p} + \delta g_\phi^{(1)} \frac{\partial \delta \bar{f}^{(1)}}{\partial \phi} \right. \\
&\qquad\qquad \left. + \delta g_p^{(2)} \frac{\partial \bar{f}}{\partial p} + \delta g_\phi^{(2)} \frac{\partial \bar{f}}{\partial \phi} \right] \\
&\quad + \frac{1}{N^{3/2}} \left[ \delta g_p^{(1)} \frac{\partial \delta \bar{f}^{(2)}}{\partial p} + \delta g_\phi^{(1)} \frac{\partial \delta \bar{f}^{(2)}}{\partial \phi} + \delta g_p^{(2)} \frac{\partial \delta \bar{f}^{(1)}}{\partial p} + \delta g_\phi^{(2)} \frac{\partial \delta \bar{f}^{(1)}}{\partial \phi} \right]
\end{aligned}$$

$$+\frac{1}{N^2}\left[g_p^{(2)}\frac{\partial\delta\bar{f}^{(2)}}{\partial p}+g_\phi^{(2)}\frac{\partial\delta\bar{f}^{(2)}}{\partial\phi}\right]=0\,, \tag{3.167}$$

where we have

$$g_p^{(0)}\equiv g_p[\bar{f}]=\sqrt{1-p^2}\left(m_y[\bar{f}]\cos\phi-m_x[\bar{f}]\sin\phi\right), \tag{3.168}$$

$$g_\phi^{(0)}\equiv g_\phi[\bar{f}]=2nDp^{2n-1}+m_x[\bar{f}]\frac{p\cos\phi}{\sqrt{1-p^2}}+m_y[\bar{f}]\frac{p\sin\phi}{\sqrt{1-p^2}}-m_z[\bar{f}]\,, \tag{3.169}$$

and

$$\delta g_p^{(\nu)}\equiv\sqrt{1-p^2}\left(m_y[\delta\bar{f}^{(\nu)}]\cos\phi-m_x[\delta\bar{f}^{(\nu)}]\sin\phi\right), \tag{3.170}$$

$$\delta g_\phi^{(\nu)}\equiv m_x[\delta\bar{f}^{(\nu)}]\frac{p}{\sqrt{1-p^2}}\cos\phi+m_y[\delta\bar{f}^{(\nu)}]\frac{p}{\sqrt{1-p^2}}\sin\phi-m_z[\delta\bar{f}^{(\nu)}]\,, \tag{3.171}$$

for $\nu\geq 1$.

We now average Eq. (3.167) with respect to the ensemble of initial microstates. Note that $\langle\delta\bar{f}^{(\nu)}\rangle=0$ implies $\langle m_x[\delta\bar{f}^{(\nu)}]\rangle=\langle m_y[\delta\bar{f}^{(\nu)}]\rangle=\langle m_z[\delta\bar{f}^{(\nu)}]\rangle=0$ for $\nu\geq 1$. In this case, we also have $\langle\delta g_p^{(\nu)}\rangle=\langle\delta g_\phi^{(\nu)}\rangle=0$ for $\nu\geq 1$. Keeping these facts in mind, we obtain from Eq. (3.167) that

$$\begin{aligned}&\frac{\partial\bar{f}}{\partial t}+g_p^{(0)}\frac{\partial\bar{f}}{\partial p}+g_\phi^{(0)}\frac{\partial\bar{f}}{\partial\phi}\\&+\frac{1}{N}\left\langle\delta g_p^{(1)}\frac{\partial\delta\bar{f}^{(1)}}{\partial p}+\delta g_\phi^{(1)}\frac{\partial\delta\bar{f}^{(1)}}{\partial\phi}+\delta g_p^{(2)}\frac{\partial\bar{f}}{\partial p}+\delta g_\phi^{(2)}\frac{\partial\bar{f}}{\partial\phi}\right\rangle\\&+\frac{1}{N^{3/2}}\left\langle\delta g_p^{(1)}\frac{\partial\delta\bar{f}^{(2)}}{\partial p}+\delta g_\phi^{(1)}\frac{\partial\delta\bar{f}^{(2)}}{\partial\phi}+\delta g_p^{(2)}\frac{\partial\delta\bar{f}^{(1)}}{\partial p}+\delta g_\phi^{(2)}\frac{\partial\delta\bar{f}^{(1)}}{\partial\phi}\right\rangle\\&+\frac{1}{N^2}\left\langle g_p^{(2)}\frac{\partial\delta\bar{f}^{(2)}}{\partial p}+g_\phi^{(2)}\frac{\partial\delta\bar{f}^{(2)}}{\partial\phi}\right\rangle=0\,.\end{aligned} \tag{3.172}$$

For finite times and in the limit $N\to\infty$ (or, for times $t\ll N$), keeping only the leading-order terms, we obtain the Vlasov equation from Eq. (3.172) as

$$\frac{\partial\bar{f}(\phi,p,t)}{\partial t}=-g_p^{(0)}\frac{\partial\bar{f}(\phi,p,t)}{\partial p}-g_\phi^{(0)}\frac{\partial\bar{f}(\phi,p,t)}{\partial\phi}\,. \tag{3.173}$$

Using Eqs. (3.168) and (3.169), Eq. (3.173) may be written in the form of Eq. (3.152).

From Eq. (3.172), the following scenario for the Heisenberg spin model (3.99) emerges: Consider for large but finite $N$ the evolution of an initial single-spin distri-

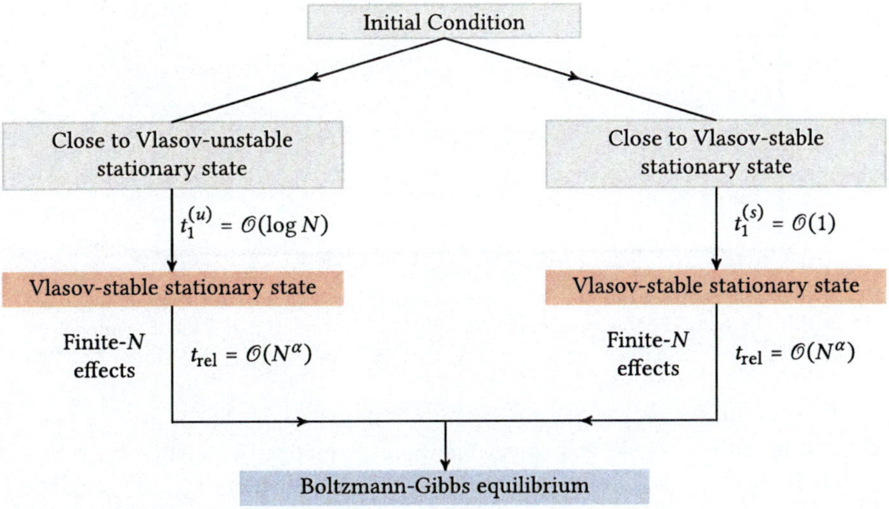


**Fig. 3.6 Schematic diagram showing the dynamical evolution of the long-range system** (3.99). As we have seen in the main text, whether the initial condition is close to an unstable or a stable Vlasov-stationary solution depends respectively on whether we prepare the initial state at an energy below or above the (linear) stability threshold $\epsilon^\star$ of the Vlasov-stationary state. Here, the exponent $\alpha$ satisfies $\alpha \geq 1$

bution $\bar{f}$ that is close to a stable stationary solution of the Vlasov equation (3.173). In this case, $\bar{f}$ evolves on a timescale $t_1^{(s)} = \mathcal{O}(1)$ given by the timescale of single-particle dynamics and settles into the stable stationary solution of the Vlasov equation. The Vlasov description is valid only for times $t \ll N$, and so the single-spin distribution $\bar{f}$ would over a longer timescale $t_{\rm rel} = \mathcal{O}(N^\alpha)$; $\alpha \geq 1$ start to however evolve away from the stable stationary solution of the Vlasov equation. The process of relaxation for the Heisenberg spin model (3.99) is shown schematically in Fig. 3.6. It has been argued [3] that finite-$N$ effects drive the system slowly across an infinite number of stationary and stable Vlasov states before the system finally attains Boltzmann-Gibbs equilibrium, which incidentally is also a Vlasov stationary state. As regards the slow evolution alluded to above, one may draw an analogy with the one-dimensional dynamics generated by an evolution equation of the form

$$\frac{\mathrm{d}x}{\mathrm{d}t} = f(x) + \epsilon g_1(x) + \epsilon^2 g_2(x) + \epsilon^3 g_3(x) + \ldots; \quad |\epsilon| \ll 1\,, \tag{3.174}$$

where $f(x)$ and the different $g$'s are known functions of $x$, and where the dynamics with $\epsilon = 0$ admits a line of stable fixed points. Then, for times $t \ll 1/\epsilon$, a typical initial condition would relax to the stable fixed point it is closest to, while subsequent evolution would involve motion across the other fixed points, with the motion invariably slowing down as $x$ evolves past the fixed points. The dynamical evolution is shown schematically in Fig. 3.7, in which we have represented the fixed points

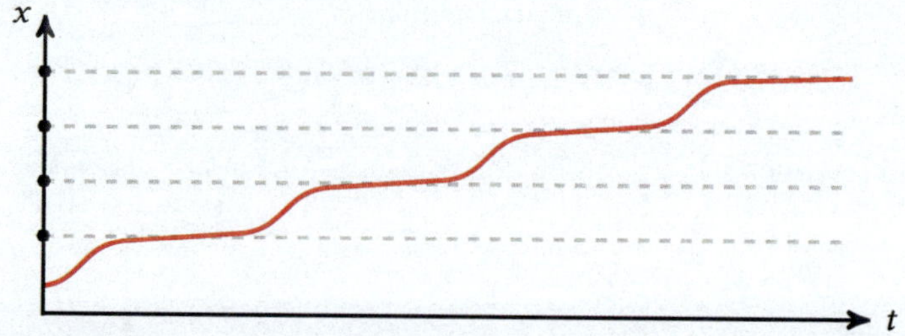


**Fig. 3.7 Schematic representation of the time evolution generated by the dynamics** (3.174), **as discussed in the text.** The filled circles denote stable fixed points of the dynamics with $\epsilon = 0$

as filled circles (the fixed points actually form a line, but are shown here as separated to bring out pictorially and somewhat more dramatically the sluggish nature of evolution past the fixed points.). From the figure, it is evident that the presence of the parameter $|\epsilon| \ll 1$ acts to slow down the eventual settling down of the initial condition into a globally-stable fixed point.

Reverting to our model system, we may conclude on the basis of Eq. (3.172) that one would expect the timescale $t_{\rm rel}$ to be at least of order $N$, or, in case the fluctuation term multiplying $1/N$ on the right hand side of Eq. (3.172) evaluates to zero, to be of order $N^{3/2}$. In general, the $N$-dependent prefactor of whichever is the first nonzero term on the right hand side of Eq. (3.172) determines the timescale $t_{\rm rel}$ over which $\bar{f}$ evolves away from the stable stationary solution of the Vlasov equation, implying possible timescales to be either $\mathcal{O}(N)$, or, $\mathcal{O}(N^{3/2})$, or, $\mathcal{O}(N^2)$. Let us remark that the foregoing picture holds if one considers $N$ to be finite and large, and observes the dynamics for longer and longer times. On the other hand, suppose one keeps the time of observation to be fixed and finite, and considers the evolution for larger and larger system sizes. Then, it is evident that the system while in the Vlasov-stable state would never attain the Boltzmann-Gibbs equilibrium state in the limit $N \to \infty$. An important implication is then that the two limits $t \to \infty$ and $N \to \infty$ do not commute for our model long-range system (3.99), a statement that holds in general for long-range interacting systems [3].

We close this section by discussing the scenario that emerges when the single-spin distribution $\bar{f}$ is close to an unstable stationary solution of the Vlasov equation (3.173). In this case, $\bar{f}$ evolves over a time $t_1^{(u)} \sim \log N$ into a stable stationary solution of the Vlasov equation. Once in a stable-stationary Vlasov state, the eventual relaxation process is expected to be similar to that discussed above, see Fig. 3.6. One would then expect relaxation to Boltzmann-Gibbs equilibrium state over a timescale that is of order $N$ or $N^{3/2}$ or $N^2$. From Fig. 3.6, it is evident that Vlasov-stable stationary states are true stationary states in an infinite system but are quasistationary for finite systems in that they show a slow relaxation over times diverging with the system size $N$. As discussed following Eq. (3.76), a dilute short-range interacting system relaxes to Boltzmann-Gibbs equilibrium over an $N$-independent timescale.

By contrast, a long-range interacting system relaxes to Boltzmann-Gibbs equilibrium over a time that depends crucially on the system size $N$, as shown in Fig. 3.6.

## References

1. A. Campa, T. Dauxois, S. Ruffo, Phys. Rep. **480**(3–6), 57 (2009). https://doi.org/10.1016/j.physrep.2009.07.001
2. F. Bouchet, S. Gupta, D. Mukamel, Phys. A **389**(20), 4389 (2010). https://doi.org/10.1016/j.physa.2010.02.024
3. A. Campa, T. Dauxois, D. Fanelli, S. Ruffo, *Physics of Long-Range Interacting Systems*, 1st edn. (Oxford University Press, Oxford, 2014). https://doi.org/10.1093/acprof:oso/9780199581931.001.0001
4. Y. Levin, R. Pakter, F.B. Rizzato, T.N. Teles, F.P.C. Benetti, Phys. Rep. **535**(1), 1 (2014). https://doi.org/10.1016/j.physrep.2013.10.001
5. S. Gupta, S. Ruffo, Int. J. Mod. Phys. A **32**(09), 1741018 (2017). https://doi.org/10.1142/S0217751X17410184
6. M. Kac, G.E. Uhlenbeck, P.C. Hemmer, J. Math. Phys. **4**(2), 216 (1963). https://doi.org/10.1063/1.1703946
7. G.F. Kventsel, J. Katriel, Phys. Rev. B **30**(5), 2828 (1984). https://doi.org/10.1103/PhysRevB.30.2828
8. S. Gupta, D. Mukamel, J. Stat. Mech: Theory Exp. **2011**(03), P03015 (2011). https://doi.org/10.1088/1742-5468/2011/03/P03015
9. Y.L. Klimontovich, *The Statistical Theory of Non-Equilibrium Processes in a Plasma* (Elsevier Science, Kent, 2014)
10. W. Feller, *An Introduction to Probability Theory and its Applications*, Vols. 1 and 2 (Wiley, New York, 2009)

# Chapter 4
# Statistical Mechanics

**Abstract** The chapter discusses the statistical physics approach to study of interacting particle systems. We start with a brief review of thermodynamics and then introduce the concept of statistical ensembles, discussing in turn the three major statistical ensembles, the microcanonical, the canonical, and the grand canonical ensemble. We recapitulate the notion of major thermodynamic potentials and how they may be used to define phase transitions. We next discuss equivalence of statistical ensembles, using the Legendre–Fenchel transformation. We work out for a model system of interacting classical Heisenberg spins the properties in canonical equilibrium, and discuss how ensemble equivalence allows to infer properties in microcanonical equilibrium.

## 4.1 Review of Thermodynamics

Here, we review briefly some foundational aspects of thermodynamics, which would be essential in appreciating ideas and concepts developed later in this chapter. As standard references on the topic, we may cite Refs. [1, 2]. The review on thermodynamics is followed by a detailed overview of the statistical mechanics approach to study of thermodynamic systems. For this part, a standard reference, which we followed, is Ref. [3].

Thermodynamics is a phenomenological theory, whose building blocks are concepts that are based purely on experimental observations and everyday experience. The subject deals exclusively with macroscopic systems (so-called thermodynamic systems) comprising a large number ($\sim 10^{23}$) of particles (more precisely, degrees of freedom) contained in volumes that are several orders of magnitude larger than molecular volumes. The theory aims to characterize the time-independent (more precisely, equilibrium) behavior of macroscopic observables of such systems. We enumerate below some of the basic and relevant elements of the theory. In the following, we will use the word system to mean a thermodynamic system.

1. **Thermodynamic parameters of a system**: These are measurable macroscopic properties that characterize the system at a macroscopic level. For a fluid contained in a closed volume $V$, the thermodynamic parameters could be $V$ itself,

 
D. Das and S. Gupta, *Facets of Noise*, Fundamental Theories of Physics 214,
https://doi.org/10.1007/978-3-031-45312-0_4

the number $N$ of particles in the system, the pressure $P$ that it exerts on the walls of the container, and the (absolute) temperature $T$ of the fluid. As already mentioned, thermodynamics deals with only equilibrium (namely, time-independent) values of thermodynamic parameters.

2. **Thermodynamic state of a system**: This is defined to be the macroscopic state of the system specified by giving the values of all the relevant thermodynamic parameters pertaining to the system. In equilibrium, the values of the thermodynamic parameters and, consequently, the thermodynamic state of the system do not change as a function of time. In this and the following section, by "state" or "thermodynamic state" will be meant the equilibrium state of the system, unless specified otherwise. Moreover, we will mostly deal with the case when the system has a fixed number of particles. Towards the end of the next section, we will discuss as to what happens when the system is allowed to have a fluctuating number of particles. The system that we will choose for the purposes of illustration will be a fluid system, for which the thermodynamic parameters will be the triplet $(P, V, T)$ (the number of $N$ of particles will not be a variable, as we will assume the fluid to have a fixed number of particles).
3. **Equation of state of a system**: A functional relationship between the thermodynamic parameters of the system in equilibrium is called the equation of state of the system. For a fluid system in equilibrium with pressure $P$, volume $V$ and temperature $T$, its equation of state is a relation of the form $f(P, V, T) = 0$, which implies that only two of the three thermodynamic parameters $P, V, T$ are independent. The function $f$ is supposed to be either given or to be determined from experiments.

   The simplest example of an equation of state is the one for one mole of an ideal gas, namely, the ideal gas law $PV = RT$, in which $R$ is the universal gas constant (one mole of any substance is that amount which contains exactly an Avogadro number of particles (atoms/molecules)). Let us recall that an ideal gas is a hypothesized gas composed of particles that are not interacting with one another. Generally, most real gases behave qualitatively like an ideal gas at high temperature and low pressure. Considering a hypothetical coordinate space with its axes spanned by the thermodynamic parameters of a system, e.g., by the values of $P, V, T$ for a fluid system, a thermodynamic state of the system is represented by a point in this space. The equation of state then defines a surface in this space, such that every point on this surface is a thermodynamic state.
4. **Thermodynamic transformation**: This is defined to be a change in the thermodynamic state of the system that is brought about by changing the external conditions. Consider a system that is in equilibrium. Then, on changing its external conditions, the system will adjust its thermodynamic parameters to adapt to the changed external environment, and in the process, it will have undergone a thermodynamic transformation.

   Quasistatic transformations are defined to be those thermodynamic transformations in which, with the system initially in equilibrium, the external conditions are changed slowly in time (i.e., appreciable change occurs over a timescale that is larger than the slowest relaxation time of the system under consideration), such that at any instant of time, the system is approximately in equilibrium. Thus, in a quasistatic transformation, the system passes through a sequence of thermodynamics states that are almost in equilibrium. Reversible transformations are those in which the transformation retraces its history in time when the change in the external conditions retraces its history in time. Reversible transformations are necessarily quasistatic, but the converse is not necessarily true. An example is that of a gas that freely expands into successive infinitesimal volume elements, which is a quasistatic but not a reversible transformation. A transformation that is not reversible is called an irreversible transformation. From now on, all thermodynamic transformations that we will talk about will be quasistatic, unless mentioned otherwise.

5. **Work and heat**: There are two distinct ways in which energy can be transferred to a system, namely, through performing work on the system or through absorption of heat by the system. For a fluid system whose volume increases by an infinitesimal amount $\mathrm{d}V$ at constant pressure $P$, one has $đW = P\mathrm{d}V$ as the infinitesimal work done by the system. When work is done on a system, its energy and, hence, its temperature increase.

   Now, temperature may increase even when no work is done on the system, but instead heat has been absorbed by the system. We have $đQ = C\mathrm{d}T$ as the infinitesimal heat absorbed by the system when its temperature increases by an infinitesimal amount $\mathrm{d}T$. Here, $C$ is the amount of heat required to increase the temperature by unity and is called the heat capacity of the system. It is an experimental fact that for the same $\mathrm{d}T$, one has different amount $đQ$ of heat absorbed depending on whether heat is being absorbed at constant pressure or at constant volume. This fact requires to define two heat capacities, $C_P$ and $C_V$, which are respectively the heat capacity at constant pressure and constant volume. Heat capacity when defined per unit mole of the substance making up the system is referred to as the molar heat capacity of the system. Standard sign convention is to consider the work done by a system (respectively, on a system) as a positive (respectively, a negative) quantity. On the other hand, heat absorbed (respectively, released) by a system is taken to be a positive (respectively, a negative) quantity.

   It is an experimental fact that the quantities $đQ$ and $đW$ are inexact differentials (which explains the use of $đ$ instead of the standard symbol d for differentials), in the sense that in a thermodynamic transformation from state $A$ to state $B$, the total work done, given by $\Delta W = \int_A^B đW$, and the total heat transferred, given by $\Delta Q = \int_A^B đQ$, depend on the precise path that the system takes in going from state $A$ to state $B$ during the transformation. Consequently, the quantities $Q$ and $W$ are not functions solely of the thermodynamic states of a system.
6. **Heat reservoir**: This is defined to be a system so large that gain or loss of any finite amount of heat does not change its temperature appreciably.
7. **Isolated systems**: A system exchanges energy with the external world via performance of work and transfer of heat. Under conditions in which no energy exchange takes place, the system is said to be isolated. This requires the system

to be thermally insulated and mechanically isolated so that no flow of heat to the system can take place and no mechanical work can be done on or by the system.

8. **Adiabatic, isothermal, isobaric, and isochoric transformation**: An adiabatic transformation is a thermodynamic transformation in which there is no heat transfer between the system and the external world. A thermodynamic transformation is said to be isothermal when the temperature of the system remains constant all through the transformation. An isobaric (respectively, an isochoric) transformation is on the other hand one in which the pressure (respectively, the volume) of the system does not change during the transformation.
9. **Extensive variable**: A macroscopic variable that is proportional to the amount of the substance making up the system is called an extensive variable; instead, it is called intensive when it is independent of the amount of the substance. Volume, mass, entropy, internal energy, heat capacity, etc. are extensive variables, while pressure, temperature, specific heat capacity, density, etc. are intensive variables.
10. **First law of thermodynamics**: The First law of thermodynamics states that in any thermodynamic transformation between states $A$ and $B$, the difference between the heat absorbed by the system and the work done by the system is a quantity $\Delta U$ that does not depend on the path that the system takes in going from state $A$ to state $B$, and is consequently the same for all paths connecting states $A$ and $B$:

$$\Delta U = \Delta Q - \Delta W \,. \tag{4.1}$$

Here, $U$, the internal energy of the system, is a function solely of the thermodynamic states of a system, and is called a state function (correspondingly, the infinitesimal quantity $\mathrm{d}U$ is a perfect differential). In the above example, we have $\Delta U = U_B - U_A$. The first law of thermodynamics is a restatement of the conservation of energy: amount of heat absorbed by a system goes into increasing the internal energy of the system and for doing work.

11. **Second law of thermodynamics**: The Second law of thermodynamics states that there exists no thermodynamic transformation whose sole effect is to extract an amount of heat from a heat reservoir and convert it entirely into work. The foregoing is the statement of the second law due to Lord Kelvin, while an equivalent statement of the law due to Rudolf Clausius states that there exists no thermodynamic transformation whose sole effect is to extract an amount of heat from a colder reservoir and deliver it to a hotter reservoir.
12. **Clausius' inequality**: In any cyclic transformation for which the temperature is defined all through, one has $\oint \frac{đQ}{T} \leq 0$, where the integral is to be computed along the complete cycle. Here, the inequality (respectively, the equality) holds for irreversible (respectively, reversible) transformations.

For a reversible transformation, it follows from the Clausius' inequality that $\int_A^B \frac{đQ}{T}$ is independent of the path connecting states $A$ and $B$ and depends only on the states $A$ and $B$.

***Proof*** Using the Clausius' inequality, and considering two arbitrary reversible paths $I$ and $II$ connecting states $A$ and $B$, we have $\left(\int_{A,I}^{B} + \int_{B,II}^{A}\right)\frac{đQ}{T} = 0$, since paths $I$ and $II$ together constitute a reversible cyclic process. Being reversible, we have $\int_{B,II}^{A}\frac{đQ}{T} = -\int_{A,II}^{B}\frac{đQ}{T}$. Hence, the desired result, namely, $\int_{A,I}^{B}\frac{đQ}{T} = \int_{A,II}^{B}\frac{đQ}{T}$, follows.

13. **Entropy**: On the basis of the foregoing, it follows that one can define a state function called the entropy $S$ such that $\int_{A}^{B}\frac{đQ}{T} = S(B) - S(A)$, where the integral on the left hand side may be computed along any reversible path connecting states $A$ and $B$.

For an irreversible transformation, we have $\int_{A}^{B}\frac{đQ}{T} < S(B) - S(A)$.

***Proof*** Consider a reversible path $I$ and an irreversible path $II$, both of which take the system from $A$ to $B$. Consider the irreversible cyclic process consisting of path $II$ taking the system from $A$ to $B$ and the reverse path of $I$ taking the system from $B$ to $A$. Using the Clausius' inequality, we get $\int_{A,II}^{B}\frac{đQ}{T} + \int_{B,I}^{A}\frac{đQ}{T} < 0$. Path $I$ being reversible, we then get $\int_{A,II}^{B}\frac{đQ}{T} < \int_{A,I}^{B}\frac{đQ}{T} = S(B) - S(A)$, yielding the desired result.

For an infinitesimal reversible transformation at temperature $T$, the entropy $S$ being a state function, the change in $S$ is given by the exact differential

$$\mathrm{d}S = \frac{đQ}{T}, \tag{4.2}$$

where $đQ$ is the amount of heat absorbed by the system. On the basis of the discussions above, we conclude that for any arbitrary thermodynamic transformation from a state $A$ to a state $B$, one has in general that

$$S(B) - S(A) \geq \int_{A}^{B}\frac{đQ}{T}, \tag{4.3}$$

where the equality (respectively, the inequality) holds for reversible (respectively, irreversible) transformations. Now, for an isolated system, according to the definition given in the earlier part of this section, one has evidently that $đQ = 0$. Consequently, Eq. (4.3) yields $S(B) - S(A) \geq 0$, suggesting that the entropy of an isolated system undergoing a thermodynamic transformation never decreases; it either remains constant when the transformation is reversible or increases when the transformation is irreversible.

For an isolated system, it is evident from its definition that the thermodynamic state of the system cannot be changed by changing the external conditions, and so the only way the state can change is via spontaneous processes. Here, by spontaneous process is meant a process that takes place on its own without any external input to the system. A spontaneous process necessarily brings about an irreversible transformation of the state of the system. This is because, in order that a system undergoing a transformation due to a spontaneous process retraces its history in time, the process itself has to retrace its history in time, which requires fine tuning of the conditions underlying the process. In the absence of any external intervention, such a fine tuning is practically impossible, and so a spontaneous process leads to an irreversible transformation.

Now, consider an isolated system in a state $A$ of maximum entropy. Any spontaneous process that brings about an irreversible transformation of the system to another state $B$, say, can according to the discussions above result in only an increase in the entropy of the system, i.e., $S(B) > S(A)$. But state $A$ has already been assumed to be the maximum-entropy state, and hence, we conclude that the state of the system cannot be changed by a spontaneous process to any other state $B$. In other words, the state $A$ is the state of equilibrium. Thus, for an isolated system, we arrive at an important conclusion that the equilibrium state of the system is the state of maximum entropy.

## 4.2 Thermodynamic Potentials

Thermodynamic potentials are a set of state functions (namely, functions of thermodynamic states) that are used to characterize the behavior of a system under different thermodynamic conditions. Some common thermodynamic potentials are the internal energy $U$, the enthalpy $H$, the Helmholtz free energy $F$, and the Gibbs free energy $G$. We discuss below each of these thermodynamic potentials.

1. **Internal energy** $U$**:** We have already defined the internal energy while discussing the First law of thermodynamics as the difference between the heat absorbed and the work done by a system, see Eq. (4.1).
2. **Enthalpy** $H$: The enthalpy of a system in a given state is defined as the sum of its internal energy and the product of its pressure $P$ and volume $V$ in the state:

$$H \equiv U + PV\,. \tag{4.4}$$

The change in enthalpy under any thermodynamic transformation is given by

$$\Delta H = \Delta U + P\Delta V + V\Delta P\,. \tag{4.5}$$

For an isobaric transformation, using $\Delta P = 0$, $\Delta W = P\Delta V$, and the First law of thermodynamics (4.1), one obtains $\Delta H = \Delta Q$. Therefore, the heat absorbed during an isobaric transformation is nothing but the change in enthalpy of the system. Considering an infinitesimal isobaric transformation, one may obtain the heat capacity at constant pressure from enthalpy by using

$$C_P = \left(\frac{\partial H}{\partial T}\right)_P\,. \tag{4.6}$$

3. **Helmholtz free energy** $F$**:** To define the Helmholtz free energy of a system, consider a system that is mechanically isolated and in contact with a heat bath that is at temperature $T$. Suppose the system is in an equilibrium state $A$, and we perform a thermodynamic transformation to take the system to another equilibrium state $B$. Being in contact with the heat bath, the temperature of the system cannot change during the transformation, which therefore qualifies as an isothermal transformation. Being mechanically isolated, no work can be done on or by the system, but heat can be transferred to the system, which is how the system may be made to undergo the thermodynamic transformation. If $\Delta Q$ is the amount of heat transferred to the system, and $\Delta U = U(B) - U(A)$ is the associated change in the internal energy of the system, the First law gives $\Delta U = \Delta Q$, while the second law gives

$$\Delta U \leq T \Delta S \,, \tag{4.7}$$

where the equality (respectively, the inequality) holds when the heat transfer is done reversibly (respectively, irreversibly). Hence, we get $\Delta U - T\Delta S \leq 0$. Let us then define a state function, the Helmholtz free energy $F$, as

$$F \equiv U - TS \,, \tag{4.8}$$

in terms of which Eq. (4.7) reads as

$$\Delta F \leq 0 \,. \tag{4.9}$$

We thus conclude that in an isothermal thermodynamic transformation occurring under mechanical isolation, the Helmholtz free energy of the system cannot increase; it either remains constant if the transformation is reversible or that it decreases if the transformation is irreversible.

In the above set-up, had the system not been mechanically isolated, work could have been done on or by the system, and so Eq. (4.7) would have been modified to

$$\Delta U + \Delta W \leq T \Delta S \,, \tag{4.10}$$

where $\Delta W$ is the work done by the system. In other words, we would have

$$-\,\Delta W \geq \Delta F \,. \tag{4.11}$$

Noting that following our sign convention, the quantity $-\Delta W$ is the work done on the system, we therefore arrive at the conclusion that in an isothermal transformation from state $A$ to state $B$, the total work done on the system goes entirely into changing the free energy of the system by the amount $\Delta F = F(B) - F(A)$, when the transformation is performed reversibly. On the other hand, when the transformation is performed irreversibly, the change in the free energy is the same (as the system goes between the same set of states, and $F$ is a state function), but not all of the work done on the system goes into increasing the free energy by the amount $\Delta F$. In this sense, $\Delta F$ has the interpretation of useful work, namely, the part of the irreversible work done on the system that goes into changing the Helmholtz free energy of the system.

Now, consider a system in a state $A$ of minimum Helmholtz free energy while being mechanically isolated and in contact with a heat bath at temperature $T$. Any heat transfer to the system done irreversibly to make the system undergo a thermodynamic transformation to another state $B$, say, can according to the discussions above result in only a decrease in the Helmholtz free energy of the system, i.e., $F(B) < F(A)$. But state $A$ has already been assumed to be the minimum-$F$ state, and hence, we conclude that the state of the system cannot be changed by an irreversible heat transfer to any other state $B$ (a reversible transfer leads to $F(B) = F(A)$ and hence cannot change the state of the system either). In other words, the state $A$ is the state of equilibrium. Thus, for a mechanically-isolated system at a constant temperature, we arrive at an important conclusion that the equilibrium state of the system is the state of minimum Helmholtz free energy.

4. **Gibbs free energy**: To define the Gibbs free energy of a system, let us now consider a system that is in contact with a heat bath at temperature $T$ and is also at a constant pressure $P$. Suppose the system is in an equilibrium state $A$, and we perform a thermodynamic transformation that takes the system to another equilibrium state $B$. Being in contact with the heat bath, the temperature of the system cannot change during the transformation, which therefore qualifies as an isothermal transformation. Being at a constant pressure, the transformation qualifies also as an isobaric transformation. Heat can be transferred to the system and work also can be performed, and these are the means through which the system may be made to undergo a thermodynamic transformation. If $\Delta Q$ is the amount of heat transferred to the system, $\Delta W = P\Delta V$ is the amount of work done in changing the volume by an amount $\Delta V$, and $\Delta U = U(B) - U(A)$ is the associated change in the internal energy of the system, the First law gives $\Delta U + P\Delta V = \Delta Q$, while the second law gives

$$\Delta U + P\Delta V \leq T\Delta S\,, \tag{4.12}$$

where the equality (respectively, the inequality) holds when the transformation takes place reversibly (respectively, irreversibly). Hence, we get $\Delta U + P\Delta V - T\Delta S \leq 0$. Let us then define a state function, the Gibbs free energy $G$, as

$$G \equiv U + PV - TS\,; \tag{4.13}$$

Equation (4.12) then reads as

$$\Delta G \leq 0\,. \tag{4.14}$$

We thus conclude that in an isothermal-isobaric thermodynamic transformation, the Gibbs free energy of the system cannot increase; it either remains constant if the transformation is reversible or that it decreases if the transformation is irreversible.

Now, consider a system in a state $A$ of minimum Gibbs free energy while being at constant pressure $P$ and in contact with a heat bath at temperature $T$. Any heat transfer to the system and any work done, both performed irreversibly to take it to another state $B$, can according to the discussions above lead to only a decrease in the Gibbs free energy of the system, i.e., $G(B) < G(A)$. But the state $A$ has already been assumed to be the minimum-$G$ state, and hence, we conclude that the state of the system cannot be changed to any other state $B$ by heat transfer and work done, either reversibly or irreversibly (a reversible transfer of heat and work done reversibly lead to $G(B) = G(A)$, and so they cannot indeed change the state of the system). In other words, the state $A$ is the state of equilibrium. Thus, for a system at constant temperature and pressure, we arrive at an important conclusion that the equilibrium state of the system is the state of minimum Gibbs free energy.

The natural arguments of the thermodynamic potentials $U$, $H$, $F$, and $G$ are identified by considering infinitesimal reversible transformations. The First law of thermodynamics for any infinitesimal transformation may be written as $\mathrm{d}U = đQ - P\mathrm{d}V$. Using Eq. (4.2), which is valid only for infinitesimal reversible transformations, one has

$$\mathrm{d}U = T\mathrm{d}S - P\mathrm{d}V\,. \tag{4.15}$$

Using the above equations and the defining equations for $H$, $F$ and $G$, one may obtain for any infinitesimal reversible transformation the exact differentials $\mathrm{d}H, \mathrm{d}F$, $\mathrm{d}G$ as

$$\mathrm{d}H = T\mathrm{d}S + V\mathrm{d}P\,, \tag{4.16}$$

$$\mathrm{d}F = -P\mathrm{d}V - S\mathrm{d}T\,, \tag{4.17}$$

$$\mathrm{d}G = V\mathrm{d}P - S\mathrm{d}T\,. \tag{4.18}$$

From Eqs. (4.15)–(4.18), one identifies the natural arguments for the various thermodynamic potentials, allowing to express them as $U(S, V)$, $H(S, P)$, $F(T, V)$, and $G(T, P)$. The thermodynamic variables $S, P, T, V$ may be obtained from the thermodynamic potentials by using Eqs. (4.15)–(4.18); This information is encoded in the so-called Maxwell relations, which are summarized in Table 4.1.

Let us note for later convenience the fact that Eq. (4.15) yields

$$\frac{1}{T} = \left(\frac{\partial S(U, V)}{\partial U}\right)_V\,, \tag{4.19}$$

**Table 4.1** Thermodynamic potentials, their natural arguments, and the Maxwell relations for a fluid $(P, V, T)$ system

| | Internal energy $U$ | Enthalpy $H = U + PV$ | Helmholtz free energy $F = U - TS$ | Gibbs free energy $G = U - TS + PV$ |
|---|---|---|---|---|
| Natural arguments | $U(S, V)$ | $H(S, P)$ | $F(T, V)$ | $G(T, P)$ |
| Maxwell relations | $P = -\left(\frac{\partial U}{\partial V}\right)_S$ | $V = \left(\frac{\partial H}{\partial P}\right)_S$ | $P = -\left(\frac{\partial F}{\partial V}\right)_T$ | $V = \left(\frac{\partial G}{\partial P}\right)_T$ |
| | $T = \left(\frac{\partial U}{\partial S}\right)_V$ | $T = \left(\frac{\partial H}{\partial S}\right)_P$ | $S = -\left(\frac{\partial F}{\partial T}\right)_V$ | $S = -\left(\frac{\partial G}{\partial T}\right)_P$ |

and

$$\frac{P}{T} = \left(\frac{\partial S(U, V)}{\partial V}\right)_U . \tag{4.20}$$

The various thermodynamic potentials may be related to one another via suitable Legendre transformations. Recall that a Legendre transformation implements the following [4]: Given a function $f(x)$ and its local derivative $p \equiv \mathrm{d}f/\mathrm{d}x$, the transformation, defined as $g \equiv f - px$, obtains a new function $g = g(p)$. Indeed, we have $\mathrm{d}g = p\mathrm{d}x - p\mathrm{d}x - x\mathrm{d}p = -x\mathrm{d}p$, implying that $g = g(p)$. Then, given the thermodynamic potential $U = U(S, V)$, suppose we want a new thermodynamic potential whose natural variables are $T$ and $V$. Using Legendre transformation, this is given as $F = U - TS$. Indeed, then, we have $\mathrm{d}F = \mathrm{d}U - T\mathrm{d}S - S\mathrm{d}T = -P\mathrm{d}V - S\mathrm{d}T$, implying thereby that $F = F(V, T)$. In the same manner, $G$ is obtained by performing a Legendre transformation on $F$, while $H$ is obtained as a Legendre transformation on $U$.

For completeness, we briefly mention the case of a magnetic system, for which the thermodynamic parameters are $(B, M, T)$, where $B$ is the magnetic field, $M$ is the magnetization, while $T$ is as usual the temperature. Here, $B$ and $T$ are intensive variables, while $M$ is extensive. In this case, the infinitesimal work done by the system in which its magnetization increases by an amount $\mathrm{d}M$ is given by $\text{đ}W = -B\mathrm{d}M$, and so the differential form of the First law reads as

$$dU = \text{đ}Q + B\mathrm{d}M , \tag{4.21}$$

which together with the Second law yields for an infinitesimal reversible transformation that

$$\mathrm{d}U = T\mathrm{d}S + B\mathrm{d}M . \tag{4.22}$$

The list of thermodynamic potentials is given in Table 4.2.

**Table 4.2** Thermodynamic potentials, their natural arguments, and the Maxwell relations for a magnetic $(B, M, T)$ system

| | Internal energy $U$ | Enthalpy $H = U - BM$ | Helmholtz free energy $F = U - TS$ | Gibbs free energy $G = U - TS - BM$ |
|---|---|---|---|---|
| Natural arguments | $U(S, M)$ | $H(S, B)$ | $F(T, M)$ | $G(T, B)$ |
| Maxwell relations | $B = \left(\frac{\partial U}{\partial M}\right)_S$ | $M = -\left(\frac{\partial H}{\partial B}\right)_S$ | $B = \left(\frac{\partial F}{\partial M}\right)_T$ | $M = -\left(\frac{\partial G}{\partial B}\right)_T$ |
| | $T = \left(\frac{\partial U}{\partial S}\right)_M$ | $T = \left(\frac{\partial H}{\partial S}\right)_B$ | $S = -\left(\frac{\partial F}{\partial T}\right)_M$ | $S = -\left(\frac{\partial G}{\partial T}\right)_B$ |

For a magnetic system, we thus see that the Helmholtz free energy is a function of the magnetization $M$ and the temperature $T$. In typical experiments involving magnets, it is the magnetic field and not the magnetization that is tuned from outside, which implies that it is $B$ and not $M$ that is the independent variable in such set-ups. The Gibbs free energy $G$ with $B$ and $T$ as its natural arguments may be obtained by performing a Legendre transformation on $F(T, M)$, as

$$G \equiv F - BM . \tag{4.23}$$

One easily checks that

$$\mathrm{d}G = -S\mathrm{d}T - M\mathrm{d}B , \tag{4.24}$$

yielding

$$M = -\left(\frac{\partial G}{\partial B}\right)_T , \quad S = -\left(\frac{\partial G}{\partial T}\right)_B . \tag{4.25}$$

Similar to a fluid system, one may establish that (i) for an isolated magnetic system, the equilibrium state of the system is the state of maximum of the entropy $S$; (ii) for a magnetic system at a constant temperature and with a fixed magnetization, the equilibrium state of the system is the state of minimum of the Helmholtz free energy $F$, and (iii) for a system at constant temperature and magnetic field, the equilibrium state of the system is the state of minimum of the Gibbs free energy $G$.

So far, we have discussed systems with a fixed number of particles. We now discuss the case when the particle number is no longer a constant. The infinitesimal work done on the system to increase its particle number by $\mathrm{d}N$ is defined to equal $\mu\mathrm{d}N$, so that the corresponding work done by the system is $đW = -\mu\mathrm{d}N$. Here, the quantity $\mu$, called the chemical potential, is defined to be the work done on the system to increase its particle number by unity. Correspondingly, the differential form of the First law reads as (we will discuss only the fluid system below):

$$dU = đQ - PdV + \mu dN\,, \tag{4.26}$$

which together with the Second law yields for an infinitesimal reversible transformation that

$$\mathrm{d}U = T\mathrm{d}S - P\mathrm{d}V + \mu\mathrm{d}N\,. \tag{4.27}$$

We thus have $U = U(S, V, N)$, inverting which one gets $S = S(N, V, U)$; moreover, Eqs. (4.19) and (4.20) continue to hold as

$$\frac{1}{T} = \left(\frac{\partial S(N, V, U)}{\partial U}\right)_{N,V}\,, \tag{4.28}$$

and

$$\frac{P}{T} = \left(\frac{\partial S(N, V, U)}{\partial V}\right)_{N,U}\,, \tag{4.29}$$

and additionally, one has

$$-\frac{\mu}{T} = \left(\frac{\partial S(N, V, U)}{\partial N}\right)_{V,U}\,. \tag{4.30}$$

In this changed set-up that we are considering now, all the definitions, $H = U + PV$, $F = U - TS$, and $G = U - TS + PV$, remain unchanged. The change in the Gibbs free energy in an infinitesimal reversible transformation is given by

$$\mathrm{d}G = -S\mathrm{d}T + V\mathrm{d}P + \mu\mathrm{d}N\,, \tag{4.31}$$

implying that one has $G = G(T, P, N)$. Now, the natural arguments of the internal energy $U$ are the extensive variables $S$, $V$ and $N$. Consequently, when the amount of substance constituting the system is increased by a factor $\alpha > 0$, all of $S$, $V$ and $N$ get rescaled by the same factor, and the internal energy, being extensive, must itself get rescaled by the same factor. Thus, we must have

$$U(\alpha S, \alpha V, \alpha N) = \alpha U(S, V, N)\,. \tag{4.32}$$

Differentiating both sides with respect to $\alpha$, we get

$$
\begin{aligned}
U &= \left(\frac{\partial U(\alpha S, \alpha V, \alpha N)}{\partial(\alpha S)}\right)_{V,N} \frac{\mathrm{d}(\alpha S)}{\mathrm{d}\alpha} + \left(\frac{\partial U(\alpha S, \alpha V, \alpha N)}{\partial(\alpha V)}\right)_{S,N} \frac{\mathrm{d}(\alpha V)}{\mathrm{d}\alpha} \\
&+ \left(\frac{\partial U(\alpha S, \alpha V, \alpha N)}{\partial(\alpha N)}\right)_{S,V} \frac{\mathrm{d}(\alpha N)}{\mathrm{d}\alpha} \\
&= TS - PV + \mu N\,,
\end{aligned}
\tag{4.33}
$$

which is the celebrated Gibbs–Duhem equation. In obtaining the last equality above, we have used Eq. (4.27). Considering the exact differential $\mathrm{d}U$ and using Eqs. (4.33) and (4.27), we get

$$
S\mathrm{d}T - V\mathrm{d}P + N\mathrm{d}\mu = 0\,, \tag{4.34}
$$

which implies that the quantities $T$, $P$, $\mu$ are not all independent. Moreover, using Eq. (4.34) in Eq. (4.31), we obtain the chemical potential $\mu$ to be equal to the Gibbs free energy per particle, as

$$
G = \mu N\,. \tag{4.35}
$$

Note that when the particle number in the system is not a constant number, a system in isolation would imply that no heat can be transferred to the system, no work can be done either on or by the system, and also, no particle can be introduced into or taken away from the system.

## 4.3 Statistical Ensembles

In Sect. 3.2, we introduced the idea of statistical ensembles, which we extend here. To this end, consider an isolated thermodynamic system comprising $N$ interacting particles, and which has total energy $E$ and is confined in volume $V$ in three spatial dimensions ($d = 3$). The system is completely isolated and is thus not interacting at all with the environment: No heat can be transferred to the system, nor can any work be done either on or by the system, and even particles cannot be added to or taken away from the system. The quantities $N$, $V$ and $E$ are thus conserved quantities that do not vary as the system evolves in time. In fact, one can be a bit more ambitious, and take into account any machine precision involved with energy measurement by stipulating that the energy of the system lies in a small range $[E, E + \Delta]$, where $0 < \Delta \ll 1$ is a measure of machine precision, and which is a pre-assigned number characteristic of the measuring apparatus that is being employed to make the measurement. The macrostate of the system is specified by the values of $N$, $V$, and $E$. The Hamiltonian of our system has the general form

$$H = \sum_{i=1}^{N} \frac{\mathbf{p}_i \cdot \mathbf{p}_i}{2m} + \sum_{\substack{i,j=1 \\ i<j}}^{N} V\left(|\mathbf{q}_i - \mathbf{q}_j|\right), \tag{4.36}$$

where symbols have their usual meaning. The microstate of the system, i.e., a full description of the state of the system, at time $t$ is specified by the values of the set of $6N$ variables $\{\mathbf{q}_i(t), \mathbf{p}_i(t)\}$. The collection of all microstates that are consistent with a given macrostate defines a statistical ensemble, which is characterized by the $N$-particle phase-space distribution function $f_N(\mathbf{q}_1, \mathbf{p}_1, \ldots, \mathbf{q}_N, \mathbf{p}_N, t)$ defined in and around Eq. 3.16. The consistent microstates evolve in time following the Hamilton's equations of motion, which for the Hamiltonian (4.36) are given by

$$\dot{\mathbf{q}}_i = \frac{\partial H}{\partial \mathbf{p}_i} = \frac{\mathbf{p}_i}{m}, \quad \dot{\mathbf{p}}_i = -\frac{\partial H}{\partial \mathbf{q}_i} = -\sum_{\substack{j=1 \\ j\neq i}}^{N} \frac{\partial V_{ij}}{\partial \mathbf{q}_i}, \tag{4.37}$$

where the dot denotes as usual the derivative with respect to time.

Now, in order to obtain the microstate of the system at time $t$, one has to solve, subject to a given initial condition $\{\mathbf{q}_i(0), \mathbf{p}_i(0)\}$, the Hamilton's equations (4.37). The latter constitute a set of $6N$ coupled (typically nonlinear) partial differential equations. Hence, solving this set of equations is not just difficult but is practically impossible, especially since our thermodynamic system comprises a very (very) large number of particles that are interacting and therefore not executing independent motion. To make headway, one is thus led to rely on statistical methods. In doing so, one is able to compute at any time instant of evolution the value of any macroscopic observable $O = O(\{\mathbf{q}_i, \mathbf{p}_i\})$ only as an average with respect to the different consistent microstates, and not able to compute its value for individual microstates. One may motivate the utility of knowing the average value thus: In real experiments, measurements yield only macroscopic properties of the system. Moreover, all measuring apparatus have an inevitable temporal resolution of finite amount $\Delta t$, however small, and as such, they are typically incapable of making measurement of instantaneous values of observables. During this time interval $\Delta t$, the system evolving according to the Hamilton's equations of motion (4.37) passes through a large number of microstates.[1] Therefore, in effect, what the apparatus provides is the value of a macroscopic observable averaged over these consistent microstates the system evolves through in time $\Delta t$. The basic assumption of statistical mechanics is that one can replace the mentioned time averaging by an averaging at a fixed instant of time with respect to a *suitably-chosen* ensemble in which the microstates occur with relative weights that are *pre-assigned* based on physical intuition.

For notational brevity and in conformity with traditional textbooks on statistical mechanics, in this chapter, we denote the $N$-particle phase-space distribution

[1] Typical $\Delta t$'s would be of the order of seconds, while particles in typical thermodynamic systems move on a timescale of order of milliseconds or less. This suggests that the system would indeed pass through really a very large number of microstates during the duration $\Delta t$.

function (which we will refer to as the phase-space density) by $\rho(q, p, t)$ instead of $f_N(\mathbf{q}_1, \mathbf{p}_1, \ldots, \mathbf{q}_N, \mathbf{p}_N, t)$, and a phase-space point by $(q, p)$ instead of the long-hand notation $\{\mathbf{q}_i, \mathbf{p}_i\} = \{\mathbf{q}_1, \mathbf{p}_1, \ldots, \mathbf{q}_N, \mathbf{p}_N\}$. Consequently, the probability density to find the system in any of the microstates contained in volume $\mathrm{d}^{3N}q\mathrm{d}^{3N}p$ centered at $(q, p)$ at time $t$ equals $\rho(q, p, t)$. We have the normalization of the phase-space density as $\int_{E<H(q,p)<E+\Delta} \mathrm{d}^{3N}q\mathrm{d}^{3N}p\ \rho(q, p, t) = 1\,\forall\, t$, where the integration is limited to consistent microstates, namely, those that have number of particles equal to $N$, volume equal to $V$, and whose energy lies between $E$ and $E + \Delta$.

For any macroscopic observable $O(q, p)$ of the system, the ensemble average $\langle O(q, p)\rangle$ at time $t$ is by definition given by

$$\langle O(q, p)\rangle(t) = \int_{E<H(q,p)<E+\Delta} \mathrm{d}^{3N}q\mathrm{d}^{3N}p\, O(q, p)\, \rho(q, p, t)\,. \tag{4.38}$$

Note that in the above equation, all the microscopic degrees of freedom of the system have been integrated out to obtain the value of a macroscopic observable. When a system is in equilibrium, a measuring apparatus will yield a time-independent value of any macroscopic observable pertaining to the system. On the basis of our discussion in one of the preceding paragraphs, we conclude that in our theoretical set-up, equilibrium would imply a time-independent value of the ensemble average $\langle O(q, p)\rangle(t)$, which by virtue of Eq. (4.38) implies a phase-space density $\rho$ that has no explicit time dependence in equilibrium, i.e., one has

$$\frac{\partial \rho(q, p, t)}{\partial t} = 0\,. \tag{4.39}$$

We are thus tasked with specifying the appropriate form of the time-independent phase-space density $\rho(q, p)$, and to bridge the gap between the microscopic description (encoded in how the different microstates are distributed in the phase space) and the macroscopic one (the one that the readings of a measuring apparatus provide us with.).

The Liouville's equation (3.15) may with the help of the Hamilton's equations (4.37) be written in the following form:

$$\frac{\partial \rho}{\partial t} = -\sum_{i=1}^{N} \frac{\partial H}{\partial \mathbf{p}_i} \cdot \frac{\partial \rho}{\partial \mathbf{q}_i} + \sum_{i=1}^{N} \frac{\partial H}{\partial \mathbf{q}_i} \cdot \frac{\partial \rho}{\partial \mathbf{p}_i} = -\,\{\rho, H\}\,, \tag{4.40}$$

where $\{\cdot\,,\cdot\}$ denotes the Poisson bracket. From Eqs. (4.39) and (4.40), it is then evident that a sufficient condition to have an equilibrium ensemble is to stipulate that the equilibrium phase-space density Poisson-commutes with the Hamiltonian:

$$\{\rho, H\} = 0\,. \tag{4.41}$$

### 4.3.1 Microcanonical Ensemble

The most straightforward route to obtain $\rho(q, p)$ is to solve Eq. (4.40) for $\rho(q, p, t)$ and to then take the limit of infinite times with the hope that we have $\rho(q, p, t \to \infty) = \rho(q, p)$. This task is equally difficult and thus as hopeless as our original task of solving the Hamilton's equations (4.37). In order to proceed, we thus need to appeal to physical intuition to guess the form of $\rho(q, p)$. In absence of any further input, a fair (and, may we say, a democratic) guess would be to stipulate that in equilibrium, all consistent microstates are equally likely.[2] In other words, we have

$$\rho(q, p) \equiv \begin{cases} \frac{1}{\Gamma(N,V,E)} & \text{if } E < H(q, p) < E + \Delta\,, \\ 0 & \text{otherwise}\,, \end{cases} \tag{4.42}$$

where the quantity $\Gamma(N, V, E)$, defined as

$$\Gamma(N, V, E) \equiv \int_{E<H(q,p)<E+\Delta} \mathrm{d}^{3N}q\, \mathrm{d}^{3N}p\,, \tag{4.43}$$

is nothing but the volume of the phase space occupied by consistent microstates, and this number is a constant of motion by virtue of the system under consideration being completely isolated. The phase-space integral in Eq. (4.43) is understood to be performed such that the total particle number is $N$ and the total volume is $V$. Note that the phase-space density in Eq. (4.42) is properly normalized to unity: $\int_{E<H(q,p)<E+\Delta} \mathrm{d}^{3N}q\, \mathrm{d}^{3N}p\, \rho(q, p) = 1$. Figure 4.1, left panel, shows a caricature of a system in microcanonical equilibrium.

The choice (4.42) evidently satisfies the criterion (4.41) for equilibrium. The ansatz that the equilibrium phase-space density $\rho(q, p)$ is constant over the allowed region of the phase space and is zero elsewhere goes under the name of the postulate of equal a priori probability. The equilibrium ensemble characterized by the phase-space density (4.42) is referred to as the microcanonical ensemble, and a system described by such an ensemble is said to be in microcanonical equilibrium. This ensemble is characterized by the conserved quantities appropriate to such an ensemble, namely, the number of particles $N$, the volume $V$, and the energy $E$ of the system. The ensemble average of an observable $O(q, p)$ in microcanonical equilibrium is then given by (see Eq. (4.38)):

$$\langle O(q, p)\rangle = \frac{\int_{E<H(q,p)<E+\Delta} \mathrm{d}^{3N}q\, \mathrm{d}^{3N}p\, O(q, p)}{\Gamma(N, V, E)}\,. \tag{4.44}$$

It is usual to refer to $\Gamma(N, V, E)$ as the microcanonical partition function. Just as in Eq. (4.43), even in Eq. (4.44) (and everywhere else in this subsection, unless specified

[2] This is only an a priori guess, and its correctness in a given situation may only be checked a posteriori by validating theoretical results based on this guess against experimental findings.

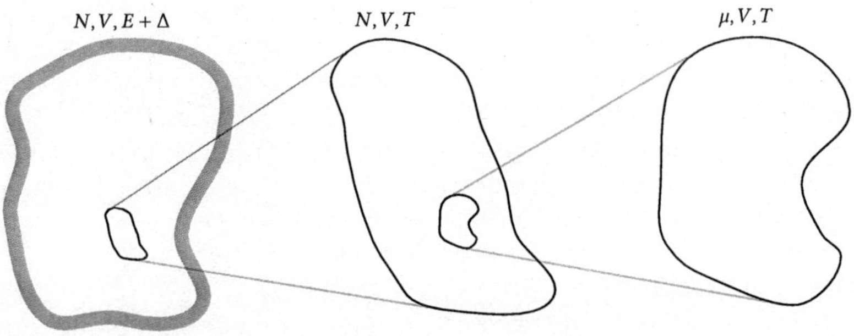


**Fig. 4.1 Caricature of set-ups for microcanonical (left panel), canonical (middle panel) and grand canonical (right panel) equilibrium.** Left panel: A thermodynamic system that is isolated can be in microcanonical equilibrium characterized by fixed values of number of particles $N$, volume $V$ and fixed energy in the range $[E, E+\Delta]$. Middle panel: A macroscopic subpart (a subsystem) of such a system, which has fixed number of particles and fixed volume but has continuous energy exchange with the rest of the system, is in canonical equilibrium characterized by the fixed values of the particle number and the volume and by the constant value of the temperature $T$ that it shares with the rest of the system. Right panel: A further macroscopic subpart of such a subsystem, which has fixed volume but has continuous energy and particle exchange with the rest of the system, is in grand canonical equilibrium characterized by the fixed value of the volume and by the constant value of the temperature $T$ and the constant value of the chemical potential $\mu$ that it shares with the rest of the system

otherwise), the phase-space integral is to be performed such that the total particle number is $N$ and the total volume is $V$.

Defining

$$\Sigma(N, V, E) \equiv \int_{H(q,p)<E} \mathrm{d}^{3N}q\, \mathrm{d}^{3N}p \tag{4.45}$$

as the volume of the phase space enclosed by the energy hyper-surface with energy $E$, we obtain on using $\Delta \ll E$ that

$$\Gamma(N, V, E) = \frac{\partial \Sigma(N, V, E)}{\partial E}\Delta = \omega(N, V, E)\Delta\,, \tag{4.46}$$

where $\omega(N, V, E) \equiv \partial \Sigma(N, V, E)/\partial E$ is known as the density of states of the system at energy $E$.

Now that we have defined microcanonical equilibrium, the connection to thermodynamics is made through the identification that the entropy of the system in microcanonical equilibrium is given by

$$S(N, V, E) \equiv k_{\mathrm{B}} \log \Gamma(N, V, E) + S_0\,, \tag{4.47}$$

where the Boltzmann constant is defined to be $k_{\mathrm{B}} = 1.380649 \times 10^{-23}\ \mathrm{JK}^{-1}$, while $S_0 \equiv -k_{\mathrm{B}} \log \Gamma_0$ is an arbitrary constant that is independent of the details of the system. Here, the quantity $\Gamma_0$ has the same dimension as that of $\Gamma(N, V, E)$, and $S_0$ defines a reference level with respect to which we may choose to measure the entropy of any system. Without loss of generality, in the rest of the chapter, we set $S_0 = 0$. Note that the above equation connects the macroscopic quantity $S$ to the quantity $\Gamma$ that depends on the microscopic description of the system through the Hamiltonian $H(q, p)$, see Eq. (4.43). One may wonder about how the degree of arbitrariness in the choice of $S_0$ affects our subsequent analysis, an issue we will address towards the end of this subsection.

In the following, we will elaborate on some important properties of the entropy that follow from the definition (4.47). To this end, consider an arbitrary partitioning of our thermodynamic system in microcanonical equilibrium into two macroscopic subsystems 1 and 2 with respective particle number being $N_1$ and $N_2$ and respective volume being $V_1$ and $V_2$. We have obviously the equalities $N_1 + N_2 = N$ and $V_1 + V_2 = V$. The partition separating the two systems is such that it allows only energy and not particle or volume to be exchanged between the two subsystems; thus, not just the sum but even the individual values of $V_1$ and $V_2$ and of $N_1$ and $N_2$ do not change as a function of time. Particles could have been exchanged if the partition had holes in it, while volume would have been exchanged if the partition was not fixed in position but rather that its position fluctuated in course of time.

Let us now specialize to the case when the system is short-ranged, i.e., the interaction potential $V(q)$ in the Hamiltonian (4.36) has the behavior $V(q) \sim 1/q^{\alpha}$, with $\alpha \geq d = 3$, see Sect. 3.3. Then, given that each subsystem is macroscopic, the total energy of the composite system is the sum of energies of individual subsystems (see the discussion on the property of additivity discussed in Sect. 3.3), with corrections that become negligibly small in the simultaneous limits $N_1 \to \infty$, $V_1 \to \infty$, keeping the ratio $\rho_1 \equiv N_1/V_1$ fixed and finite, and $N_2 \to \infty$, $V_2 \to \infty$, keeping the ratio $\rho_2 \equiv N_2/V_2$ fixed and finite. These considerations lead to the result that the energies of the two subsystems, denoted by $E_1$ and $E_2$, of course fluctuate as a function of time while satisfying the inequality $E < E_1 + E_2 < E + \Delta$. Moreover, being short-ranged, the property of extensivity holds, that is, we have $E_1 \propto N_1$, and $E_2 \propto N_2$. Note that the quantities $\rho_1$ and $\rho_2$ mentioned above correspond to the particle density of the two subsystems in the thermodynamic limit, see Sect. 3.4.

The presence of the finite machine precision $\Delta$ implies a natural "discretization" of the energy of the two subsystems, that is, we may say that the energy levels of subsystem 1 are $E_1^{(1)} = 0$, $E_1^{(2)} = \Delta/2, \ldots, E_1^{(E/(\Delta/2))} = E - \Delta/2$, and hence, the energy of subsystem 1 lies in the following range of values: $[0, \Delta/2]$, $[\Delta/2, \Delta]$,...,$[E - \Delta/2, E]$. Correspondingly, the energy levels of subsystem 2 are $E_2^{(1)} = E$, $E_2^{(2)} = E - \Delta/2, \ldots, E_2^{(E/(\Delta/2))} = \Delta/2$, and hence, the energy of subsystem 2 can lie in the following range of values: $[E, E + \Delta/2]$, $[E - \Delta/2, E]$,...,$[\Delta/2, \Delta]$. The total number of consistent microstates of the composite system is then given by

$$\Gamma(E) = \sum_{\alpha=1}^{E/(\Delta/2)} \Gamma_1(E_1^{(\alpha)})\Gamma_2(E - E_1^{(\alpha)}), \tag{4.48}$$

where we have suppressed the dependence of $\Gamma$ on the volume and particle number. The rationale behind writing down the above equation is that in the thermodynamic limit, the two subsystems are effectively independent in view of the system under consideration being a short-ranged interacting system. Consequently, the entropy of the composite system is obtained as

$$S(N, V, E) = k_B \log \sum_{\alpha=1}^{E/(\Delta/2)} \Gamma_1(E_1^{(\alpha)})\Gamma_2(E - E_1^{(\alpha)}) . \tag{4.49}$$

The sum on the right hand side is a sum of positive terms. Let the largest term in the sum be $\Gamma_1(\overline{E_1})\Gamma_2(E - \overline{E_1})$. It is then evidently true that one has $\Gamma_1(\overline{E_1})\Gamma_2(E - \overline{E_1}) \leq \Gamma(E) \leq (E/(\Delta/2))\Gamma_1(\overline{E_1})\Gamma_2(E - \overline{E_1})$, that is, one has

$$\begin{aligned} &S_1(N_1, V_1, \overline{E_1}) + S_2(N_2, V_2, E - \overline{E_1}) \leq S(N, V, E) \\ &\leq S_1(N_1, V_1, \overline{E_1}) + S_2(N_2, V_2, E - \overline{E_1}) + k_B \log\left(\frac{2E}{\Delta}\right) . \end{aligned} \tag{4.50}$$

Since we are dealing with macroscopic systems, both the energy and the entropy of each subsystem and hence of the composite system will grow linearly with respect to the corresponding particle number.[3] Consequently, the last term on the right hand side of the above equation is of order $\log N$ and may be neglected in comparison to the other terms. We then finally arrive at the equation

$$S(N, V, E) = S_1(N_1, V_1, \overline{E_1}) + S_2(N_2, V_2, \overline{E_2}); \quad \overline{E_1} + \overline{E_2} = E = \text{constant} . \tag{4.51}$$

The above treatment may be generalized to any number of partitioning of our thermodynamic system into macroscopic subsystems, leading to the following result: The entropy of a thermodynamic system is additive in the sense that the entropy equals the sum of entropies of each macroscopic subsystem making up the system, provided that the interaction between the particles constituting the system is short-ranged.

The maximization condition, namely, that the value $\overline{E_1}$ is such that the product $\Gamma_1(E_1)\Gamma_2(E - E_1)$ is maximum may be written as

$$\frac{\partial}{\partial E_1}\left[\log \Gamma_1(E_1) + \log \Gamma_2(E - E_1)\right]\Big|_{\overline{E_1}} = 0 , \tag{4.52}$$

which may be rewritten as

---

[3] We have already remarked that we have $E_1 \propto N_1$ and $E_2 \propto N_2$, which implies that we have $E \propto N$. Computation of entropy for an arbitrary system is not an easy task because of the difficulties in computing the constrained integral (4.43) defining the quantity $\Gamma(N, V, E)$. Nevertheless, at least for an ideal gas, one can compute $\Gamma$ exactly and thence, show that one has $S \propto N$ [3].

$$\frac{\partial}{\partial E_1}\left[\log \Gamma_1(E_1)\right]\Big|_{\overline{E_1}} = \frac{\partial}{\partial E_2}\left[\log \Gamma_2(E_2)\right]\Big|_{\overline{E_2}}, \tag{4.53}$$

subject to the constraint $\overline{E_1} + \overline{E_2} = E$, a constant. Let us define the microcanonical temperature as

$$\frac{1}{T} \equiv \frac{\partial S(N, V, E)}{\partial E}\Big|_{N,V}, \tag{4.54}$$

and then it follows from Eq. (4.53) that

$$T_1 = T_2\,. \tag{4.55}$$

In the above discussions, we have considered a partitioning of the system into two subsystems, such that only energy can be exchanged between the two subsystems. One may instead consider the partitioning to be such that only volume is being exchanged between the subsystems (vibrating partition), and repeat the same line of arguments to conclude that in microcanonical equilibrium, the quantity $P/T$ of the different subsystems will be equal, with

$$\frac{P}{T} \equiv \frac{\partial S(N, V, E)}{\partial V}\Big|_{N,E}. \tag{4.56}$$

The partitioning may as well be taken to be such that only particles are being exchanged between the subsystems (permeable partition), and then one may repeat the analysis to arrive at the result that in microcanonical equilibrium, the quantity $\mu/T$ of the different subsystems will be equal, with

$$-\frac{\mu}{T} \equiv \frac{\partial S(N, V, E)}{\partial N}\Big|_{V,E}. \tag{4.57}$$

Now, one may imagine partitioning of any thermodynamic system into different macroscopic subsystems between which energy, volume and particles may all be simultaneously exchanged. Let us consider such a partitioning of our isolated thermodynamic system into two subsystems. Moreover, let the system be initially in microcanonical equilibrium in which the subsystems 1 and 2 have respective temperature, pressure and chemical potential given by $(T_1, P_1, \mu_1)$ and $(T_2, P_2, \mu_2)$, and then, the system is subject to a thermodynamic transformation. We have already seen in Sect. 4.1 that the equilibrium state of such a system is the one that maximizes the entropy. Viewing this statement in the light of current analysis, we may say that the state of microcanonical equilibrium should be such that in an infinitesimal transformation, the entropy should not change: $\mathrm{d}S = 0$, which on using Eq. (4.51) yields $\mathrm{d}S_1(N_1, V_1, E_1) = -\mathrm{d}S_2(N_2, V_2, E_2)$, with $N_1 + N_2 =$ constant, $V_1 + V_2 =$ constant and $E_1 + E_2 =$ constant. We therefore obtain

$$\left(\frac{\partial S_1(N_1, V_1, E_1)}{\partial N_1}\bigg|_{V_1,E_1} - \frac{\partial S_2(N_2, V_2, E_2)}{\partial N_2}\bigg|_{V_2,E_2}\right) \mathrm{d}N_1$$
$$+ \left(\frac{\partial S_1(N_1, V_1, E_1)}{\partial V_1}\bigg|_{N_1,E_1} - \frac{\partial S_2(N_2, V_2, E_2)}{\partial V_2}\bigg|_{N_2,E_2}\right) \mathrm{d}V_1$$
$$+ \left(\frac{\partial S_1(N_1, V_1, E_1)}{\partial E_1}\bigg|_{N_1,V_1} - \frac{\partial S_2(N_2, V_2, E_2)}{\partial E_2}\bigg|_{N_2,V_2}\right) \mathrm{d}E_1 = 0\,. \tag{4.58}$$

Using Eqs. (4.54), (4.56) and (4.57), the above equation may be rewritten as

$$\left(\frac{\mu_2}{T_2} - \frac{\mu_1}{T_1}\right) \mathrm{d}N_1 + \left(\frac{P_1}{T_1} - \frac{P_2}{T_2}\right) \mathrm{d}V_1 + \left(\frac{1}{T_1} - \frac{1}{T_2}\right) \mathrm{d}E_1 = 0\,. \tag{4.59}$$

Now, the variables $N$, $V$, $E$ occurring as arguments of entropy are independent variables and thus may be varied independently of one another. Then, Eq. (4.59) may be satisfied only with having $T_1 = T_2$, $P_1 = P_2$ and $\mu_1 = \mu_2$ simultaneously satisfied. The above argument may be generalized to any number of partitioning of our thermodynamic system into macroscopic subsystems. We thus conclude that when an isolated thermodynamic system is in microcanonical equilibrium, different macroscopic parts of the system will have the same temperature, same pressure and same chemical potential.

Comparing Eqs. (4.28), (4.29) and (4.30) with Eqs. (4.54), (4.56) and (4.57), respectively, suggests the identification:

$$U = E \equiv U(S, V, N)\,. \tag{4.60}$$

An inspection of Eqs. (4.54), (4.56) and (4.57) convinces oneself that relevant thermodynamic quantities such as the temperature, the pressure and the chemical potential are obtained as derivatives of the entropy, and hence, the arbitrary constant $S_0$ in Eq. (4.47) has no role to play in so far as observable properties are concerned.

In this subsection, we have discussed how the various thermodynamics quantities of relevance such as the entropy, the temperature, and the internal energy may be derived from the microscopic details of a system through computation of its microcanonical partition function $\Gamma(N, V, E)$. However, such a computation is not an easy affair as the integral in the defining Eq. (4.43) is a constrained multi-dimensional (the dimension is as large as $10^{23}$ !!) integral

### 4.3.2 *Canonical Ensemble*

We now change perspectives and ask: suppose we take our thermodynamic system and consider one of its subparts that is macroscopically large and is moreover separated from the rest of the system by an imaginary partition that allows only transfer of energy and not volume or particle between the subsystem and the rest of the system.

Let us focus on this subsystem, which evidently is not isolated from the rest of the system owing to transfer of energy (since the volume of the subsystem is held fixed, no work can be done and so only heat can be transferred) between the two, and ask the following question. When the full system is isolated and is in microcanonical equilibrium, we have already seen in the previous subsection that the subsystem will also be in equilibrium and will be at the same temperature as the rest of the system. But what is the form of the phase-space density for this subsystem alone? The subsystem, although macroscopically large, is yet smaller than the whole system, and so we may take the rest of the system to be serving the role of a heat bath for the subsystem. In contrast to the equilibrium of an isolated system, which we had named microcanonical equilibrium, the term "canonical equilibrium" is given to the equilibrium state of a system that is exchanging energy with a heat bath. Figure 4.1, middle panel, shows a caricature of a system in canonical equilibrium.

The subsystem (call it system 1) and the rest of the system (call it system 2) together constitute an isolated system in microcanonical equilibrium specified by the total energy in $[E, E+\Delta]$, total volume $V$, and total number of particles $N$, all of which have given values that do not vary with time. These quantities for the respective systems satisfy the equalities

$$E = E_1 + E_2\,, \tag{4.61}$$

$$V = V_1 + V_2\,, \tag{4.62}$$

$$N = N_1 + N_2\,, \tag{4.63}$$

where the quantities $N_1$, $N_2$, $V_1$, $V_2$ do not vary in time, and moreover, one has $V_2 \gg V_1$, $N_2 \gg N_1$. As we have allowed heat exchange between system 1 and the much-larger system 2, both $E_1$ and $E_2$ have values that vary in time, while satisfying (4.61) and the condition $E_2 \gg E_1$ (implying $E \approx E_2$). Note that in writing Eq. (4.61), we have used the fact that our thermodynamic system comprises particles interacting with each other via short-range interactions. This latter feature, in the light of the fact that both the systems are macroscopically large, allows to neglect in the thermodynamic limit the interaction between the particles in systems 1 and 2 across the dividing partition. This leads to the Hamiltonian of systems 1 and 2 taken together to be written as[4]

$$H = H_1(q_1, p_1) + H_2(q_2, p_2)\,. \tag{4.64}$$

Here, $(q_1, p_1)$ and $(q_2, p_2)$ are short-hand notations, respectively, for the set of $6N_1$ and $6N_2$ variables denoting the coordinates and the momenta of particles constituting systems 1 and 2.

Our goal now is to obtain the form of the phase-space density $\rho_1(q_1, p_1)$ of microstates in the phase space of system 1, subject to the conditions mentioned

---

[4] Note that system 1 and system 2 are parts of the same system, and so the quantities $H_1$ and $H_2$ have the same mathematical form; the subscripts 1 and 2 are being added here for ease of comprehension by the uninitiated.

above. Let us say that corresponding to the microstate $(q_1, p_1)$, system 1 has energy $H_1(q_1, p_1) = E_1$, so that system 2 has energy in $[E - E_1, E - E_1 + \Delta]$. Using the results discussed in Sect. 1.3, the phase-space density $\rho_1(q_1, p_1)$ is obtained as a marginal of the microcanonical phase-space density $\rho(q, p) = \rho(q_1, p_1, q_2, p_2)$ of systems 1 and 2 taken together, in which we integrate over all possible values of $(q_2, p_2)$, subject to the constraints that $H_2(q_2, p_2) = E - E_1$ and that system 2 has $N_2$ number of particles contained in volume $V_2$. In other words, we have

$$\begin{aligned}\rho_1(q_1, p_1) &\propto \int_{E-E_1<H_2(q_2,p_2)<E-E_1+\Delta} \mathrm{d}^{3N_2}q_2\mathrm{d}^{3N_2}p_2\, \rho(q_1, p_1, q_2, p_2)\\ &= \frac{\Gamma_2(N_2, V_2, E - E_1)}{\Gamma(N, V, E)},\end{aligned} \tag{4.65}$$

where we have used Eqs. (4.42) and (4.43). Note that as elaborated in Sect. 1.3 (see the discussion around Eq. (1.19)), the marginal density $\rho_1(q_1, p_1)$ as expressed in Eq. (4.65) is not normalized, which explains the occurrence of the proportionality sign in the equation.

Now, using the relations $E - E_1 \gg E_1$ implying $E \approx E_2$ (the energy of system 2 is almost a constant, which justifies its role as a heat bath for system 1) and Eq. (4.47) (with $S_0 = 0$, as discussed in the paragraph following Eq. (4.47)), one obtains

$$\begin{aligned}\Gamma_2(N_2, V_2, E - E_1) &= \exp\left[\frac{S_2(N_2, V_2, E - E_1)}{k_\mathrm{B}}\right]\\ &\approx \exp\left[\frac{S_2(N_2, V_2, E)}{k_\mathrm{B}} - \frac{E_1}{k_\mathrm{B}}\left(\frac{\partial S_2(N_2, V_2, E_2)}{\partial E_2}\right)_{E_2=E}\right].\end{aligned} \tag{4.66}$$

Denoting the microcanonical-equilibrium temperature of our thermodynamic system (which is also the temperature of each of its macroscopic subparts) by $T$, and using Eqs. (4.28) and (4.60), implying

$$\left(\frac{\partial S_2(N_2, V_2, E_2)}{\partial E_2}\right)_{E_2=E} = \left(\frac{\partial S_2(N_2, V_2, E_2)}{\partial E_2}\right)_{N_2,V_2,E_2=E} = \frac{1}{T}, \tag{4.67}$$

we obtain from Eq. (4.66) that

$$\frac{\Gamma_2(N_2, V_2, E - E_1)}{\Gamma(N, V, E)} \approx \left(\frac{1}{\Gamma(N, V, E)} \exp\left[\frac{S_2(N_2, V_2, E)}{k_\mathrm{B}}\right]\right) \exp\left[-\frac{E_1}{k_\mathrm{B}T}\right], \tag{4.68}$$

where the bracketed factor on the right hand side may be treated as constant when only system 1 (the system of interest) is being considered. Therefore, using Eqs. (4.65) and (4.68) and identifying $E_1 = H_1(q_1, p_1)$, one finally obtains, up to a proportionality constant, the phase-space density of a system in thermal equilibrium with a much-larger heat bath at temperature $T$ as

$$\rho(q,p) \propto \exp\left(-\frac{H(q,p)}{k_B T}\right), \tag{4.69}$$

where the subscript 1 has been dropped with the understanding that from now on, we are interested primarily in system 1, and will therefore refer to it as the system and not any more as subsystem or system 1.

Equation (4.69) gives the un-normalized phase-space density of microstates in canonical equilibrium at temperature $T$. These microstates make up the canonical ensemble of our system of interest. This ensemble is characterized by the conserved quantities appropriate to such an ensemble, namely, the number of particles $N$, the volume $V$, and the temperature $T$ of the system. The normalized canonical phase-space density reads as

$$\rho(q,p) = \frac{\exp(-\beta H(q,p))}{Z(N,V,T)}, \tag{4.70}$$

where we have defined the quantity $\beta$ as

$$\beta \equiv \frac{1}{k_B T}, \tag{4.71}$$

while the normalizing factor is the canonical partition function given by

$$Z(N,V,T) \equiv \int \mathrm{d}^{3N}q\, \mathrm{d}^{3N}p\, \mathrm{e}^{-\beta H(q,p)}, \tag{4.72}$$

which gives the volume of the phase space of the system occupied by consistent microstates in canonical equilibrium; here, the phase-space integral is understood to be evaluated subject to the constraint of total number of particles equal to $N$ and total volume equal to $V$. Moreover, the integral should in principle be restricted to those microstates that have energy in the range $E < H(q,p) < E + \Delta$, but which, for a reason we will explain later, may be extended to all consistent microstates irrespective of their energy values. In canonical equilibrium, the ensemble average of an observable $O(q,p)$ is given by (see Eq. (4.38))

$$\langle O(q,p)\rangle = \frac{\int \mathrm{d}^{3N}q\, \mathrm{d}^{3N}p\, O(q,p)\, \mathrm{e}^{-\beta H(q,p)}}{Z(N,V,T)}. \tag{4.73}$$

Two important remarks are in order. Firstly, Eq. (4.70) gives the canonical phase-space density of (i) any macroscopic subsystem of a much-larger system that is in microcanonical equilibrium, provided the system is a short-range interacting one, but also (ii) of any system, short- or long-range interacting, that is in contact with a heat bath in equilibrium at fixed temperature $T$, provided the coupling between the system and the bath is short-ranged. The latter fact should be evident from the manner in which we obtained the result (4.70). Secondly, it does not seem straightforward to obtain the result (4.70) as the canonical-equilibrium density of a subsystem of a much-larger long-range interacting system in microcanonical equilibrium. This is because the very premise on which the derivation was founded, namely, Eq. (4.64), would not hold in such a case. Hence, in this book, whenever we talk about canonical equilibrium of a long-range interacting system, we would imply the canonical equilibrium reached when the system is kept in contact, via a short-ranged coupling, with an external heat bath in equilibrium at a fixed temperature.

Now, we need to make the appropriate connection between the macroscopic properties of the system and the canonical partition function $Z(N, V, T)$ obtained from our microscopic description of the system in terms of a suitable phase-space density. To this end, one defines the Helmholtz free energy $F(N, V, T)$ up to an additive constant[5] as

$$F(N, V, T) \equiv -k_B T \log Z(N, V, T). \tag{4.74}$$

The internal energy of the system is obtained as the ensemble average of the energy of the system:

$$U \equiv \langle H(q, p) \rangle ; \tag{4.75}$$

one obtains from Eq. (4.73) that

$$U = -\frac{\partial}{\partial \beta} \log Z(N, V, T). \tag{4.76}$$

To show that the definition (4.74) of the Helmholtz free energy $F$ is indeed consistent with its thermodynamic definition, one may use this equation along with Eqs. (4.72) and (4.76) to straightforwardly obtain the desired answer:

$$F = U + T\left(\frac{\partial F}{\partial T}\right)_V = U - TS, \tag{4.77}$$

where, in obtaining the last equality, we have used the Maxwell relation $S = -(\partial F/\partial T)_{V,N}$, see Table 4.1. Using $\mathrm{d}F = \mathrm{d}U - T\mathrm{d}S - S\mathrm{d}T$ and Eq. (4.27), we

[5] Customarily, while defining $Z$, one divides the right hand side of Eq. (4.72) by $N!\,h^{3N}$, which would only add a constant term in $F$. The factor $h^{3N}$ makes $Z$ dimensionless and the factor of $N!$, known as the correct Boltzmann counting, takes into account the indistinguishability of particles in quantum mechanics. However, there is no way to derive this within the ambit of classical mechanics, in which no consistent way exists to regard the particles indistinguishable. We remark that these constant factors would prove irrelevant in what concerns observable properties of the system, which would all be defined in terms of derivatives of the free energy, in much the same way as the constant factor $S_0$ in the definition of the entropy in Eq. (4.47) had no relevance.

obtain the thermodynamic relation $\mathrm{d}F = -S\mathrm{d}T - P\mathrm{d}V + \mu\mathrm{d}N$, which yields the pressure $P$ and the chemical potential $\mu$ as

$$S = -\left(\frac{\partial F}{\partial T}\right)_{N,V}, \quad P = -\left(\frac{\partial F}{\partial V}\right)_{N,T}, \quad \mu = \left(\frac{\partial F}{\partial N}\right)_{V,T}, \tag{4.78}$$

We thus see that corresponding to the three variables $N, V, T$ characterizing the canonical ensemble, we have obtained the corresponding conjugate variables (meaning a pair of variables whose product has the dimension of energy, as in $\mu N$, $PV$, $ST$) as suitable derivatives of the free energy, and the element of indeterminacy in defining the free energy in Eq. (4.74) has no relevance in this matter. Therefore, once the canonical partition function is known, computation of the Helmholtz free energy and then the rest of the thermodynamic variables is very straightforward.

As we have already seen, the energy of the system in canonical equilibrium does not have a fixed value. Instead, the energy fluctuates in time due to heat transfer with the bath with which it is in contact. While the ensemble average of the energy is given by Eq. (4.76), it behooves us to compute its variance in order to have a sense of the magnitude of fluctuations. From Eq. (4.76), one may obtain

$$\begin{aligned}\frac{\partial U}{\partial \beta} &= -\frac{\int \mathrm{d}^{3N}q\,\mathrm{d}^{3N}p\, H^2\,\mathrm{e}^{-\beta H(q,p)}}{\int \mathrm{d}^{3N}q\,\mathrm{d}^{3N}p\,\mathrm{e}^{-\beta H(q,p)}} + \frac{\left(\int \mathrm{d}^{3N}q\,\mathrm{d}^{3N}p\, H\,\mathrm{e}^{-\beta H(q,p)}\right)^2}{\left(\int \mathrm{d}^{3N}q\,\mathrm{d}^{3N}p\,\mathrm{e}^{-\beta H(q,p)}\right)^2} \\ &= -\langle H^2\rangle + \langle H\rangle^2 .\end{aligned} \tag{4.79}$$

Using $\partial U/\partial\beta = -k_{\mathrm{B}}T^2\partial U/\partial T$ and the heat capacity at constant volume $C_V = (\partial U/\partial T)_V$, we get the variance of energy of the system in canonical equilibrium as

$$\langle H^2\rangle - \langle H\rangle^2 = k_{\mathrm{B}}T^2 C_V . \tag{4.80}$$

For a short-range interacting system (and even for a long-range interacting system that is rendered extensive by means of the Kac's trick, see Sect. 3.3), both the quantities $\langle H\rangle$ and $C_V$ are proportional to the number of particle $N$. Equation (4.80) then suggests that one has

$$\frac{\sqrt{\langle H^2\rangle - \langle H\rangle^2}}{\langle H\rangle} \propto \frac{1}{\sqrt{N}} \ll 1 , \tag{4.81}$$

i.e., the relative fluctuation in energy is negligible by virtue of the fact that our thermodynamic system has number of particles $N \gg 1$.[6] In other words, despite the fact that the consistent microstates in canonical ensemble can have distributed energy values, effectively all the microstates possess the same energy $\langle H\rangle$, which is the internal energy of the system. This justifies why we had remarked that following

[6] At a second-order phase transition point, one observes a scaling $\left(\sqrt{\langle H^2\rangle - \langle H\rangle^2}\right)/\langle H\rangle \propto N^{-\alpha}$ with $1/2 < \alpha < 1$. Thus, the relative fluctuation in energy is still negligibly small for $N \gg 1$.

Eq. (4.72) that integration over the phase-space variables, which should in principle be restricted to those microstates that have energy in the range $E < H(q, p) < E + \Delta$, may well be extended to all consistent microstates irrespective of their energy values. This is because one has only negligibly-small microstates that have energy outside the range $[E, E + \Delta]$, where $E$ and $T$ are related via Eq. (4.67), and $U$ is related to $T$ via Eq. (4.76).

Equation (4.81) also leads us to establish equivalence between the microcanonical and the canonical ensemble in the following sense. One may do a microcanonical computation of thermodynamic quantities at a given energy value $E$ and expect that a canonical computation done at a temperature $T$ that may be uniquely related to the energy value $E$ via Eq. (4.67) should yield the same answer, provided Eq. (4.81) holds. As remarked earlier, the latter equation generally works, a notable exception being a second-order phase transition point.

For systems in equilibrium, while thermodynamics talks about only the average of macroscopic observables, statistical mechanics not only yields consistent thermodynamics results but also enables us to compute the corresponding variance. Equation (4.80) is one example that directly connects the energy dispersion of a system in thermal equilibrium with a heat reservoir to a conveniently measurable quantity, namely, heat capacity at constant volume.

### 4.3.3 Grand Canonical Ensemble

The canonical ensemble has been obtained by relaxing the restriction of having a constant energy value in microcanonical ensemble, while retaining the restrictions on fixed particle number and fixed volume. The grand canonical ensemble, the object of study in this subsection, is obtained from the canonical ensemble by relaxing the restriction of having a fixed particle number. Thus, the scenario we have in mind is: we have our thermodynamic system in canonical equilibrium with fixed values of the corresponding thermodynamic parameters $N, V, T$, and we consider a macroscopic subsystem that forms its part and which is separated from the rest of the system by an imaginary partition that allows transfer of heat and particles between the subsystem and the rest of the system. Figure 4.1, right panel, shows a caricature of a system in grand canonical equilibrium. We refer to the subsystem as system 1 and the rest of the system as system 2. Now, the composite system may be considered a part of an even bigger system that is in microcanonical equilibrium, see Fig. 4.1. We have proved in Sect. 4.3.1 that different macroscopic subparts of a system in microcanonical equilibrium have the same temperature, pressure and chemical potential. Hence, denoting by $(N_1, N_2)$, $(V_1, V_2)$ and $(T_1, T_2)$, the particle number, the volume and the temperature, respectively, of systems 1 and 2, we have

$$N = N_1 + N_2\,, \tag{4.82}$$

$$V = V_1 + V_2\,, \tag{4.83}$$

$$T_1 = T_2 = T\,, \tag{4.84}$$

where the quantities $V_1$ and $V_2$ do not vary in time, and moreover, one has $V_2 \gg V_1$. As we have allowed particle exchange between system 1 and the much-larger system 2, both $N_1$ and $N_2$ have values that vary in time, while satisfying Eq. (4.82) and the condition $N_2 \gg N_1$ (implying $N \approx N_2$). The chemical potential of systems 1 and 2 are equal: $\mu_1 = \mu_2$. Similar to what we had assumed in treating microcanonical and canonical ensembles earlier in the chapter, we will take our system to be short-range interacting, allowing the decomposition

$$H(q, p, N) = H_1(q_1, p_1, N_1) + H_2(q_2, p_2, N_2)\,, \tag{4.85}$$

where, as regards use of the subscripts 1 and 2 for the Hamiltonian, the footnote for Eq. (4.64) also applies here. Here, $(q_1, p_1, N_1)$ and $(q_2, p_2, N_2)$ are short-hand notations, respectively, for the set of $6N_1 + 1$ and $6N_2 + 1$ variables denoting the coordinates, the momenta, and the number of particles constituting systems 1 and 2. Note that to define the phase space of the system of our interest, namely, system 1, we need in contrast to the situation in canonical ensemble an extra variable denoting the number of particles in the system. In this sense, the phase-space dimension of system 1 (and also system 2) keeps on fluctuating as a function of time.

We would like to investigate the form of the phase-space density $\rho_1(q_1, p_1, N_1)$. Our stratagem will have an exact parallel in the way we derived canonical-equilibrium phase-space density: $\rho_1(q_1, p_1, N_1)$ is obtained as a marginal of the canonical phase-space density $\rho(q_1, p_1, N_1, q_2, p_2, N_2)$ of systems 1 and 2 taken together, in which we integrate over all possible values of $(q_2, p_2)$ and sum over all possible values of $N_2$ subject to the constraint that $N_2 = N - N_1$ and $V_2 = V - V_1$. In other words, we have

$$\begin{aligned}\rho_1(q_1, p_1, N_1) &\propto \sum_{N_2} \int \mathrm{d}^{3N_2} q_2 \mathrm{d}^{3N_2} p_2\, \delta_{N_2,(N-N_1)}\, \rho(q_1, p_1, N_1, q_2, p_2, N_2) \\ &= \mathrm{e}^{-\beta H_1(q_1, p_1, N_1)} \frac{Z_2(N - N_1, V_2, T)}{Z(N, V, T)}\,, \end{aligned} \tag{4.86}$$

where $\beta = 1/(k_\mathrm{B} T)$ and we have used Eqs. (4.70) and (4.72). The marginal density $\rho_1(q_1, p_1, N_1)$ as expressed in Eq. (4.86) is not normalized, which explains the occurrence of the proportionality sign in the equation. Now, using $N - N_1 \gg N_1$ implying $N \approx N_2$ (the particle number in system 2 is almost a constant, which makes it play the role of a particle reservoir for system 1), one obtains

$$\frac{Z_2(N-N_1,V_2,T)}{Z(N,V,T)}=\frac{\exp\left[-\beta F_2(N-N_1,V_2,T)\right]}{Z(N,V,T)}$$
$$\approx\left(\frac{\exp\left[-\beta F_2(N,V_2,T)\right]}{Z(N,V,T)}\right)\exp\left[\beta N_1\left(\frac{\partial F_2(N_2,V_2,T)}{\partial N_2}\right)_{N_2=N}\right], \tag{4.87}$$

where we have set to zero the arbitrary additive constant with respect to which the Helmholtz free energy is measured. Denoting the microcanonical-equilibrium chemical potential of our thermodynamic system (which is also the chemical potential of each of its macroscopic subparts) by $\mu$, that is, $\mu_1=\mu_2=\mu$, and using Eq. (4.78), implying

$$\left(\frac{\partial F_2(N_2,V_2,T)}{\partial N_2}\right)_{N_2=N}=\left(\frac{\partial F_2(N_2,V_2,T)}{\partial N_2}\right)_{N_2=N,V_2,T}=\mu\,, \tag{4.88}$$

we obtain from Eq. (4.87) that

$$\frac{Z_2(N-N_1,V_2,T)}{Z(N,V,T)}\approx\left(\frac{\exp\left[-\beta F_2(N,V_2,T)\right]}{Z(N,V,T)}\right)\exp(\mu\beta N_1)\,, \tag{4.89}$$

where the bracketed factor on the right hand side may be treated as constant when only system 1 (the system of interest) is being considered. Therefore, using Eqs. (4.86) and (4.89), one finally obtains, up to a proportionality constant, the phase-space density of a system in chemical and thermal equilibrium with a much-larger heat and particle reservoir at temperature $T$ and with chemical potential $\mu$ as

$$\rho(q,p,N)\propto\exp\left(\beta\mu N-\beta H(q,p,N)\right)\,, \tag{4.90}$$

where the subscript 1 has been dropped with the understanding that from now on, we are interested primarily in system 1, and will therefore refer to it as the system and not any more as subsystem or system 1.

Equation (4.90) gives the un-normalized phase-space density of microstates in grand canonical equilibrium at temperature $T$ and chemical potential $\mu$. These microstates make up the grand canonical ensemble of our system of interest, and this ensemble is characterized by the conserved quantities appropriate to such an ensemble, namely, the chemical potential $\mu$, the volume $V$, and the temperature $T$ of the system. The normalized grand canonical phase-space density reads as

$$\rho(q,p,N)=\frac{\exp\left(\beta\mu N-\beta H(q,p,N)\right)}{Q(\mu,V,T)}\,, \tag{4.91}$$

where the normalizing factor is the grand canonical partition function given by

$$Q(\mu, V, T) \equiv \sum_{N=0}^{\infty} e^{\beta\mu N} \int d^{3N}q d^{3N}p \, e^{-\beta H(q,p,N)} = \sum_{N=0}^{\infty} e^{\beta\mu N} Z(N, V, T), \tag{4.92}$$

where $Z(N, V, T)$ is the canonical partition function. Here (and everywhere else in this subsection, unless specified otherwise), the phase-space integral is understood to be evaluated subject to the constraint of total volume equal to $V$. Moreover, the integral should in principle be restricted to those microstates that have particle number between a lower cut-off $N^\star \gg 1$ and $N$, but which, for a reason we will explain later, may be extended to all positive values of $N$. The lower cut-off arises because our system of interest, although a subpart of a much larger system, is yet a macroscopic system and hence would contain a macroscopic number of particles. Defining a new quantity $z$, called the fugacity, by

$$z \equiv \exp(\beta\mu), \tag{4.93}$$

the grand canonical partition function may also be written as

$$\widetilde{Q}(z, V, T) = \sum_{N=0}^{\infty} z^N Z(N, V, T). \tag{4.94}$$

The above equation shows that the grand canonical partition function acts as a generating function for the canonical partition function, in the sense that knowing the former generates the latter as

$$Z(N, V, T) = \frac{1}{N!} \frac{\partial^N}{\partial z^N} \widetilde{Q}(z, V, T). \tag{4.95}$$

It may be noted that the grand canonical distribution (4.91) may be derived directly by starting with systems 1 and 2 together to be in microcanonical equilibrium and then allowing system 1 to exchange both particle and energy with system 2. Accordingly, Eq. (4.65) is modified to

$$\begin{aligned}
\rho_1(q_1, p_1, N_1) &\propto \sum_{N_2} \int_{E-E_1<H_2(q_2,p_2,N_2)<E-E_1+\Delta} d^{3N_2}q_2 d^{3N_2}p_2 \, \delta_{N_2,(N-N_1)} \\
&\quad \times \rho(q_1, p_1, N_1, q_2, p_2, N_2) \\
&= \frac{\Gamma_2(N-N_1, V_2, E-E_1)}{\Gamma(N, V, E)} = \frac{\exp\left[\frac{S_2(N-N_1, V_2, E-E_1)}{k_B}\right]}{\Gamma(N, V, E)} \\
&\approx \frac{\exp\left[\frac{S_2(N, V_2, E)}{k_B} - \frac{N_1}{k_B}\frac{\partial S_2(N_2, V_2, E_2)}{\partial N_2}\Big|_{V_2,E_2} - \frac{E_1}{k_B}\frac{\partial S_2(N_2, V_2, E_2)}{\partial E_2}\Big|_{N_2,E_2}\right]}{\Gamma(N, V, E)} .
\end{aligned} \tag{4.96}$$

Using the relations $N_1 \ll N_2 \approx N$, $E_1 \ll E_2 \approx E$ and the definition of temperature and chemical potential, given by Eqs. (4.54) and (4.57), respectively, we obtain from Eq. (4.96) that

$$\begin{aligned}\rho_1(q_1, p_1, N_1) &\propto \frac{1}{\Gamma(N, V, E)} \exp\left[\frac{S_2(N, V_2, E)}{k_B} + \frac{\mu N_1}{k_B T} - \frac{E_1}{k_B T}\right] \\ &\propto \exp\left(\beta\mu N_1 - \beta E_1\right) ,\end{aligned} \tag{4.97}$$

from which directly follows the distribution (4.91).

Along the lines of Eq. (4.38), the ensemble average of an observable $O(q, p, N)$ in grand canonical equilibrium is then given by

$$\langle O(q, p, N)\rangle = \frac{\sum_{N=0}^{\infty} \int \mathrm{d}^{3N}q\mathrm{d}^{3N}p\, O(q, p, N)\, \mathrm{e}^{\beta\mu N - \beta H(q,p,N)}}{Q(\mu, V, T)} . \tag{4.98}$$

Using Eqs. (4.98) and (4.92), one straightforwardly finds the average number of particles $\overline{N}$ in volume $V$ to be

$$\overline{N} \equiv \langle N\rangle = \frac{\sum_{N=0}^{\infty} \int \mathrm{d}^{3N}q\mathrm{d}^{3N}p\, N\, \mathrm{e}^{\beta\mu N - \beta H(q,p,N)}}{Q(\mu, V, T)} = k_B T \frac{\partial}{\partial\mu} \log Q(\mu, V, T) . \tag{4.99}$$

Similarly, the internal energy $U$ in grand canonical equilibrium is obtained as

$$U \equiv \langle H(q, p, N)\rangle = -\frac{\partial}{\partial\beta} \log Q(\mu, V, T) + \mu\overline{N} . \tag{4.100}$$

Now, we will make an assumption regarding fluctuations of particle number, whose validity will be seen only a posteriori from the consequences we derive based on the assumption. To this end, we assume that the relative fluctuations in the particle number, given by $\sqrt{\langle N^2\rangle - \langle N\rangle^2}/\langle N\rangle$, is a number much smaller than unity. It then follows that in the sum on the right hand side of Eq. (4.92), there is only one significant term that corresponds to $\overline{N}$, which allows to write

$$\log Q(\mu, V, T) \approx \log\left\{\mathrm{e}^{\beta\mu\overline{N}} Z(\overline{N}, V, T)\right\} = \beta\mu\overline{N} - \beta F(\overline{N}, V, T) , \tag{4.101}$$

where we have used Eq. (4.74). It then follows that

$$-k_B T \log Q(\mu, V, T) = -\mu\overline{N} + F(\overline{N}, V, T) = -G(\mu, P, T) + F(\overline{N}, V, T) , \tag{4.102}$$

where we have used Eq. (4.35). Let us use the definition of the Gibbs free energy as $G = F + PV$, so that on the right hand side of the above equation, we have $-G(\mu, P, T) + F(\overline{N}, V, T) = -PV$. We thus obtain

$$\frac{PV}{k_B T} = \log \mathcal{Q}(\mu, V, T), \tag{4.103}$$

which yields the connection of the grand canonical ensemble to thermodynamics. Corresponding to the variables $\mu$, $V$ characterizing the grand canonical ensemble, Eqs. (4.99) and (4.103) determine from the grand canonical partition function the corresponding conjugate variables that together form quantities with the dimension of energy.

Just as we had obtained the variance of the energy in canonical equilibrium, it is instructive to compute the variance in $N$ in grand canonical ensemble. Differentiating the last equality of Eq. (4.99) with respect to $\mu$, we obtain

$$\langle N^2 \rangle - \langle N \rangle^2 = (k_B T)^2 \frac{\partial}{\partial \mu} \log \mathcal{Q}(\mu, V, T). \tag{4.104}$$

Next, we note from Eq. (4.103) that the quantities $P$ and $T$ being intensive variables, one has the scaling $\log \mathcal{Q}(\mu, V, T) \sim V$. It then follows from Eq. (4.104) that $\langle N^2 \rangle - \langle N \rangle^2 \propto V \propto \overline{N}$, since in order that the system has a finite density in the thermodynamic limit, we must have $\overline{N} \propto V$ so that the density $\overline{N}/V$ remains finite. We thus have

$$\frac{\sqrt{\langle N^2 \rangle - \langle N \rangle^2}}{\langle N \rangle} \propto \frac{1}{\sqrt{\overline{N}}} \ll 1, \tag{4.105}$$

which is perfectly consistent with what we had assumed earlier regarding number fluctuations. The above equation tells us that the relative fluctuation in particle number is negligible by virtue of the fact that our thermodynamic system has number of particles $N \gg 1$. We arrive at a remarkable result that despite the fact that the consistent microstates in the grand canonical ensemble can have distributed particle number, effectively all the microstates possess the same number of particles equal to $\overline{N}$. It thus makes it possible that in Eq. (4.92), the integration over the phase-space variables is well extended to all consistent microstates irrespective of their particle number, as only a negligibly-small number of microstates are going to have particle number different from $\overline{N}$ given by Eq. (4.99).

Equation (4.105) establishes equivalence between the canonical and the grand canonical ensemble. In other words, one may compute thermodynamic quantities within either a canonical ensemble characterized by the triplet $(N, V, T)$ or a grand canonical ensemble characterized by the triplet $(\mu, V, T)$ and obtain the same answer, provided $\mu$ is chosen such that the average particle number given by Eq. (4.99) coincides with the value $N$. As we will discuss latter, the equivalence holds excepting at a second-order phase transition point.

Dividing both sides of Eq. (4.104) by the volume $V$, one obtains the variance in particle density as

$$\langle n^2\rangle - \langle n\rangle^2 = \frac{(k_B T)^2}{V^2}\frac{\partial}{\partial\mu}\log Q(\mu, V, T) = \frac{k_B T}{V}\frac{\partial P}{\partial\mu}, \tag{4.106}$$

where we have $n \equiv N/V$, and we have used Eq. (4.103). Defining the volume and the Helmholtz free energy per particle as $v \equiv V/N$ and $f(v, T) = F(N, V, T)/N$,[7] respectively and using Eq. (4.74), we obtain from the Maxwell relations (4.78) that

$$P = -\frac{\partial f(T, v)}{\partial v}, \tag{4.107}$$

$$\mu = f(T, v) + N\frac{\partial f(T, v)}{\partial N} = f(T, v) - v\frac{\partial f(T, v)}{\partial v} = f(T, v) + Pv, \tag{4.108}$$

where we have used $\partial f(T, v)/\partial N = (\partial v/\partial N)\partial f(T, v)/\partial v = -(V/N^2)\partial f(T, v)/\partial v$. From Eqs. (4.107) and (4.108), one has $\partial\mu/\partial v = v\partial P/\partial v$. A measure of the instantaneous relative volume change of a system when subject to a pressure change at constant temperature is provided by the so-called isothermal compressibility $\kappa_T$, defined as

$$\kappa_T \equiv -\frac{1}{v}\left(\frac{\partial v}{\partial P}\right)_T; \tag{4.109}$$

in terms of $\kappa_T$, one gets from Eq. (4.108) that

$$\frac{\partial P}{\partial\mu} = \frac{\partial P/\partial v}{\partial\mu/\partial v} = \frac{1}{v}, \quad \frac{\partial P}{\partial\mu} = -\frac{1}{v^2}\frac{\partial v}{\partial\mu} = \frac{\kappa_T}{v^2}, \tag{4.110}$$

yielding on using Eq. (4.106) that

$$\frac{\langle n^2\rangle - \langle n\rangle^2}{\langle n\rangle^2} = \frac{k_B T\kappa_T}{V}. \tag{4.111}$$

The above equation relates the density fluctuation in a system in grand canonical equilibrium to the conveniently measurable quantity, namely, the isothermal compressibility $\kappa_T$.

## 4.4 Phase Transitions

Phase transition is a physical phenomenon that occurs when a system undergoes a qualitative change (sudden or gradual) in its macroscopic properties as a result of changes in external fields such as temperature, pressure, or magnetic field. As the

[7] Since we are dealing with short-range interacting systems, energy and hence, free energy will grow linearly in $N$. For long-range interacting systems, such a linear dependence will be guaranteed once the Kac's trick is invoked, see Sect. 3.3.

system transitions from one phase to another at a phase transition point, the associated change in the macroscopic behavior manifests in a change in the observable properties of the system, such as density, heat capacity, magnetic susceptibility, etc. Phase transitions are encountered frequently in many areas of physics, chemistry, and materials science. Common examples of phase transitions include the melting of ice, the boiling of water, and the transition from a ferromagnetic to a paramagnetic phase in a magnetic material. Depending on the behavior of an appropriate thermodynamic potential (e.g., the Gibbs free energy $G$, the Helmholtz free energy $F$), phase transitions are commonly classified into two types: (i) first-order phase transitions and (ii) second-order or continuous phase transitions.

In first-order phase transitions, one or more of the first derivatives of the appropriate thermodynamic potential exhibits a finite discontinuity. For a fluid system, the Gibbs free energy $G(T, P)$ is the appropriate thermodynamic potential, and the volume and the entropy show discontinuity at the first-order transition point, which are respectively given by (see Table 4.1)

$$V = \left(\frac{\partial G}{\partial P}\right)_T , \quad \text{and} \quad S = -\left(\frac{\partial G}{\partial T}\right)_P . \tag{4.112}$$

On the other hand, for a magnetic system, the Gibbs free energy $G(T, B)$ is the appropriate thermodynamic potential, and it is the magnetization that shows discontinuity at the first-order transition point.

The most common example of a first-order phase transition is the liquid-gas transition. The phase diagram of a typical fluid in the $(P, T)$-plane is shown in Fig. 4.2. At the first-order phase transition point marked $A$ in the figure, both the liquid and the gas phase coexist in equilibrium and have the same temperature and the same pressure. The common value of the pressure at the transition point is known as the vapor pressure and is denoted as $P = P(T)$. Let us then envisage a situation in which at this common pressure $P$ and common temperature $T$, one starts with a given mass of liquid and follow its evolution in time, and one would see that the initial state with only liquid changes to a state in which liquid and gas coexist. Despite $T$

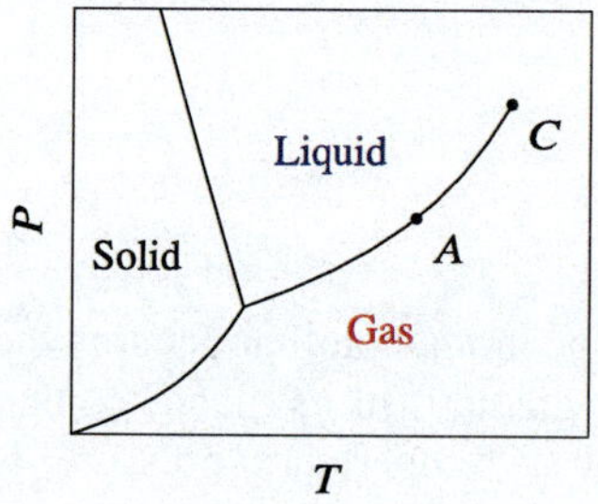


**Fig. 4.2 Phase diagram of a typical fluid in the $(P, T)$-plane.** The lines correspond to coexistence of two phases (namely, liquid-gas, solid-liquid, solid-gas), and define lines of first-order phase transitions. The point marked $C$ is the critical point or the point of second-order phase transition

being constant during this process, one requires heat to be supplied to the system so that part of the initial amount of liquid could convert into gas. On the other hand, had we started with a gas-only initial state, heat would have been released by the system in order to effect the conversion of a part of the initial amount of gas into liquid. At the common temperature and pressure, let $s_l$ and $s_g$ denote the entropy per unit mass of the liquid and the gas, respectively, in terms of which the amount of heat that is either absorbed or released during the aforementioned liquid-gas transformation is given by

$$l \equiv T(s_g - s_l) . \tag{4.113}$$

Here, the quantity $l$ is commonly referred to as the latent heat of liquid-gas transformation.

As the temperature and the pressure remain constant during the mentioned liquid-gas transformation into the state of phase coexistence, it must be that the latter state, which is the equilibrium state at the given temperature $T$ and the given pressure $P$, is the one that minimizes the Gibbs free energy of the system, see Sect. 4.2. Then, denoting the mass and the Gibbs free energy per unit mass of the liquid and the gas phase by $(m_l, m_g)$ and $(g_l, g_g)$, respectively, we may write the total Gibbs potential of the system as

$$G = m_l g_l + m_g g_g , \tag{4.114}$$

where we have considered the bulk effects and have neglected the surface effects. Note that the total mass $m = m_l + m_g$ remains constant during the transformation. In the state of phase coexistence, the Gibbs free energy $G(T, P)$ being minimum, any transformation should be such that the change in $G$ is zero. In other words, we have $\delta G = g_l \delta m_l + g_g \delta m_g = 0$. Then, using $\delta m_l = -\delta m_g$, we obtain $(g_g - g_l)\delta m_g = 0$, or, given that $\delta m_g$ is arbitrary, we have

$$\Delta g \equiv g_g - g_l = 0 . \tag{4.115}$$

Equation (4.115) suggests that the coexisting phases must have equal Gibbs free energy per particle, or, using Eq. (4.35), that they must have the same chemical potential. The specific volume (volume per unit mass) and the specific entropy (entropy per unit mass) of the gas phase are given by

$$v_g = \left(\frac{\partial g_g}{\partial P}\right)_T , \quad \text{and} \quad s_g = -\left(\frac{\partial g_g}{\partial T}\right)_P , \tag{4.116}$$

respectively, see Eq. (4.112). One has similar expressions for the liquid phase:

$$v_l = \left(\frac{\partial g_l}{\partial P}\right)_T , \quad \text{and} \quad s_l = -\left(\frac{\partial g_l}{\partial T}\right)_P . \tag{4.117}$$

As the specific volume and the specific entropy of the liquid and the gas phase are different (indeed, for the same mass, liquid has higher density and lower entropy than gas), one has

$$\left(\frac{\partial\,(\Delta g)}{\partial P}\right)_T = v_\mathrm{g} - v_\mathrm{l} \equiv \Delta v > 0\,, \quad -\left(\frac{\partial\,(\Delta g)}{\partial T}\right)_P = s_\mathrm{g} - s_\mathrm{l} \equiv \Delta s > 0\,. \tag{4.118}$$

Equation (4.118) shows that the first derivatives of Gibbs free energy have finite discontinuities at the point of phase coexistence, which defines the phase boundary between the liquid and the gas phase.

Note that the quantities $\Delta g$, $\Delta v$, and $\Delta s$ considered above are all functions of $T$ and $P$. Using the relation $\Delta g = \Delta g(T, P)$ implying $f(\Delta g, T, P) = 0$ and the chain rule of partial differentiation, we may write

$$\frac{(\partial(\Delta g)/\partial T)_P}{(\partial(\Delta g)/\partial P)_T} = -\left(\frac{\partial P}{\partial T}\right)_{\Delta g}\,. \tag{4.119}$$

From Eqs. (4.118) and (4.119), one then obtains for the phase boundary that

$$\frac{\mathrm{d}P(T)}{\mathrm{d}T} \equiv \left(\frac{\partial P}{\partial T}\right)_{\Delta g=0} = \frac{\Delta s}{\Delta v}\,, \tag{4.120}$$

where we have used the equilibrium condition $\Delta g = 0$. The above equation gives the derivative of the vapor pressure $P(T)$ with respect to temperature. Combining Eqs. (4.113) and (4.120), one obtains the celebrated Clapeyron equation for the vapor-pressure curve:

$$\frac{\mathrm{d}P(T)}{\mathrm{d}T} = \frac{l}{T\Delta v}\,. \tag{4.121}$$

In second-order or continuous phase transitions, while the first derivatives of the appropriate thermodynamic potential are continuous, one or more of its higher derivatives may show discontinuity. In liquid-gas systems, there exists a special point in the $(P, T)$-plane (marked $C$ in Fig. 4.2) at which the specific volume and the specific entropy of the liquid and the gas phase are equal, i.e., $\Delta v = \Delta s = 0$. At this point, known as the critical point, the first derivatives of the Gibbs free energy become continuous (see Eq. (4.118)), and consequently, the transition between the liquid and the gas becomes second order.

## 4.5 Ensemble Equivalence

In this section, we discuss briefly the notion of equivalence of equilibrium ensembles for short and long-range interacting systems. For a more extensive discussion, the reader is referred to Ref. [5]. A macroscopic short-range interacting system may be divided into two or more macroscopic subsystems while keeping the total energy of the system fixed. Because of this additive nature of short-range systems, it is then guaranteed that in microcanonical equilibrium, the entropy $S(E)$ is a concave function of energy $E$. Suppose we start with an entropy $S(E)$ that is non-concave in a region $E_1 < E < E_2$, as shown in Fig. 4.3a. For a short-range system, we now argue that this curve cannot represent the observed entropy in microcanonical equilibrium: a state of the system that is lying on the curve in the energy interval $E_1 < E < E_2$ would not be observed in microcanonical equilibrium. To understand this, let us recall that for isolated system in equilibrium (i.e., the microcanonical equilibrium), the observed state of a system for a given energy $E$ is that, which maximizes the entropy $S(E)$, see Sect. 4.1. Because of the property of additivity, any short-range system with total energy $E$ in the interval $E_1 < E < E_2$ would separate into two coexisting subsystems with energies $E_1$ and $E_2$ such that the total energy $E$ is given by the weighted average of the energies of the subsystems:

$$E = xE_1 + (1 - x)E_2; \quad 0 \leq x \leq 1 . \tag{4.122}$$

This is because in doing so, the system would owing to the non-concavity of $S(E)$ in the region $E_1 < E < E_2$ gain in entropy with respect to states lying on the non-concave region of the curve:

$$xS(E_1) + (1 - x)S(E_2) > S(E) ; \quad E_1 < E < E_2 . \tag{4.123}$$

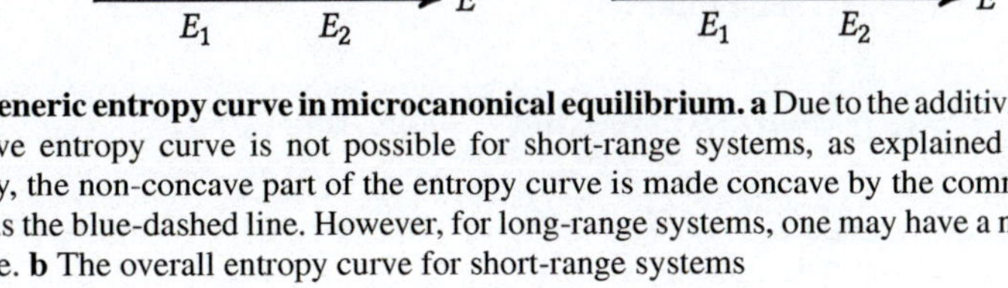


**Fig. 4.3 A generic entropy curve in microcanonical equilibrium. a** Due to the additivity property, a non-concave entropy curve is not possible for short-range systems, as explained in the text. Consequently, the non-concave part of the entropy curve is made concave by the common tangent line, shown as the blue-dashed line. However, for long-range systems, one may have a non-concave entropy curve. **b** The overall entropy curve for short-range systems

Therefore, the correct entropy curve in the non-concave region is given by a common tangent line shown in Fig. 4.3a as a blue-dashed line. Consequently, the overall entropy curve becomes concave in microcanonical equilibrium for short-range systems, as shown in Fig. 4.3b. On the contrary, long-range systems are generically non-additive [5], see Sect. 3.3. Therefore, for these systems, the system would not gain in entropy by undergoing a separation into two parts with energies satisfying Eq. (4.122). As a result, for long-range systems in microcanonical equilibrium, one may very well have a non-concave entropy curve as shown by the red line in Fig. 4.3a.

We now discuss how equivalence of microcanonical and canonical ensembles may be attributed to the concavity of the entropy and the free energy. Let us consider a system of $N$ particles contained in volume $V$ with total energy in $[E, E+\Delta]$, the set-up considered in Sect. 4.3.1. We rewrite the microcanonical partition function $\Gamma(N, V, E)$ defined in Eq. (4.43) in a slightly different way, namely,

$$\Gamma(N, V, E) = \int \mathrm{d}^{3N}q\, \mathrm{d}^{3N}p\, \delta\left(E - H(q, p)\right), \tag{4.124}$$

where $\delta(x)$ is the Dirac delta function. Note that both the definitions (4.43) and (4.124) of $\Gamma(N, V, E)$ are equivalent as we have the tolerance $\Delta \ll E$. The entropy of the system in microcanonical equilibrium is related as discussed earlier via

$$S(N, V, E) = k_{\mathrm{B}} \log \Gamma(N, V, E). \tag{4.125}$$

Now, let us invoke the thermodynamic limit, namely, the simultaneous limit $N \to \infty$, $V \to \infty$, and $E \to \infty$, while keeping intensive quantities like particle density $n \equiv N/V$, energy density $\epsilon \equiv E/N$, and entropy density $s(\epsilon) \equiv S/N$ fixed and finite. In this limit, the Helmholtz free energy per particle $f(\beta)$ in canonical equilibrium obtained from Eq. (4.74) may be written as

$$f(\beta) \equiv - \lim_{N\to\infty} \frac{1}{N\beta} \log Z(N, V, T), \tag{4.126}$$

where, as usual, $\beta = 1/(k_{\mathrm{B}}T)$, and $Z(N, V, T)$ is the canonical partition function defined in Eq. (4.72). Note that in defining the entropy density as well as the Helmholtz free energy density in the foregoing, we have suppressed the dependence of these quantities on $n$. For later convenience, let us define a rescaled Helmholtz free energy density $\phi(\beta)$ such that

$$\phi(\beta) \equiv \beta f(\beta) = - \lim_{N\to\infty} \frac{1}{N} \log Z(N, V, T). \tag{4.127}$$

Ensemble equivalence is a consequence of concavity of $s(\epsilon)$ and $\phi(\beta)$, as we show in the following. The canonical partition function $Z(N, V, T)$ may be recast as

$$Z(N,V,T) = \int \mathrm{d}^{3N}q\, \mathrm{d}^{3N}p\, \mathrm{d}E\, \delta\left(E - H(q,p)\right) \mathrm{e}^{-\beta E}, \tag{4.128}$$

which along with Eq. (4.124) yields

$$Z(N,V,T) = \int \mathrm{d}E\; \Gamma(N,V,E)\,\mathrm{e}^{-\beta E}. \tag{4.129}$$

By using the definition of $S$ given in Eqs. (4.125), (4.129) may be written as

$$Z(N,V,T) = \int \mathrm{d}E\; \exp\Big[ -\Big\{\beta E - S(N,V,E)/k_{\mathrm{B}}\Big\}\Big]. \tag{4.130}$$

In the thermodynamic limit, one obtains from Eq. (4.130) that

$$Z(N,V,T) = \int \mathrm{d}(N\epsilon)\; \exp\Big[ -N\Big\{\beta\epsilon - s(\epsilon)/k_{\mathrm{B}}\Big\}\Big]. \tag{4.131}$$

For large values of $N$, the integral on the right hand side of Eq. (4.131) may be performed by using the saddle-point method. Then, using Eq. (4.127) and the fact that $\lim_{N\to\infty} \log N/N = 0$, we get

$$\phi(\beta) = \inf_{\epsilon} \left[\beta\epsilon - s(\epsilon)/k_{\mathrm{B}}\right]. \tag{4.132}$$

Here, inf stands for infimum.[8] The right hand side of Eq. (4.132) is known as the Legendre–Fenchel Transformation (LFT) of the microcanonical entropy $s(\epsilon)$. We thus see that the LFT of the entropy $s(\epsilon)$ in microcanonical equilibrium yields the Helmholtz free energy $\phi(\beta)$ in canonical equilibrium.

### 4.5.1 *The Legendre–Fenchel Transformation*

The Legendre–Fenchel transformation of a generic function is always concave [4]. As a result, the free energy density $\phi(\beta)$ is a concave function of $\beta$. Corresponding to an LFT, its inverse always exists [4], so that in the present case, the LFT of $\phi(\beta)$ yields

$$s^{\star}(\epsilon)/k_{\mathrm{B}} \equiv \inf_{\beta} \left[\beta\epsilon - \phi(\beta)\right]. \tag{4.133}$$

[8] The infimum of a set $A$ is its greatest lower bound, defined as a quantity $m$ such that there exists no member of the set with a value less than $m$. When the infimum belongs to $A$, it is considered a minimum of $A$.

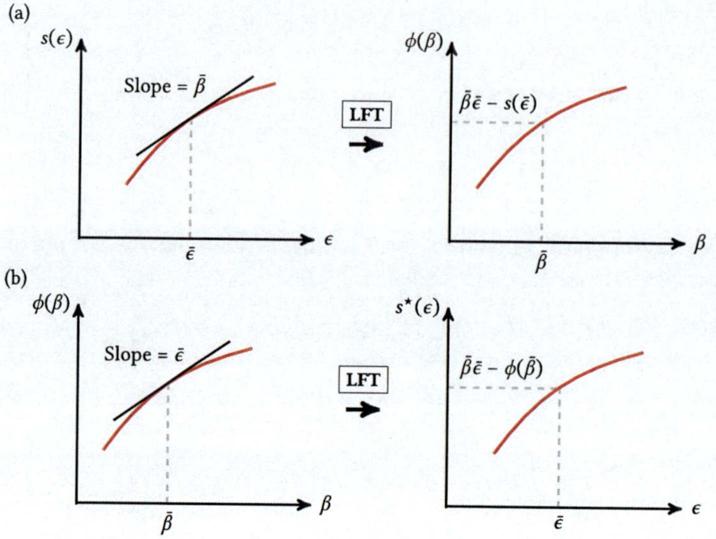


**Fig. 4.4** We present here a schematic diagram for a given particle density $n$ of the Legendre–Fenchel transformation (LFT) of a strictly concave function, namely, the entropy $s(\epsilon)$ (in units of $k_{\rm B}$) to **a** the rescaled free energy density $\phi(\beta)$, and subsequently, **b** to the function $s^\star(\epsilon)$ (again in units of $k_{\rm B}$), see Eqs. (4.132) and (4.133)

It then follows from the property of the LFT that the function $s^\star(\epsilon)$ is a concave function of $\epsilon$. In Fig. 4.4, we show a graphical representation for the LFT and the inverse LFT given in Eqs. (4.132) and (4.133), respectively.

- For a given value $\bar{\epsilon}$ of the microcanonical ensemble parameter $\epsilon$, one may obtain the corresponding canonical ensemble parameter $\beta = \bar{\beta}$ by solving Eq. (4.132), as $\bar{\beta} = (1/k_{\rm B})[\partial s(\epsilon)/\partial\epsilon]_{\epsilon=\bar{\epsilon}}$ (namely, $\bar{\beta}$ is the slope of $s(\epsilon)/k_{\rm B}$ versus $\epsilon$ curve at $\epsilon = \bar{\epsilon}$), see Fig. 4.4a. Then the corresponding free energy density is obtained as $\phi(\bar{\beta}) = \bar{\beta}\bar{\epsilon} - s(\bar{\epsilon})/k_{\rm B}$.
- Similarly, for a given value $\bar{\beta}$, one obtains $\bar{\epsilon}$ as the value of $\epsilon$ satisfying Eq. (4.133), i.e., satisfying $\bar{\epsilon} = [\partial\phi(\beta)/\partial\beta]_{\beta=\bar{\beta}}$ (namely, the slope of the $\phi(\beta)$ versus $\beta$ curve at $\beta = \bar{\beta}$), see Fig. 4.4b. One then obtains the function $s^\star(\bar{\epsilon})$ as $s^\star(\bar{\epsilon})/k_{\rm B} = \bar{\beta}\bar{\epsilon} - \phi(\bar{\beta})$.

We would regard the microcanonical and canonical equilibrium ensembles to be equivalent if it so turns out that for a generic $s(\epsilon)$, one has

$$s^\star(\epsilon) = s(\epsilon)\,; \tag{4.134}$$

otherwise, the two ensembles would be considered inequivalent. For a concave $s(\epsilon)$ in microcanonical equilibrium, Eq. (4.134) is always satisfied leading to equivalence of ensembles.

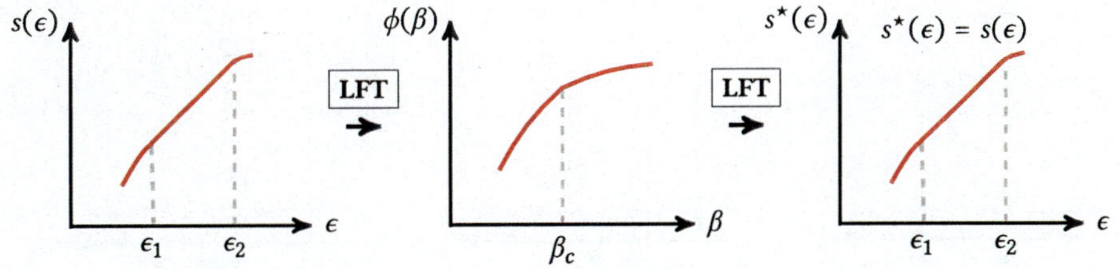


**Fig. 4.5 For a generic short-range system, we show here a schematic diagram of Legendre–Fenchel transformation (LFT) between entropy density $s(\epsilon)$ (in units of $k_B$) and rescaled free energy density $\phi(\beta)$ for given particle density $n$, in the case of a first-order phase transition in canonical ensemble.** The LFT when applied to the concave function $s(\epsilon)$ (in units of $k_B$) yields the concave free energy density $\phi(\beta)$. The inverse LFT of the free energy density yields back $s(\epsilon)$ (in units of $k_B$), i.e., we have $s^\star(\epsilon) = s(\epsilon)$. This fact corresponds to ensemble equivalence

In presence of a first-order phase transition in canonical equilibrium, the first derivative of the free energy $\phi(\beta)$ has by definition a discontinuity at the critical point $\beta = \beta_c$. In addition, we may recall that the free energy, obtained via the LFT of the function $s(\epsilon)/k_B$, is always a concave function irrespective of whether $s(\epsilon)$ is concave or not. Then, the microcanonical entropy $s(\epsilon)$ from which such a free energy may have been obtained through the LFT is either concave with a flat region (see Fig. 4.3b), or, non-concave (see Fig. 4.3a). As we have already discussed, non-concave entropy in microcanonical equilibrium is not possible for short-range systems due to the property of additivity. Therefore, in case of a first-order phase transition in canonical equilibrium, the microcanonical entropy of short-range systems is necessarily concave with a flat region. Consequently, Eq. (4.134) is satisfied in this case, leading to ensemble equivalence. In Fig. 4.5, we schematically show the equality between $s(\epsilon)$ and $s^\star(\epsilon)$ in this case. The other option of having a non-concave entropy is possible only for long-range systems. Hence, for long-range systems, Eq. (4.134) is not satisfied in presence of first-order phase transitions in canonical equilibrium, and consequently one observes ensemble inequivalence for such systems. In this case, Fig. 4.6 depicts schematically the fact that we have $s^\star(\epsilon) \neq s(\epsilon)$.

For continuous phase transitions in canonical equilibrium, the first derivative of the free energy $\phi(\beta)$ is continuous. We may then conclude that the microcanonical entropy from which the free energy may have been obtained through the LFT is a concave function. Hence, for continuous phase transitions in canonical equilibrium, Eq. (4.134) is always satisfied. The scenario is as depicted in Fig. 4.4, where since $s(\epsilon)$ is a strictly concave function to start with, it is evidently true that $s(\epsilon) = s^\star(\epsilon)$. In this case, then, for both short and long-range systems, the microcanonical and canonical ensembles are equivalent.

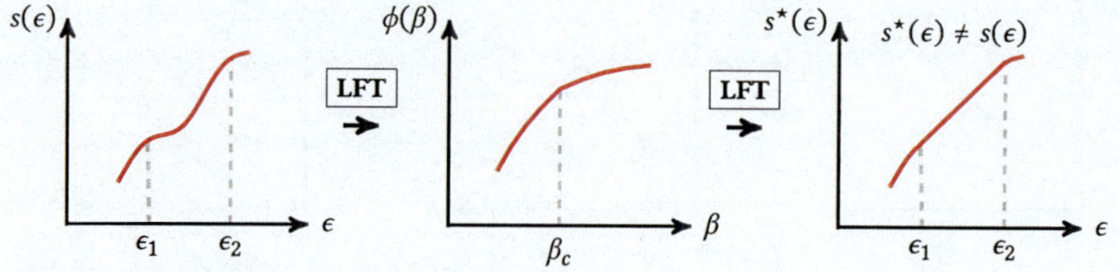


**Fig. 4.6 For a generic long-range system, we show here a schematic diagram of Legendre–Fenchel transformation (LFT) between entropy density $s(\epsilon)$ (in units of $k_B$) and rescaled free energy density $\phi(\beta)$ for given particle density $n$, in the case of a first-order phase transition in canonical ensemble.** The LFT when applied to the non-concave function $s(\epsilon)$ (in units of $k_B$) yields the concave free energy density $\phi(\beta)$. However, the inverse LFT of the free energy density yields a concave function $s^\star(\epsilon)$ different from $s(\epsilon)$ (both in units of $k_B$), i.e., we have $s^\star(\epsilon) \neq s(\epsilon)$. This fact corresponds to ensemble inequivalence

## 4.6 An Example: Interacting Classical Heisenberg Spins

We consider the same $N$ globally-coupled classical Heisenberg spins introduced in Sect. 3.8 with the Hamiltonian

$$H = -\frac{J}{2N} \sum_{i,j=1}^{N} \mathbf{S}_i \cdot \mathbf{S}_j + D \sum_{i=1}^{N} S_{iz}^{2n}, \tag{4.135}$$

where $n > 0$ is an integer. Without loss of generality, we set $J = 1$.

### *4.6.1 Canonical Equilibrium*

Let us now discuss the canonical equilibrium properties of the system (4.135) in the thermodynamic limit $N \to \infty$. We denote the temperature of the system in canonical equilibrium by $T = 1/\beta$, where we set the Boltzmann constant $k_B$ to unity, as was done in writing Eq. (4.125). One may consider that model (4.135) as a mean-field system such that it describes the dynamics of a single spin moving in a self-consistent mean-field generated by its interaction with all the spins. Then, the corresponding single-spin Hamiltonian is given by

$$h(\theta, \phi, m_x, m_y, m_z) \equiv -m_x \sin\theta \cos\phi - m_y \sin\theta \sin\phi - m_z \cos\theta + D \cos^{2n}\theta\,. \tag{4.136}$$

Consequently, in the thermodynamic limit, one may straightforwardly write down exact expressions for the average magnetization and the average energy in equilibrium. For example, the single-spin equilibrium distribution is given by

$$f_{\mathrm{eq}}(\theta,\phi) \propto \exp(-\beta h(\theta,\phi,m_x^{\mathrm{eq}},m_y^{\mathrm{eq}},m_z^{\mathrm{eq}})), \tag{4.137}$$

where the equilibrium magnetization components $(m_x^{\mathrm{eq}}, m_y^{\mathrm{eq}}, m_z^{\mathrm{eq}})$ now obey the self-consistent relation

$$(m_x^{\mathrm{eq}}, m_y^{\mathrm{eq}}, m_z^{\mathrm{eq}}) = \frac{\int \mathrm{d}\theta\, \mathrm{d}\phi\ \sin\theta\ (m_x^{\mathrm{eq}}, m_y^{\mathrm{eq}}, m_z^{\mathrm{eq}})\ f_{\mathrm{eq}}(\theta,\phi)}{\int \mathrm{d}\theta\, \mathrm{d}\phi\ \sin\theta\ f_{\mathrm{eq}}(\theta,\phi)}. \tag{4.138}$$

We recall that for $D > 0$ the system (4.135) orders in the $xy$-plane. We may then choose the equilibrium ordering direction to be along $x$ without loss of generality. For such a choice, one has $m_x^{\mathrm{eq}} \neq 0$ and $m_y^{\mathrm{eq}} = m_z^{\mathrm{eq}} = 0$. From Eq. (4.138), one thus obtains for $m_{\mathrm{eq}} \equiv m_x^{\mathrm{eq}}$ the equation [6]

$$m_{\mathrm{eq}} = \frac{\int \mathrm{d}\theta\, \mathrm{d}\phi\ \sin^2\theta\ \cos\phi\ \mathrm{e}^{\beta m_{\mathrm{eq}} \sin\theta\cos\phi - \beta D \cos^{2n}\theta}}{\int \mathrm{d}\theta\, \mathrm{d}\phi\ \sin\theta\ \mathrm{e}^{\beta m_{\mathrm{eq}} \sin\theta\cos\phi - \beta D \cos^{2n}\theta}}. \tag{4.139}$$

The average energy density, i.e., the average energy per spin is given by [6]

$$\epsilon = -\frac{1}{2} m_{\mathrm{eq}}^2 + D\langle \cos^{2n}\theta\rangle_{\mathrm{eq}}, \tag{4.140}$$

where one has

$$\langle \cos^{2n}\theta\rangle_{\mathrm{eq}} = \frac{\int \mathrm{d}\theta\, \mathrm{d}\phi\ \sin\theta \cos^{2n}\theta\ \mathrm{e}^{\beta m_{\mathrm{eq}} \sin\theta\cos\phi - \beta D \cos^{2n}\theta}}{\int \mathrm{d}\theta\, \mathrm{d}\phi\ \sin\theta\ \mathrm{e}^{\beta m_{\mathrm{eq}} \sin\theta\cos\phi - \beta D \cos^{2n}\theta}}. \tag{4.141}$$

It is known that the model (4.135) with $n = 1$ shows a continuous phase transition in magnetization across critical inverse temperature $\beta_c = 1/T_c$ [6]. From this fact, one may anticipate that the model exhibits continuous phase transition for general $n$. One may then consider Eq. (4.139) close to the critical point, i.e., for $\beta \gtrsim \beta_c$, when $m_{\mathrm{eq}}$ is small so that the equation may be expanded to leading order in $m_{\mathrm{eq}}$, as

$$m_{\mathrm{eq}}\left(\int \mathrm{d}\theta\, \mathrm{d}\phi\ \sin\theta\ \mathrm{e}^{-\beta D\cos^{2n}\theta} - \beta \int \mathrm{d}\theta\, \mathrm{d}\phi\ \sin^3\theta\cos^2\phi\ \mathrm{e}^{-\beta D\cos^{2n}\theta}\right) = 0. \tag{4.142}$$

Using the fact that $m_{\mathrm{eq}}$ is small but nonzero, we obtain the critical value $\beta_c$ as the value of $\beta$ that sets the bracketed quantity to zero. On computing the integrals, we obtain $\beta_c$ to be satisfying

$$1 - \frac{2}{\beta_c} = \frac{\Gamma(3/(2n)) - \Gamma(3/(2n), \beta_c D)}{(\beta_c D)^{2/(2n)}\left[\Gamma(1/(2n)) - \Gamma(1/(2n), \beta_c D)\right]}, \tag{4.143}$$

where $\Gamma(s)$ is the Gamma function and $\Gamma(s, x)$ is the upper incomplete Gamma function. At the critical point, when one has $m_{\text{eq}} = 0$, we obtain the critical energy density $\epsilon_c$ from Eq. (4.140) as

$$\begin{aligned}\epsilon_c = D\langle\cos^{2n}\theta\rangle_{\text{eq}} &= D\,\frac{\int \mathrm{d}\theta\,\mathrm{d}\phi\ \sin\theta\cos^{2n}\theta\ \mathrm{e}^{-\beta_c D\cos^{2n}\theta}}{\int \mathrm{d}\theta\,\mathrm{d}\phi\ \sin\theta\ \mathrm{e}^{-\beta_c D\cos^{2n}\theta}} \\ &= \frac{\Gamma(1/(2n)) - 2n\Gamma(1+1/(2n), \beta_c D)}{2n\beta_c\big[\Gamma(1/(2n)) - \Gamma(1/(2n), \beta_c D)\big]}\,. \end{aligned} \tag{4.144}$$

It may be easily checked that for $n = 1$, Eqs. (4.143) and (4.144) yield, respectively, $1 - 2/\beta_c = 1/(2\beta_c D) - \mathrm{e}^{-\beta_c D}/(\sqrt{\pi\beta_c D}\,\mathrm{Erf}[\sqrt{\beta_c D}])$ and $\epsilon_c = D\,(1 - 2/\beta_c)$, where $\mathrm{Erf}[x] \equiv (2/\sqrt{\pi})\int_0^x \mathrm{d}t\ \mathrm{e}^{-t^2}$ is the error function, as was reported in Ref. [6].

### 4.6.2 Ensemble Equivalence

Since the phase transition exhibited by the model (4.135) is a continuous one in canonical equilibrium, the canonical and microcanonical ensemble properties in equilibrium would be equivalent. Hence, Eq. (4.144) may also be interpreted as the conserved microcanonical energy density at the transition point. Figure 4.7 shows for $n = 1, 2$ the energy density $\epsilon_c$ as a function of $D$, obtained by first solving numerically for a given $D$ the transcendental Eq. (4.143) for $\beta_c$ and then using the obtained value of $\beta_c$ in Eq. (4.144). Moreover, one may construct a one-to-one mapping between a value of microcanonical equilibrium energy density $\epsilon$ and canonical equilibrium temperature $T$ by first solving Eq. (4.139) at a given $T$ to obtain the equilibrium magnetization $m_{\text{eq}}$, then substituting in Eq. (4.140) to obtain the corresponding average energy in canonical equilibrium, and finally demanding that the latter is the conserved energy density in microcanonical equilibrium. On carrying out this program for $n = 1$ and

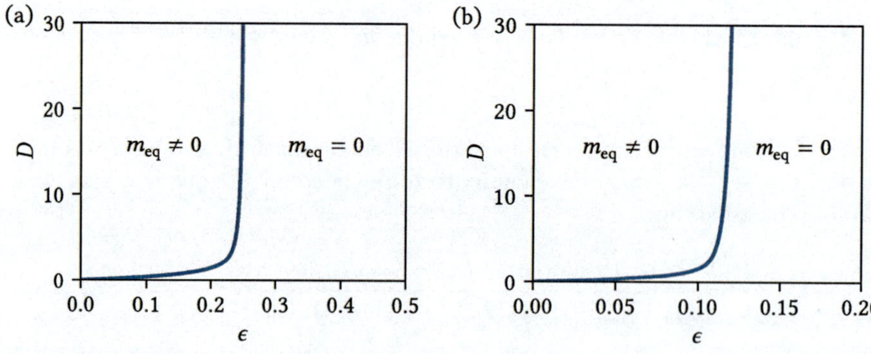


**Fig. 4.7 Phase diagram of the model (4.135): a** $n = 1$, **b** $n = 2$. Both the diagrams show the equilibrium phase boundary $\epsilon_c$

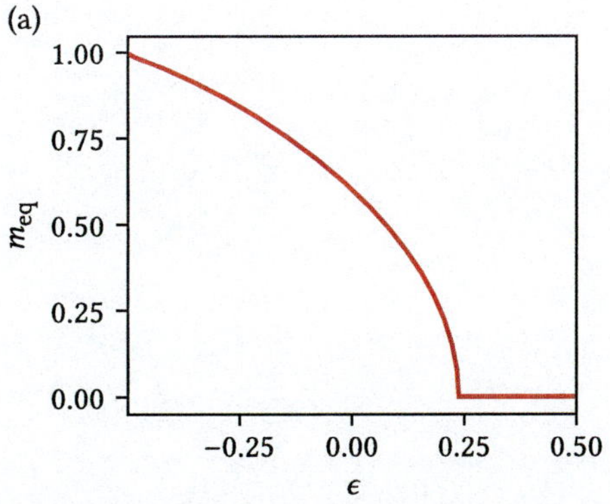


**Fig. 4.8 Magnetization $m_{eq}$ and temperature $T$ as a function of energy density $\epsilon$ in microcanonical equilibrium for the model (4.135) with $n = 1$ and $D = 5.0$. a** The magnetization decreases continuously from unity to zero at the critical energy density $\epsilon_c$, obtained from Eq. (4.144) as $\epsilon_c \approx 0.2381$, and remains zero at higher energies. **b** Correspondingly, the $T$ versus $\epsilon$ curve (the so-called caloric curve) is non-differentiable at the critical energy $\epsilon_c$, namely, $\mathrm{d}\epsilon/\mathrm{d}T|_{\epsilon_c-} \neq \mathrm{d}\epsilon/\mathrm{d}T|_{\epsilon_c+}$

$D = 5.0$, one obtains the results shown in Fig. 4.8, where we also show $m_{eq}$ as a function of microcanonical energy density $\epsilon$.

## References

1. M.W. Zemansky, R.H. Dittman, *Heat and Thermodynamics: An intermediate textbook*, 7th edn. (McGraw-Hill, New York, 1997)
2. F.W. Sears, G.L. Salinger, *Thermodynamics, Kinetic Theory, and Statistical Thermodynamics*, 3rd edn. (Addison-Wesley, Reading, 1975)
3. K. Huang, *Statistical Mechanics*, 2nd edn. (Wiley, New York, 1987)
4. R.K.P. Zia, E.F. Redish, S.R. McKay, Am. J. Phys. **77**(7), 614 (2009). https://doi.org/10.1119/1.3119512
5. A. Campa, T. Dauxois, D. Fanelli, S. Ruffo, *Physics of Long-Range Interacting Systems*, 1st edn. (Oxford University Press, Oxford, 2014). https://doi.org/10.1093/acprof:oso/9780199581931.001.0001
6. S. Gupta, D. Mukamel, J. Stat. Mech.: Theory Exp. **2011**(03), P03015 (2011). https://doi.org/10.1088/1742-5468/2011/03/P03015

# Chapter 5
# Nonlinear Dynamics

**Abstract** We provide an introduction to some basic concepts of nonlinear dynamics, focusing on those required in understanding noise effects in nonlinear systems dealt with later in the book. We start with defining a nonlinear dynamical system and discussing the notion of fixed points, namely, values of dynamical variables such that if the dynamics is initiated at these values, the variables do not evolve in time. We next turn to types of fixed points in one and two-dimensional systems. A particular type of fixed point in two dimensions is the so-called limit cycle, which is next discussed. It may happen as one tunes the dynamical parameters that the nature of the fixed points changes, new fixed points are generated, or existing fixed points are destroyed. This phenomenon of bifurcation and its various prototypical types in one dimension are then discussed. We proceed to detail characterization of one limit cycle as well as two and more than two interacting limit cycles. As an example of the latter, we discuss the celebrated Kuramoto model of interacting limit-cycle oscillators of distributed frequencies and discuss spontaneous synchronization, whereby a macroscopic number of the oscillators exhibits collective oscillations at a common frequency in the stationary state.

## 5.1 A Brief Introduction

In this chapter, we provide an introduction to nonlinear dynamical systems without being exhaustive in any way. For a more detailed treatment of nonlinear dynamical systems, the reader is referred to standard texts on nonlinear dynamics, e.g., [1, 2]. A dynamical system is any system whose state evolves in time. For a system with $d$ degrees of freedom, the state of the system at a given time is specified by specifying the values of all the degrees of freedom.[1] In mathematical terms, the state at time

[1] In textbooks on classical mechanics, the number of degrees of freedom of a system is defined to be the number of generalized coordinates. In this case, one needs to specify all the generalized coordinates and the generalized velocities (or momenta) to completely specify the state of the system. In this setting of dynamical systems, however, all the variables needed to specify completely the state of the system are collectively called the degrees of freedom.



D. Das and S. Gupta, *Facets of Noise*, Fundamental Theories of Physics 214,
https://doi.org/10.1007/978-3-031-45312-0_5

$t$ is defined by a set of $d$ dynamical variables denoted by $\{x_1(t), x_2(t), \ldots, x_d(t)\}$, where $x_i(t)$ gives the value of the $i$th variable at time $t$. Our job is to predict the state at a future time for a given initial state $\{x_1(0), x_2(0), \ldots, x_d(0)\}$. The time evolution of a dynamical system is typically given by a set of $d$ coupled differential equations:

$$\frac{\mathrm{d}x_i}{\mathrm{d}t} = f_i(x_1, x_2, \ldots, x_d, t)\,; \qquad i = 1, 2, \ldots, d\,, \tag{5.1}$$

where $f_i$'s are functions of the degrees of freedom and may also depend explicitly on time $t$. When none of the functions $f_i$ show explicit time dependence, the corresponding dynamical system is referred to as an autonomous system, for which Eq. (5.1) becomes

$$\frac{\mathrm{d}x_i}{\mathrm{d}t} = f_i(x_1, x_2, \ldots, x_d)\,; \qquad i = 1, 2, \ldots, d\,. \tag{5.2}$$

In this chapter, we restrict ourselves only to autonomous dynamical systems. The $d$-dimensional space spanned by the variables $x_1, x_2, \ldots, x_d$ is called the phase space of the system, where the state at a given time is denoted by a single point. As time goes on, the state starts to evolve according to the dynamical rules set by Eq. (5.2), thereby tracing out a path in the phase space. This path is called a phase-space trajectory. A dynamical system is linear when all the $f_i$'s are linear in the variables $x_i$. On the contrary, it is considered to be a nonlinear system if there exists at least one $f_i$ that is not linear even in one $x_i$.

Now, there could be special points $\{\overline{x}_1, \overline{x}_2, \ldots, \overline{x}_d\}$ in the phase-space, such that if the system is set to evolve starting from such points, it does not evolve at all. Such points are known as fixed or equilibrium points. From Eq. (5.2), we see that a fixed point is obtained as the solution of the set of equations

$$f_i(\overline{x}_1, \overline{x}_2, \ldots, \overline{x}_d) = 0 \;\forall\, i\,. \tag{5.3}$$

There could be several fixed points existing in the phase space for a given dynamical system. On starting slightly away from a fixed point, if the trajectories eventually approach the point in time, then the fixed point is said to be stable. If on the other hand the trajectories move away from it, the fixed point is said to be unstable. In presence of a stable fixed point, the dynamical system is expected to settle into this point at long times. We are typically interested in the long-time behavior of the dynamics (5.2), and hence, much of studies of dynamical systems is focused on fixed-point analysis. One may classify the fixed points according to their stability. To this end, we will restrict ourselves to one and two-dimensional phase spaces.

### 5.1.1 Fixed Points in One Dimension

In one dimension, the dynamical Eq. (5.2) reduces to

$$\frac{dx}{dt} = f(x), \tag{5.4}$$

for which the fixed points $\overline{x}$ are defined by $f(\overline{x}) = 0$. To analyze the stability of a given fixed point $\overline{x}$, one may study the dynamical evolution of a small perturbation $\delta x(t)$ around it. Using linearization in the neighborhood of $\overline{x}$:

$$x(t) = \overline{x} + \delta x(t), \tag{5.5}$$

with $|\delta x(t)| \ll 1$, one obtains from Eq. (5.4) that

$$\frac{d\delta x}{dt} = f(\overline{x} + \delta x) = f'(\overline{x})\,\delta x + \text{ higher-order terms in } \delta x\,, \tag{5.6}$$

where $f'(x) \equiv df/dx$. For $f'(\overline{x}) \neq 0$, one may neglect the higher-order term in the above equation, and then the solution of the equation is given by $\delta x(t) = \delta x(0)\exp(f'(\overline{x})t)$, which clearly shows that the perturbation around the fixed point $\overline{x}$ decays (respectively, grows) exponentially with time for $f'(\overline{x}) < 0$ (respectively, for $f'(\overline{x}) > 0$). Therefore, all trajectories in the neighborhood of the fixed point $\overline{x}$ eventually approach it when $f'(\overline{x}) < 0$ and move away from it when $f'(\overline{x}) > 0$. Consequently, the fixed point at $x = \overline{x}$ is said to be stable for $f'(\overline{x}) < 0$ and unstable for $f'(\overline{x}) > 0$. For $f'(\overline{x}) = 0$, the linearization method of deciding the stability evidently fails, and one needs to examine the higher-order nonlinear terms in Eq. (5.6).

### 5.1.2 Fixed Points in Two Dimensions

In two dimensions, we obtain the dynamical equations from Eq. (5.2) as

$$\begin{aligned} \frac{dx_1}{dt} &= f_1(x_1, x_2)\,, \\ \frac{dx_2}{dt} &= f_2(x_1, x_2)\,. \end{aligned} \tag{5.7}$$

The fixed points $(\overline{x}_1, \overline{x}_2)$ of the above dynamics satisfy

$$f_1(\overline{x}_1, \overline{x}_2) = f_2(\overline{x}_1, \overline{x}_2) = 0\,. \tag{5.8}$$

Once again, in order to analyze the stability of a given fixed point $(\overline{x}_1, \overline{x}_2)$, we consider dynamical evolution of small perturbations $\delta x_1(t)$ and $\delta x_2(t)$ around it, i.e., write $x_1(t)$ and $x_2(t)$ as

$$x_1(t) = \overline{x}_1 + \delta x_1(t)\,, \tag{5.9}$$

$$x_2(t) = \overline{x}_2 + \delta x_2(t)\,, \tag{5.10}$$

with $|\delta x_1(t)| \ll 1$, $|\delta x_2(t)| \ll 1$. Substituting Eqs. (5.9) and (5.10) in Eq. (5.7), and using Eq. (5.8), one obtains

$$\frac{\mathrm{d}\delta x_1}{\mathrm{d}t} = \delta x_1 \left.\frac{\partial f_1}{\partial x_1}\right|_{(\overline{x}_1,\overline{x}_2)} + \delta x_2 \left.\frac{\partial f_1}{\partial x_2}\right|_{(\overline{x}_1,\overline{x}_2)} + \text{ higher-order terms in } \delta x_1 \text{ and } \delta x_2\,, \tag{5.11}$$

and a similar equation for $\mathrm{d}\delta x_2/\mathrm{d}t$. In obtaining the last equation, we have expanded $f_1(x_1, x_2)$ and $f_2(x_1, x_2)$ around the point $(\overline{x}_1, \overline{x}_2)$. Keeping terms only linear in $\delta x_1$ and $\delta x_2$ and defining

$$a \equiv \left.\frac{\partial f_1}{\partial x_1}\right|_{(\overline{x}_1,\overline{x}_2)}; \quad b \equiv \left.\frac{\partial f_1}{\partial x_2}\right|_{(\overline{x}_1,\overline{x}_2)}; \quad c \equiv \left.\frac{\partial f_2}{\partial x_1}\right|_{(\overline{x}_1,\overline{x}_2)}; \quad d \equiv \left.\frac{\partial f_2}{\partial x_2}\right|_{(\overline{x}_1,\overline{x}_2)}, \tag{5.12}$$

one obtains

$$\frac{\mathrm{d}\delta x_1}{\mathrm{d}t} = a\,\delta x_1 + b\,\delta x_2\,, \tag{5.13}$$

$$\frac{\mathrm{d}\delta x_2}{\mathrm{d}t} = c\,\delta x_1 + d\,\delta x_2\,, \tag{5.14}$$

which may also be written as a matrix equation as

$$\frac{\mathrm{d}}{\mathrm{d}t}\begin{bmatrix}\delta x_1\\ \delta x_2\end{bmatrix} = \begin{bmatrix}a & b\\ c & d\end{bmatrix}\begin{bmatrix}\delta x_1\\ \delta x_2\end{bmatrix} \equiv \boldsymbol{M}\begin{bmatrix}\delta x_1\\ \delta x_2\end{bmatrix}. \tag{5.15}$$

The matrix $\boldsymbol{M}$ is called the Jacobian matrix at the fixed point $(\overline{x}_1, \overline{x}_2)$.

Differentiating Eq. (5.13) with respect to $t$, and then eliminating $\delta x_2$ with the help of Eq. (5.14), we get

$$\frac{\mathrm{d}^2\delta x_1}{\mathrm{d}t^2} - (a+d)\frac{\mathrm{d}\delta x_1}{\mathrm{d}t} + (ad - bc)\delta x_1 = 0\,. \tag{5.16}$$

The solution of Eq. (5.16) is obtained following standard procedure as

$$\delta x_1(t) = A\,\mathrm{e}^{\lambda_1 t} + B\,\mathrm{e}^{\lambda_2 t}\,, \tag{5.17}$$

with

$$\lambda_{1,2} \equiv \frac{1}{2}[(a+d) \pm \sqrt{(a+d)^2 - 4(ad-bc)}], \tag{5.18}$$

and $A$, $B$ are constants to be fixed by initial conditions $\delta x_1(0)$, $\delta x_2(0)$. Note that $\lambda_1$, $\lambda_2$ are the two eigenvalues of the matrix $\boldsymbol{M}$ defined in Eq. (5.15). Here, $A$, $B$, $\lambda_1$, $\lambda_2$ are in general complex constants. Substituting Eq. (5.17) in Eq. (5.14), one obtains the solution

$$\delta x_2(t) = C\, e^{\lambda_1 t} + D\, e^{\lambda_2 t}, \tag{5.19}$$

where the constants $C$ and $D$ are respectively related to $A$ and $B$ as

$$C \equiv (\lambda_1 - a)A/b; \qquad D \equiv (\lambda_2 - a)B/b. \tag{5.20}$$

The fixed point $(\overline{x}_1, \overline{x}_2)$ is linearly stable if the perturbations $\delta x_1$, $\delta x_2$ decay exponentially in time. Contrarily, the fixed point is unstable if even one of the perturbations grows in time. Then, it is seen from the solutions (5.17) and (5.19) that the stability of the fixed point is solely determined by the eigenvalues $\lambda_1$, $\lambda_2$. However, if $ad - bc = 0$, then at least one of the eigenvalues vanishes. In this case, one needs to consider the higher-order terms on the right hand side of Eq. (5.11) (and the corresponding equation for $\delta x_2$) to decide on the stability. We will not consider such cases in the following.

Let us now discuss different cases that may occur depending on the nature of the eigenvalues $\lambda_1$, $\lambda_2$.

**Case I: $\lambda_1$, $\lambda_2$ real and $\lambda_1 = \lambda_2$: Star**

1. $\lambda_1 = \lambda_2 < 0$: From Eqs. (5.17) and (5.19), we see that both $\delta x_1(t)$ and $\delta x_2(t)$ decay to zero with time. Therefore, all the neighboring trajectories approach the fixed point $(\overline{x}_1, \overline{x}_2)$ in the long-time limit. Hence, the fixed point is stable. In order to find the nature of the trajectories, we note that the slope in the $(x_1, x_2)$-phase space in the neighborhood of the fixed point is given by

$$\frac{dx_2}{dx_1} = \frac{dx_2}{dt}\left[\frac{dx_1}{dt}\right]^{-1} = \frac{d\delta x_2}{dt}\left[\frac{d\delta x_1}{dt}\right]^{-1} = \frac{C\,\lambda_1 e^{(\lambda_1-\lambda_2)t} + D\,\lambda_2}{A\,\lambda_1 e^{(\lambda_1-\lambda_2)t} + B\,\lambda_2}, \tag{5.21}$$

where we have used Eqs. (5.9), (5.10), (5.17) and (5.19). For $\lambda_1 = \lambda_2$, one obtains from Eq. (5.21) that $dx_2/dx_1 = (C+D)/(A+B) =$ constant, which suggests that in the $(x_1, x_2)$-phase space, the neighboring trajectories approach the fixed point along straight lines. This type of fixed point is called a stable star (Fig. 5.1a).
2. $0 < \lambda_1 = \lambda_2$: Here, both $\delta x_1(t)$ and $\delta x_2(t)$ grow with time. Therefore, the fixed point is unstable. The slope of the trajectories is again a constant. Hence, the trajectories in the neighborhood of the fixed point diverge from it along straight lines. This type of fixed point is called an unstable star (Fig. 5.1b).

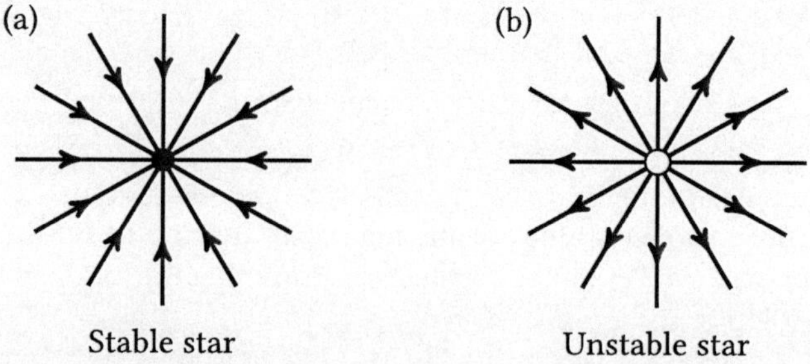


**Fig. 5.1 Stars with neighboring trajectories. a** Stable star. The stable fixed point is shown by a filled circle. **b** Unstable star. The unstable fixed point is depicted as an open circle

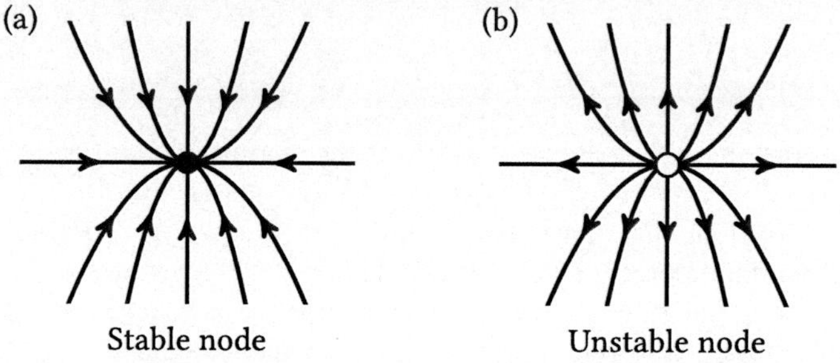


**Fig. 5.2 Nodes with neighboring trajectories. a** Stable node. The stable fixed point is shown by a filled circle. **b** Unstable node. The unstable fixed point is depicted as an open circle

**Case II: $\lambda_1, \lambda_2$ real and $\lambda_1 < \lambda_2 < 0$ or $0 < \lambda_2 < \lambda_1$: Node**

1. $\lambda_1 < \lambda_2 < 0$: Since both the eigenvalues are negative, the perturbations $\delta x_1(t)$ and $\delta x_2(t)$ decay with time. Hence, the fixed point is stable. It is seen from Eq. (5.21) that in this case, the slope is not constant, but that it changes with time and approaches the value $D/B$ as $t \to \infty$. It may be shown that the trajectories approach the fixed point along parabolic paths [2]. A fixed point of this type is called a stable node. We have shown a typical stable node in Fig. 5.2a.
2. $0 < \lambda_2 < \lambda_1$: Here, the perturbations grow with time. Hence, the neighboring trajectories diverge from the fixed point, making it unstable. The slope of the trajectories changes with time and approaches $C/A$ as $t \to \infty$, see Eq. (5.21). Such fixed points are called unstable nodes. Figure 5.2b depicts an unstable node.

**Case III: $\lambda_1, \lambda_2$ real and $\lambda_1 < 0 < \lambda_2$: Saddle**
We see from Eqs. (5.17) and (5.19) that both $\delta x_1$ and $\delta x_2$ diverge as $t \to \infty$. Along the eigendirection corresponding to $\lambda_1$, the trajectories approach the fixed point. On the other hand, along the eigendirection corresponding to $\lambda_2$, the trajectories move away from the fixed point. Figure 5.3 depicts a typical saddle point. As we have shown in the figure, the fixed point is stable only for trajectories along the vertical axis. For all other nearby trajectories, it is unstable. Hence, a saddle is considered an unstable fixed point.

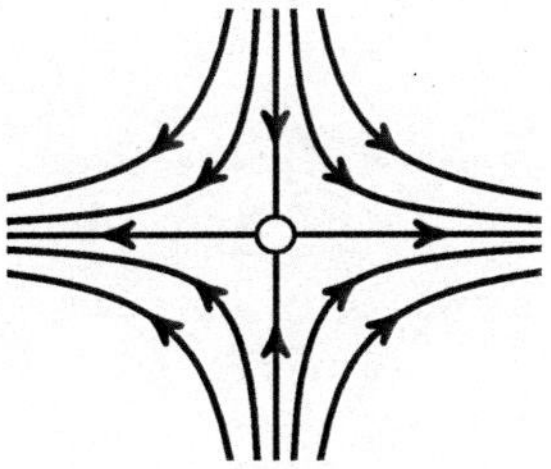


**Fig. 5.3 A saddle point with neighboring trajectories.** In this schematic figure, the saddle point is stable only along the vertical axis. For all other neighboring trajectories, it is unstable, and this is why we depict the fixed point as an open circle

**Case IV: $\lambda_1,\ \lambda_2$ forming a complex-conjugate pair: Spiral and Center**
Let us say that the eigenvalues are

$$\lambda_1 = \alpha + i\beta\,, \qquad \lambda_2 = \alpha - i\beta\,, \tag{5.22}$$

where $\alpha$ and $\beta$ are real constants. Substituting these expressions in Eq. (5.17), one obtains

$$\delta x_1(t) = e^{\alpha t}\left[(A+B)\cos\beta t + i(A-B)\sin\beta t\right]\,. \tag{5.23}$$

Invoking the condition that $\delta x_1(t)$ is real, and using the fact that $\cos\beta t$ and $\sin\beta t$ are functions that are linearly independent, one obtains $A = B^\star$, where $\star$ denotes complex conjugation. Then, using $A = A_r + iA_i$ with $A_r,\ A_i$ real, Eq. (5.23) is reduced to

$$\delta x_1(t) = 2e^{\alpha t}\left[A_r\cos\beta t - A_i\sin\beta t\right] = 2|A|e^{\alpha t}\cos(\beta t + \phi)\,, \tag{5.24}$$

with $|A| = \sqrt{A_r^2 + A_i^2}$ and $\phi = \tan^{-1}(A_i/A_r) = \arg(A)$. Similarly, from Eq. (5.19), one obtains

$$\delta x_2(t) = 2\,|C|\,e^{\alpha t}\cos(\beta t + \phi')\,, \tag{5.25}$$

where $C$ is given by Eq. (5.20), and $\phi' = \arg(C)$.

1. $\alpha < 0$: In this case, as seen from Eqs. (5.24) and (5.25), both $|\delta x_1|$ and $|\delta x_2|$ decay to zero as $t \to \infty$. Such perturbations represent decaying oscillations, and the trajectories approach the fixed point along spiral paths. The corresponding fixed points are stable spiral. Figure 5.4a shows a stable spiral.
2. $\alpha > 0$: Here, both the perturbations grow, and the trajectories move away from the fixed point along spiral paths. Therefore, this fixed point is called an unstable spiral. Such a fixed point is shown in Fig. 5.4b.

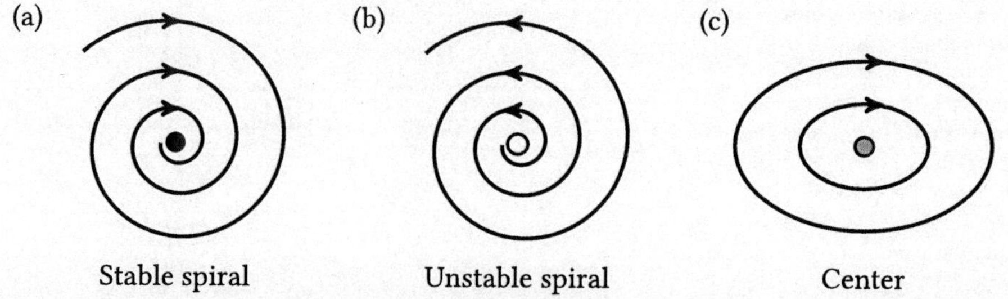


**Fig. 5.4 Spiral and center with neighboring trajectories. a** Stable spiral. The filled circle denotes the stable fixed point. **b** Unstable spiral. The open circle denotes the unstable fixed point. **c** Center. The fixed point is denoted as the gray-filled circle

3. $\alpha = 0$: In this case, $\delta x_1$ and $\delta x_2$ neither decay nor diverge, but vary periodically with time. Consequently, neighboring trajectories are neither attracted to nor repelled from the fixed point. The trajectories form a family of closed orbits around the fixed point. This is called a center type fixed point. Figure 5.4c depicts a center type fixed point.

In order to develop a clearer understanding of fixed points in two dimensions, we now move on to discuss a representative example, namely, the Van der Pol oscillator.[2] The equation of motion for the oscillator is given by

$$\frac{\mathrm{d}^2x}{\mathrm{d}t^2} + \mu\,(x^2 - 1)\frac{\mathrm{d}x}{\mathrm{d}t} + x = 0\,, \tag{5.26}$$

where $\mu$ is a real parameter that indicates nonlinearity and damping (depending on the value of $x$ with respect to unity, the damping could be either positive or negative). In terms of the variables $x_1 \equiv x$ and $x_2 \equiv \mathrm{d}x/\mathrm{d}t$, we may rewrite the second-order differential Eq. (5.26) as two coupled first-order differential equations:

$$\begin{aligned} \frac{\mathrm{d}x_1}{\mathrm{d}t} &= x_2\,, \\ \frac{\mathrm{d}x_2}{\mathrm{d}t} &= -x_1 - \mu\, x_2\,(x_1^2 - 1)\,. \end{aligned} \tag{5.27}$$

It is evident that the form of Eq. (5.27) is exactly the same as Eq. (5.7) with $f_1(x_1, x_2) = x_2$, and $f_2(x_1, x_2) = -x_1 - \mu\, x_2\,(x_1^2 - 1)$. Clearly, the fixed point of dynamics (5.27) is given by $(\overline{x}_1 = 0,\ \overline{x}_2 = 0)$.

[2] The Van der Pol oscillator model was originally proposed by Balthasar van der Pol in 1920. He was then working as an electrical engineer for the multinational conglomerate Philips in Netherlands. The model describes a nonlinear electrical circuit used in early radios.

**Table 5.1** Different types of fixed points appearing in the Van der Pol dynamics (5.27)

| Range of $\mu$ | Eigenvalues $\lambda_1, \lambda_2$ | Fixed point type |
|---|---|---|
| $\mu < -2$ | $\lambda_1 < \lambda_2 < 0$ | Stable node |
| $\mu = -2$ | $\lambda_1 = \lambda_2 < 0$ | Stable star |
| $-2 < \mu < 0$ | $\lambda_1 = \lambda_2^\star = \alpha + \mathrm{i}\beta,\ \alpha < 0$ | Stable spiral |
| $\mu = 0$ | $\lambda_1 = \lambda_2^\star = 2\mathrm{i}$ | Center |
| $0 < \mu < 2$ | $\lambda_1 = \lambda_2^\star = \alpha + \mathrm{i}\beta,\ \alpha > 0$ | Unstable spiral |
| $\mu = 2$ | $0 < \lambda_2 = \lambda_1$ | Unstable star |
| $2 < \mu$ | $0 < \lambda_2 < \lambda_1$ | Unstable node |

In order to study the stability of the fixed point ($\overline{x}_1 = 0,\ \overline{x}_2 = 0$), we linearize Eq. (5.27) around it and obtain the Jacobian matrix with the help of Eqs. (5.15) and (5.12) as

$$\boldsymbol{M} = \begin{bmatrix} 0 & 1 \\ -1 & \mu \end{bmatrix}, \tag{5.28}$$

for which the eigenvalues are given by

$$\lambda_1 = \frac{1}{2}\left[\mu + \sqrt{\mu^2 - 4}\right], \qquad \lambda_2 = \frac{1}{2}\left[\mu - \sqrt{\mu^2 - 4}\right]. \tag{5.29}$$

The eigenvalues change their nature depending on the value of $\mu$, thereby changing the type of the fixed point. We have enumerated in Table 5.1 all possible cases that may arise if one varies $\mu$.

In the long-time limit, the phase-space trajectories approach the stable fixed point. From Table 5.1, it is seen that the phase-space origin is stable if $\mu < 0$. For $\mu = 0$, the Van der Pol equation (5.26) reduces to the equation of motion of an undamped simple harmonic oscillator with unit angular frequency, and one has a center type fixed point. In this case, the phase space trajectories for two different initial conditions are shown in Fig. 5.5a. From the figure, we see that the trajectories originating from two different phase points form two closed orbits. In fact, in this case, every distinct initial condition forms its own closed orbit around the center.

## 5.2 Limit Cycle

Apart from fixed points discussed in the previous section, one may also have isolated closed orbits in two-dimensional phase spaces. Suppose for a given initial condition, the phase-space trajectory of dynamics (5.7) forms a closed orbit. In mathematical terms, this means that if $(x_1(t), x_2(t))$ is the solution of Eq. (5.7) at time $t$, then there exists a time $T > 0$, such that $x_1(t + T) = x_1(t);\ x_2(t + T) = x_2(t)\ \forall\, t$. Then, the

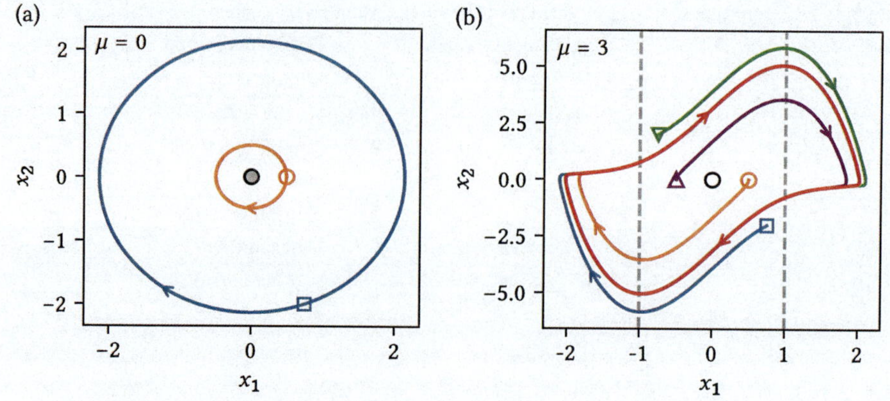


**Fig. 5.5 Phase-space trajectories for the Van der Pol oscillator dynamics** (5.27). **a** For $\mu = 0$, we have a center. The orange trajectory corresponds to the initial condition $x_1(0) = 0.5,\ x_2(0) = 0.0$, while the blue trajectory is for $x_1(0) = 0.75,\ x_2(0) = -2.0$. The initial conditions are shown as the orange circle and the blue square, respectively. **b** For $\mu = 3$, the fixed point at the origin is an unstable node shown as the black circle. Here, we see that trajectories originating from different phase points approach a closed orbit as $t \to \infty$. The orange and the blue trajectory correspond to the same initial condition as in (**a**). The green and the purple trajectory have corresponding initial conditions $x_1(0) = -0.5,\ x_2(0) = 0.0$, and $x_1(0) = -0.75,\ x_2(0) = 2.0$, denoted respectively by a triangle and an inverted triangle. It is seen that as $x_1$ tries to escape the region $[-1, 1]$, marked by the dashed lines, $|x_2|$ decreases significantly, thereby keeping the trajectories confined to a smaller region of phase space. In integrating the dynamics (5.27) numerically, we have used the fourth-order Runge–Kutta method discussed in Appendix M.1, with time step $\Delta t = 0.01$

solution is called a periodic solution with time period $T$. We have already seen a family of closed orbits in the neighborhood of the center type fixed point for the Van der Pol oscillator dynamics (5.27), and these are necessarily periodic orbits. A periodic orbit that is isolated, i.e., there exists no other periodic orbit in its neighborhood, is called a limit cycle. Every nearby trajectory in course of dynamical evolution either approaches the limit cycle or moves away from it. If all the neighboring trajectories approach the limit cycle, we consider it to be a stable limit cycle. On the contrary, if all of them deviate from it, it is said to be an unstable limit cycle.

As seen in Table 5.1, the phase-space origin for the Van der Pol dynamics becomes unstable for all positive values of $\mu$. In this case, nearby trajectories diverge from the origin as time increases. But, do they diverge all the way to infinity? In order to find an answer to the question, we note from Eq. (5.27) that with $\mu > 0$, the nonlinear term $\mu x_2(x_1^2 - 1)$ damps down $x_2$ and eventually $x_1$ if $|x_1|$ tends to become too large ($|x_1| > 1$). The converse is true if $|x_1|$ tends to become too small ($|x_1| < 1$): here, the nonlinear term pumps up $x_2$ and eventually $x_1$. Because of these two inherently opposite dynamical tendencies, the system in the long-time limit settles into a characteristic closed curve, a limit cycle, irrespective of the initial conditions. In Fig. 5.5b, we have shown the trajectories originating from four different phase points for $\mu = 3$. It is seen from the figure that as soon as $x_1$ tries to escape the region

$[-1, 1]$, marked by the dashed lines, $|x_2|$ decreases significantly, thereby keeping the diverging trajectories confined to a smaller region of the phase space. In the limit $t \to \infty$, all of them approach a closed orbit. We see from the figure that neighboring trajectories from inside as well as from outside of the closed orbit gradually approach it. We thus conclude that the closed orbit is a stable limit cycle, shown as the closed red curve in Fig. 5.5b.

## 5.3 Bifurcations

From the foregoing, we note that as one tunes the value of $\mu$, the type and stability of the fixed point, and concomitantly, the long-time dynamical behavior of the Van der Pol oscillator changes abruptly (see Table 5.1). Such sudden qualitative changes in the dynamical behavior of a system because of tuning of one or more dynamical parameters are called bifurcations. The critical value of the parameter at which a bifurcation occurs is called the bifurcation value. We now discuss some simple examples of bifurcations in one dimension. These examples constitute the so-called normal forms of bifurcation in one-dimensional nonlinear dynamical systems. It may be shown that all one-dimensional nonlinear systems exhibiting a certain type of bifurcation are locally (namely, around a fixed point) equivalent in a topological sense to one of these normal forms of bifurcation [1].

### *5.3.1 Saddle-Node Bifurcation*

Let us consider the dynamical system

$$\frac{\mathrm{d}x}{\mathrm{d}t} = f(x) = \mu - x^2 , \tag{5.30}$$

where $\mu$ is a dynamical parameter. For the dynamics (5.30), we have the following scenarios:

- For $\mu < 0$, there are no real fixed points.
- For $\mu = 0$, there exists a fixed point at $x = 0$.
- For $\mu > 0$, there are two fixed points at $x = \pm\sqrt{\mu}$.

From the linear stability analysis discussed in Sect. 5.1.1, one may easily see that the fixed points at $x = \sqrt{\mu}$ and $x = -\sqrt{\mu}$ are stable and unstable, respectively, whereas the method of linearization fails for the fixed point at $x = 0$.

There are no fixed points for $\mu < 0$. As one increases the parameter $\mu$ from the negative side, a fixed point (the one at $x = 0$) of the dynamics suddenly appears at $\mu = 0$. At the bifurcation value $\mu = 0$, one has $\mathrm{d}x/\mathrm{d}t = -x^2 < 0$, and therefore, the fixed point at $x = 0$ is a half-stable fixed point, namely, it acts as an unstable

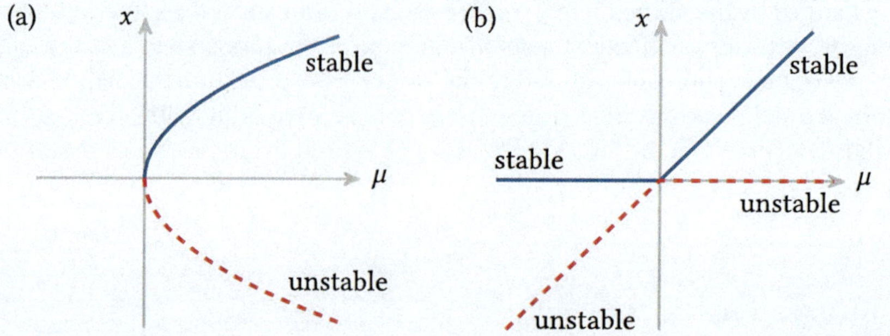


**Fig. 5.6 Bifurcation diagrams in one dimensional nonlinear dynamical systems: a** Saddle-node bifurcation **b** Transcritical bifurcation

fixed point for $x < 0$ (as $f'(x) = -2x > 0$) and as a stable fixed point for $x > 0$ (as $f'(x) = -2x < 0$). With further increase in $\mu$, the half-stable fixed point at $x = 0$ bifurcates into one stable and one unstable fixed point. This type of bifurcation is known as the Saddle-node bifurcation. One may show the dependence of the flow of the dynamical variable $x(t)$ on the parameter $\mu$ on an $x$-vs-$\mu$ diagram, known as the bifurcation diagram. Figure 5.6a depicts the bifurcation diagram for the Saddle-node bifurcation.

### 5.3.2 *Transcritical Bifurcation*

Now, consider the dynamical system

$$\frac{dx}{dt} = f(x) = \mu x - x^2 . \tag{5.31}$$

For the above dynamics, we have the following scenarios:

- For $\mu < 0$, there exist a stable fixed point at $x = 0$ and an unstable fixed point at $x = \mu$.
- For $\mu = 0$, there is a half-stable fixed point at $x = 0$.
- For $\mu > 0$, there exist an unstable fixed point at $x = 0$ and a stable fixed point at $x = \mu$.

In this case, at the bifurcation value $\mu = 0$, the stability of the half-stable fixed point at $x = 0$ is exactly similar to that of the saddle-node case, since then the governing dynamical equation becomes the same. As one tunes $\mu$ from negative to positive values, the stability of the fixed points at $x = 0$ and $x = \mu$ gets interchanged at $\mu = 0$. This type of bifurcation is known as transcritical bifurcation. In contrast to the saddle-node bifurcation, here fixed points do not disappear at the bifurcation point; they rather switch their stability. The corresponding bifurcation diagram is shown in Fig. 5.6b.

### 5.3.3 Pitchfork Bifurcation

We now consider a third dynamical system

$$\frac{\mathrm{d}x}{\mathrm{d}t} = f(x) = \mu x - x^3 , \tag{5.32}$$

which has the following fixed points:

- For $\mu < 0$, there exists a stable fixed point at $x = 0$,
- For $\mu = 0$, there is a fixed point at $x = 0$; note that the method of linearization for deciding its stability fails.
- For $\mu > 0$, there exist an unstable fixed point at $x = 0$ and two stable fixed points at $x = \pm\sqrt{\mu}$.

The bifurcation diagram in this case is shown in Fig. 5.7a, which clearly depicts the "pitchfork" nature. Such a bifurcation behavior is known as the supercritical pitchfork bifurcation. The negative $x^3$ term in Eq. (5.32) acts as a stabilizer, that is, it plays the role of a restoring force directed to the origin $x = 0$ and attracts $x(t)$ towards it. When the cubic term becomes positive and therefore destabilizing, one has the dynamics

$$\frac{\mathrm{d}x}{\mathrm{d}t} = f(x) = \mu x + x^3 , \tag{5.33}$$

that has:

- For $\mu < 0$, there exist an unstable fixed point at $x = 0$ and two stable fixed points at $x = \pm\sqrt{-\mu}$.
- For $\mu > 0$, there is an unstable fixed point at $x = 0$.

The bifurcation behavior of the dynamics (5.33) is known as the subcritical pitchfork bifurcation, and is shown in Fig. 5.7b.

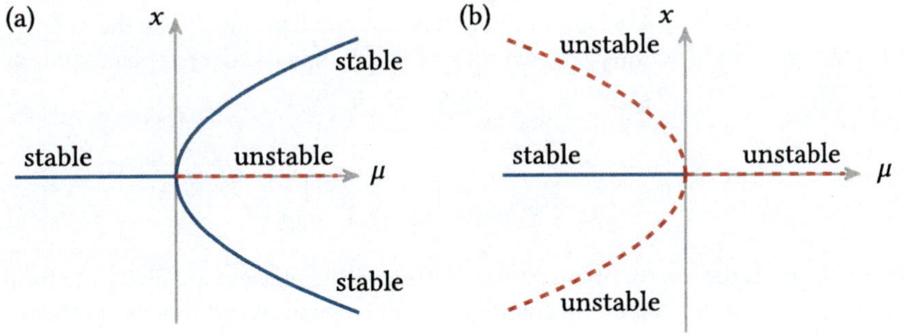


**Fig. 5.7 Bifurcation diagrams in one-dimensional nonlinear dynamical systems: a** Supercritical pitchfork bifurcation **b** Subcritical pitchfork bifurcation

## 5.4 Characterization of Limit Cycles

We will discuss later in this chapter the phenomenon of spontaneous synchronization, whereby a large collection of interacting nonlinear oscillators with distributed natural frequencies spontaneously evolves to a state in which a macroscopic number of them oscillate with a common frequency. The building blocks of a synchronizing system are self-sustained oscillators that can oscillate on their own without the need of any external source of energy. Moreover, each constituent unit is capable of exhibiting oscillations that have a characteristic waveform, amplitude, and frequency of oscillations depending only on the internal dynamical parameters of the unit and independent of any typical initial condition of its dynamics. The characteristic waveform is stable under small perturbations: if we perturb slightly the motion of the individual units, it would soon return to its characteristic waveform. On the basis of these features, one may anticipate that the dynamics of such self-sustained oscillators must have the two opposing tendencies of energy-pumping and dissipation in such a way that oscillations that tend to become too large are damped down by dissipation. On the other hand, it is the energy-pumping mechanism in the dynamics that pumps up oscillations that tend to become too small. The characteristic waveform would then correspond to a balance between energy pumping and dissipation, and would be the one sustained by the dynamics at long times. It turns out that the limit-cycle dynamics is the ideal candidate to model such self-sustained oscillators. Indeed, the Van der Pol oscillator discussed previously in this chapter has for example the aforementioned characteristics required of a self-sustained oscillator constituting a synchronizing system.

A single self-sustained oscillator may have a large number of internal degrees of freedom. When the oscillator dynamics settles into a limit cycle, it forms a one-dimensional isolated closed curve in the phase space. This enables us to do a huge simplification in studying the dynamics of the self-sustained oscillator, i.e., to map the large set of internal degrees of freedom into a single monotonically increasing variable $s(t)$ [3, 4]. The variable $s(t)$ has to satisfy the periodicity condition $s(t + nT) = s(t) + ns(T)\ \forall\ t$, where $T$ is the period of the limit-cycle oscillator, and $n$ is an integer. For an illustration, we consider the example of the stable limit cycle occurring in the Van der Pol oscillator dynamics discussed previously in the chapter. For this example, we have only two degrees of freedom: $x_1$ and $x_2$. Then, one may map these two variables into a single variable $s$ by defining

$$s(t) \equiv \int_0^t \mathrm{d}t \sqrt{\left(\frac{\mathrm{d}x_1}{\mathrm{d}t}\right)^2 + \left(\frac{\mathrm{d}x_2}{\mathrm{d}t}\right)^2}, \tag{5.34}$$

such that $s(t)$ gives the length of the path traversed by the phase point along the limit cycle in time $t$. Note that the definition (5.34) may be generalized to any number of degrees of freedom. For a self-sustained oscillator with $d$ degrees of freedom, one has

$$s(t) \equiv \int_0^t \mathrm{d}t \sqrt{\sum_{i=1}^{d} \left( \frac{\mathrm{d}x_i}{\mathrm{d}t} \right)^2} . \tag{5.35}$$

In general, the variation of $s(t)$ along the limit cycle would be nonuniform: in equal intervals of time, $s(t)$ would not change by an equal amount. Nevertheless, there is a convenient way to transform the variable $s(t)$ into a $2\pi$-periodic phase variable $\theta \equiv \theta(s)$ that changes uniformly along the limit cycle, by defining [4]

$$\theta \equiv \frac{2\pi}{T} \int_0^s \mathrm{d}s \left[ \frac{\mathrm{d}s}{\mathrm{d}t} \right]^{-1} . \tag{5.36}$$

From Eq. (5.36), we have

$$\frac{\mathrm{d}\theta}{\mathrm{d}t} = \frac{\mathrm{d}\theta}{\mathrm{d}s} \frac{\mathrm{d}s}{\mathrm{d}t} = \frac{2\pi}{T} \left[ \frac{\mathrm{d}s}{\mathrm{d}t} \right]^{-1} \frac{\mathrm{d}s}{\mathrm{d}t} = \frac{2\pi}{T} \equiv \omega , \tag{5.37}$$

where the constant $\omega$ is the natural frequency of the oscillator. We thus see that the variable $\theta$ changes uniformly in time.

In Fig. 5.8, we have shown the limit cycle of the Van der Pol oscillator dynamics (5.27). Figure 5.8a shows that the phase point takes the same time $\Delta t$ to traverse from $A$ to $B$ as well as from $C$ to $D$. During this time interval, changes in $s(t)$ from $A$ to $B$ and from $C$ to $D$ are different; yet, the phase $\theta(t)$ changes by the same amount in both the cases. In Fig. 5.8b, we have shown the evolution of $s(t)$ and $\theta(t)$ as a function of time $t$, with both renormalized to zero after every time interval $t = T$. Here, the periodic behavior of $s(t)$ and $\theta(t)$ is obtained by numerically integrating Eq. (5.27) at times long enough that the trajectories $s(t)$ and $\theta(t)$ are essentially the same (up to numerical precision) as that on the limit cycle. It may be seen from the figure that $s(t)$ changes at a nonuniform rate, while $\theta(t)$ changes uniformly. The dashed line in gray denoting the value $2\pi$ implies the $2\pi$-periodicity in $\theta(t)$.

To observe spontaneous collective synchronization, we need many weakly-interacting limit-cycle oscillators. Considering a single oscillator, its interactions with others act as small perturbations that try to deviate its motion away from the limit cycle. The stability of the limit cycle nullifies the effect of any possible motion of the phase point in the direction transverse to the limit cycle. However, there is no such restriction for the perturbations in driving the phase along the limit cycle. Thus, the stability of the limit cycle enables us to characterize the dynamics in its neighborhood also in terms of the phase $\theta(t)$.

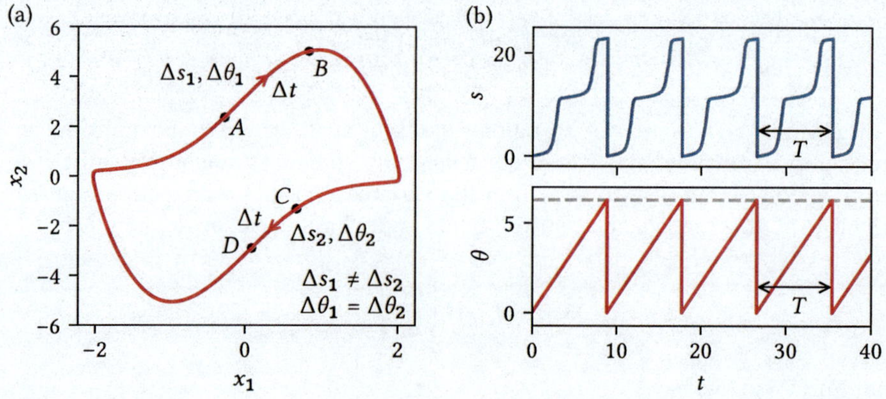


**Fig. 5.8 The limit cycle corresponding to the Van der Pol oscillator dynamics** (5.27). The parameter $\mu$ has the same value as in Fig. 5.5a. The phase point takes the same time $\Delta t$ to traverse from $A$ to $B$ as well as from $C$ to $D$. Within this time interval, changes in $s$ are different, while $\theta$ changes by the same amount. **b** The periodic behavior of $s(t)$ and $\theta(t)$ is obtained by numerically integrating Eq. (5.27) at times long enough that the trajectories $s(t)$ and $\theta(t)$ are essentially the same (up to numerical precision) as that on the limit cycle. Here, we have used the fourth-order Runge–Kutta method discussed in Appendix M.1 with time step $\Delta t = 0.01$. In the figure, $T$ denotes the time period of motion. After every time interval $t = T$, we reset $s(t)$ and $\theta(t)$ to zero. The path length $s(t)$ does not change uniformly with time. However, the rate of change of the phase $\theta(t)$ is uniform. The gray-dashed line corresponds to the value $2\pi$. It is seen that indeed, $\theta$ is $2\pi$-periodic

## 5.5 Two Interacting Limit Cycles

In the previous section, we have seen how to characterize a single limit cycle. We now discuss the case of two interacting limit cycles. To proceed, let us consider two similar weakly-interacting limit-cycle oscillators with phases $\theta_1$ and $\theta_2$ and frequencies $\omega_1$ and $\omega_2$, with $\theta_1, \theta_2 \in [0, 2\pi)$. By similar oscillators, we mean $\omega_1 \sim \omega_2$. The dynamics is given by the following set of equations:

$$\frac{\mathrm{d}\theta_k}{\mathrm{d}t} = \omega_k + \epsilon\, \mathcal{G}_k(\theta_1, \theta_2)\,; \quad k = 1, 2\,, \tag{5.38}$$

where $\epsilon$ is the coupling constant. To take account of the fact that the oscillators are weakly interacting, we assume $\epsilon$ to be small, i.e., $|\epsilon| \ll 1$. The interaction term $\mathcal{G}_k$ is assumed to be $2\pi$-periodic in both $\theta_1$ and $\theta_2$. In terms of the center-of-mass coordinates defined by $\Theta \equiv (\theta_1 + \theta_2)/2$ and $\Phi \equiv (\theta_1 - \theta_2)/2$, Eq. (5.38) becomes

$$\frac{\mathrm{d}\theta_k}{\mathrm{d}t} = \omega_k + \epsilon G_k(\Theta, \Phi)\,; \quad k = 1, 2\,. \tag{5.39}$$

Using Eq. (5.39), one may obtain the equations of motion for $\Theta(t)$ and $\Phi(t)$ as

$$\begin{aligned}\frac{\mathrm{d}\Theta}{\mathrm{d}t} &= \frac{1}{2}(\omega_1+\omega_2)+\frac{\epsilon}{2}\Big[G_1(\Theta,\Phi)+G_2(\Theta,\Phi)\Big],\\ \frac{\mathrm{d}\Phi}{\mathrm{d}t} &= \frac{1}{2}(\omega_1-\omega_2)+\frac{\epsilon}{2}\Big[G_1(\Theta,\Phi)-G_2(\Theta,\Phi)\Big].\end{aligned} \tag{5.40}$$

Since $|\epsilon| \ll 1$, one obtains from Eq. (5.40) the behavior of $\Theta$ and $\Phi$ to leading order in $\epsilon$ as

$$\Theta(t)=\frac{(\omega_1+\omega_2)}{2}t\,,\quad \Phi(t)=\frac{(\omega_1-\omega_2)}{2}t\,. \tag{5.41}$$

The definition of the center-of-mass coordinate system ensures that $G_k$'s are also $2\pi$-periodic in both $\Theta$ and $\Phi$. Therefore, we may expand them in Fourier series as

$$G_k(\Theta,\Phi)=\sum_{n,m=-\infty}^{\infty}\widetilde{G}^{(k)}_{n,m}\,\mathrm{e}^{\mathrm{i}n\Theta}\,\mathrm{e}^{\mathrm{i}m\Phi}\,. \tag{5.42}$$

Substituting Eq. (5.42) in Eq. (5.39), we get

$$\frac{\mathrm{d}\theta_k}{\mathrm{d}t}=\omega_k+\epsilon\sum_{n,m=-\infty}^{\infty}\widetilde{G}^{(k)}_{n,m}\,\mathrm{e}^{\mathrm{i}n\Theta}\,\mathrm{e}^{\mathrm{i}m\Phi}\,. \tag{5.43}$$

As the interaction term in Eq. (5.43) is already suppressed because of the presence of the multiplying factor $\epsilon$, we may use Eq. (5.41) in Eq. (5.43) to obtain the equation of motion for $\theta_k$ to leading order in $\epsilon$ as

$$\frac{\mathrm{d}\theta_k}{\mathrm{d}t}=\omega_k+\epsilon\sum_{n,m=-\infty}^{\infty}\widetilde{G}^{(k)}_{n,m}\,\mathrm{e}^{\mathrm{i}n\frac{(\omega_1+\omega_2)}{2}t}\,\mathrm{e}^{\mathrm{i}m\frac{(\omega_1-\omega_2)}{2}t}\,. \tag{5.44}$$

From Eq. (5.44), it may be seen that in the interaction term, there are two timescales of order $1/(\omega_1+\omega_2)$ and $1/|\omega_1-\omega_2|$ owing to the presence of the terms $\exp[\mathrm{i}n(\omega_1+\omega_2)t/2]$ and $\exp[\mathrm{i}m(\omega_1-\omega_2)t/2]$, respectively. Since we have $\omega_1 \sim \omega_2$, the first timescale is much shorter than the second one. Consequently, the interaction term shows fast oscillations on the former timescale and slow oscillations on the latter. In Fig. 5.9, we show schematically the behavior of the fast and slow oscillations. We are interested in the long-time behavior of the dynamics, and consequently, the slowly-varying terms would be most important to our interest. The fast-oscillating terms would induce changes of $\mathcal{O}(\epsilon)$ in $\theta_k$ over short times of order $1/(\omega_1+\omega_2)$. Over the (long) time of our interest, these fast-oscillating terms would change their sign many times, thus having no appreciable effect on $\theta_k$. On the contrary, the slowly-oscillating terms do not change their sign over short times. As a result, their contributions, although of $\mathcal{O}(\epsilon)$, would keep accruing until times $\sim 1/|\omega_1-\omega_2|$ over which they undergo a change of sign. Therefore, the slowly-oscillating terms would have appreciable influence on $\theta_k$ at long times of order $1/|\omega_1-\omega_2|$.

On the basis of the above discussions, we conclude that in order to obtain the long-time dynamics, i.e., dynamics over a time $\tau$ satisfying $1/(\omega_1+\omega_2) \ll \tau \ll$

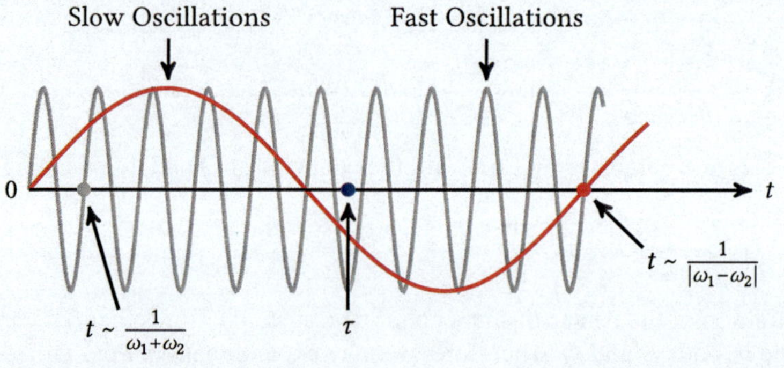


**Fig. 5.9 Schematic diagram to show the time variation of the fast and slowly-oscillating terms in Eq. (5.44).** The fast oscillations have a timescale $\sim 1/(\omega_1+\omega_2)$. Contrarily, the slow oscillations have a timescale $\sim 1/|\omega_1-\omega_2|$. We see over a time $\tau$ satisfying $1/(\omega_1+\omega_2) \ll \tau \ll 1/|\omega_1-\omega_2|$ that the fast oscillations change sign many times unlike the slow oscillations

$1/|\omega_1-\omega_2|$, we can average out the fast-oscillating terms over such a time. Hence, we get

$$\left\langle \frac{\mathrm{d}\theta_k}{\mathrm{d}t} \right\rangle \equiv \frac{1}{\tau}\int_0^\tau \mathrm{d}t \frac{\mathrm{d}\theta_k}{\mathrm{d}t} = \omega_k + \frac{\epsilon}{\tau}\sum_{n,m=-\infty}^{\infty} \widetilde{G}_{n,m}^{(k)} \int_0^\tau \mathrm{d}t\, \mathrm{e}^{\mathrm{i}n\frac{(\omega_1+\omega_2)}{2}t}\, \mathrm{e}^{\mathrm{i}m\frac{(\omega_1-\omega_2)}{2}t}, \tag{5.45}$$

where, as denoted, the angular brackets stand for the mentioned time averaging. Now, among the fast-oscillating terms, the term with $n = 0$ is independent of time. Therefore, only this term has a non-vanishing contribution to the integral on the right hand side of Eq. (5.45). Then, from Eq. (5.45), we obtain

$$\left\langle \frac{\mathrm{d}\theta_k}{\mathrm{d}t} \right\rangle = \omega_k + \frac{\epsilon}{\tau}\sum_{m=-\infty}^{\infty} \widetilde{G}_{0,m}^{(k)} \int_0^\tau \mathrm{d}t\, \mathrm{e}^{\mathrm{i}m\frac{(\omega_1-\omega_2)}{2}t}. \tag{5.46}$$

Over the time $\tau$, the slowly-oscillating terms $\exp(\mathrm{i}m(\omega_1-\omega_2)t/2)$ do not change appreciably, so that we may take the corresponding terms outside of the integral in Eq. (5.46). Thus, we get

$$\begin{aligned}\left\langle \frac{\mathrm{d}\theta_k}{\mathrm{d}t} \right\rangle &= \omega_k + \frac{\epsilon}{\tau}\sum_{m=-\infty}^{\infty} \widetilde{G}_{0,m}^{(k)}\, \mathrm{e}^{\mathrm{i}m\frac{(\omega_1-\omega_2)}{2}t} \int_0^\tau \mathrm{d}t \\ &= \omega_k + \epsilon \sum_{m=-\infty}^{\infty} \widetilde{G}_{0,m}^{(k)}\, \mathrm{e}^{\mathrm{i}m\frac{(\omega_1-\omega_2)}{2}t} \\ &= \omega_k + \epsilon \sum_{m=-\infty}^{\infty} \widetilde{G}_{0,m}^{(k)}\, \mathrm{e}^{\mathrm{i}m\Phi t}, \end{aligned} \tag{5.47}$$

where we have used Eq. (5.41). Introducing $Q_k(\Phi) \equiv \sum_{m=-\infty}^{\infty} \widetilde{G}_{0,m}^{(k)} \, \mathrm{e}^{\mathrm{i}m\Phi t}$, and dropping the angular brackets for brevity, we thus obtain the long-time behavior of Eq. (5.38) to leading order in $\epsilon$ as

$$\frac{\mathrm{d}\theta_k}{\mathrm{d}t} = \omega_k + \epsilon \, Q_k(\theta_1 - \theta_2); \qquad k = 1, 2\,. \tag{5.48}$$

Equation (5.48) evidently shows that in the long-time limit, only the difference between the phases contributes to the interaction term [3, 4].

This section was devoted to a discussion of two interacting limit cycles. In the next section, we generalize the discussion to a population of $N$ interacting limit cycles.

## 5.6 Many Interacting Limit Cycles: An Example

We may now generalize the above discussion to $N \gg 1$ weakly-interacting oscillators [4]. To this end, we only need to find a suitable time $\tau$ that satisfies $1/(\omega_i + \omega_j) \ll \tau \ll 1/|\omega_i - \omega_j| \; \forall \; i$ and $j$. For large $N$, the set of natural frequencies $\{\omega_i\}$ may be considered sampled from a given distribution $\overline{g}(\omega)$. In situations of practical relevance, the oscillators will be similar, so that we may assume $\overline{g}(\omega)$ to be appreciably different from zero over a narrow range $[\omega_{\min}, \omega_{\max}]$. Then, it may be difficult in general to find a single $\tau$ that satisfies the above condition for every pair $(i, j)$. In order to proceed, we seek for a more stringent constraint on $\tau$, namely, $1/\min(|\omega_{\max}|, |\omega_{\min}|) \ll \tau \ll 1/(\omega_{\max} - \omega_{\min})$. Note that the range $[\omega_{\min}, \omega_{\max}]$ being narrow, the upper and the lower limit of this constraint are widely separated, so that we may always find a suitable $\tau$ that satisfies this constraint. By definition, an oscillator must have a nonzero frequency. Therefore, we may take the range $[\omega_{\min}, \omega_{\max}]$ to not include the origin. Then, introducing $Q_{ij}$ as the interaction between the pair $(i, j)$, and following Eq. (5.48), one may obtain in the long-time limit the equations of motion for a system of $N$ weakly-interacting oscillators as

$$\frac{\mathrm{d}\theta_i}{\mathrm{d}t} = \omega_i + \epsilon \sum_{\substack{j=1 \\ i \neq j}}^{N} Q_{ij}(\theta_i - \theta_j)\,; \quad i = 1, 2, \ldots, N\,. \tag{5.49}$$

### 5.6.1 *The Kuramoto Model*

The Kuramoto model serves as a minimal framework to study the phenomenon of spontaneous synchronization among a population of coupled oscillating units of diverse natural frequencies [3]. Spontaneous synchrony is commonly observed in nature, e.g., in yeast cell suspensions [5], flashing fireflies [6], arrays of Josephson junctions [7], laser arrays [8], power-grid networks [9], and others. The model involves a large collection of limit-cycle oscillators equally coupled to each other through an interaction that depends sinusoidally on the difference of the phases between the oscillators. The model may be derived from Eq. (5.49) by assuming all pair interactions to be the same, i.e., $Q_{ij}(\theta) = Q(\theta)\ \forall\ i$ and $j$, making the choice $Q(\theta) = -\sin\theta$, and considering $\epsilon > 0$. Thus, we get

$$\frac{\mathrm{d}\theta_i}{\mathrm{d}t} = \omega_i - \epsilon \sum_{j=1}^{N} \sin(\theta_i - \theta_j)\,; \quad i = 1, 2, \ldots, N. \tag{5.50}$$

The inequality $i \neq j$ in the summation in Eq. (5.49) is already taken care of by the sine term in Eq. (5.50). Note that the second term on the right hand side of Eq. (5.50) scales as $N$, whereas the first term does not scale with $N$. Consequently, in the limit $N \to \infty$, we have the undesirable situation in which the second term dominates over the first term, so that the latter may be neglected to a good approximation. In order to have an effective competition between the two terms on the right hand side of Eq. (5.50), we scale down the interaction term by $N$. Then, from Eq. (5.50), we obtain the dynamics of the Kuramoto model as [4, 10–13]

$$\frac{\mathrm{d}\theta_i}{\mathrm{d}t} = \omega_i - \frac{K}{N} \sum_{j=1}^{N} \sin(\theta_i - \theta_j)\,; \quad i = 1, 2, \ldots, N, \tag{5.51}$$

where we have introduced $K \geq 0$ as the coupling constant with $\theta_i \in [0, 2\pi)\ \forall i$. The natural frequencies $\omega_i$ are taken to be sampled independently from a common distribution $\overline{g}(\omega)$ with the normalization $\int_{-\infty}^{\infty} \mathrm{d}\omega\, \overline{g}(\omega) = 1$. The average frequency $\omega_0$ is defined as

$$\omega_0 \equiv \int_{-\infty}^{\infty} \mathrm{d}\omega\, \omega\, \overline{g}(\omega). \tag{5.52}$$

As is usual in studies of the Kuramoto model, we consider $\overline{g}(\omega)$ to be symmetric about $\omega_0$, and unimodal, i.e., it decreases monotonically and continuously to zero with increasing $|\omega - \omega_0|$. A representative example of such a distribution would be the Gaussian distribution given by

$$\overline{g}(\omega) = \frac{1}{\sqrt{2\pi\sigma^2}} \mathrm{e}^{-\frac{(\omega-\omega_0)^2}{2\sigma^2}}, \tag{5.53}$$

where $\sigma > 0$ is the standard deviation.

It would prove convenient for further analysis to scale out the mean $\omega_0$ of the natural frequencies, which may be easily done by viewing the dynamics (5.51) in a frame rotating uniformly with the frequency $\omega_0$ with respect to the laboratory frame. Indeed, in this rotating frame, Eq. (5.51) in terms of the redefined phase variable $\theta_i'(t) \equiv \theta_i(t) - \omega_0 t$ becomes

$$\frac{\mathrm{d}\theta_i'}{\mathrm{d}t} = \omega_i' - \frac{K}{N}\sum_{j=1}^{N}\sin(\theta_i' - \theta_j')\,; \quad i = 1, 2, \ldots, N. \tag{5.54}$$

Here, the natural frequencies $\omega_i' \equiv (\omega_i - \omega_0)$ may be considered sampled from the distribution $g(\omega') \equiv \overline{g}(\omega - \omega_0)$, thereby having zero mean. Dropping the primes in Eq. (5.54) for simplicity, we obtain the Kuramoto dynamics in the rotating frame as

$$\frac{\mathrm{d}\theta_i}{\mathrm{d}t} = \omega_i - \frac{K}{N}\sum_{j=1}^{N}\sin(\theta_i - \theta_j)\,; \quad i = 1, 2, \ldots, N, \tag{5.55}$$

where the distribution $g(\omega)$ now satisfies $g(\omega) = g(-\omega)$, and we have $\omega_0 = 0$.

### 5.6.2 Spontaneous Synchronization

We now discuss the dynamics (5.55) in the limit $N \to \infty$. In this limit, the system at long times ($t \to \infty$) settles into a stationary state. In the stationary state, the dynamics (5.55) shows qualitatively different behaviors depending on the value of the coupling constant $K$. Clearly, for small $K$, it is the first term on the right hand side of (5.55) that dominates the dynamics, while for large $K$, it is the other term that is dominant. Indeed, there exists a critical value $K_c$, such that for larger $K$, a macroscopic number (of order $N$) of the oscillators oscillate with a common frequency, thereby maintaining time-independent phase differences among themselves. Such a coherent state characterizes a synchronized phase. Contrarily, for $K < K_c$, the system exhibits an incoherent behavior in which there is no macroscopic number of oscillators that oscillate with a common frequency. In this case, the oscillators have phase differences among themselves that are explicitly time dependent. Consequently, the phase angles at every instant may be considered to be distributed uniformly over the range $[0, 2\pi)$. An incoherent state of this sort exemplifies an incoherent phase. To avoid any confusion, we will from now on use the word 'angle' for $\theta_i$ and the word 'phase' to refer to the state of the system.

In order to distinguish between the synchronized and the incoherent phase, it is customary to define a complex order parameter $r(t)\mathrm{e}^{\mathrm{i}\psi(t)}$ as [10]

$$r(t)\mathrm{e}^{\mathrm{i}\psi(t)} \equiv \frac{1}{N}\sum_{j=1}^{N}\mathrm{e}^{\mathrm{i}\theta_j(t)}, \tag{5.56}$$

where $r(t);\ 0 \le r(t) \le 1$ measures the amount of synchrony present in the system at time $t$, while $\psi(t);\ 0 \le \psi(t) < 2\pi$ is the average of the oscillator angles. In Fig. 5.10, we have shown a geometric representation of the order parameter. The quantities $r(t)$ and $\psi(t)$ are obtained as

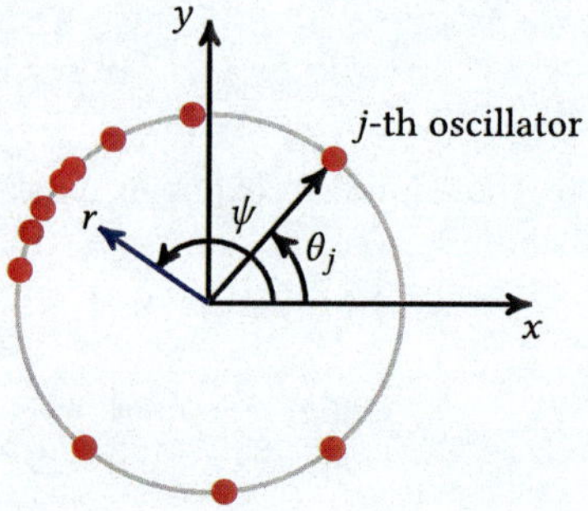


**Fig. 5.10 Geometric interpretation of the Kuramoto order parameter** (5.56). A given configuration of angles $\theta_j$ of different oscillators are shown as red dots distributed over a unit circle. The centroid of the oscillator angles is shown as a blue arrow

$$r(t) = \sqrt{r_x^2(t) + r_y^2(t)}\,, \quad \psi(t) = \tan^{-1}(r_y(t)/r_x(t))\,, \tag{5.57}$$

where we have

$$r_x(t) \equiv \frac{1}{N}\sum_{j=1}^{N}\cos\theta_j(t)\,, \quad r_y(t) \equiv \frac{1}{N}\sum_{j=1}^{N}\sin\theta_j(t)\,. \tag{5.58}$$

In terms of the Kuramoto order parameter, Eq. (5.55) may be written as

$$\frac{\mathrm{d}\theta_i}{\mathrm{d}t} = \omega_i - Kr(t)\sin(\theta_i - \psi(t))\,; \quad i = 1, 2, \ldots, N. \tag{5.59}$$

The complex order parameter (5.56) being an average of complex numbers with unit modulus may also be understood in the complex $\mathcal{Z}$-plane as a sum of individual vectors pointing from the origin to points on a unit circle. In the incoherent phase, when the oscillator angles are distributed uniformly over the range $[0, 2\pi)$, the contribution of any such vector would be canceled by the one due to a vector pointing in the opposite direction (this would strictly be true as $N \to \infty$). This results in $r(t) = 0$. In the synchronized phase, however, such a cancellation would not take place for a macroscopic number of vectors, thereby resulting in $r(t) > 0$. In the case of perfect synchrony, achieved as $K \to \infty$, when all the oscillators attain the same angle, one has $r(t) = 1$. In Fig. 5.11, we have shown the stationary-state behavior of the dynamics (5.59) for three different values of the coupling constant $K$.

Equation (5.59) shows the mean-field nature of the dynamics, i.e., every oscillator experiences a mean-field of magnitude $r(t)$ produced by all the oscillators. In the stationary state, the quantities $r(t)$ and $\psi(t)$ both become time independent, and we denote the corresponding stationary values by $r_{\mathrm{st}}$ and $\psi_{\mathrm{st}}$, respectively. In the stationary state, it makes sense to measure all the angles with respect to $\psi_{\mathrm{st}}$, which is tantamount to setting $\psi_{\mathrm{st}}$ to zero. Then, Eq. (5.59) simplifies to

$$\frac{\mathrm{d}\theta_i}{\mathrm{d}t} = \omega_i - Kr_{\mathrm{st}}\sin\theta_i\,; \quad i = 1, 2, \ldots, N. \tag{5.60}$$

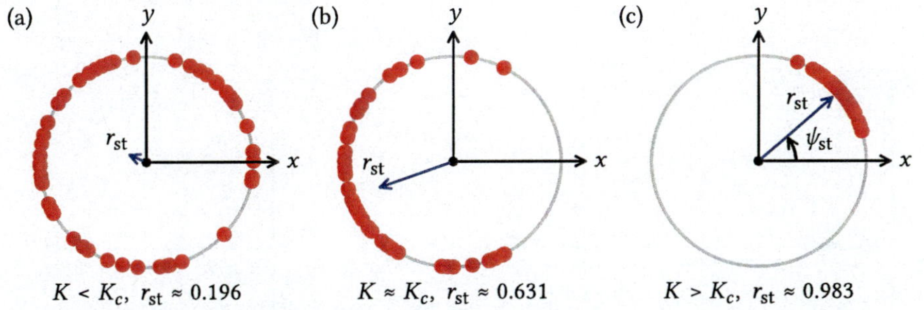


**Fig. 5.11 Stationary-state behavior of the Kuramoto dynamics** (5.59). The angles of different oscillators are shown as red dots. In order to obtain the figure, we integrate the dynamics (5.59) numerically by using the Euler method discussed in Appendix M.2 with time step $\Delta t = 0.01$ and $N = 50$. The frequencies $\omega_i$ are sampled from a Gaussian distribution with zero mean and standard deviation equal to 0.2. The initial configuration $\{\theta_i\}$ at $t = 0$ is taken to be distributed uniformly in the range $[0, 2\pi)$. **a** Incoherent phase. Here, $K = 0.1$. The angles are distributed all over the unit circle, yielding a small value for $r_{st}$. In fact, in the limit $N \to \infty$, one has $r_{st} \to 0$. **b** We consider $K = 0.32 \approx K_c$. The critical value $K_c$ is obtained by using Eq. (5.87). It may be seen that the oscillators have formed a few clusters, and the value of $r_{st}$ is sufficiently larger than in (**a**). **c** Synchronized phase. Here, we have $K = 1$. In this case, a macroscopic number of the oscillators are phase locked

Setting $\psi_{st} = 0$ also yields $r_{st,y} = 0$, hence, $r_{st} = r_{st,x}$ , see Eq. (5.57).

### *5.6.3 Stationary-State Distribution*

The dynamics (5.60) has fixed points at those values of $\theta_i$ that satisfy $\mathrm{d}\theta_i/\mathrm{d}t = 0$. In Figs. 5.12 and 5.13, we have shown the dependence of $\mathrm{d}\theta_i/\mathrm{d}t$ as a function of $\theta_i$ for $\omega_i > 0$ and $\omega_i < 0$, respectively. For given values of $K$ and $r_{st}$, as shown in Figs. 5.12 and 5.13, we have the following fixed points [1]

1. For oscillators satisfying $|\omega_i| < Kr_{st}$: As shown in Fig. 5.12a and 5.13a, in this case, we have two fixed points. For later convenience, we denote them as

$$\text{Point } A : \overline{\theta}_i^A(\omega_i) \equiv \sin^{-1}[\omega_i/(Kr_{st})]; \ \cos\overline{\theta}_i^A = +\sqrt{1 - \omega_i^2/(Kr_{st})^2}\,; \tag{5.61a}$$

$$\text{Point } B : \overline{\theta}_i^B(\omega_i) \equiv \sin^{-1}[\omega_i/(Kr_{st})]; \ \cos\overline{\theta}_i^B = -\sqrt{1 - \omega_i^2/(Kr_{st})^2}\,. \tag{5.61b}$$

2. For oscillators satisfying $|\omega_i| = Kr_{st}$: In this case, we have a single fixed point at $\pi/2$ for $\omega_i > 0$, and at $3\pi/2$ for $\omega_i < 0$, see Figs. 5.12b and 5.13b.
3. For oscillators satisfying $|\omega_i| > Kr_{st}$: In this case, no fixed point exists, as seen from Figs. 5.12c and 5.13c.

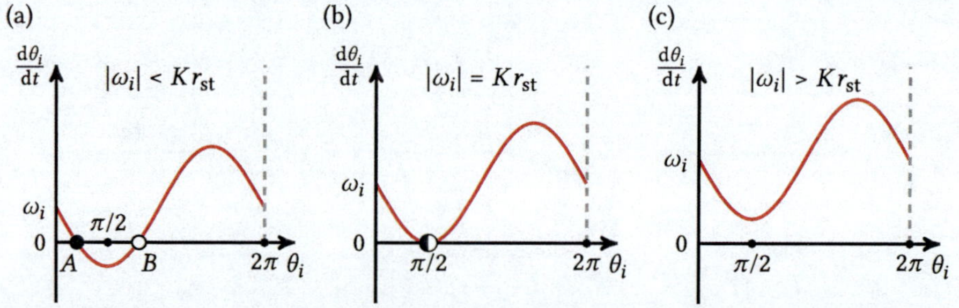


**Fig. 5.12 Behavior of** $d\theta_i/dt$ **from Eq. (5.60) as a function of** $\theta_i$ **in different regimes of** $\omega_i > 0$**, and for given values of** $K$ **and** $r_{st}$. The solid and void circles in **a** show the stable and unstable fixed points, respectively. In **b**, we show the half-stable fixed point at $\theta_i = \pi/2$ as a half-filled circle. It shows that the point is stable for $\theta_i < \pi/2$ and unstable for $\theta_i > \pi/2$. In **c**, we see that there is no fixed point

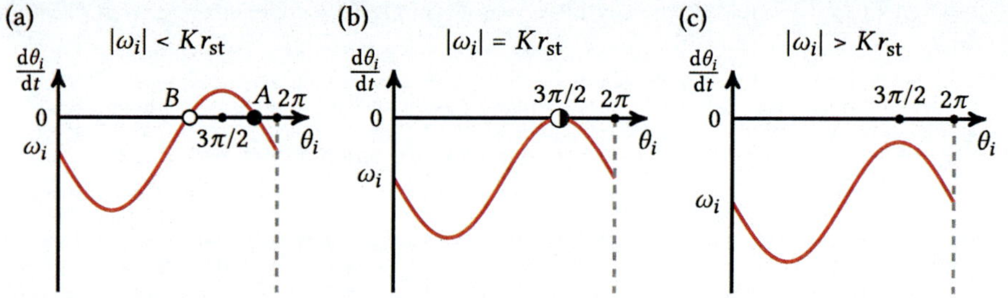


**Fig. 5.13 Behavior of** $d\theta_i/dt$ **from Eq. (5.60) as a function of** $\theta_i$ **in different regimes of** $\omega_i < 0$**, and for given values of** $K$ **and** $r_{st}$. The solid and void circles in **a** show the stable and unstable fixed points, respectively. In **b**, we show the half-stable fixed point at $\theta_i = 3\pi/2$ as a half-filled circle. It shows that the point is stable for $\theta_i > 3\pi/2$ and unstable for $\theta_i < 3\pi/2$. In **c**, we see that there is no fixed point.

The stability of the fixed points $\overline{\theta}_i$ may be investigated by linearizing Eq. (5.60) about $\overline{\theta}_i$. To this end, writing $\theta_i$ as $\overline{\theta}_i + \delta\theta_i$, with $|\delta\theta_i| \ll 1$ and substituting in Eq. (5.60), we obtain the linearized equation

$$\frac{d\delta\theta_i}{dt} = -Kr_{st}\,\delta\theta_i \cos\overline{\theta}_i\,, \tag{5.62}$$

for which the solution is given by

$$\delta\theta_i(t) = \delta\theta_i(0)\,e^{-(Kr_{st}\cos\overline{\theta}_i)t}\,. \tag{5.63}$$

From Eq. (5.63), it may be seen that the perturbation $\delta\theta_i(t)$ decays to zero in time if $\cos\overline{\theta}_i > 0$. Therefore, the fixed point with $\cos\overline{\theta}_i > 0$ is linearly stable. Now, for the oscillators with $|\omega_i| < Kr_{st}$, it may be seen from Eq. (5.61a) that $\overline{\theta}_i^A(\omega_i)$ is the stable fixed point. Contrarily, as seen from Eq. (5.61b), $\cos\overline{\theta}_i^B(\omega_i)$ is negative, from which one may infer that the point $\overline{\theta}_i^B(\omega_i)$ is an unstable fixed point. However, for those very few oscillators with $|\omega_i| = Kr_{\rm st}$, we see that $\cos\overline{\theta}_i$ vanishes at the fixed point, which means that the linear stability analysis does not let us infer anything about the stability property of the fixed point.

To have a more lucid understanding of stability, we now switch to the graphical way of thinking about it. Let us focus on the case with $\omega_i > 0$ as shown in Fig. 5.12. From Fig. 5.12a, we see that the quantity $\mathrm{d}\theta_i/\mathrm{d}t$ changes its sign from positive to negative at $\theta_i = \overline{\theta}_i^A(\omega_i)$. In the neighborhood of the point $A$, one has $\mathrm{d}\theta_i/\mathrm{d}t$ positive for those angles that satisfy $\theta_i < \overline{\theta}_i^A(\omega_i)$, so that the long-time dynamics starting from these $\theta_i$'s would be governed by (namely, attracted towards) the value $\overline{\theta}_i^A(\omega_i)$. Contrarily, the quantity $\mathrm{d}\theta_i/\mathrm{d}t$ is negative for angles that satisfy $\theta_i > \overline{\theta}_i^A(\omega_i)$ in the neighborhood of $A$. Thus, nearby trajectories originating from these $\theta_i$'s would also be attracted to the point $A$ during evolution. As the neighboring angles from both sides of $A$ are being attracted to it, the point $A$ becomes a stable fixed point. Let us now consider the case of the fixed point $B$ at which $\mathrm{d}\theta_i/\mathrm{d}t$ changes its sign from negative to positive. By similar arguments as above, one may see that all the angles in the neighborhood of $B$ would move away from it as time goes on. Therefore, $B$ is an unstable fixed point. In the particular case of $\omega_i = Kr_{\rm st}$, the fixed points $A$ and $B$ coalesce at $\theta_i = \pi/2$, as shown in Fig. 5.12b. Here, $\mathrm{d}\theta_i/\mathrm{d}t$ is positive on both sides of $\theta_i = \pi/2$, thus making the point stable from the left side and unstable from the right side. Hence, this point becomes a half-stable fixed point. In fact, the number of oscillators with such a fixed point is so small (since they have to satisfy a very stringent condition: $|\omega_i| = Kr_{\rm st}$ ) that they can be neglected in the limit $N \to \infty$. Depending on the values of the parameters $\omega_i, K, r_{\rm st}$, we see from Fig. 5.12 that (i) there is no fixed point for $|\omega_i| > Kr_{\rm st}$, (ii) a half-stable fixed point is born at $\theta_i = \pi/2$ for $|\omega_i| = Kr_{\rm st}$, and (iii) the half-stable fixed point splits into one stable and one unstable fixed point for $|\omega_i| < Kr_{\rm st}$. Therefore, here we have a saddle-node bifurcation at $\theta_i = \pi/2$. By arguing along the same lines as above, one may ascertain the nature of the fixed points and conclude about the existence of a saddle-node bifurcation at $\theta_i = 3\pi/2$, for the case $\omega_i < 0$.

To characterize the dynamics (5.59) in the limit $N \to \infty$, we define a single-oscillator distribution function $\rho(\theta, \omega, t)$, such that $\rho(\theta, \omega, t)\mathrm{d}\theta$ gives among the oscillators with frequency $\omega$ the fraction that have angles between $\theta$ and $\theta + \mathrm{d}\theta$ at time $t$. The distribution is $2\pi$-periodic, i.e., $\rho(\theta, \omega, t) = \rho(\theta + 2\pi, \omega, t)$, and satisfies the normalization $\int_0^{2\pi} \mathrm{d}\theta\ \rho(\theta, \omega, t) = 1\ \forall\ \omega$ and $t$. In the limit $N \to \infty$, Eq. (5.56) reads

$$r(t)\mathrm{e}^{\mathrm{i}\psi(t)} = \int_{-\infty}^{\infty} \mathrm{d}\omega\ g(\omega) \int_0^{2\pi} \mathrm{d}\theta\ \mathrm{e}^{\mathrm{i}\theta} \rho(\theta, \omega, t)\,. \tag{5.64}$$

The dynamics (5.59) conserves the number of oscillators with a given frequency $\omega$, implying thereby a continuity equation for the time evolution of $\rho(\theta, \omega, t)$:

$$\frac{\partial \rho(\theta, \omega, t)}{\partial t} = -\frac{\partial J(\theta, \omega, t)}{\partial \theta}, \tag{5.65}$$

where $J(\theta, \omega, t)$ is the probability current. Using the flow velocity of $\theta$, the probability current may be written as

$$J(\theta, \omega, t) = \rho(\theta, \omega, t) \frac{\mathrm{d}\theta}{\mathrm{d}t} = \left[\omega - Kr(t) \sin\left(\theta - \psi(t)\right)\right] \rho(\theta, \omega, t). \tag{5.66}$$

In obtaining the second equality, we have used Eq. (5.59). Then, the continuity Eq. (5.65) reduces to

$$\frac{\partial \rho(\theta, \omega, t)}{\partial t} = -\frac{\partial}{\partial \theta}\Big[\Big\{\omega - Kr(t) \sin\left(\theta - \psi(t)\right)\Big\} \rho(\theta, \omega, t)\Big]. \tag{5.67}$$

In general, it does not prove easy to solve the partial differential Eq. (5.67) for arbitrary initial conditions. Let us look for the stationary solutions. In the stationary state, $\rho(\theta, \omega, t)$ settles into a time-independent form $\rho_{\mathrm{st}}(\theta, \omega)$. The latter quantity solves the equation

$$\frac{\partial}{\partial \theta}\Big[\Big(\omega - Kr_{\mathrm{st}} \sin(\theta - \psi_{\mathrm{st}})\Big) \rho_{\mathrm{st}}(\theta, \omega; r_{\mathrm{st}})\Big] = 0. \tag{5.68}$$

From Eq. (5.68), it is evident that the form of $\rho_{\mathrm{st}}$ depends explicitly on $r_{\mathrm{st}}$, which by virtue of Eq. (5.64) satisfies the self-consistent equation

$$r_{\mathrm{st}} e^{\mathrm{i}\psi_{\mathrm{st}}} = \int_{-\infty}^{\infty} \mathrm{d}\omega\, g(\omega) \int_0^{2\pi} \mathrm{d}\theta\, e^{\mathrm{i}\theta} \rho_{\mathrm{st}}(\theta, \omega; r_{\mathrm{st}}), \tag{5.69}$$

where we have introduced $r_{\mathrm{st}}$ as an argument to $\rho_{\mathrm{st}}$ to show the explicit dependence of the latter on the former. From Eq. (5.66), we see that the probability current in the stationary state is given by $J_{\mathrm{st}}(\theta, \omega) = (\omega - Kr_{\mathrm{st}} \sin(\theta - \psi_{\mathrm{st}})) \rho_{\mathrm{st}}(\theta, \omega; r_{\mathrm{st}})$. However, Eq. (5.68) suggests that the stationary current does not depend on $\theta$. Considering this fact, one obtains the probability current in the stationary state as

$$J_{\mathrm{st}}(\omega) = (\omega - Kr_{\mathrm{st}} \sin(\theta - \psi_{\mathrm{st}})) \rho_{\mathrm{st}}(\theta, \omega; r_{\mathrm{st}}). \tag{5.70}$$

As was done following Eq. (5.59), we now set $\psi_{\mathrm{st}} = 0$ to obtain from Eqs. (5.69) and (5.70) that

$$r_{\mathrm{st}} = \int_{-\infty}^{\infty} \mathrm{d}\omega\, g(\omega) \int_0^{2\pi} \mathrm{d}\theta\, e^{\mathrm{i}\theta} \rho_{\mathrm{st}}(\theta, \omega; r_{\mathrm{st}}), \tag{5.71}$$

$$J_{\mathrm{st}}(\omega) = (\omega - Kr_{\mathrm{st}} \sin\theta) \rho_{\mathrm{st}}(\theta, \omega; r_{\mathrm{st}}). \tag{5.72}$$

The stationary distribution $\rho_{\rm st}(\theta, \omega; r_{\rm st})$ would have two contributions owing to the drifting and the phase-locked oscillators. We denote these two contributions as $\rho_{\rm st}^{\rm drift}(\theta, \omega; r_{\rm st})$ and $\rho_{\rm st}^{\rm lock}(\theta, \omega; r_{\rm st})$, respectively. Then, Eq. (5.71) becomes

$$r_{\rm st} = \int_0^{2\pi} {\rm d}\theta \left[ \int_{|\omega|>Kr_{\rm st}} {\rm d}\omega\, g(\omega)\, {\rm e}^{{\rm i}\theta} \rho_{\rm st}^{\rm drift}(\theta, \omega; r_{\rm st}) + \int_{|\omega|<Kr_{\rm st}} {\rm d}\omega\, g(\omega)\, {\rm e}^{{\rm i}\theta} \rho_{\rm st}^{\rm lock}(\theta, \omega; r_{\rm st}) \right]. \quad (5.73)$$

For the phase-locked oscillators with $|\omega| < Kr_{\rm st}$, we have already seen that there exists a stable fixed point $\overline{\theta}_i^A(\omega) = \sin^{-1}[\omega/(Kr_{\rm st})]$; $\cos\overline{\theta}_i^A = +\sqrt{1 - \omega_i^2/(Kr_{\rm st})^2}$. Therefore, in the stationary state, the distribution for these oscillators is given by [4]

$$\rho_{\rm st}^{\rm lock}(\theta, \omega; r_{\rm st}) = Kr_{\rm st} \cos\theta\, \delta(\omega - Kr_{\rm st}\sin\theta)\, \Theta(\cos\theta)\,; \quad |\omega| < Kr_{\rm st}\,, \quad (5.74)$$

where $\Theta(\cdot)$ is the Heaviside step function standing for the fact that the stable fixed point has $\cos\theta > 0$. In Eq. (5.74), the prefactor $Kr_{\rm st}\cos\theta$ takes care of the normalization of $\rho_{\rm st}^{\rm lock}(\theta, \omega; r_{\rm st})$. Substituting Eq. (5.74) in Eq. (5.72), one obtains the stationary probability current for the phase-locked oscillators as

$$J_{\rm st}^{\rm lock}(\omega) = 0\,; \quad |\omega| < Kr_{\rm st}\,. \quad (5.75)$$

For the drifting oscillators with $|\omega| > Kr_{\rm st}$, the stationary distribution may be obtained from Eq. (5.72), as [4, 11]

$$\rho_{\rm st}^{\rm drift}(\theta, \omega; r_{\rm st}) = \frac{J_{\rm st}^{\rm drift}(\omega)}{|\omega - Kr_{\rm st}\sin\theta|}\,; \quad |\omega| > Kr_{\rm st}\,, \quad (5.76)$$

where the modulus ensures that the probability is positive, as required. The normalization condition $\int_0^{2\pi} {\rm d}\theta\, \rho_{\rm st}^{\rm drift}(\theta, \omega; r_{\rm st}) = 1$ may be used to obtain the probability current in the stationary state for these oscillators. Then, one obtains

$$J_{\rm st}^{\rm drift}(\omega) = \frac{1}{2\pi}\sqrt{\omega^2 - K^2 r_{\rm st}^2}\,; \quad |\omega| > Kr_{\rm st}\,. \quad (5.77)$$

That Eq. (5.76) is indeed the correct stationary-state distribution for the drifting oscillators may be checked by comparing it with results obtained from numerical integration of the dynamics (5.59). In Fig. 5.14, we have made such a comparison for three different values of $\omega$ such that $|\omega| > Kr_{\rm st}$. The figure shows a very good match between theory and numerical integration results, thereby implying the correctness of the form (5.76).

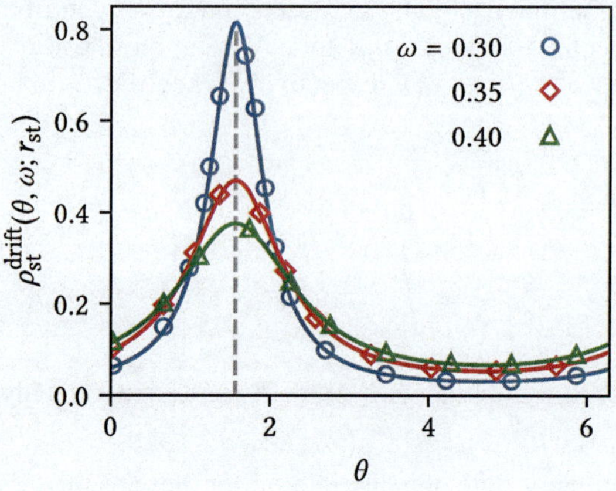


**Fig. 5.14 Stationary-state distribution for the drifting oscillators in the Kuramoto model** (5.59). The figure shows $\rho_{\rm st}^{\rm drift}(\theta, \omega; r_{\rm st})$ as a function of $\theta$ for three different values of $\omega > 0$, such that $|\omega| > K r_{\rm st}$. As implied by Fig. 5.12c, the most probable value for the angles is seen to be $\pi/2$, which is denoted by the dashed line in gray. The continuous lines are obtained by using the analytical form, Eqs. (5.76) and (5.77), whereas the points are obtained by numerically integrating the dynamics (5.59) by using the Euler method discussed in Appendix M.2 with $\Delta t = 0.01$, $N = 2000$, and $K = 0.4$. The frequencies $\omega_i$ are sampled from a Gaussian distribution with zero mean and standard deviation equal to 0.2. The initial $\theta_i$'s are taken to be distributed independently and uniformly in the range $[0, 2\pi)$. We first let the system evolve until it reaches the stationary state. In the stationary state, we obtain $r_{\rm st} \approx 0.695$. Then, we start the dynamics all over again from the initial state, and monitor the motion of three tagged oscillators with frequencies set at 0.30, 0.35, and 0.40, so that $|\omega| > K r_{\rm st}$ is satisfied for them. After the system reaches the stationary state, we check that the condition $|\omega| > K r_{\rm st}$ is satisfied for these three oscillators. Then, we recorded the angles of these three oscillators to obtain the displayed probability distribution. We see a very good agreement between theory and numerical results

## 5.6.4 *Synchronization Phase Transition*

Now that we know the forms of $\rho_{\rm st}^{\rm drift}(\theta, \omega; r_{\rm st})$ and $\rho_{\rm st}^{\rm lock}(\theta, \omega; r_{\rm st})$, Eqs. (5.76) and (5.74), respectively, let us evaluate the first integral on the right hand side of Eq. (5.73). Denoting the mentioned integral by $I_1$, we have

$$\begin{aligned} I_1 &\equiv \int_0^{2\pi} \mathrm{d}\theta \int_{|\omega|>K r_{\rm st}} \mathrm{d}\omega\, g(\omega)\, \mathrm{e}^{\mathrm{i}\theta} \rho_{\rm st}^{\rm drift}(\theta, \omega; r_{\rm st}) \\ &= \int_0^{2\pi} \mathrm{d}\theta\, \mathrm{e}^{\mathrm{i}\theta} \left[ \int_{-\infty}^{-K r_{\rm st}} \mathrm{d}\omega\, g(\omega)\, \rho_{\rm st}^{\rm drift}(\theta, \omega; r_{\rm st}) + \int_{K r_{\rm st}}^{\infty} \mathrm{d}\omega\, g(\omega)\, \rho_{\rm st}^{\rm drift}(\theta, \omega; r_{\rm st}) \right]. \end{aligned} \tag{5.78}$$

By changing $\omega$ to $-\omega$ and using $g(-\omega) = g(\omega)$ in the first integral on the right hand side of Eq. (5.78), we get

$$I_1 = \int_0^{2\pi} \mathrm{d}\theta\, \mathrm{e}^{\mathrm{i}\theta} \left[ \int_{Kr_{\mathrm{st}}}^{\infty} \mathrm{d}\omega\, g(\omega)\, \rho_{\mathrm{st}}^{\mathrm{drift}}(\theta, -\omega; r_{\mathrm{st}}) + \int_{Kr_{\mathrm{st}}}^{\infty} \mathrm{d}\omega\, g(\omega)\, \rho_{\mathrm{st}}^{\mathrm{drift}}(\theta, \omega; r_{\mathrm{st}}) \right]. \tag{5.79}$$

Using a new integration variable $\theta'$, such that $\theta' \equiv \theta - \pi$, in the first term on the right hand side of Eq. (5.79), one obtains

$$\begin{aligned} I_1 = &- \int_{-\pi}^{\pi} \mathrm{d}\theta' \int_{Kr_{\mathrm{st}}}^{\infty} \mathrm{d}\omega\, g(\omega)\, \mathrm{e}^{\mathrm{i}\theta'} \rho_{\mathrm{st}}^{\mathrm{drift}}(\theta' + \pi, -\omega; r_{\mathrm{st}}) \\ &+ \int_0^{2\pi} \mathrm{d}\theta \int_{Kr_{\mathrm{st}}}^{\infty} \mathrm{d}\omega\, g(\omega)\, \mathrm{e}^{\mathrm{i}\theta} \rho_{\mathrm{st}}^{\mathrm{drift}}(\theta, \omega; r_{\mathrm{st}}). \end{aligned} \tag{5.80}$$

From Eq. (5.76), it is seen that there exists a symmetry relation for the stationary-state distribution of the drifting oscillators, namely, $\rho_{\mathrm{st}}^{\mathrm{drift}}(\theta + \pi, -\omega; r_{\mathrm{st}}) = \rho_{\mathrm{st}}^{\mathrm{drift}}(\theta, \omega; r_{\mathrm{st}})$. Using this relation in Eq. (5.80), we obtain

$$\begin{aligned} I_1 &= \int_{Kr_{\mathrm{st}}}^{\infty} \mathrm{d}\omega\, g(\omega) \left[ \int_0^{2\pi} \mathrm{d}\theta - \int_{-\pi}^{\pi} \mathrm{d}\theta \right] \mathrm{e}^{\mathrm{i}\theta} \rho_{\mathrm{st}}^{\mathrm{drift}}(\theta, \omega; r_{\mathrm{st}}) \\ &= \int_{Kr_{\mathrm{st}}}^{\infty} \mathrm{d}\omega\, g(\omega) \left[ \int_{\pi}^{2\pi} \mathrm{d}\theta - \int_{-\pi}^{0} \mathrm{d}\theta \right] \mathrm{e}^{\mathrm{i}\theta} \rho_{\mathrm{st}}^{\mathrm{drift}}(\theta, \omega; r_{\mathrm{st}}). \end{aligned} \tag{5.81}$$

The integrand in Eq. (5.81) is $2\pi$-periodic in $\theta$. Using this fact, Eq. (5.81) may be reduced to

$$I_1 = \int_{Kr_{\mathrm{st}}}^{\infty} \mathrm{d}\omega\, g(\omega) \left[ \int_{\pi}^{2\pi} \mathrm{d}\theta - \int_{\pi}^{2\pi} \mathrm{d}\theta \right] \mathrm{e}^{\mathrm{i}\theta} \rho_{\mathrm{st}}^{\mathrm{drift}}(\theta, \omega; r_{\mathrm{st}}) = 0. \tag{5.82}$$

Thus, the first integral on the right hand side of Eq. (5.73) vanishes. Since $I_1$ gives the contribution to $r_{\mathrm{st}}$ from the drifting oscillators, it is physically meaningful that $I_1$ evaluates to zero: (1) when there are only drifting oscillators, we have $r_{\mathrm{st}} = 0$, and consistently with Eq. (5.73), we have both $I_1$ and the second integral on the right hand side vanishing; (2) for $r_{\mathrm{st}} \neq 0$, one has $I_1$ still equal to zero, and Eq. (5.73) shows that as expected, only the phase-locked oscillators contribute to $r_{\mathrm{st}}$.

Now, substituting the stationary distribution for the phase-locked oscillators (5.74) in Eq. (5.73), we obtain

$$
\begin{aligned}
r_{\rm st} &= \int_0^{2\pi} {\rm d}\theta \int_{|\omega|<Kr_{\rm st}} {\rm d}\omega \; g(\omega)\, {\rm e}^{{\rm i}\theta}\; Kr_{\rm st}\cos\theta\; \delta(\omega - Kr_{\rm st}\sin\theta)\; \Theta(\cos\theta) \\
&= Kr_{\rm st}\int_0^{2\pi} {\rm d}\theta\; g(Kr_{\rm st}\sin\theta)\, {\rm e}^{{\rm i}\theta} \cos\theta\; \Theta(\cos\theta) \\
&= Kr_{\rm st}\left[\int_0^{\frac{\pi}{2}} {\rm d}\theta\; g(Kr_{\rm st}\sin\theta)\, {\rm e}^{{\rm i}\theta}\cos\theta + \int_{\frac{3\pi}{2}}^{2\pi} {\rm d}\theta\; g(Kr_{\rm st}\sin\theta)\, {\rm e}^{{\rm i}\theta}\cos\theta\right].
\end{aligned}
\tag{5.83}
$$

By changing $\theta$ to $(\theta - 2\pi)$ in the second integral on the right hand side of Eq. (5.83), one obtains

$$
r_{\rm st} = Kr_{\rm st}\int_{-\pi/2}^{\pi/2} {\rm d}\theta\; g(Kr_{\rm st}\sin\theta)\; \cos\theta\, {\rm e}^{{\rm i}\theta}\,. \tag{5.84}
$$

As the distribution $g(\omega)$ is symmetric about zero, the imaginary part of the integral on the right hand side of Eq. (5.84) vanishes. Then, the self-consistent Eq. (5.84) becomes

$$
r_{\rm st} = Kr_{\rm st}\int_{-\pi/2}^{\pi/2} {\rm d}\theta\; g(Kr_{\rm st}\sin\theta)\; \cos^2\theta\,. \tag{5.85}
$$

There is a trivial solution to Eq. (5.85), i.e., $r_{\rm st} = 0$, which refers to the incoherent phase of the system. Any nonzero solution of $r_{\rm st}$ warrants complete or partial synchrony. Therefore, considering Eq. (5.85) in the limit $r_{\rm st} \to 0^+$, one may obtain the critical value of the coupling constant $K_c$ to be satisfying

$$
1 = K_c\int_{-\pi/2}^{\pi/2} {\rm d}\theta\; g(0)\; \cos^2\theta + \mathcal{O}(r_{\rm st}^2)\,. \tag{5.86}
$$

To leading order in $r_{\rm st}$, the solution to Eq. (5.86) reads [4, 10, 11]

$$
K_c = \frac{2}{\pi g(0)}\,. \tag{5.87}
$$

With increase of the coupling strength $K$, more oscillators satisfy the condition $|\omega_i| < Kr_{\rm st}$, which increases the number of oscillators in the phase-locked group. Hence, $r_{\rm st}$ also increases, which further increases the number of phase-locked oscillators (see Ref. [13] for a mathematical proof). Variation of $r_{\rm st}$ with $K$ is shown in Fig. 5.15. In the limit $K \to \infty$, one has $r_{\rm st}$ approaching unity. It may be noted that $r_{\rm st} = 0$ is always a solution to Eq. (5.85) whatever be the value of $K$. In fact, the incoherent solution always exists, but for $K > K_c$, it is rendered unstable. The stable stationary solution corresponding to the synchronized phase bifurcates from the unstable solution at $K = K_c$.

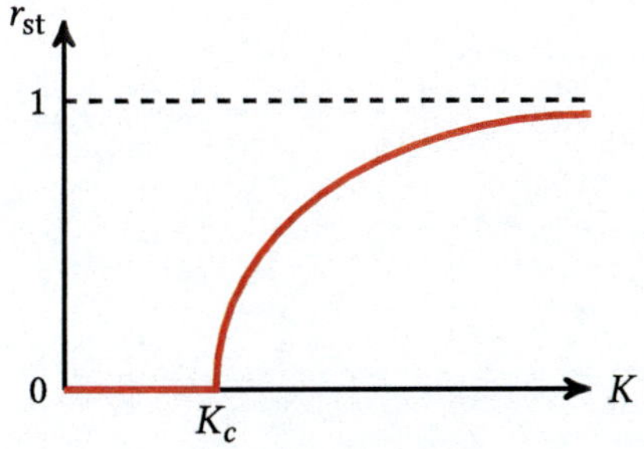


**Fig. 5.15** Schematic diagram to show the dependence of the stationary-state order parameter $r_{st}$ on the coupling constant $K$ in the Kuramoto dynamics (5.60)

# References


1. S.H. Strogatz, *Nonlinear Dynamics and Chaos* (CRC Press, Boca Raton, 2018). https://doi.org/10.1201/9780429492563
2. M. Lakshmanan, S. Rajasekar, *Nonlinear Dynamics* (Springer, Berlin, 2003). https://doi.org/10.1007/978-3-642-55688-3
3. A. Pikovsky, M. Rosenblum, J. Kurths, *Synchronization: A Universal Concept in Nonlinear Sciences* (Cambridge University Press, Cambridge, 2001)
4. S. Gupta, A. Campa, S. Ruffo, *Statistical Physics of Synchronization* (Springer, Cham, 2018). https://doi.org/10.1007/978-3-319-96664-9
5. M. Bier, B.M. Bakker, H.V. Westerhoff, Biophys. J. **78**(3), 1087 (2000). https://doi.org/10.1016/S0006-3495(00)76667-7
6. J. Buck, Q. Rev. Biol. **63**(3), 265 (1988). https://doi.org/10.1086/415929
7. K. Wiesenfeld, P. Colet, S. Strogatz, Phys. Rev. E **57**(2), 1563 (1998). https://doi.org/10.1103/PhysRevE.57.1563
8. K. Hirosawa, S. Kittaka, Y. Oishi, F. Kannari, T. Yanagisawa, Opt. Express **21**(21), 24952 (2013). https://doi.org/10.1364/OE.21.024952
9. M. Rohden, A. Sorge, M. Timme, D. Witthaut, Phys. Rev. Lett. **109**(6), 064101 (2012). https://doi.org/10.1103/PhysRevLett.109.064101
10. Y. Kuramoto, *Chemical Oscillations, Waves, and Turbulence* (Springer, Berlin, 1984). https://doi.org/10.1007/978-3-642-69689-3
11. S.H. Strogatz, Phys. D: Nonlinear Phenom. **143**(1–4), 1 (2000). https://doi.org/10.1016/S0167-2789(00)00094-4
12. J.A. Acebrón, L.L. Bonilla, C.J. Pérez Vicente, F. Ritort, R. Spigler, Rev. Mod. Phys. **77**(1), 137 (2005). https://doi.org/10.1103/RevModPhys.77.137
13. S. Gupta, A. Campa, S. Ruffo, J. Stat. Mech. **2014**(8), R08001 (2014). https://doi.org/10.1088/1742-5468/14/08/R08001

# Part II
# Applications

# Chapter 6
# Stationary Correlations in the Noisy Kuramoto Model

**Abstract** In this chapter, we study the facet of noise arising in the dynamics of a physical system due to its interaction with its environment. We consider the system of Kuramoto oscillators evolving in presence of Gaussian, white noise. Introducing noise into the bare Kuramoto dynamics models its interaction with the environment. For this noisy Kuramoto model, we obtain exact results on autocorrelation of the order parameter in the nonequilibrium stationary state that the dynamics attains at long times. In obtaining the results, we employ a method based on an exact mapping of the stationary-state dynamics of the model in the thermodynamic limit to the noisy dynamics of a single, nonuniform oscillator. The method also allows to obtain the autocorrelation in the equilibrium stationary state of a related model of long-range interactions, the Brownian mean-field model. Both the models exhibit a phase transition between a synchronized and an incoherent phase at a critical value of the noise strength. We show that in the two phases as well as at the critical point, the autocorrelation for both the model decays as an exponential with a rate that increases continuously with the noise strength. This chapter is based on our published work, Ref. [1].

## 6.1 Introduction

One of the primary objectives of statistical mechanics is to characterize the stationary state of a many-body interacting system evolving according to a given dynamics [2]. Such computation often turns out to be formidable due to the many-body nature of the dynamics, and furthermore when the stationary state is out of equilibrium [3]. Indeed, the phase-space distribution in an equilibrium stationary state is given unequivocally by the Gibbs–Boltzmann weight independent of the underlying dynamics, while that in a nonequilibrium stationary state does not have a universal form but has to be obtained from an explicit consideration of the dynamics. The phase-space distribution is a characterization of a one-time snapshot of the possible values of the dynamical variables in the stationary state, while it is obviously of interest to ask about how do the values at one time relate to those at another time. The autocorrelation function provides a measure of similarity of the values of dynamical variables at two different

 
D. Das and S. Gupta, *Facets of Noise*, Fundamental Theories of Physics 214,
https://doi.org/10.1007/978-3-031-45312-0_6

times as a function of the time lag between them. Autocorrelations in equilibrium may be deduced from the response of the system to small external perturbations by invoking the framework of the linear response theory [4]. On the other hand, there is as yet no general procedure that allows for computation of autocorrelations in generic nonequilibrium stationary states in a similar manner. This warrants the need to study model nonequilibrium systems for which explicit formulas may be derived for autocorrelation. In this backdrop, we obtain in this chapter exact results on the autocorrelation of a paradigmatic model showing spontaneous order in a nonequilibrium stationary state, the Kuramoto model. We have already encountered the model in Sect. 5.6.1, and here we study a noisy variant of the model. The method we employ is based on an exact mapping of the stationary-state dynamics of the model in the thermodynamic limit to the noisy dynamics of a single, nonuniform oscillator. Remarkably, the method also allows to obtain the autocorrelation in the equilibrium stationary state of a related model of long-range interactions.

The chapter is organized as follows. In Sect. 6.2, we provide a brief reminder of the Kuramoto model. In the following section, we discuss in detail the noisy Kuramoto model, and define our quantity of interest, namely, the stationary correlations. The analysis of the single noisy nonuniform oscillator pursued in Sect. 6.4 is then used in Sect. 6.5 to obtain the core results of the chapter, namely, stationary correlations of the noisy Kuramoto model. Conclusions and perspectives are summarized in Sect. 6.6.

## 6.2 A Brief Reminder of the Kuramoto Model

In this chapter, we study a nontrivial variant of the Kuramoto model introduced and discussed in Sect. 5.6.1. Before we embark on a discussion of this variant, it would prove convenient to recall briefly the model. The Kuramoto model comprises $N$ phase-only oscillators that are globally coupled through the sine of their phase differences. The phase $\theta_i \in [0, 2\pi)$; $i = 1, 2, \ldots, N$ evolves in time according to the dynamics

$$\frac{\mathrm{d}\theta_i}{\mathrm{d}t} = \omega_i - \frac{K}{N}\sum_{j=1}^{N} \sin(\theta_i - \theta_j)\,; \quad i = 1, 2, \ldots, N, \tag{6.1}$$

where $\omega_i$ is the natural frequency of the $i$th oscillator. The $\omega_i$'s are quenched-disordered random variables sampled from a common distribution $\overline{g}(\omega)$ that is taken to be unimodal, with mean $\omega_0 \equiv \int \mathrm{d}\omega\, \omega\, \overline{g}(\omega)$. One may scale out the effect of the mean $\omega_0$ of the natural frequencies from the dynamics (6.1) by viewing the latter in a frame rotating uniformly with frequency $\omega_0$ with respect to the laboratory frame. Effecting such a transformation, we will from now on consider $\omega_i$'s to be sampled from a common distribution $g(\omega)$ with zero mean. The degree of ordering between the phases of the oscillators is effectively characterized by the order parameter

$$r(t) \equiv \frac{1}{N}\left|\sum_{j=1}^{N} \mathrm{e}^{\mathrm{i}\theta_j(t)}\right|, \tag{6.2}$$

such that $r(t) = 0$ (respectively, $0 < r(t) \leq 1$) corresponds to the incoherent (respectively, synchronized) phase. In the limit $N \to \infty$, the system (6.1) in the stationary state undergoes a phase transition in terms of the stationary value $r_{\mathrm{st}}$ of the order parameter: as one tunes $K$, one observes a phase transition from an incoherent phase ($r_{\mathrm{st}} = 0$) for $K < K_c$ to a synchronized phase ($0 < r_{\mathrm{st}} \leq 1$) for $K > K_c$. Here, the critical $K_c$ is given by

$$K_c = \frac{2}{\pi g(0)} \, . \tag{6.3}$$

Armed with the above background, we now discuss a variant of the model (6.1), in which the dynamics involves evolution in presence of noise due to interaction with the environment.

## 6.3 The Noisy Kuramoto Model

The natural frequencies of the Kuramoto oscillators may in practice be not time independent but undergoing stochastic fluctuations in time as an effect of interaction with the environment. The effect of these fluctuations on the dynamics of the oscillators was studied by Hidetsugu Sakaguchi [5]. In this case, the Kuramoto dynamics (5.51) is replaced by a set of $N$ coupled overdamped Langevin equations of the form

$$\frac{\mathrm{d}\theta_i}{\mathrm{d}t} = \omega_i - \frac{K}{N}\sum_{j=1}^{N} \sin(\theta_i - \theta_j) + \eta_i(t)\,; \qquad i = 1, 2, \ldots, N. \tag{6.4}$$

We remark that in the spirit of the so-called Zwanzig approach [6], the above set of Langevin equations may be obtained by considering bilinear interactions between the Kuramoto oscillators and a harmonic oscillator heat bath, and finally, taking the overdamped limit [7]. The quantity $\eta_i(t)$ in Eq. (6.4) models the aforementioned stochastic fluctuations. Moreover, it is taken to be a Gaussian, white noise, independent for every oscillator, and satisfying

$$\langle \eta_i(t) \rangle = 0, \quad \langle \eta_i(t)\eta_j(t') \rangle = 2\,D\,\delta_{i,j}\,\delta(t - t')\,, \tag{6.5}$$

with $D > 0$ setting the noise strength, while here, the angular brackets denote averaging with respect to noise realizations. The Gaussian, white noise $\eta_i(t)$ may also be interpreted as the effect of interaction between the Kuramoto oscillators and the environment. In terms of the Kuramoto order parameter (5.56), Eq. (6.4) may be written as

$$\frac{\mathrm{d}\theta_i}{\mathrm{d}t} = \omega_i - Kr(t)\sin\big(\theta_i - \psi(t)\big) + \eta_i(t)\,; \qquad i = 1, 2, \ldots, N. \tag{6.6}$$

In the case of noisy Kuramoto dynamics (6.6), there are two types of randomness:

1. We may consider the sum $\omega_i + \eta_i(t)$ to be representing the time-dependent natural frequencies of the oscillators; these frequencies are changing randomly in time due to the presence of the Gaussian, white noise $\eta_i(t)$. The latter being time dependent may be interpreted as annealed disorder, while the constant $\omega_i$ may be considered as the time-independent average frequency of the $i$th oscillator.
2. As we have discussed earlier, the $\omega_i$'s themselves are changing from one oscillator to another as their values are taken to be sampled from the distribution $g(\omega)$, and that this type of time-independent randomness is called quenched disorder. Indeed, once the frequencies have been assigned to the different oscillators, their values do not change during the dynamical evolution.

In the particular case, when the frequency term on the right hand side of the dynamics (6.4) is absent (i.e., $\omega_i = 0\ \forall\ i$), or, when all the oscillators have the same frequency equal to $\overline{\omega}$ (that is, $g(\omega) = \delta(\omega - \overline{\omega})$) and one observes the dynamics (6.4) in a frame rotating uniformly with frequency $\overline{\omega}$ with respect to the laboratory frame, the corresponding equations of motion are given by

$$\frac{\mathrm{d}\theta_i}{\mathrm{d}t} = -\frac{K}{N}\sum_{j=1}^{N}\sin(\theta_i - \theta_j) + \eta_i(t)\,; \qquad i = 1, 2, \ldots, N. \tag{6.7}$$

Equation (6.7) has the same form as that of the so-called Brownian mean-field (BMF) model [8]. The model mimics the canonical ensemble dynamics of a paradigmatic model of long-range interactions,[1] namely, the Hamiltonian mean-field (HMF) model [9] in the overdamped limit. The HMF model comprises $N$ particles moving on a unit circle and interacting through a mean-field pair potential. The Hamiltonian of the model system is given by (see Appendix C)

$$H(\{p_i, \theta_i\}) = \mathcal{T} + \mathcal{V}, \tag{6.8}$$

$$\mathcal{T} \equiv \mathcal{T}(\{p_i\}) = \sum_{i=1}^{N}\frac{p_i^2}{2m}, \tag{6.9}$$

$$\mathcal{V} \equiv \mathcal{V}(\{\theta_i\}, \overline{K}) = \frac{\overline{K}}{2N}\sum_{i,j=1}^{N}\big[1 - \cos(\theta_i - \theta_j)\big], \tag{6.10}$$

[1] For detailed discussions on long-range interactions, see Chap. 3.

where $\theta_i \in [0, 2\pi)$ and $p_i \equiv m\,\mathrm{d}\theta_i/\mathrm{d}t$ are, respectively, the coordinate and the conjugate momentum of the $i$th particle, $m$ is the common mass, while $\overline{K} \geq 0$ is the coupling constant. In Eq. (6.8), $\mathcal{T}$ and $\mathcal{V}$ represent respectively the kinetic energy and the potential energy of the system. The equations of motion of the system are given by the Hamilton's equations corresponding to the Hamiltonian (6.8):

$$\frac{\mathrm{d}\theta_i}{\mathrm{d}t} = \frac{\partial H}{\partial p_i} = \frac{p_i}{m}, \tag{6.11}$$

$$\frac{\mathrm{d}p_i}{\mathrm{d}t} = -\frac{\partial H}{\partial \theta_i} = -\frac{\overline{K}}{N}\sum_{j=1}^{N}\sin(\theta_i - \theta_j). \tag{6.12}$$

In order to model the canonical ensemble dynamics of the HMF system when in contact with a heat bath at temperature $T$, one writes down a set of Langevin equations corresponding to the equations of motion (6.11) and (6.12), such that

$$\frac{\mathrm{d}\theta_i}{\mathrm{d}t} = \frac{p_i}{m}, \tag{6.13}$$

$$\frac{\mathrm{d}p_i}{\mathrm{d}t} = -\gamma\frac{p_i}{m} - \frac{\overline{K}}{N}\sum_{j=1}^{N}\sin(\theta_i - \theta_j) + \xi_i(t), \tag{6.14}$$

where the real constant $\gamma \geq 0$ is the friction coefficient, $\xi_i(t)$ is a Gaussian, white noise satisfying

$$\langle \xi_i(t)\rangle = 0, \quad \langle \xi_i(t)\xi_j(t')\rangle = 2\,\overline{D}\,\delta_{i,j}\,\delta(t-t'), \tag{6.15}$$

with $\overline{D} > 0$ setting the strength of $\xi_i(t)$. From Eqs. (6.13) and (6.14), one may obtain

$$\frac{m}{\gamma}\frac{\mathrm{d}^2\theta_i}{\mathrm{d}t} = -\frac{\mathrm{d}\theta_i}{\mathrm{d}t} - \frac{\overline{K}}{N\gamma}\sum_{j=1}^{N}\sin(\theta_i - \theta_j) + \frac{1}{\gamma}\xi_i(t). \tag{6.16}$$

Considering the dynamics (6.16) in the overdamped limit, i.e., $(m/\gamma) \to 0$, Eq. (6.16) may be reduced to

$$\frac{\mathrm{d}\theta_i}{\mathrm{d}t} = -\frac{\overline{K}}{N\gamma}\sum_{j=1}^{N}\sin(\theta_i - \theta_j) + \frac{1}{\gamma}\xi_i(t). \tag{6.17}$$

Redefining the variables as $K \equiv \overline{K}/\gamma$ and $\eta_i(t) \equiv \xi_i(t)/\gamma$, such that

$$\langle \eta_i(t)\rangle = \langle \xi_i(t)\rangle = 0, \tag{6.18}$$

$$\langle \eta_i(t)\eta_j(t')\rangle = \frac{1}{\gamma^2}\langle \xi_i(t)\xi_j(t')\rangle = \frac{2\overline{D}}{\gamma^2}\delta_{i,j}\delta(t-t') \equiv 2D\,\delta_{i,j}\delta(t-t'), \tag{6.19}$$

one obtains Eq. (6.7). In obtaining the last equality of Eq. (6.19), we have defined $D \equiv \overline{D}/\gamma^2$. Using the form of the potential energy (6.10), one may also write Eq. (6.7) as

$$\frac{\mathrm{d}\theta_i}{\mathrm{d}t} = -\frac{\partial}{\partial \theta_i}\Big[\mathcal{V}(\{\theta_j\}, K)\Big] + \eta_i(t)\,, \tag{6.20}$$

which suggests that Eq. (6.7) indeed exhibits the noisy overdamped dynamics of the HMF model, alternatively, canonical ensemble dynamics at temperature $T$ and in the overdamped limit. In passing, let us note down the dimensions of various quantities encountered in this part of the section. Since 'length scales' in the system are set by the angle variable $\theta$ that is dimensionless, we have

$$\big[p\big] = \big[\gamma\big] = \mathrm{M\,T^{-1}}\,, \quad \big[E\big] = \big[\overline{K}\big] = \big[\xi_i\big] = \mathrm{M\,T^{-2}}\,, \quad \big[\overline{D}\big] = \mathrm{M^2\,T^{-3}}\,, \tag{6.21}$$

$$\big[\omega\big] = \big[K\big] = \big[\eta_i\big] = \big[D\big] = \mathrm{T^{-1}}\,, \tag{6.22}$$

where $E$ is the energy: $E \equiv H(\{p_i, \theta_i\})$.

### 6.3.1 Stationary-State Distribution

In the limit $N \to \infty$, we may characterize the dynamics (6.6) in terms of the single-oscillator distribution function $\rho(\theta, \omega, t)$, as was done for the Kuramoto model in Sect. 5.6.3. For the Langevin dynamics with Gaussian, white noise, the time evolution of such a distribution function is given in terms of a partial differential equation, namely, the Fokker–Planck equation. In our case, for the Langevin dynamics (6.6), the corresponding Fokker–Planck equation may be derived by following standard procedures [5], and is given by

$$\frac{\partial \rho(\theta, \omega, t)}{\partial t} = -\frac{\partial}{\partial \theta}\Big[\Big\{\omega - Kr(t)\sin\big(\theta - \psi(t)\big)\Big\}\,\rho(\theta, \omega, t)\Big] + D\,\frac{\partial^2 \rho(\theta, \omega, t)}{\partial \theta^2}\,. \tag{6.23}$$

In the absence of noise, i.e., with $D = 0$, Eq. (6.23) reduces to the continuity Eq. (5.67) for the Kuramoto model. Equation (6.23) shows that $\rho(\theta, \omega, t)$ has an explicit dependence on $r(t)$, which is to be obtained by solving the self-consistent equation

$$r(t)\mathrm{e}^{\mathrm{i}\psi(t)} = \int_{-\infty}^{\infty} \mathrm{d}\omega\, g(\omega) \int_0^{2\pi} \mathrm{d}\theta\, \mathrm{e}^{\mathrm{i}\theta(t)}\rho(\theta, \omega, t; r)\,. \tag{6.24}$$

In the limit $t \to \infty$, the dynamics (6.6) settles into a stationary state.[2] Correspondingly, the quantities $r(t)$ and $\psi(t)$ assume the time-independent values $r_{\text{st}}$ and $\psi_{\text{st}}$, respectively, and the single-oscillator distribution $\rho(\theta, \omega, t; r)$ assumes a time-independent form $\rho_{\text{st}}(\theta, \omega; r_{\text{st}})$. In the stationary state, setting $\psi_{\text{st}} = 0$, we obtain the stationary-state dynamics as

$$\frac{\mathrm{d}\theta_i}{\mathrm{d}t} = \omega_i - K r_{\text{st}} \sin\theta_i + \eta_i(t)\,, \tag{6.25}$$

while the Fokker–Planck equation (6.23) reduces to

$$-\frac{\partial}{\partial\theta}\Big[\big(\omega - K r_{\text{st}} \sin\theta\big)\, \rho_{st}(\theta, \omega; r_{\text{st}})\Big] + D\frac{\partial^2 \rho_{\text{st}}(\theta, \omega; r_{\text{st}})}{\partial\theta^2} = 0\,, \tag{6.26}$$

thereby implying that

$$-D\frac{\partial \rho_{\text{st}}(\theta, \omega; r_{\text{st}})}{\partial\theta} + \big(\omega - K r_{\text{st}} \sin\theta\big)\, \rho_{\text{st}}(\theta, \omega; r_{\text{st}}) = J_{\text{st}}(\omega)\,. \tag{6.27}$$

Here, $J_{\text{st}}(\omega)$, which is independent of $\theta$, denotes the probability current in the stationary state. The first and the second term on the left hand side of Eq. (6.27) account for the contribution due to diffusion and drift, respectively.

For given values of $\omega$ and $r_{\text{st}}$, the first-order differential Eq. (6.27) may be solved by using the integrating factor

$$I(\theta) = \exp\left[\int_0^\theta \mathrm{d}\theta' \,\frac{K r_{\text{st}} \sin\theta' - \omega}{D}\right]. \tag{6.28}$$

Using the periodicity condition $\rho_{\text{st}}(0, \omega; r_{\text{st}}) = \rho_{\text{st}}(2\pi, \omega; r_{\text{st}})$, one obtains the solution for $\rho_{\text{st}}(\theta, \omega; r_{\text{st}})$ as [5]

$$\rho_{\text{st}}(\theta, \omega; r_{\text{st}}) = A(\omega; r_{\text{st}})\, \mathrm{e}^{\frac{-K r_{\text{st}} + K r_{\text{st}} \cos\theta + \omega\theta}{D}} \times \left[1 + \frac{(\mathrm{e}^{-\frac{2\pi\omega}{D}} - 1)\int_0^\theta \mathrm{d}\theta' \mathrm{e}^{(-\omega\theta' - K r_{\text{st}} \cos\theta')/D}}{\int_0^{2\pi} \mathrm{d}\theta' \mathrm{e}^{(-\omega\theta' - K r_{\text{st}} \cos\theta')/D}}\right], \tag{6.29}$$

where $A(\omega; r_{\text{st}})$ may be found from the normalization $\int_0^{2\pi} \mathrm{d}\theta\, \rho_{\text{st}}(\theta, \omega; r_{\text{st}}) = 1$.

From Eq. (6.29), by substituting $r_{\text{st}} = 0$, i.e., considering the incoherent stationary state, we obtain

$$\rho_{\text{st}}(\theta, \omega; r_{\text{st}} = 0) = A(\omega; r_{\text{st}} = 0)\,. \tag{6.30}$$

Then, using the normalization condition $\int_0^{2\pi} \mathrm{d}\theta\, \rho_{\text{st}}(\theta, \omega; r_{\text{st}} = 0) = 1$, we see from Eq. (6.30), that the angles distribute themselves uniformly over the unit circle:

[2] It is assumed that the dynamics (6.6) has a well-defined steady state. For that, one presumably needs the condition of ergodicity [10], which is also assumed to hold here.

$$\rho_{\rm st}(\theta, \omega; r_{\rm st} = 0) = A(\omega; r_{\rm st} = 0) = \frac{1}{2\pi}\ \forall\, \omega\,. \tag{6.31}$$

For the BMF model, i.e., with $\omega = 0$, one may find the stationary distribution from Eq. (6.29), as

$$\rho_{\rm st}(\theta, \omega = 0; r_{\rm st}) \propto \exp\left[(-Kr_{\rm st} + Kr_{\rm st}\cos\theta)/D\right]\,. \tag{6.32}$$

Substituting the distribution (6.32) in Eq. (6.27), one finds that the probability current $J_{\rm st}(0)$ is identically zero for all $\theta$. Therefore, one concludes that the dynamics of the BMF model settles into an equilibrium stationary state.[3] This is contrary to what happens in the case of the noisy Kuramoto model (6.4), where the dynamics relaxes to a nonequilibrium stationary state, as we show now. We define the $N$-oscillator distribution function $f_N(\{\theta_l\}, t)$, such that the quantity $f_N(\{\theta_l\}, t)\prod_{l=1}^{N} {\rm d}\theta_l$ gives the probability to observe the system around the state $\{\theta_l\}_{1\le l\le N}$ at time $t$. At any instant $t$, the distribution $f_N(\{\theta_l\}, t)$ has to satisfy two conditions:

Normalization: $$\int \left(\prod_{l=1}^{N} {\rm d}\theta_l\right) f_N(\{\theta_l\}, t) = 1\,, \tag{6.33}$$

Periodicity: $$f_N(\{\theta_l + 2\pi\}, t) = f_N(\{\theta_l\}, t) \quad \text{for any number of } \theta_l\text{'s}. \tag{6.34}$$

For the noisy Kuramoto dynamics (6.4), the time evolution of such a distribution is given by the $N$-dimensional Fokker–Planck equation:

$$\begin{aligned}
&\frac{\partial f_N(\{\theta_l\}, t)}{\partial t} \\
&= -\sum_{i=1}^{N} \frac{\partial}{\partial \theta_i}\left[\left(\omega_i - \frac{K}{N}\sum_{j=1}^{N}\sin(\theta_i - \theta_j)\right) f_N(\{\theta_l\}, t)\right] + D\sum_{i=1}^{N}\frac{\partial^2 f_N(\{\theta_l\}, t)}{\partial \theta_i^2} \\
&= -\sum_{i=1}^{N} \frac{\partial}{\partial \theta_i}\left[A_i(\{\theta_l\})\, f_N(\{\theta_l\}, t) - \frac{1}{2}\sum_{j=1}^{N}\frac{\partial}{\partial \theta_j}\left\{B_{ij}(\{\theta_l\})\, f_N(\{\theta_l\}, t)\right\}\right],
\end{aligned} \tag{6.35}$$

using the definitions:

Drift vector: $$A_i(\{\theta_l\}) \equiv \omega_i - \frac{K}{N}\sum_{j=1}^{N}\sin(\theta_i - \theta_j)\,, \tag{6.36}$$

Diffusion matrix: $$B_{ij}(\{\theta_l\}) \equiv 2D\delta_{i,j}\,. \tag{6.37}$$

Equation (6.35) may be written in the form of a continuity equation as

---

[3] Unlike a nonequilibrium stationary state, an equilibrium stationary state does not support any net probability current [11].

$$\frac{\partial f_N(\{\theta_l\},t)}{\partial t} = -\sum_{i=1}^{N} \frac{\partial J_i(\{\theta_l\})}{\partial \theta_i}, \tag{6.38}$$

where the $i$th component of the $N$-particle probability current is given by

$$J_i(\{\theta_l\},t) \equiv A_i(\{\theta_l\})\, f_N(\{\theta_l\},t) - \frac{1}{2}\sum_{j=1}^{N} \frac{\partial}{\partial \theta_j} \left\{ B_{ij}(\{\theta_l\})\, f_N(\{\theta_l\},t) \right\}. \tag{6.39}$$

Here, the first term on the right hand side corresponds to the drift current, while the second term gives the diffusion current. Denoting the time-independent stationary solution of the Fokker–Planck equation (6.35) as $f_N^{\mathrm{st}}(\{\theta_l\})$, and the stationary current as $J_i^{\mathrm{st}}(\{\theta_l\})$, one obtains from Eq. (6.39) in the stationary state that

$$J_i^{\mathrm{st}}(\{\theta_l\}) = A_i(\{\theta_l\})\, f_N^{\mathrm{st}}(\{\theta_l\}) - \frac{1}{2}\sum_{j=1}^{N} \frac{\partial}{\partial \theta_j} \left\{ B_{ij}(\{\theta_l\})\, f_N^{\mathrm{st}}(\{\theta_l\}) \right\}. \tag{6.40}$$

The probability current in an equilibrium stationary state always vanishes [11]. Therefore, considering such an equilibrium state, from Eqs. (6.40) and (6.37), we obtain

$$D\,\frac{\partial f_N^{\mathrm{st}}(\{\theta_l\})}{\partial \theta_i} = A_i(\{\theta_l\})\, f_N^{\mathrm{st}}(\{\theta_l\}), \tag{6.41}$$

solving which, we get

$$f_N^{\mathrm{st}}(\{\theta_l\}) = C\, \exp\left[ \sum_{i=1}^{N} \frac{\omega_i \theta_i}{D} + \frac{K}{2ND} \sum_{i,j=1}^{N} \cos(\theta_i - \theta_j) \right], \tag{6.42}$$

where $C$ is the normalization constant. Using the normalization condition (6.33), one may find the value of $C$, which will be independent of $\theta_l$'s. Now, we are in trouble. Note that the first term in the exponential in Eq. (6.42) does not respect the periodicity condition (6.34). We would not be able to fix this discrepancy unless $\omega_i = 0\ \forall\, i$. Therefore, to have an equilibrium stationary state, we need to necessarily set all the oscillator frequencies to zero; otherwise, one would have a nonequilibrium stationary state. Hence, for the noisy Kuramoto model (6.4), in which we cannot set all $\omega_i$'s to zero, the dynamics relaxes to a nonequilibrium stationary state. For the BMF model, in which case we have $\omega_i = 0\ \forall\, i$, the $N$-particle distribution function in the stationary state is obtained from Eq. (6.42) as

$$f_N^{\mathrm{st}}(\{\theta_l\}) \propto \exp\left[ \frac{K}{2ND} \sum_{i,j=1}^{N} \cos(\theta_i - \theta_j) \right] \propto \exp\Big[ -\mathcal{V}(\{\theta_i\},K)/D \Big], \tag{6.43}$$

where in the last step, we have used Eq. (6.10) for the definition of $\mathcal{V}$. On the other hand, noting that the BMF dynamics (6.20) is nothing but the canonical ensemble dynamics at temperature $T$, one would write down the $N$-particle distribution function in the stationary state as the canonical equilibrium distribution

$$f_N^{\rm st}(\{\theta_l\}) \propto \exp\Big[-\mathcal{V}(\{\theta_i\}, K)/T\Big], \tag{6.44}$$

with the Boltzmann constant set to unity. Comparing Eqs. (6.43) and (6.44) allows to identify the parameter $D$ with the temperature $T$ of canonical equilibrium.

### 6.3.2 Synchronization Phase Transition: Critical Noise Strength

In the stationary state of the noisy Kuramoto dynamics (6.4), for given $g(\omega)$ and $K$ and on tuning $D$, one observes a continuous phase transition from a low-$D$ synchronized ($r_{\rm st} \neq 0$) to a high-$D$ incoherent ($r_{\rm st} = 0$) phase at the critical value $D_c$ [5]. Our aim here is to obtain an analytical expression for $D_c$. To this end, let us start with the self-consistent equation for $r_{\rm st}$, obtained from Eq. (6.24) and the discussions following that equation, as

$$r_{\rm st} = \int_{-\infty}^{\infty} {\rm d}\omega\ g(\omega) \int_0^{2\pi} {\rm d}\theta\ {\rm e}^{{\rm i}\theta} \rho_{\rm st}(\theta, \omega; r_{\rm st}), \tag{6.45}$$

where $\rho_{\rm st}(\theta, \omega; r_{\rm st})$ is given by Eq. (6.29). Denoting the imaginary part of the integral on the right hand side of Eq. (6.45) as $I_2$, one has

$$I_2 = \int_{-\infty}^{\infty} {\rm d}\omega\ g(\omega) \int_0^{2\pi} {\rm d}\theta\ \sin\theta\ \rho_{\rm st}(\theta, \omega; r_{\rm st}). \tag{6.46}$$

With the change $\omega$ to $-\omega$, one obtains from Eq. (6.46) that

$$I_2 = \int_{-\infty}^{\infty} {\rm d}\omega\ g(-\omega) \int_0^{2\pi} {\rm d}\theta\ \sin\theta\ \rho_{\rm st}(\theta, -\omega; r_{\rm st}). \tag{6.47}$$

Using $g(-\omega) = g(\omega)$ and changing $\theta$ to $-\theta$ in Eq. (6.47), we get

$$I_2 = \int_{-\infty}^{\infty} {\rm d}\omega\ g(\omega) \int_0^{-2\pi} {\rm d}\theta\ \sin\theta\ \rho_{\rm st}(-\theta, -\omega; r_{\rm st}). \tag{6.48}$$

From Eq. (6.26), it is seen that $\rho_{\rm st}(-\theta, -\omega; r_{\rm st}) = \rho_{\rm st}(\theta, \omega; r_{\rm st})$. Then, from Eq. (6.48), one obtains

$$I_2 = \int_{-\infty}^{\infty} \mathrm{d}\omega\, g(\omega) \int_0^{-2\pi} \mathrm{d}\theta\, \sin\theta\, \rho_{\rm st}(\theta, \omega; r_{\rm st})\,, \tag{6.49}$$

which after using the periodicity condition $\rho_{\rm st}(\theta - 2\pi, \omega; r_{\rm st}) = \rho_{\rm st}(\theta, \omega; r_{\rm st})\ \forall\ \omega$ yields

$$I_2 = \int_{-\infty}^{\infty} \mathrm{d}\omega\, g(\omega) \int_{2\pi}^{0} \mathrm{d}\theta\, \sin\theta\, \rho_{\rm st}(\theta, \omega; r_{\rm st}) = -I_2\,. \tag{6.50}$$

Evidently, from Eq. (6.50), one obtains $I_2 = 0$. Thus, the imaginary part of the integral in Eq. (6.45) vanishes. Then, the self-consistent equation (6.45) reduces to

$$r_{\rm st} = \int_{-\infty}^{\infty} \mathrm{d}\omega\, g(\omega) \int_0^{2\pi} \mathrm{d}\theta\, \cos\theta\, \rho_{\rm st}(\theta, \omega; r_{\rm st})\,. \tag{6.51}$$

Substituting Eq. (6.31) in Eq. (6.51), one obtains the trivial solution of Eq. (6.51), namely, $r_{\rm st} = 0$, corresponding to the incoherent phase. In the synchronized phase, we will have a nonzero $r_{\rm st}$. Therefore, studying the behavior of Eq. (6.51) in the limit $r_{\rm st} \to 0^+$, we may obtain the critical value $D_c$ below which the system will be synchronized. Then, expanding $\rho_{\rm st}(\theta, \omega; r_{\rm st})$ about the point $r_{\rm st} = 0$ in Eq. (6.51), we obtain

$$r_{\rm st} = \int_{-\infty}^{\infty} \mathrm{d}\omega\, g(\omega) \int_0^{2\pi} \mathrm{d}\theta\, \cos\theta \left[\rho_{\rm st}(\theta, \omega; 0) + r_{\rm st} \left.\frac{\partial \rho_{\rm st}(\theta, \omega; r_{\rm st})}{\partial r_{\rm st}}\right|_{r_{\rm st}\to 0^+} + O(r_{\rm st}^2)\right]. \tag{6.52}$$

As we have already argued in obtaining the incoherent solution, the contribution to the integral from the first term within the brackets vanishes. Then, Eq. (6.52) reduces to

$$r_{\rm st} = \int_{-\infty}^{\infty} \mathrm{d}\omega\, g(\omega) \int_0^{2\pi} \mathrm{d}\theta\, \cos\theta \left[r_{\rm st} \left.\frac{\partial \rho_{\rm st}(\theta, \omega; r_{\rm st})}{\partial r_{\rm st}}\right|_{r_{\rm st}\to 0^+} + O(r_{\rm st}^2)\right]. \tag{6.53}$$

In the limit $r_{\rm st} \to 0^+$, using $A(\omega; r_{\rm st} \to 0^+) \to 1/(2\pi)$ from Eq. (6.31), and denoting $B(\omega) \equiv \partial A(\omega; r_{\rm st})/\partial r_{\rm st}|_{r_{\rm st}\to 0^+}$, one obtains from Eq. (6.29) that

$$\begin{aligned}\left.\frac{\partial \rho_{\rm st}(\theta, \omega; r_{\rm st})}{\partial r_{\rm st}}\right|_{r_{\rm st}\to 0^+} &= B(\omega) + \frac{K\cos\theta - K}{2\pi D} + \frac{K}{2\pi D}\, \mathrm{e}^{\frac{\omega\theta}{D}} \left(\mathrm{e}^{-\frac{2\pi\omega}{D}} - 1\right) \\ &\times \left[-\frac{\int_0^\theta \mathrm{d}\theta' \mathrm{e}^{\frac{-\omega\theta'}{D}} \cos\theta'}{\int_0^{2\pi} \mathrm{d}\theta' \mathrm{e}^{\frac{-\omega\theta'}{D}}} + \frac{\int_0^\theta \mathrm{d}\theta' \mathrm{e}^{\frac{-\omega\theta'}{D}} \left(\int_0^{2\pi} \mathrm{d}\theta'' \mathrm{e}^{\frac{-\omega\theta''}{D}} \cos\theta''\right)}{\left(\int_0^{2\pi} \mathrm{d}\theta' \mathrm{e}^{\frac{-\omega\theta'}{D}}\right)^2}\right].\end{aligned} \tag{6.54}$$

In order to proceed, we note that

$$\int_0^\theta \mathrm{d}\theta' \mathrm{e}^{-\frac{\omega\theta'}{D}} = \frac{D}{\omega}\left(1 - \mathrm{e}^{-\frac{\omega\theta}{D}}\right), \tag{6.55}$$

$$\int_0^\theta \mathrm{d}\theta' \mathrm{e}^{-\frac{\omega\theta'}{D}} \cos\theta' = \frac{D}{D^2+\omega^2}\left(\omega - \omega\mathrm{e}^{-\frac{\omega\theta}{D}}\cos\theta + D\mathrm{e}^{-\frac{\omega\theta}{D}}\sin\theta\right). \tag{6.56}$$

Substituting Eqs. (6.55) and (6.56) in Eq. (6.54), we obtain

$$\left.\frac{\partial \rho_{\mathrm{st}}(\theta, \omega; r_{\mathrm{st}})}{\partial r_{\mathrm{st}}}\right|_{r_{\mathrm{st}}\to 0^+} = B(\omega) + \frac{K(\cos\theta - 1)}{2\pi D} + \frac{K(\omega^2 - \omega^2\cos\theta + D\omega\sin\theta)}{2\pi D(D^2+\omega^2)}. \tag{6.57}$$

Considering Eq. (6.53) to leading order in $r_{\mathrm{st}}$ and using Eq. (6.57), we obtain the critical value of noise strength $D_c$ to be satisfying

$$1 = \int_{-\infty}^{\infty} \mathrm{d}\omega\, g(\omega)\left[\frac{K}{2D_c} - \frac{K\omega^2}{2D_c(D_c^2+\omega^2)}\right]. \tag{6.58}$$

Simplifying Eq. (6.58) further, we obtain the critical value $D_c$ to be solving the equation [5]

$$\int_{-\infty}^{\infty} \mathrm{d}\omega\, g(\omega)\,\frac{D_c}{D_c^2+\omega^2} = \frac{2}{K}. \tag{6.59}$$

In Fig. 6.1, we have shown schematically in the $(D, K)$ parameter space the synchronized and the incoherent phase as well as the boundary between them, to be obtained from Eq. (6.59). In the Kuramoto model, i.e., with $D = 0$, the critical coupling strength is given by $K_c(D = 0)$, such that for $K > K_c(D = 0)$ one has a synchronized phase in the stationary state. However, such a value of $K$ does not

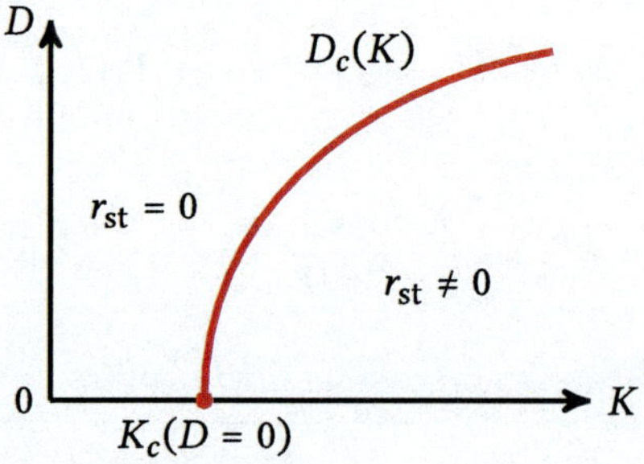


**Fig. 6.1 Schematic diagram of different stationary-state phases in the parameter space $(D, K)$ for the noisy Kuramoto dynamics (6.6).** The two regions denoted by $r_{\mathrm{st}} \neq 0$ and $r_{\mathrm{st}} = 0$ refer to the synchronized and the incoherent phase, respectively, while $D_c(K)$ shows the boundary between these two phases

ensure the existence of a synchronized phase in the stationary state for the noisy Kuramoto dynamics. Indeed, for every value of $K$ ($> K_c(D = 0)$), there is a critical value $D_c$, such that for $D < D_c$ the system relaxes to a synchronized phase, while for $D > D_c$, it shows incoherent behavior in the stationary state. In particular, for the BMF model, one obtains by substituting $g(\omega) = \delta(\omega)$ in Eq. (6.59) the critical noise strength as $D_c = K/2$. In both the noisy Kuramoto and the BMF model, the quantity $r_{\text{st}}$ decreases continuously from the value of unity at $D = 0$ to zero at $D = D_c$, and remains zero at higher values of $D$.

### 6.3.3 *Our Object of Study: Stationary Correlations*

In our published work [1], we have characterized in the thermodynamic limit the stationary-state dynamics (6.25) in terms of autocorrelation of the dynamical variable $\cos\theta$. The reason behind choosing such a dynamical variable is that the stationary order parameter is indeed $r_{\text{st}} = \langle\langle\cos\theta\rangle\rangle$, see Eq. (6.51), and so through such a choice, we would be able to monitor the correlation between the phase coherence at two times in the stationary state. Here, and in the following, double angular brackets are used to denote averaging over both the sources of randomness, namely, the annealed randomness $\eta_i(t)$ and the quenched randomness $\omega_i$. To study the correlation, we define the autocorrelation function as [1]

$$C(\tau) \equiv \lim_{\substack{t'\to\infty \\ \tau=\text{finite}}} \langle\langle\cos\theta(t')\cos\theta(t'+\tau)\rangle\rangle\,. \tag{6.60}$$

In order to understand better the physical meaning of $C(\tau)$, we will in the next section discuss the dynamics of a much simpler system, namely, a single noisy nonuniform oscillator. The purpose of this discussion will be two fold:

1. Understanding the physical meaning of the stationary autocorrelation.
2. Obtaining exact results for the autocorrelation, which will be used in the following section to derive stationary autocorrelation for the noisy Kuramoto model.

## 6.4 Single Noisy Nonuniform Oscillator: Stationary Correlation

The simplest form of a one-dimensional oscillator is the one for which an angle $\theta \in [0, 2\pi)$ changes uniformly in time.[4] For such an oscillator, we have $\mathrm{d}\theta/\mathrm{d}t = \omega$, where the constant $\omega$ is the frequency of the oscillator. However, in this section, we

[4] Here, we are considering a one-dimensional phase space. In one dimension, one has oscillations only if the motion is periodic [12]. The motion of a simple harmonic oscillator is two dimensional. The reason is that the second-order equation of motion of a simple harmonic oscillator reduces to two coupled first-order equations, thereby resulting in a two-dimensional phase space.

will talk about the nonuniform oscillator for which the angle $\theta$ has a nonuniform variation in time: at times, it speeds up, while at other times, it slows down [12].

### 6.4.1 Model Definition

We consider a single nonuniform oscillator characterized by its angle $\theta \in [0, 2\pi)$, whose time evolution in the presence of a Gaussian, white noise $\eta(t)$ is given by the following Langevin dynamics [1]:

$$\frac{\mathrm{d}\theta}{\mathrm{d}t} = \omega - \mathcal{K} \sin\theta + \eta(t) . \tag{6.61}$$

Here, $\mathcal{K} \geq 0$ and the natural frequency $\omega$ are real constants, while $\eta(t)$ satisfies

$$\langle \eta(t) \rangle = 0, \quad \langle \eta(t)\eta(t') \rangle = 2D\delta(t - t') , \tag{6.62}$$

with $D > 0$, and angular brackets denoting average over noise realizations. The dynamics (6.61) may be thought of in terms of motion of a point on a circle of unit radius, with $\theta$ characterizing the position of the point on the circle. Analogous to the discussion given in Sect. 6.3 on obtaining the BMF model as the overdamped limit of the HMF model, one may argue for the single oscillator (6.61) with $\omega = 0$ that the equation of motion corresponds to the noisy, overdamped dynamics of $\theta$ in a potential

$$\mathcal{V}_{\text{s.o.}}(\theta) \equiv -\mathcal{K} \cos\theta , \tag{6.63}$$

such that

$$\frac{\mathrm{d}\theta}{\mathrm{d}t} = -\frac{\mathrm{d}}{\mathrm{d}\theta}\Big[\mathcal{V}_{\text{s.o.}}(\theta)\Big] + \eta(t) . \tag{6.64}$$

### 6.4.2 Fixed Points of the Noiseless Dynamics

In the absence of noise (i.e., with $D = 0$), the dynamics of the nonuniform oscillator (6.61) has the same form as that of the Kuramoto dynamics (5.60). The local velocity $\mathrm{d}\theta/\mathrm{d}t$ is minimum at $\pi/2$ and maximum at $3\pi/2$. On the circle on which the dynamics is taking place, Fig. 6.2 shows the fixed points occurring in the dynamics for different values of $\omega > 0$ at a fixed $\mathcal{K}$; this figure may be obtained from Fig. 5.12.

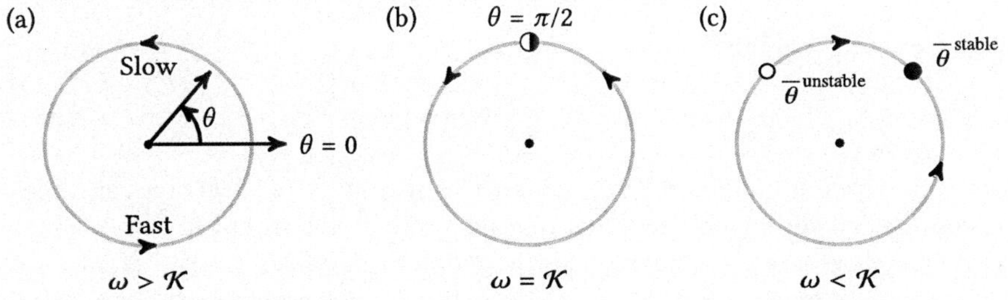


**Fig. 6.2 Different fixed points for the nonuniform-oscillator dynamics** (6.61) **in the absence of noise, i.e., with** $D = 0$. **a** No fixed point. **b** A half-stable fixed point at $\pi/2$. **c** Two fixed points: one unstable, and another stable. The unstable fixed point has $\cos\overline{\theta}^{\text{unstable}} < 0$, while the stable fixed point has $\cos\overline{\theta}^{\text{stable}} > 0$, see Eqs. (6.65) and (6.66). The arrows on the circle denote the direction of flow due to the dynamical evolution

The corresponding figure for $\omega < 0$ may be obtained from Fig. 5.13. The fixed points are as follows [12]:

1. For $\omega > \mathcal{K}$: No fixed point.
2. For $\omega = \mathcal{K}$: A half-stable fixed point at $\pi/2$.
3. For $\omega < \mathcal{K}$: Two fixed points, with one unstable and another stable.

$$\text{Unstable}: \overline{\theta}^{\text{unstable}}(\omega) \equiv \sin^{-1}[\omega/\mathcal{K}]; \;\; \cos\overline{\theta}^{\text{unstable}} = -\sqrt{1-\omega^2/\mathcal{K}^2}\,, \tag{6.65}$$

$$\text{Stable}: \overline{\theta}^{\text{stable}}(\omega) \equiv \sin^{-1}[\omega/\mathcal{K}]; \;\; \cos\overline{\theta}^{\text{stable}} = +\sqrt{1-\omega^2/\mathcal{K}^2}\,. \tag{6.66}$$

In the long-time limit, the dynamics (6.61) without noise and for $\omega < \mathcal{K}$ results in $\theta$ settling into the stable fixed-point value $\overline{\theta}^{\text{stable}}$. In presence of weak noise ($D \to 0$), we expect $\theta$ in the long-time limit to have a (narrow) distribution of values around $\overline{\theta}^{\text{stable}}$.

### 6.4.3 Stationary-State Distribution

To characterize the stationary state of the dynamics (6.61), we define a one-time probability distribution $\mathcal{P}(\theta, t)$, such that $\mathcal{P}(\theta, t)\mathrm{d}\theta$ gives the probability to observe a value of the angle in the interval $[\theta, \theta + \mathrm{d}\theta]$ at time $t$. One has the normalization

$$\int_0^{2\pi} \mathrm{d}\theta\, \mathcal{P}(\theta, t) = 1 \;\; \forall\, t\,, \tag{6.67}$$

and the periodicity condition

$$\mathcal{P}(\theta + 2\pi, t) = \mathcal{P}(\theta, t) \,. \tag{6.68}$$

For a given initial condition $\mathcal{P}(\theta, t = 0)$, the quantity $\mathcal{P}(\theta, t)$ evolves in time according to a Fokker–Planck equation that one may derive from Eq. (6.61) using standard procedures [13], see Chap. 2. The equation reads

$$\frac{\partial \mathcal{P}(\theta, t)}{\partial t} = -\frac{\partial}{\partial \theta}\Big[\Big(\omega - \mathcal{K}\sin\theta\Big)\mathcal{P}(\theta, t)\Big] + D\frac{\partial^2 \mathcal{P}(\theta, t)}{\partial \theta^2} \,. \tag{6.69}$$

From Eq. (6.69), it is evident that the form of $\mathcal{P}(\theta, t)$ depends on the frequency $\omega$. Therefore, introducing the frequency as an argument, the distribution is denoted as $\mathcal{P}(\theta, \omega, t)$.

In the stationary state, when $\mathcal{P}(\theta, \omega, t)$ relaxes to a time-independent form $\mathcal{P}_{\rm st}(\theta, \omega)$, Eq. (6.69) gives

$$0 = -\frac{\partial}{\partial \theta}\big[(\omega - \mathcal{K}\sin\theta)\mathcal{P}_{\rm st}(\theta, \omega)\big] + D\frac{\partial^2 \mathcal{P}_{\rm st}(\theta, \omega)}{\partial \theta^2} \,. \tag{6.70}$$

This last equation has the same form as Eq. (6.26). Then, it may be solved by using the same technique of integrating factor as was employed there. One gets

$$\mathcal{P}_{\rm st}(\theta, \omega) = \mathcal{A}(\omega)\, {\rm e}^{\frac{-\mathcal{K}+\mathcal{K}\cos\theta+\omega\theta}{D}}\left[1 + \frac{({\rm e}^{-2\pi\omega/D} - 1)\int_0^\theta {\rm d}\theta'\, {\rm e}^{(-\omega\theta' - \mathcal{K}\cos\theta')/D}}{\int_0^{2\pi} {\rm d}\theta'\, {\rm e}^{(-\omega\theta' - \mathcal{K}\cos\theta')/D}}\right], \tag{6.71}$$

where $\mathcal{A}(\omega)$ is a constant whose value may be fixed by employing the normalization condition: $\int_0^{2\pi} {\rm d}\theta\, \mathcal{P}_{\rm st}(\theta, \omega) = 1$. In Fig. 6.3, we have compared the analytical form of the stationary distribution (6.71) with that obtained by numerically integrating the dynamics (6.61) for three different values of noise strength $D$ with $\omega = 0.4$ and $\mathcal{K} = 1$. We see a very good agreement between theory and numerical results. For these values of $\omega$ and $\mathcal{K}$, the noiseless limit of dynamics (6.61) has a stable fixed point at $\overline{\theta}^{\rm stable} = \sin^{-1}(0.4) \approx 0.4115$. From the figure, it is seen that as the strength of the noise decreases, the distribution gets increasingly peaked around $\overline{\theta}^{\rm stable}$ shown as a gray dashed line.

For $\omega = 0$, we obtain the stationary distribution from Eq. (6.71), as

$$\mathcal{P}_{\rm st}(\theta, \omega = 0) \propto {\rm e}^{(-\mathcal{K}+\mathcal{K}\cos\theta)/D} \sim {\rm e}^{-\mathcal{V}_{\rm s.o.}(\theta)/D} \,, \tag{6.72}$$

where we have used Eq. (6.63). Equation (6.72) shows that the stationary state has the same form as in the Gibbs–Boltzmann canonical equilibrium at temperature $T$ given by $T = D$, with the Boltzmann constant set to unity. The stationary distribution in equilibrium, Eq. (6.72), is shown in Fig. 6.4 for three different values of $D$ along with the results obtained by numerically integrating the dynamics (6.61) with $\omega = 0$. We

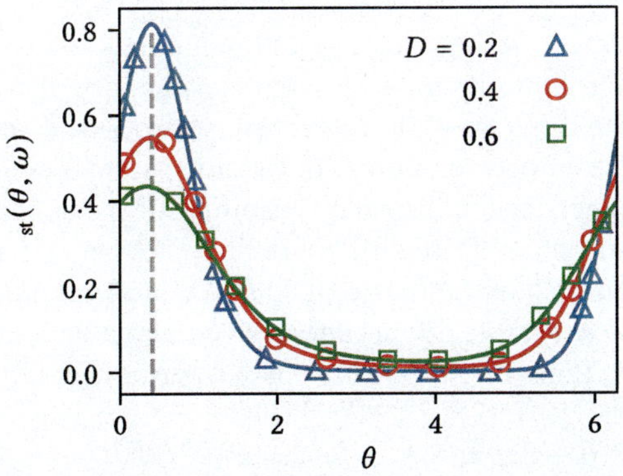


**Fig. 6.3 Stationary-state distribution for the noisy nonuniform oscillator** (6.61) **with** $\omega = 0.4$. Here, $\mathcal{K} = 1$. The dashed line in gray shows the stable fixed point $\overline{\theta}^{\text{stable}}$ for the dynamics (6.61) in the absence of noise. As the value of $D$ decreases, we see that the distribution gets increasingly peaked around $\overline{\theta}^{\text{stable}}$. The lines are obtained by using the analytical form (6.71), while the points are obtained by numerically integrating the stochastic dynamics (6.61), by using the stochastic version of the Euler method discussed in Appendix M.2 with $\Delta t = 0.01$. The initial value of $\theta$ at $t = 0$ is chosen randomly from the range $[0, 2\pi)$. After the system reaches the stationary state, we record the value of $\theta$. In order to draw the displayed distribution, we repeat these steps $10^5$ times. We see a very good agreement between theory and numerical results

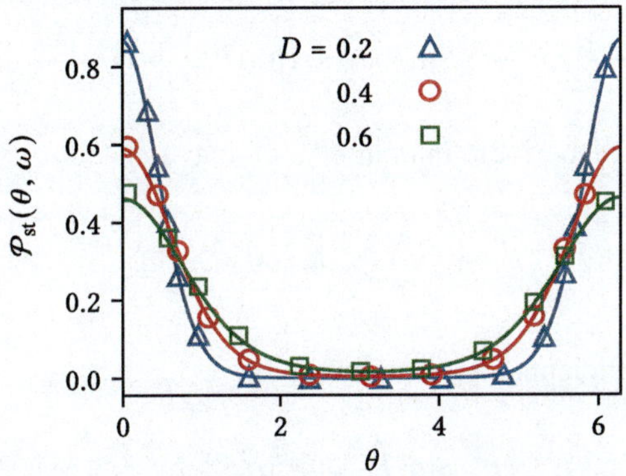


**Fig. 6.4 Stationary-state distribution for the noisy nonuniform oscillator** (6.61) **with** $\omega = 0$. Here, $\mathcal{K} = 1$. We see that the stationary distributions are peaked around the value zero that is the minimum of the potential $\mathcal{V}_{\text{s.o.}}(\theta) = -\mathcal{K} \cos \theta$. The lines are obtained by using the analytical form (6.72), while the points are obtained by numerically integrating the stochastic dynamics (6.61), by using the stochastic version of the Euler method discussed in Appendix M.2 with $\Delta t = 0.01$. The initial value of $\theta$ at $t = 0$ is chosen randomly from the range $[0, 2\pi)$. After the system reaches the stationary state, we record the value of $\theta$. In order to draw the displayed distribution, we repeat these steps $10^5$ times. We see a very good agreement between theory and numerical results

find a very good agreement between theory and numerical results. As the potential $\mathcal{V}_{\text{s.o.}}(\theta) = -\mathcal{K}\cos\theta$ has a minimum at $\theta = 0$, the stationary distribution is peaked around this value. In the limit $D \to 0$, we expect a sharper peak around $\theta = 0$.

In order to obtain the autocorrelation in the stationary state, to be discussed later in this chapter, we will need a conditional probability for the dynamics (6.61) along with the stationary distribution $\mathcal{P}_{\text{st}}(\theta, \omega)$. We define $P(\theta, t|\theta', t')$ as the conditional probability distribution, such that $P(\theta, t|\theta', t')\mathrm{d}\theta$ gives the probability that the angle has a value in the interval $[\theta, \theta + \mathrm{d}\theta]$ at time $t$, given that it had the value $\theta'$ at time $t'$, such that $t \geq t'$. The function $P(\theta, t|\theta', t')$ is $2\pi$-periodic in both $\theta$ and $\theta'$:

$$P(\theta + 2\pi, t|\theta' + 2\pi, t') = P(\theta, t|\theta', t'), \tag{6.73}$$

and satisfies the normalization condition

$$\int_0^{2\pi} \mathrm{d}\theta\; P(\theta, t|\theta', t') = 1\; \forall\, \theta', t' \text{ and } \forall\, t \geq t'. \tag{6.74}$$

The time evolution of $P(\theta, t|\theta', t')$ also follows a Fokker–Planck equation that may be written down by using the Langevin equation (6.61). The equation is given by [1]

$$\frac{\partial P(\theta, t|\theta', t')}{\partial t} = -\frac{\partial}{\partial \theta}\Big[(\omega - \mathcal{K}\sin\theta)P(\theta, t|\theta', t')\Big] + D\frac{\partial^2 P(\theta, t|\theta', t')}{\partial \theta^2}. \tag{6.75}$$

Since $P(\theta, t|\theta', t')$ is $2\pi$-periodic in $\theta$ and $\theta'$, we may expand it in a Fourier series, as

$$P(\theta, t|\theta', t') = \sum_{n,m=-\infty}^{\infty} \widetilde{P}_{n,m}(t|t')\, \mathrm{e}^{\mathrm{i}(n\theta + m\theta')}, \tag{6.76}$$

where the Fourier coefficients are given by

$$\widetilde{P}_{n,m}(t|t') = \frac{1}{4\pi^2}\int_0^{2\pi} \mathrm{d}\theta \int_0^{2\pi} \mathrm{d}\theta'\; P(\theta, t|\theta', t')\, \mathrm{e}^{-\mathrm{i}(n\theta + m\theta')}. \tag{6.77}$$

Since $P(\theta, t|\theta', t)$ is real, we have $\widetilde{P}_{-n,-m}(t|t') = \widetilde{P}^{\star}_{n,m}(t|t')$, where $\star$ denotes complex conjugation; also, the condition $P(\theta, t'|\theta', t') = \delta(\theta - \theta')$ implies that

$$\widetilde{P}_{n,m}(t'|t') = \frac{\delta_{n,-m}}{2\pi}. \tag{6.78}$$

In the following discussions, we will see that only a few Fourier components are relevant to find the correlation. But, let us first give a precise meaning of the stationary-state correlation.

### 6.4.4 Stationary-State Correlation

#### 6.4.4.1 Definition and Physical Interpretation

In the context of the model (6.61), the autocorrelation of the quantity $\cos\theta$ between times $t$ and $t'$ is defined as

$$C(t,t',\omega) \equiv \langle \cos\theta \cos\theta' \rangle\,;\ \theta \equiv \theta(t)\,;\ \ \theta' \equiv \theta(t')\,;\ \ t \geq t'\,, \tag{6.79}$$

where we have indicated the explicit dependence of the autocorrelation on the parameter $\omega$. In order to understand the physical meaning of $C(t,t',\omega)$, let us introduce a two-time joint probability distribution $\mathcal{P}(\theta',t';\theta,t)$, such that the quantity $\mathcal{P}(\theta',t';\theta,t)\,\mathrm{d}\theta'\,\mathrm{d}\theta$ denotes the joint probability to observe the angle value around $\theta'$ at time $t'$ and around $\theta$ at time $t \geq t'$. In Fig. 6.5, we have shown a few trajectories $\theta(t)$ for the dynamics (6.61) and for different initial conditions. The joint probability is obtained by considering an infinite number of such trajectories and counting the fraction that have the angle value around $\theta'$ at time $t'$ and around $\theta$ at time $t$, for specified values of $\theta'$, $\theta$, $t'$, and $t$. For the representative case shown in the figure, we see that out of five trajectories, only two satisfy this condition for the values of $\theta'$, $\theta$, $t'$, and $t$ indicated in the figure. In terms of the joint probability, one may rewrite the autocorrelation (6.79) as

$$C(t,t',\omega) = \int_0^{2\pi} \mathrm{d}\theta \int_0^{2\pi} \mathrm{d}\theta'\ \cos\theta\ \cos\theta'\,\mathcal{P}(\theta',t';\theta,t)\,. \tag{6.80}$$

The two-time joint probability distribution may be obtained from the one-time probability distribution and the two-time conditional probability distribution, by invoking the Bayes' theorem (see Chap. 1)

**Fig. 6.5 A few trajectories of $\theta(t)$ for different initial conditions evolving under the dynamics (6.61).** We numerically integrate the stochastic dynamics (6.61), by using the stochastic version of the Euler method discussed in Appendix M.2 with $\Delta t = 0.01$, $\mathcal{K} = 1$, $D = 0.2$, and $\omega = 0.8$

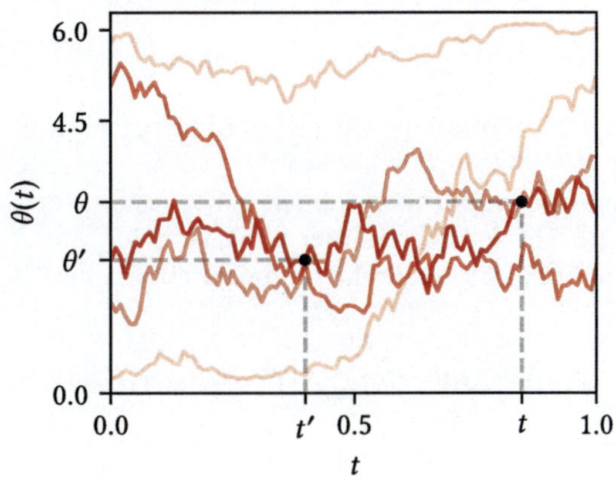

$$P(A \mid B) = \frac{P(A \cap B)}{P(B)}. \tag{6.81}$$

Hence, we get

$$\mathcal{P}(\theta', t'; \theta, t)\,\mathrm{d}\theta'\,\mathrm{d}\theta = P(\theta, t|\theta', t')\,\mathrm{d}\theta\,\mathcal{P}(\theta', t')\,\mathrm{d}\theta', \tag{6.82}$$

where $\mathcal{P}(\theta', t')\,\mathrm{d}\theta'$ denotes the one-time probability to observe a value around $\theta'$ at time $t'$. The quantity $P(\theta, t|\theta', t')$ has already been defined in the paragraph preceding Eq. (6.73) as the conditional probability distribution, such that $P(\theta, t|\theta', t')\mathrm{d}\theta$ gives the probability that the angle has a value in the interval $[\theta, \theta + \mathrm{d}\theta]$ at time $t$, given that it had the value $\theta'$ at time $t'$ satisfying $t \geq t'$. It may be seen from Fig. 6.5 that for the representative cases displayed therein, only three out of five trajectories take the value around the indicated $\theta'$ at time $t'$; out of these three, only two are observed to have a value around the indicated $\theta$ at time $t$. In terms of the conditional probability distribution, Eq. (6.80) reads

$$C(t, t', \omega) = \int_0^{2\pi} \mathrm{d}\theta \int_0^{2\pi} \mathrm{d}\theta'\, \cos\theta\, \cos\theta'\, P(\theta, t|\theta', t')\, \mathcal{P}(\theta', t'). \tag{6.83}$$

We are interested in the autocorrelation in the stationary state. The dynamics of the nonuniform oscillator (6.61) reaches the stationary state in the long-time limit $t' \to \infty$. Then, the one-time probability distribution $\mathcal{P}(\theta', t')$ relaxes to a time-independent form $\mathcal{P}_{\mathrm{st}}(\theta', \omega)$. Introducing $\tau$ as the finite time delay between $t'$ and $t$, such that $\tau \equiv t - t'$, one obtains from Eq. (6.83) the stationary correlation for the noisy nonuniform oscillator as

$$C(\tau, \omega) = \lim_{\substack{t' \to \infty \\ \tau = \text{finite}}} \int_0^{2\pi} \mathrm{d}\theta \int_0^{2\pi} \mathrm{d}\theta'\, \cos\theta\, \cos\theta'\, P(\theta, t' + \tau|\theta', t')\, \mathcal{P}_{\mathrm{st}}(\theta', \omega). \tag{6.84}$$

#### 6.4.4.2 Stationary Correlation: Exact Results

Now, we move on to obtain exact results for the stationary correlation. Using Eq. (6.71) and the Fourier expansion for $P(\theta, t|\theta', t')$ given in Eq. (6.76) with $t \equiv t' + \tau$, we obtain the stationary correlation $C(\tau, \omega)$ from Eq. (6.84) as [1]

$$C(\tau, \omega) = \lim_{\substack{t' \to \infty \\ \tau = \text{finite}}} \pi \sum_{m=-\infty}^{\infty} \left[\widetilde{P}_{1,m}(t|t') + \widetilde{P}_{-1,m}(t|t')\right] \int_0^{2\pi} \mathrm{d}\theta\, \mathrm{e}^{\mathrm{i}m\theta} \cos\theta\, \mathcal{P}_{\mathrm{st}}(\theta, \omega). \tag{6.85}$$

Therefore, the stationary correlation depends only on $\widetilde{P}_{1,m}(t|t')$ and $\widetilde{P}_{-1,m}(t|t')$.

Now, Eq. (6.75) gives the time evolution of $\widetilde{P}_{n,m}(t|t')$ as

$$\frac{\partial \widetilde{P}_{n,m}(t|t')}{\partial t} = -(in\omega + Dn^2)\widetilde{P}_{n,m}(t|t') + \frac{n\mathcal{K}}{2}\left(\widetilde{P}_{n-1,m}(t|t') - \widetilde{P}_{n+1,m}(t|t')\right). \tag{6.86}$$

It then follows from Eq. (6.86) that $\widetilde{P}_{0,m}(t|t')$ is independent of $t$, and thus, we have

$$\widetilde{P}_{0,m}(t|t') = \widetilde{P}_{0,m}(t'|t') = \frac{\delta_{0,-m}}{2\pi}. \tag{6.87}$$

Let us consider separately the cases $\mathcal{K} \neq 0$ and $\mathcal{K} = 0$.

**The case $\mathcal{K} \neq 0$ :**
For any pair of values $(n, m)$, the system of Eq. (6.86) is not closed, and in fact, involves an infinite hierarchy: to solve for $\widetilde{P}_{n,m}$ requires knowing $\widetilde{P}_{n+1,m}$ whose solution in turn requires knowing $\widetilde{P}_{n+2,m}$, and so on. Nevertheless, for a given $m$, only one of the $\widetilde{P}_{n,m}$'s is nonzero at the initial time, i.e., $\widetilde{P}_{n,m}(t'|t') = \delta_{n,-m}/(2\pi)$, see Eq. (6.78). Therefore, the system of equations may be solved by truncating it at a given value $n = n_{\text{max}}$, i.e., by stipulating that $\widetilde{P}_{n,m}(t|t') = 0$ for $n > n_{\text{max}}$, where $n_{\text{max}}$ may be chosen to be as large as possible [1]. In practice, in evaluating the correlation (6.85), we choose the same $n_{\text{max}}$ for different $m$'s and restrict the values of $m$ to be in $[-n_{\text{max}}, n_{\text{max}}]$, checking that a larger value of $n_{\text{max}}$ does not lead to any significant change in the results obtained. In Fig. 6.6a, we have shown that the stationary autocorrelation obtained using Eq. (6.85) and applying the aforementioned numerical steps do not change significantly for $n_{\text{max}} > 10$, thereby allowing us to choose $n_{\text{max}} = 10$ to compute $C(\tau, \omega)$. With such a choice, in Fig. 6.6b, we have shown the comparison between the analytical form of the stationary correlation (6.85) and the results obtained by numerically integrating the dynamics (6.61) for different values of $D$. We see a very good agreement between theory and numerical results.

From the system of Eq. (6.86), one may obtain closed form expressions for $\widetilde{P}_{n,m}$ for the particular case of small $\mathcal{K}$ when the equations may be solved perturbatively. To this end, we expand $\widetilde{P}_{n,m}(t|t')$ as a power series in $\mathcal{K}$ as [1]

$$\widetilde{P}_{n,m}(t|t') = \widetilde{P}^{(0)}_{n,m}(t|t') + \mathcal{K}\widetilde{P}^{(1)}_{n,m}(t|t') + \mathcal{K}^2\widetilde{P}^{(2)}_{n,m}(t|t') + \cdots, \tag{6.88}$$

where we have $\widetilde{P}^{(\alpha)}_{n,m}(t|t') = O(\mathcal{K}^0)$ for $\alpha \geq 0$. Substituting Eq. (6.88) in Eq. (6.86), and comparing terms of the same order in $\mathcal{K}$ from both sides, we get

$$\mathcal{K}^0 : \quad \frac{\partial \widetilde{P}^{(0)}_{n,m}(t|t')}{\partial t} = -(in\omega + Dn^2)\widetilde{P}^{(0)}_{n,m}(t|t'), \tag{6.89}$$

$$\mathcal{K}^1 : \quad \frac{\partial \widetilde{P}^{(1)}_{n,m}(t|t')}{\partial t} = -(in\omega + Dn^2)\widetilde{P}^{(1)}_{n,m}(t|t') + \frac{n}{2}\left(\widetilde{P}^{(0)}_{n-1,m}(t|t') - \widetilde{P}^{(0)}_{n+1,m}(t|t')\right), \tag{6.90}$$

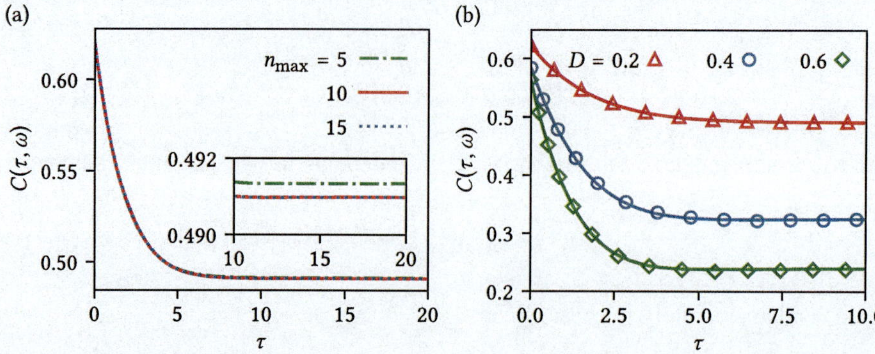


**Fig. 6.6** **Stationary correlation for the noisy nonuniform oscillator** (6.61). **a** We see that the stationary correlation $C(\tau, \omega)$ obtained from Eq. (6.85) do not change significantly with increasing $n_{\max}$. The figure suggests that we may choose $n_{\max} = 10$ for computing $C(\tau, \omega)$. Here, we set $\omega = 0.5$, $D = 0.2$, and $\mathcal{K} = 1.0$. In order to obtain $C(\tau, \omega)$ from Eq. (6.85), we first find the solutions for $\widetilde{P}_{1,m}(t|t')$ and $\widetilde{P}_{-1,m}(t|t')$ by numerically integrating the Eq. (6.86) truncated at $n = n_{\max}$ by using the Euler method discussed in Appendix M.2 with $\Delta t = 0.01$. **b** For three different values of the noise strength $D$, we show the behavior of stationary correlation for an oscillator with frequency $\omega = 0.5$. Here, we set $\mathcal{K} = 1$. The lines are given by the analytical form (6.85) with $n_{\max} = 10$, while the points are obtained by numerically integrating the stochastic dynamics (6.61) by using the stochastic version of the Euler method discussed in Appendix M.2 with $\Delta t = 0.01$. The initial value of $\theta$ at $t = 0$ is chosen randomly from the range $[0, 2\pi)$. The data are averaged over $10^5$ noise realizations. We see a very good agreement between theory and numerical results

$$\mathcal{K}^2: \quad \frac{\partial \widetilde{P}^{(2)}_{n,m}(t|t')}{\partial t} = -(\mathrm{i}n\omega + Dn^2)\widetilde{P}^{(2)}_{n,m}(t|t') + \frac{n}{2}\left(\widetilde{P}^{(1)}_{n-1,m}(t|t') - \widetilde{P}^{(1)}_{n+1,m}(t|t')\right), \tag{6.91}$$

and so on. Since $\widetilde{P}_{n,m}(t'|t') = \delta_{n,-m}/(2\pi)$, we may take

$$\widetilde{P}^{(0)}_{n,m}(t'|t') = \frac{\delta_{n,-m}}{2\pi}, \qquad \widetilde{P}^{(\alpha)}_{n,m}(t'|t') = 0 \;\forall\; \alpha \geq 1\,. \tag{6.92}$$

Solving Eq. (6.89), we get

$$\widetilde{P}^{(0)}_{n,m}(t|t') = \frac{\delta_{n,-m}}{2\pi} \exp\left[-\left(\mathrm{i}n\omega + Dn^2\right)(t - t')\right], \tag{6.93}$$

which when used in Eq. (6.90) yields the solution

$$\widetilde{P}^{(1)}_{n,m}(t|t') = \frac{n\mathrm{e}^{-(\mathrm{i}n\omega+Dn^2)(t-t')}}{4\pi}\left[\frac{\delta_{n-1,-m}}{2nD - D + \mathrm{i}\omega}\left\{\mathrm{e}^{-(-2nD+D-\mathrm{i}\omega)(t-t')} - 1\right\} + \frac{\delta_{n+1,-m}}{2nD + D + \mathrm{i}\omega}\left\{\mathrm{e}^{-(2nD+D+\mathrm{i}\omega)(t-t')} - 1\right\}\right]. \tag{6.94}$$

Substituting Eq. (6.94) in Eq. (6.91), one obtains

$$\widetilde{P}^{(2)}_{n,m}(t|t') = \frac{n e^{-(\mathrm{i}n\omega + Dn^2)(t-t')}}{8\pi} \left[ \frac{(n-1)\,\delta_{n,-m}}{2nD - D + \mathrm{i}\omega} \left( t - t' + \frac{e^{(2nD-D+\mathrm{i}\omega)(t-t')} - 1}{-2nD + D - \mathrm{i}\omega} \right) \right.$$
$$+ \frac{(n-1)\,\delta_{n-2,-m}}{2nD - 3D + \mathrm{i}\omega} \left( \frac{e^{(2nD-D+\mathrm{i}\omega)(t-t')} - 1}{-2nD + D - \mathrm{i}\omega} - \frac{e^{-(-4nD+4D-2\mathrm{i}\omega)(t-t')} - 1}{-4nD + 4D - 2\mathrm{i}\omega} \right)$$
$$- \frac{(n+1)\,\delta_{n+2,-m}}{2nD + 3D + \mathrm{i}\omega} \left( \frac{e^{-(2nD+D\mathrm{i}\omega)(t-t')} - 1}{2nD + D + \mathrm{i}\omega} - \frac{e^{-(4nD+4D+2\mathrm{i}\omega)(t-t')} - 1}{4nD + 4D + 2\mathrm{i}\omega} \right)$$
$$\left. - \frac{(n+1)\,\delta_{n,-m}}{2nD + D + \mathrm{i}\omega} \left( t - t' + \frac{e^{-(2nD+D+\mathrm{i}\omega)(t-t')} - 1}{2nD + D + \mathrm{i}\omega} \right) \right]. \tag{6.95}$$

For later convenience we introduce

$$Q_m(\omega) \equiv \int_0^{2\pi} d\theta \, \cos\theta \, e^{\mathrm{i}m\theta} \, \mathcal{P}_{\mathrm{st}}(\theta, \omega). \tag{6.96}$$

It is seen from the last equation that $Q_{-m}(\omega) = Q^\star_m(\omega)$. With the help of Eqs. (6.96), (6.85) may be reduced to

$$C(\tau, \omega) = \lim_{\substack{t' \to \infty \\ \tau = \text{finite}}} \pi \sum_{m=-\infty}^{\infty} Q_m(\omega) \left[ \widetilde{P}_{1,m}(t|t') + \widetilde{P}_{-1,m}(t|t') \right]. \tag{6.97}$$

To order $\mathcal{K}^2$, substituting $\widetilde{P}_{n,m}(t|t') = \widetilde{P}^{(0)}_{n,m}(t|t') + \mathcal{K}\widetilde{P}^{(1)}_{n,m}(t|t') + \mathcal{K}^2\widetilde{P}^{(2)}_{n,m}(t|t')$ in Eq. (6.97) and using the solutions for $\widetilde{P}^{(0)}_{n,m}(t|t')$, $\widetilde{P}^{(1)}_{n,m}(t|t')$ and $\widetilde{P}^{(2)}_{n,m}(t|t')$ given in Eqs. (6.93), (6.94), and (6.95), respectively, we obtain $C(\tau, \omega)$ for the model (6.61) in the limit of small $\mathcal{K}$, as [1]

$$C(\tau, \omega) = \left[ \frac{e^{-(D+\mathrm{i}\omega)\tau} Q_{-1}}{2} \right.$$
$$+ \frac{\mathcal{K}}{4} \left\{ \frac{(1 - e^{-(D+\mathrm{i}\omega)\tau}) Q_0}{(D + \mathrm{i}\omega)} + \frac{\left( e^{-(4D+2\mathrm{i}\omega)\tau} - e^{-(D+\mathrm{i}\omega)\tau} \right) Q_{-2}}{(3D + \mathrm{i}\omega)} \right\}$$
$$+ \frac{\mathcal{K}^2 e^{-(D+\mathrm{i}\omega)\tau}}{4} \left\{ \frac{Q_{-1}}{3D + \mathrm{i}\omega} \left( \frac{1 - e^{-(3D+\mathrm{i}\omega)\tau}}{3D + \mathrm{i}\omega} - \tau \right) \right.$$
$$\left.\left. - \frac{Q_{-3}}{5D + \mathrm{i}\omega} \left( \frac{e^{-(3D+\mathrm{i}\omega)\tau} - 1}{3D + \mathrm{i}\omega} - \frac{e^{-(8D+2\mathrm{i}\omega)\tau} - 1}{8D + 2\mathrm{i}\omega} \right) \right\} \right] + \text{c.c.}, \tag{6.98}$$

where c.c. denotes complex conjugate of the bracketed term; here, we need to evaluate the terms $Q_m$ appearing on the right hand side of the above equation, which we now set out to do.

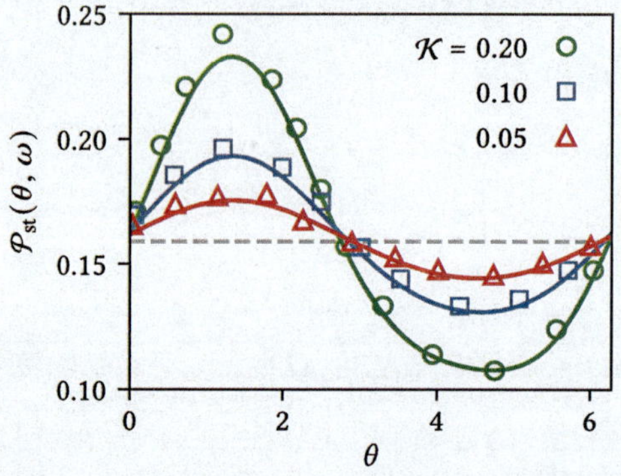


**Fig. 6.7 Stationary distribution for the noisy nonuniform oscillator (6.61) with small $\mathcal{K}$.** Here, we set $\omega = 0.5$, and $D = 0.1$. The continuous lines are given by the analytical form (6.99), while the points are obtained by numerically integrating the stochastic dynamics (6.61), by using stochastic version of the Euler method discussed in Appendix M.2 with $\Delta t = 0.01$. The initial value of $\theta$ at $t = 0$ is chosen randomly from the range $[0, 2\pi)$. After the system reaches the stationary state, we record the value of $\theta$. In order to draw the displayed distribution, we repeat these steps $10^6$ times. The figure shows a very good agreement between theory and numerical results as $\mathcal{K}$ is decreased. The gray-dashed line denotes the uniform distribution with $\mathcal{P}_{st}(\theta, \omega) = 1/(2\pi)$ for $\mathcal{K} = 0$

The stationary distribution $\mathcal{P}_{st}(\theta, \omega)$ up to order $\mathcal{K}^2$ is obtained from Eq. (6.71) as (see Appendix D)

$$\mathcal{P}_{st}(\theta, \omega) = \frac{1}{2\pi}\Bigg[1 + \frac{\mathcal{K}(D\cos\theta + \omega\sin\theta)}{D^2+\omega^2} + \frac{\mathcal{K}^2\{(2D^2-\omega^2)\cos 2\theta + 3D\omega\sin 2\theta\}}{2(D^2+\omega^2)(4D^2+\omega^2)}\Bigg]. \tag{6.99}$$

In Fig. 6.7, the stationary distribution $\mathcal{P}_{st}(\theta, \omega)$ as a function of $\theta$ has been shown for $\omega = 0.50$, and $D = 0.10$. In this figure, we compare the analytical form (6.99) of $\mathcal{P}_{st}(\theta, \omega)$, valid for small $\mathcal{K}$, with the numerical form of the distribution obtained by numerically integrating the dynamics (6.61). The figure shows that in the small-$\mathcal{K}$ limit, Eq. (6.99) does correctly represent the numerical form of $\mathcal{P}_{st}(\theta, \omega)$.

Substituting Eq. (6.99) in Eq. (6.96), we obtain

$$\mathcal{Q}_0(\omega) = \frac{\mathcal{K}D}{2\,(D^2+\omega^2)}, \quad \mathcal{Q}_{-1}(\omega) = \frac{1}{2} + \frac{\mathcal{K}^2}{8\,(D+\mathrm{i}\omega)\,(2D+\mathrm{i}\omega)}, \tag{6.100}$$

$$\mathcal{Q}_{-2}(\omega) = \frac{\mathcal{K}}{4\,(D+\mathrm{i}\omega)}, \quad \mathcal{Q}_{-3}(\omega) = \frac{\mathcal{K}^2}{8\,(D+\mathrm{i}\omega)\,(2D+\mathrm{i}\omega)}. \tag{6.101}$$

Using the above expressions for $\mathcal{Q}_0(\omega)$, $\mathcal{Q}_{-1}(\omega)$, $\mathcal{Q}_{-2}(\omega)$ and $\mathcal{Q}_{-3}(\omega)$ in Eq. (6.98), we obtain the stationary correlation $C(\tau, \omega)$ up to order $\mathcal{K}^2$ as

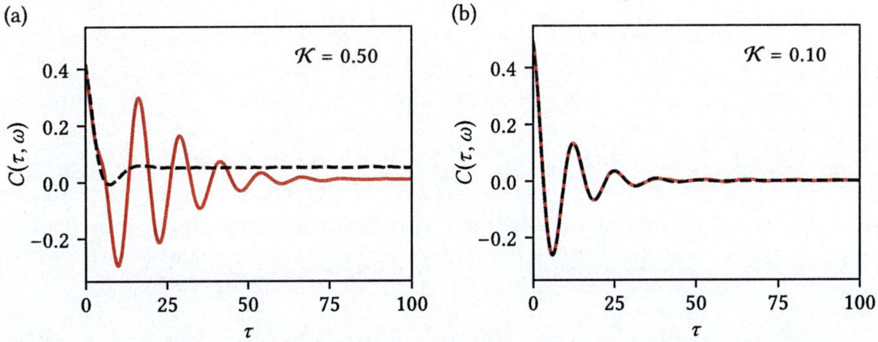


**Fig. 6.8 Correctness of the analytical form (6.102) of the stationary correlation for the noisy nonuniform oscillator (6.61) for small $\mathcal{K}$.** Here, we set $\omega = 0.5$, and $D = 0.1$. The continuous lines in red are given by the analytical form (6.102), valid for small $\mathcal{K}$, while the black-dashed lines are obtained by numerically integrating the stochastic dynamics (6.61), by using the stochastic version of the Euler method discussed in Appendix M.2 with $\Delta t = 0.01$. The initial value of $\theta$ at $t = 0$ is chosen randomly from the range $[0, 2\pi)$. The data are averaged over $10^5$ noise realizations. **a** For $\mathcal{K} = 0.50$, the match between theory and numerical results is not good. However, as we decrease the value of $\mathcal{K}$, we see in **b** a very good agreement between theory and numerics

$$C(\tau, \omega) = \left[\frac{1}{4}\,\mathrm{e}^{-(D+\mathrm{i}\omega)\tau} + \mathcal{K}^2 \left\{ \frac{\mathrm{e}^{-(D+\mathrm{i}\omega)\tau}}{16(D+\mathrm{i}\omega)\,(2D+\mathrm{i}\omega)} + \frac{D\big(1-\mathrm{e}^{-(D+\mathrm{i}\omega)\tau}\big)}{8\,(D+\mathrm{i}\omega)\,(D^2+\omega^2)} + \frac{\mathrm{e}^{-(4D+2\mathrm{i}\omega)\tau} - \mathrm{e}^{-(D+\mathrm{i}\omega)\tau}}{16(D+\mathrm{i}\omega)(3D+\mathrm{i}\omega)} + \frac{\mathrm{e}^{-(D+\mathrm{i}\omega)\tau}}{8(3D+\mathrm{i}\omega)}\left(\frac{1-\mathrm{e}^{-(3D+\mathrm{i}\omega)\tau}}{3D+\mathrm{i}\omega} - \tau\right)\right\}\right] + \text{c.c.}\,. \tag{6.102}$$

In Fig. 6.8, we have shown comparisons between the results obtained on the basis of the analytical form (6.102) and those obtained by numerically integrating the noisy nonuniform oscillator dynamics (6.61). The figure shows a very good agreement for small $\mathcal{K}$.

For $\omega = 0$, the autocorrelation for the dynamics (6.61) in the equilibrium stationary state is obtained from Eq. (6.102): to order $\mathcal{K}^2$, one has

$$C(\tau, \omega = 0) = \frac{\mathrm{e}^{-D\tau}}{4}\left[1 + \frac{\mathcal{K}^2}{72D^2}\left(36\mathrm{e}^{D\tau} + 2\mathrm{e}^{-3D\tau} - 12D\tau - 29\right)\right] + \text{c.c.}\,. \tag{6.103}$$

**The case $\mathcal{K} = 0$:**
For $\mathcal{K} = 0$, Eq. (6.71) reduces to

$$\mathcal{P}_{\mathrm{st}}(\theta, \omega) = \mathcal{A}(\omega)\,\mathrm{e}^{\omega\theta/D}\left[1 + \frac{(\mathrm{e}^{-2\pi\omega/D} - 1)\int_0^\theta \mathrm{d}\theta'\,\mathrm{e}^{-\omega\theta'/D}}{\int_0^{2\pi} \mathrm{d}\theta'\,\mathrm{e}^{-\omega\theta'/D}}\right]. \tag{6.104}$$

Using Eq. (6.55), we obtain from the last equation that

$$\mathcal{P}_{\rm st}(\theta, \omega) = \mathcal{A}(\omega). \tag{6.105}$$

Now, demanding normalization $\int_0^{2\pi} d\theta\, \mathcal{P}_{\rm st}(\theta, \omega) = 1$, the stationary distribution is found from Eq. (6.105) to be $\mathcal{P}_{\rm st}(\theta, \omega) = 1/(2\pi)$. Therefore, in the stationary state, all the values of $\theta \in [0, 2\pi)$ are equally likely. In order to find the Fourier components $\widetilde{P}_{n,m}(t|t')$, one observes that for this particular case, Eq. (6.86) reduces to

$$\frac{\partial \widetilde{P}_{n,m}(t|t')}{\partial t} = -\left(in\omega + Dn^2\right) \widetilde{P}_{n,m}(t|t'). \tag{6.106}$$

This set of uncoupled differential equations may be solved easily with the initial condition $\widetilde{P}_{n,m}(t'|t') = \delta_{n,-m}/(2\pi)$. The solutions are given by

$$\widetilde{P}_{n,m}(t|t') = \frac{\delta_{n,-m}}{2\pi} \exp\left[-\left(in\omega + Dn^2\right)(t - t')\right]. \tag{6.107}$$

Then, substituting the above results for $\mathcal{K} = 0$ in Eq. (6.85), we obtain [1]

$$\begin{aligned} C(\tau, \omega) &= \frac{1}{4\pi} \int_0^{2\pi} d\theta \cos\theta \sum_{m=-\infty}^{\infty} e^{im\theta} \left[\delta_{1,-m}\, e^{-(i\omega+D)\tau} + \delta_{-1,-m}\, e^{(i\omega-D)\tau}\right] \\ &= \frac{1}{2\pi} e^{-D\tau} \int_0^{2\pi} d\theta \cos\theta \cos(\theta + \omega\tau) = \frac{1}{2} \cos(\omega\tau)\, e^{-D\tau}, \end{aligned} \tag{6.108}$$

which is an exact result for the stationary autocorrelation. Note that the result shown in Eq. (6.108) may directly be obtained from Eq. (6.102) by putting $\mathcal{K} = 0$. We thus obtain the equal-time stationary correlation $C(\tau = 0, \omega) = \langle \cos^2\theta \rangle_{\rm st} = 1/2$. Putting $\omega = 0$, Eq. (6.108) gives the equilibrium correlation as

$$C(\tau, \omega = 0) = \frac{1}{2} e^{-D\tau}. \tag{6.109}$$

Armed with our exact results derived in this section for stationary correlation of the single noisy nonuniform oscillator, we are now in a position to derive the stationary correlation for the noisy Kuramoto model. We will focus on this computation in the next section.

## 6.5 The Noisy Kuramoto Model: Stationary Correlation

### 6.5.1 *Mapping to a Single Noisy Nonuniform Oscillator*

For the noisy Kuramoto dynamics in the stationary state, Eq. (6.25), we have $N$ interacting oscillators with distributed frequencies. Comparing the dynamics (6.25) and (6.61), we arrive at the following useful analogy between the system of Kuramoto oscillators and the single nonuniform oscillator. First, let us club the Kuramoto oscillators into groups that have the same natural frequency $\omega$. Then, the stationary-state dynamics of oscillators within each group is that of a nonuniform oscillator, Eq. (6.61), with the constant $\mathcal{K}$ equal to $Kr_{\rm st}$, where the macroscopic quantity $r_{\rm st}$ is the global order parameter obtained from the stationary-state dynamics of oscillators across all groups.

As defined in Eq. (6.60), the stationary autocorrelation for the noisy Kuramoto dynamics is given by

$$C(\tau) \equiv \lim_{\substack{t'\to\infty \\ \tau=\text{finite}}} \langle\langle \cos\theta(t')\cos\theta(t'+\tau)\rangle\rangle\,, \tag{6.110}$$

which is a measure of similarity of $\cos\theta$-values at two different times. We reiterate that the rationale of choosing $\cos\theta$ as the dynamical variable is that the order parameter $r_{\rm st}$ itself is given by $\langle\langle\cos\theta\rangle\rangle$ in the stationary state, see Eq. (6.51). The autocorrelation $C(\tau)$ measures the possibility of observing a given $\cos\theta$-value at one time and another given value at another time. Consequently, $C(\tau)$ will be given by an appropriate joint probability distribution. As we have clubbed the Kuramoto oscillators into groups of same frequency, the autocorrelation $C(\tau,\omega)$ for a group with frequency $\omega$ is the same as the autocorrelation of the noisy nonuniform oscillator with the same frequency. Then, according to Eq. (6.84), for each such group of Kuramoto oscillators, one has the autocorrelation

$$C(\tau,\omega) = \lim_{\substack{t'\to\infty \\ \tau=\text{finite}}} \int_0^{2\pi} \mathrm{d}\theta \int_0^{2\pi} \mathrm{d}\theta'\, \cos\theta\, \cos\theta'\, P(\theta,t'+\tau|\theta',t')\,\rho_{\rm st}(\theta',\omega;r_{\rm st})\,. \tag{6.111}$$

Note that the stationary distribution $\rho_{\rm st}(\theta',\omega;r_{\rm st})$ for the noisy Kuramoto oscillators that satisfies Eq. (6.26) and given by Eq. (6.29) is the same as the stationary distribution $\mathcal{P}_{\rm st}(\theta',\omega)$ given in Eq. (6.71) for the noisy nonuniform oscillator, with the identification $\mathcal{K} = Kr_{\rm st}$. In the stationary state, each group among the Kuramoto oscillators interacts with every other group only through the time-independent mean-field order parameter $r_{\rm st}$. Then, considering all the groups with different frequencies, one obtains the stationary autocorrelation $C(\tau)$ for the noisy Kuramoto model as

$$C(\tau) = \int_{-\infty}^{\infty} \mathrm{d}\omega\; g(\omega)\; C(\tau,\omega)\,. \tag{6.112}$$

### 6.5.2 Stationary Correlation: Exact Results

In order to derive our results for the stationary dynamics of the noisy Kuramoto model, Eq. (6.25), we employ the aforementioned exact analogy that exists between it and the dynamics (6.61) of the single nonuniform oscillator on setting the constant $\mathcal{K}$ in the latter to the value $Kr_{\rm st}$, where $K$ is as usual the coupling constant of the Kuramoto model and $r_{\rm st}$ is the stationary Kuramoto order parameter. For a given value of $\omega$, we rewrite here by using Eq. (6.29) the single-oscillator probability distribution in the stationary state of the noisy Kuramoto model:

$$\rho_{\rm st}(\theta,\omega;r_{\rm st}) = A(\omega;r_{\rm st})\,{\rm e}^{(-Kr_{\rm st}+Kr_{\rm st}\cos\theta+\omega\theta)/D} \times\left[1+\frac{({\rm e}^{-\frac{2\pi\omega}{D}}-1)\int_0^\theta {\rm d}\theta'{\rm e}^{(-\omega\theta'-Kr_{\rm st}\cos\theta')/D}}{\int_0^{2\pi}{\rm d}\theta'{\rm e}^{(-\omega\theta'-Kr_{\rm st}\cos\theta')/D}}\right], \tag{6.113}$$

with the constant $A(\omega;r_{\rm st})$ to be fixed by the normalization $\int_0^{2\pi}{\rm d}\theta\;\rho_{\rm st}(\theta,\omega;r_{\rm st}) = 1\;\forall\,\omega$, and $r_{\rm st}$ to be determined from the self-consistent equation

$$r_{\rm st} = \int_{-\infty}^{\infty}{\rm d}\omega\;g(\omega)\int_0^{2\pi}{\rm d}\theta\;\cos\theta\;\rho_{\rm st}(\theta,\omega;r_{\rm st})\,. \tag{6.114}$$

In order to obtain the value $r_{\rm st}$ from Eq. (6.114), one needs to perform a double integration with respect to $\omega$ and $\theta$; this cannot be carried out for generic $g(\omega)$, and one has to rely on numerical evaluation. Instead of performing the double integration numerically, we use a different prescription to reduce the computational cost [14]. We recall the Fokker–Planck equation (6.27) in the stationary state:

$$-D\frac{\partial\rho_{\rm st}(\theta,\omega;r_{\rm st})}{\partial\theta}+\big(\omega-Kr_{\rm st}\sin\theta\big)\,\rho_{\rm st}(\theta,\omega;r_{\rm st}) = J_{\rm st}(\omega)\,. \tag{6.115}$$

Multiplying Eq. (6.115) by $\exp({\rm i}n\theta)$, where $n$ is an integer, and integrating over $\theta$, one obtains

$$\begin{aligned}-D\int_0^{2\pi}{\rm d}\theta\;{\rm e}^{{\rm i}n\theta}\,\frac{\partial\rho_{\rm st}(\theta,\omega;r_{\rm st})}{\partial\theta}+\omega\int_0^{2\pi}{\rm d}\theta\;{\rm e}^{{\rm i}n\theta}\;\rho_{\rm st}(\theta,\omega;r_{\rm st})\\ -Kr_{\rm st}\int_0^{2\pi}{\rm d}\theta\;{\rm e}^{{\rm i}n\theta}\sin\theta\;\rho_{\rm st}(\theta,\omega;r_{\rm st}) = J_{\rm st}(\omega)\int_0^{2\pi}{\rm d}\theta\;{\rm e}^{{\rm i}n\theta}\,.\end{aligned} \tag{6.116}$$

In the stationary state, denoting the expectation value of the quantity $\exp({\rm i}n\theta)$ (which depends on $\omega$ and $r_{\rm st}$) as

$$\langle{\rm e}^{{\rm i}n\theta}\rangle(\omega;r_{\rm st}) \equiv \int_0^{2\pi}{\rm d}\theta\;{\rm e}^{{\rm i}n\theta}\;\rho_{\rm st}(\theta,\omega;r_{\rm st})\,, \tag{6.117}$$

one obtains from Eq. (6.116) that

$$- D \int_0^{2\pi} \mathrm{d}\theta \left[ \frac{\partial}{\partial \theta} \{ \mathrm{e}^{\mathrm{i}n\theta} \rho_{\mathrm{st}}(\theta, \omega; r_{\mathrm{st}}) \} - \mathrm{i}n \mathrm{e}^{\mathrm{i}n\theta} \rho_{\mathrm{st}}(\theta, \omega; r_{\mathrm{st}}) \right] + \omega \, \langle \mathrm{e}^{\mathrm{i}n\theta} \rangle (\omega; r_{\mathrm{st}})$$
$$+ \frac{\mathrm{i}K r_{\mathrm{st}}}{2} \int_0^{2\pi} \mathrm{d}\theta \, \left[ \mathrm{e}^{\mathrm{i}(n+1)\theta} - \mathrm{e}^{\mathrm{i}(n-1)\theta} \right] \rho_{\mathrm{st}}(\theta, \omega; r_{\mathrm{st}}) = 2\pi \, J_{\mathrm{st}}(\omega) \, \delta_{n,0} \, . \quad (6.118)$$

Using the condition $\rho_{\mathrm{st}}(2\pi, \omega; r_{\mathrm{st}}) = \rho_{\mathrm{st}}(0, \omega; r_{\mathrm{st}})$ evaluates the first integral to zero. Then, Eq. (6.118) reduces to

$$(\omega + \mathrm{i}nD) \, \langle \mathrm{e}^{\mathrm{i}n\theta} \rangle (\omega; r_{\mathrm{st}}) + \frac{\mathrm{i}K r_{\mathrm{st}}}{2} \left[ \langle \mathrm{e}^{\mathrm{i}(n+1)\theta} \rangle (\omega; r_{\mathrm{st}}) - \langle \mathrm{e}^{\mathrm{i}(n-1)\theta} \rangle (\omega; r_{\mathrm{st}}) \right]$$
$$= 2\pi \, J_{\mathrm{st}}(\omega) \, \delta_{n,0} \, . \quad (6.119)$$

From the self-consistent equation (6.114), one observes that

$$r_{\mathrm{st}} = \int_{-\infty}^{\infty} \mathrm{d}\omega \; g(\omega) \; \mathrm{Re}[\langle \mathrm{e}^{\mathrm{i}\theta} \rangle (\omega; r_{\mathrm{st}})] \, . \quad (6.120)$$

Equation (6.119) may be written as a system of linear algebraic equations for different values of $n$. Then, omitting the arguments of the expectation values for brevity, one obtains

$$\begin{aligned}
n = 1: &\; (\omega + \mathrm{i}D) \, \langle \mathrm{e}^{\mathrm{i}\theta} \rangle + \frac{\mathrm{i}K r_{\mathrm{st}}}{2} \langle \mathrm{e}^{2\mathrm{i}\theta} \rangle = \frac{\mathrm{i}K r_{\mathrm{st}}}{2} \, , \\
n = 2: &\; - \frac{\mathrm{i}K r_{\mathrm{st}}}{2} \langle \mathrm{e}^{\mathrm{i}\theta} \rangle + (\omega + 2\mathrm{i}D) \, \langle \mathrm{e}^{2\mathrm{i}\theta} \rangle + \frac{\mathrm{i}K r_{\mathrm{st}}}{2} \langle \mathrm{e}^{3\mathrm{i}\theta} \rangle = 0 \, , \\
n = 3: &\; - \frac{\mathrm{i}K r_{\mathrm{st}}}{2} \langle \mathrm{e}^{2\mathrm{i}\theta} \rangle + (\omega + 3\mathrm{i}D) \, \langle \mathrm{e}^{3\mathrm{i}\theta} \rangle + \frac{\mathrm{i}K r_{\mathrm{st}}}{2} \langle \mathrm{e}^{4\mathrm{i}\theta} \rangle = 0 \, , \\
&\; \vdots
\end{aligned} \quad (6.121)$$

We truncate the above set of infinite equations at a finite value $n = n_{\mathrm{max}}$. Next, for given values of $\omega$ and $r_{\mathrm{st}}$, we solve the set of truncated equations to obtain $\langle \mathrm{e}^{\mathrm{i}\theta} \rangle (\omega; r_{\mathrm{st}})$. The latter value is then used on the right hand side of the self-consistent equation (6.120), following which we check whether the value of $r_{\mathrm{st}}$ so obtained is within numerical accuracy the same as the one chosen in solving the truncated set of equations. If the answer is in the affirmative, we conclude that we have obtained numerically the self-consistent value of $r_{\mathrm{st}}$. If not, we repeat the whole process until we have obtained the self-consistent $r_{\mathrm{st}}$. This prescription helps us to replace the costlier computation of double integration in Eq. (6.114) with that involving solution of a set of linear equations (6.121) and a single integration (6.120). However, we pay a price in the form of the approximation involving truncation of the set of Eq. (6.121). We choose $n_{\mathrm{max}}$ to be a conveniently large integer such that a larger $n_{\mathrm{max}}$ causes no appreciable change to the self-consistent $r_{\mathrm{st}}$. Typically, we choose $n_{\mathrm{max}}$ to be fifty.

Once we obtain $r_{\mathrm{st}}$, we replace $\mathcal{K}$ by $Kr_{\mathrm{st}}$ in the expressions for the single-oscillator autocorrelation $C(\tau, \omega)$ reported in Sect. 6.4.4. Then, using Eq. (6.112), we obtain the stationary correlation $C(\tau)$ for the noisy Kuramoto oscillators. For a given choice of $g(\omega)$ and a given value of $K$, the explicit steps involved in obtaining the correlation $C(\tau)$ in the synchronized and incoherent phases of the Kuramoto model are as follows [1].

1. $D \geq D_c$: For $D \geq D_c$, when $\mathcal{K} = Kr_{\mathrm{st}} = 0$, Eq. (6.108) gives

$$C(\tau) = \frac{\mathrm{e}^{-D\tau}}{2} \int_{-\infty}^{\infty} \mathrm{d}\omega\, g(\omega)\, \cos(\omega\tau)\,. \tag{6.122}$$

2. $D < D_c$: For $D < D_c$, when $\mathcal{K} = Kr_{\mathrm{st}} \neq 0$, we first obtain the value of $r_{\mathrm{st}}$ by solving the self-consistent equation (6.120). We then solve for every value of $\omega$ in the support of $g(\omega)$ the system of Eq. (6.86) with the substitution $\mathcal{K} = Kr_{\mathrm{st}}$. We then use the solution $\widetilde{P}_{n,m}$ in Eq. (6.85) in which $\mathcal{P}_{\mathrm{st}}(\theta, \omega)$ is given by Eq. (6.71) with $\mathcal{K} = Kr_{\mathrm{st}}$, to obtain $C(\tau, \omega)$. Finally, Eq. (6.112) yields the desired correlation.
3. $D \lesssim D_c$: For the particular case of $D \lesssim D_c$, so that $\mathcal{K} = Kr_{\mathrm{st}}$ is small, we may use Eq. (6.102) to get to order $(Kr_{\mathrm{st}})^2$ the result

$$\begin{aligned} C(\tau) = \int_{-\infty}^{\infty} \mathrm{d}\omega\, g(\omega) \Bigg[ \frac{1}{4}\, \mathrm{e}^{-(D+\mathrm{i}\omega)\tau} + (Kr_{\mathrm{st}})^2 \Bigg\{ & \frac{\mathrm{e}^{-(D+\mathrm{i}\omega)\tau}}{16\,(D+\mathrm{i}\omega)\,(2D+\mathrm{i}\omega)} \\ & + \frac{D(1-\mathrm{e}^{-(D+\mathrm{i}\omega)\tau})}{8(D+\mathrm{i}\omega)(D^2+\omega^2)} + \frac{\mathrm{e}^{-(4D+2\mathrm{i}\omega)\tau} - \mathrm{e}^{-(D+\mathrm{i}\omega)\tau}}{16(D+\mathrm{i}\omega)(3D+\mathrm{i}\omega)} \\ & + \frac{\mathrm{e}^{-(D+\mathrm{i}\omega)\tau}}{8(3D+\mathrm{i}\omega)} \left( \frac{1-\mathrm{e}^{-(3D+\mathrm{i}\omega)\tau}}{3D+\mathrm{i}\omega} - \tau \right) \Bigg\} \Bigg] + \mathrm{c.c.}\,. \end{aligned} \tag{6.123}$$

Following the above procedure, and considering $g(\omega)$ to be a Gaussian distribution centered at zero and with standard deviation $\sigma = 0.2$, we show in Fig. 6.9 our analytical results for the noisy Kuramoto model compared against those obtained from direct numerical integration of the dynamics (6.4) for very large $N$. Note that all the analytical computations were done in the limit $N \to \infty$. The figure shows a very good agreement between theory and numerical results.

### 6.5.3 *Results for the Brownian Mean-Field Model*

Results for the BMF model, which corresponds to the case $g(\omega) = \delta(\omega)$, may be obtained by using Eqs. (6.113) and (6.114), as

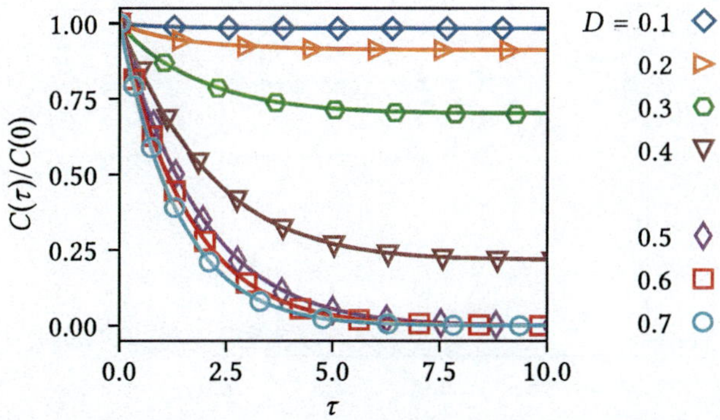


**Fig. 6.9 Stationary autocorrelation $C(\tau)$ normalized by its value at $\tau = 0$ in the noisy Kuramoto model** (6.4). The frequency distribution $g(\omega)$ is taken to be a Gaussian centered at zero and with standard deviation given by $\sigma = 0.2$, thus yielding the critical noise strength $D_c \approx 0.43$, see Eq. (6.59). The coupling constant is $K = 1$. Here, the continuous lines refer to analytical results obtained in the text, while the points are obtained by numerical integration of the stochastic dynamics (6.4), by using the stochastic version of the Euler method discussed in Appendix M.2 with $\Delta t = 0.01$ and $N = 10^5$. The set of angles $\{\theta_i\}$ at $t = 0$ is taken to be distributed independently and uniformly in the range $[0, 2\pi)$. The data are averaged over 100 realizations. The figure shows a very good agreement between theory and numerical results

$$\rho_{\rm st}(\theta; r_{\rm st}) = \frac{\mathrm{e}^{(K r_{\rm st} \cos\theta)/D}}{\int_0^{2\pi} \mathrm{d}\theta\ \mathrm{e}^{(K r_{\rm st} \cos\theta)/D}}; \qquad r_{\rm st} = \frac{I_1(K r_{\rm st}/D)}{I_0(K r_{\rm st}/D)}, \tag{6.124}$$

where for obvious reasons, we have dropped the dependence of $\rho_{\rm st}$ on $\omega$. Here, $I_n(x)$ is the modified Bessel function of the first kind, which for integer $n$ is defined as

$$I_n(x) \equiv \frac{1}{2\pi} \int_0^{2\pi} \mathrm{d}\theta\ \cos(n\theta)\ \mathrm{e}^{x \cos\theta}. \tag{6.125}$$

Then, for the BMF model, we obtain the correlation in the equilibrium stationary state as follows [1].

1. $D \geq D_c$: For $D \geq D_c$, the autocorrelation is obtained from Eq. (6.108) as

$$C(\tau) = \frac{\mathrm{e}^{-D\tau}}{2} \int_{-\infty}^{\infty} \mathrm{d}\omega\ \delta(\omega)\ \cos(\omega\tau) = \frac{1}{2}\ \mathrm{e}^{-D\tau}. \tag{6.126}$$

2. $D < D_c$: For $D < D_c$, we first solve the self-consistent equation for $r_{\rm st}$ given in Eq. (6.124). Then, using this value of $r_{\rm st}$, we solve the system of Eq. (6.86) with the substitution $\mathcal{K} = K r_{\rm st}$ and $\omega = 0$. Finally, using (6.85) with $\omega = 0$ and $\mathcal{P}_{\rm st}(\theta, \omega)$ replaced with $\rho_{\rm st}(\theta; r_{\rm st})$ given in Eq. (6.124), we obtain the correlation.
3. $D \lesssim D_c$: In particular, the result for $D \lesssim D_c$ is obtained from Eq. (6.103) with the substitution $\mathcal{K} = K r_{\rm st}$.

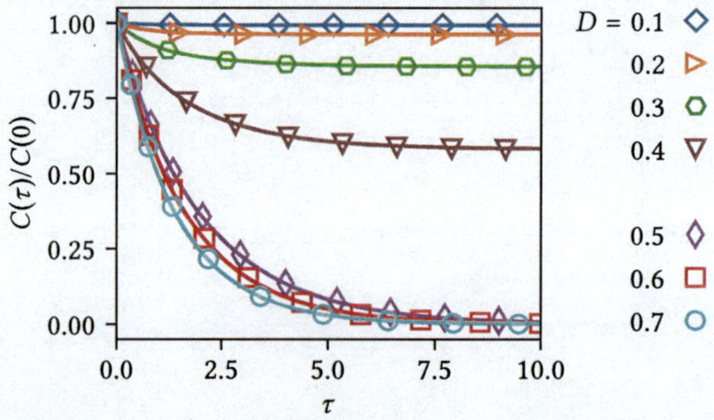


**Fig. 6.10 Stationary autocorrelation $C(\tau)$ normalized by its value at $\tau = 0$ in the BMF model** (6.7). Here, we set $K = 1$. The critical noise strength $D_c$ is obtained from Eq. (6.59) with $g(\omega) = \delta(\omega)$. Then, one finds that $D_c = 0.5$. The continuous lines refer to analytical results obtained in the text, while the points are obtained by numerical integration of the stochastic dynamics (6.7), by using the stochastic version of the Euler method discussed in Appendix M.2 with $\Delta t = 0.01$ and $N = 10^6$. The set of angles $\{\theta_i\}$ at $t = 0$ is taken to be distributed independently and uniformly in the range $[0, 2\pi)$. The data are averaged over 10 realizations. The figure shows a very good agreement between theory and numerical results

Figure 6.10 shows the correlation for the BMF model. In this figure, we have shown our analytical results for the BMF model compared against those obtained from direct numerical integration of the dynamics (6.7) for very large $N$. The figure shows a very good agreement between theory and numerical results.

The results for $D \lesssim D_c$ are shown in Fig. 6.11 for both the noisy Kuramoto model and the BMF model. We have taken the coupling constant to be $K = 1$. In Figs. 6.9, 6.10, and 6.11, the function $C(\tau)$ indeed decays to the value $r_{\rm st}^2$ for large $\tau$, as it should. In all cases, a very good agreement between theory and numerical results is evident from the plots. The autocorrelation $C(\tau)$ in all cases decays as an exponential with a rate that increases continuously with the noise strength. An implication of this result is that the more noisy the dynamics is, the faster it takes to generate uncorrelated configurations.

## 6.6 Conclusions and Perspectives

In this chapter, we obtained exact analytical results on correlations in the nonequilibrium stationary state of a paradigmatic many-body interacting system showing spontaneous order, the Kuramoto model of coupled oscillators, evolving in presence of Gaussian, white noise. The method relies on an exact mapping of the stationary-state dynamics of the model to the noisy dynamics of a single, nonuniform oscillator, which could be possible due to the mean-field nature of the dynamics of the Kuramoto model. Namely, the dynamics may be thought as that of a single oscillator evolv-

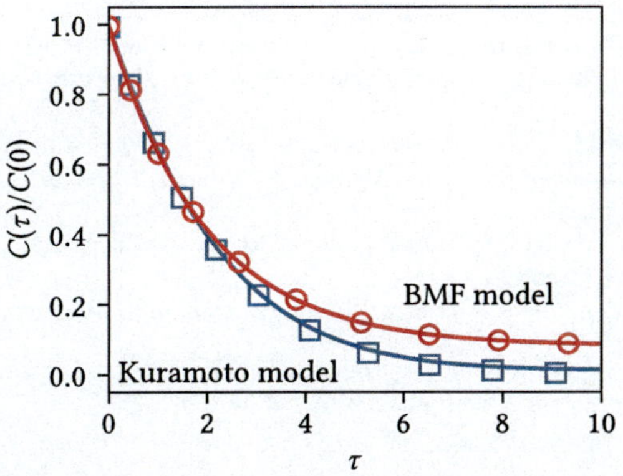


**Fig. 6.11 Stationary autocorrelation $C(\tau)$ normalized by its value at $\tau = 0$ in the special case of $D \lesssim D_c$.** For the noisy Kuramoto model (6.4), we take here the same parameter values as used in Fig. 6.9. Similarly, for the BMF model (6.7), the parameter values used here are identical to those in Fig. 6.10. For the noisy Kuramoto model, we have $D_c \approx 0.43$. The continuous blue line given by the analytical result (6.123) with $D = 0.427$ corresponds to the noisy Kuramoto model. The critical noise strength $D_c$ for the BMF model equals 0.5. The continuous red line shows the analytical result for the BMF model obtained from Eq. (6.103) with the substitution $\mathcal{K} = K r_{\rm st}$ and $D = 0.49$. In both cases, the points are obtained by numerically integrating the corresponding stochastic dynamics (6.4) and (6.7) by using the same numerical scheme used in Figs. 6.9 and 6.10, respectively. We see a good agreement between theory and numerical results in both cases

ing in a mean-field due to its interaction with all the other oscillators. Although we considered the Kuramoto model that involves a sinusoidal interaction between the oscillators, the method is designed to work for any form of interaction between the oscillators so long as it is of the mean-field type. It would be interesting to obtain correlations in the stationary state of the Kuramoto model with inertia [14]. As is well known, inertia significantly changes the nature of the synchronization transition of the model, and is thus expected to affect also the stationary-state correlation. On the analytical side, inertia brings in significant complexity in obtaining even the stationary state of the model [14].

## References


1. D. Das, S. Gupta, J. Phys. A: Math. Theor. **51**(44), 445003 (2018). https://doi.org/10.1088/1751-8121/aae2c2
2. K. Huang, *Statistical Mechanics*, 2nd edn. (Wiley, New York, 1987)
3. R. Livi, P. Politi, *Nonequilibrium Statistical Physics: A Modern Perspective* (Cambridge University Press, Cambridge, 2017)
4. R. Kubo, Rep. Prog. Phys. **29**(1), 255 (1966). https://doi.org/10.1088/0034-4885/29/1/306
5. H. Sakaguchi, Prog. Theor. Exp. Phys. **79**(1), 39 (1988). https://doi.org/10.1143/PTP.79.39
6. R. Zwanzig, *Nonequilibrium Statistical Mechanics* (Oxford University Press, Oxford, 2001)
7. I.H.d. Mendoza, L.A. Pachón, J. Gómez-Gardeñes, D. Zueco, Phys. Rev. E **90**(5), 052904 (2014). https://doi.org/10.1103/PhysRevE.90.052904

8. P.H. Chavanis, Eur. Phys. J. B **87**(5), 120 (2014). https://doi.org/10.1140/epjb/e2014-40586-6
9. M. Antoni, S. Ruffo, Phys. Rev. E **52**(3), 2361 (1995). https://doi.org/10.1103/PhysRevE.52.2361
10. R.K. Pathria, P.D. Beale, *Statistical Mechanics*, 3rd edn. (Elsevier, Amsterdam, 2011)
11. C.W. Gardiner, *Handbook of Stochastic Methods for Physics, Chemistry and the Natural Sciences* (Springer, Berlin, 1983)
12. S.H. Strogatz, *Nonlinear Dynamics and Chaos* (CRC Press, Boca Raton, 2018). https://doi.org/10.1201/9780429492563
13. H. Risken, *The Fokker-Planck Equation: Methods of Solution and Applications* (Springer, Berlin, 1996)
14. S. Gupta, A. Campa, S. Ruffo, *Statistical Physics of Synchronization* (Springer, Cham, 2018). https://doi.org/10.1007/978-3-319-96664-9

# Chapter 7
# Bifurcation Behavior of a Nonlinear System by Introducing Noise

**Abstract** In this chapter, we unveil a useful facet of noise: we show how one may obtain the bifurcation behavior of a nonlinear dynamical system by introducing noise into its dynamics. This is done by adding a Gaussian, white noise term to the dynamics and then studying the resulting Langevin dynamics in the weak-noise limit. We show that the behavior of the noisy dynamics can be effectively captured via a conditional probability of observing microscopic configurations at a given instant, conditioned on having observed a given configuration at an earlier instant. We study this conditional probability for our model system by using two distinct approaches, namely, the Fokker-Planck and the path integral approach. An exact closed-form expression for the conditional probability can be obtained within the latter approach. By studying the long-time behavior of the conditional probability in the weak-noise limit, we show how to reconstruct the bifurcation diagram of the noiseless dynamics. All our predictions are validated with direct numerical integration of the dynamical equations of motion. This chapter is based on our published work, Ref. [1].

## 7.1 Introduction

Nonlinear dynamical systems exhibit a variety of physical phenomena that are fascinating and at the same time counterintuitive and intriguing, e.g., chaos [2], pattern formation [3], occurrence of solitons [4], and many more. In this chapter, we address the issue of how one may obtain the bifurcation diagram of a nonlinear dynamical system by introducing noise into its dynamics and studying the resulting noisy dynamics using tools of stochastic processes. The work reported in this chapter provides a theoretical framework to systematically obtain the stability properties of a nonlinear dynamical system from a suitable analysis of the corresponding noisy dynamics, and as an illustration of how one may derive analytical expressions of the quantities involved in the latter analysis.

The chapter is laid out as follows. We start off with a definition of a model system in Sect. 7.2. Sections 7.3 and 7.4 are devoted respectively to an analysis of the noiseless and the noisy dynamics of our model system. Section 7.5 forms the heart of this chapter, in which we discuss our main results on the reconstruction of the

 
D. Das and S. Gupta, *Facets of Noise*, Fundamental Theories of Physics 214,
https://doi.org/10.1007/978-3-031-45312-0_7

bifurcation diagram of the noiseless dynamics using the noisy one. The chapter ends with conclusions and perspectives in Sect. 7.6.

## 7.2 The Model

We consider a nonlinear dynamical system described by a single phase-like variable $\theta \in [-\pi, \pi]$. The dynamical equation for $\theta(t)$ is given by

$$\frac{d\theta}{dt} = A \sin\theta - B \sin 2\theta\,, \tag{7.1}$$

where $A$ and $B$ are two real parameters. Using a simple rescaling of time as $\tau \equiv Bt$, one may get rid of one of the parameters from the dynamics. Then, denoting the rescaled time $\tau$ as $t$, the dynamics (7.1) may be recast as

$$\frac{d\theta}{dt} = a \sin\theta - \sin 2\theta\,, \tag{7.2}$$

where $a \equiv A/B$ is the only real parameter in the dynamics. From now on, we consider Eq. (7.2) as our nonlinear dynamical equation of study. The dynamics (7.2) may also be viewed as the evolution of $\theta(t)$ in presence of a potential $V(\theta)$:

$$\frac{d\theta}{dt} = -V'(\theta)\,, \tag{7.3}$$

with

$$V(\theta) \equiv a\cos\theta - \frac{1}{2}\cos 2\theta\,. \tag{7.4}$$

In Eq. (7.3), the prime denotes the first derivative with respect to $\theta$.

## 7.3 Noiseless Dynamics

The dynamics (7.2) is fully deterministic, i.e., once an initial value $\theta(t = 0)$ is given, we are, in principle, able to predict $\theta(t)$ for all later times. As one tunes the parameter $a$, the behavior of the potential landscape $V(\theta)$ changes. Consequently, the fixed points of the dynamics (7.2) and their stability change with change of $a$, yielding a bifurcation diagram that we discuss below.

### *7.3.1 Analysis of the Potential Landscape*

With the aim to locate the fixed points of the dynamics (7.2) in the next subsection, we analyze in this section the form of the potential $V(\theta)$, which evidently satisfies

$$V_{a<0}(\theta) = V_{a>0}(\theta - \pi)\,. \tag{7.5}$$

In order to find the extrema of $V(\theta)$, one solves $V'(\theta) = 0$ and obtains

$$\theta = \begin{cases} 0\,, \pm\pi & \text{for all values of } a, \\ \cos^{-1}(a/2) & -2 < a < 2\,. \end{cases} \tag{7.6}$$

Computing $V''(\theta)$ at these points, one may easily check that

- $\theta = 0$ maximizes $V(\theta)$ for $a > 2$ and minimizes it for $a < 2$,
- $\theta = \pm\pi$ maximize $V(\theta)$ for $a < -2$ and minimize it for $a > -2$,
- $\theta = \cos^{-1}(a/2)$ maximizes $V(\theta)$ for $-2 < a < 2$.

We also have

- $V''(0) = 0$ for $a = 2$,
- $V''(\pm\pi) = 0$ for $a = -2$,
- $V''(\cos^{-1}(a/2)) = 0$ for $a = \pm 2$.

In Fig. 7.1, we show the potential landscape $V(\theta)$ for a few representative values of $a$.

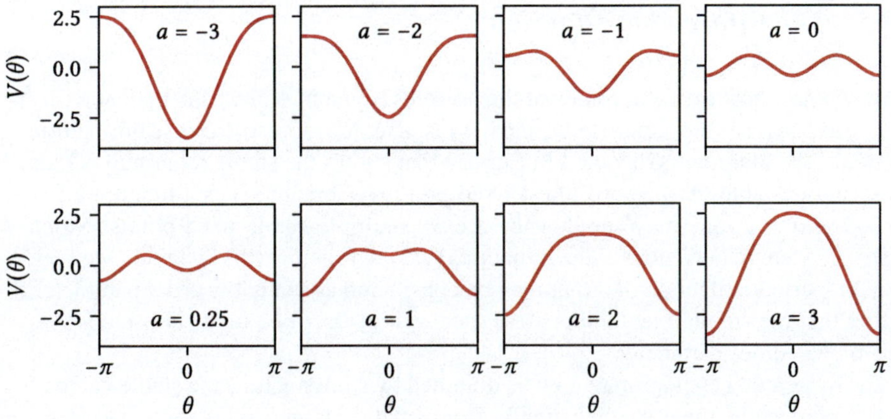


**Fig. 7.1 Potential** $V(\theta) = a\cos\theta - (1/2)\cos 2\theta$ **for representative values of the parameter** $a$

### 7.3.2 Fixed Points and Their Stability

From the discussions in Sect. 5.1.1, the fixed points $\bar{\theta}$ of the noiseless dynamics (7.2) are obtained as $\bar{\theta} = 0, \pm\pi$ for all values of $a$, and $\bar{\theta} = \cos^{-1}(a/2)$ or $\sin\bar{\theta} = \pm\sqrt{1 - a^2/4}$ for $-2 < a < 2$. These fixed points evidently satisfy $V'(\bar{\theta}) = 0$. We now discuss the linear stability of these fixed points. To this end, one substitutes in Eq. (7.2) the expansion $\theta = \bar{\theta} + \Delta\theta$, where $\Delta\theta$ is a small perturbation around the fixed point $\bar{\theta}$. Keeping terms upto linear order in $\Delta\theta$, one obtains

$$\frac{\mathrm{d}\Delta\theta}{\mathrm{d}t} = -V''(\bar{\theta})\Delta\theta\,. \tag{7.7}$$

The above equation suggests that $\Delta\theta$ decays (respectively, grows) exponentially in time for $V''(\bar{\theta}) > 0$ (respectively, $V''(\bar{\theta}) < 0$). Consequently, the fixed point $\bar{\theta}$ is linearly stable (respectively, unstable) for $V''(\bar{\theta}) > 0$ (respectively, $V''(\bar{\theta}) < 0$). From the analysis of the potential $V(\theta)$ given in Sect. 7.3.1, one concludes that

- For $a > 2$, the linearly-stable fixed points are $\bar{\theta} = \pm\pi$, while the fixed point $\bar{\theta} = 0$ is linearly unstable. For $a < -2$, the stability of these fixed points gets exchanged.
- For $a = 2$, the fixed point $\bar{\theta} = 0$ is linearly neutrally stable, while the fixed points $\bar{\theta} = \pm\pi$ are linearly stable. For $a = -2$, the fixed points $\bar{\theta} = \pm\pi$ are linearly neutrally stable, while the fixed point $\bar{\theta} = 0$ is linearly stable.
- For $-2 < a < 2$, the fixed points $\bar{\theta} = 0, \pm\pi$ are linearly stable, while the fixed points $\bar{\theta} = \cos^{-1}(a/2)$; $\sin\bar{\theta} = \pm\sqrt{1 - a^2/4}$ are linearly unstable.

### 7.3.3 The Bifurcation Diagram

We now know how the fixed points of the noiseless dynamics (7.2) and their stability change as one tunes the parameter $a$, and it is just one step more to draw the bifurcation diagram. The diagram is depicted in Fig. 7.2 that shows the stable (continuous blue lines) and unstable (dashed red lines) fixed points (or branches) as a function of $a$. For $-2 < a < 2$, one may see that there exist multiple stable fixed points, which implies hysteretic behavior for the dynamics (7.2). To capture this behavior, we seek for a $2\pi$-periodic function of $\theta$ that may be considered as a suitable order parameter. Noting that the stable fixed points are either zero or $\pm\pi$, one may choose $\cos\theta$ as the simplest order parameter.

The hysteretic behavior may now be obtained by studying how the stable value of $\cos\theta$ varies as one tunes $a$ adiabatically from small to large values and back. Here, the adiabatic tuning of $a$ is important because it allows the system enough time to relax to the stable state while starting from an initial state, before the value of $a$ changes appreciably. If we start with $a < -2$, any initial $\theta$ will relax at long times to the stable fixed point at $\bar{\theta} = 0$, see Fig. 7.1. Now, if $a$ is increased adiabatically,

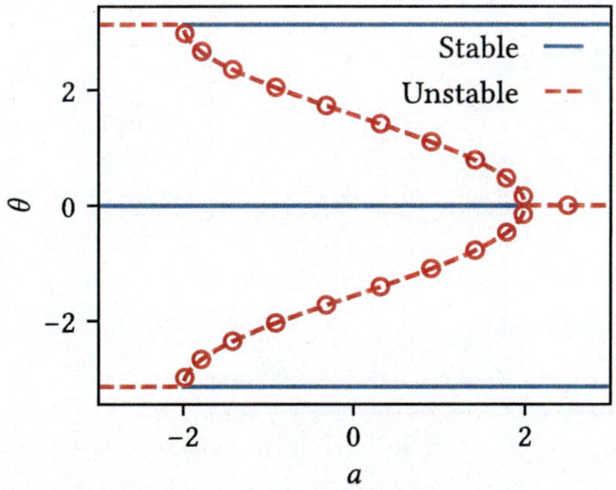


**Fig. 7.2 Bifurcation diagram of the noiseless dynamics (7.2).** For all values of $a$, the fixed points are $\theta = 0, \pm\pi$, while for $a$ in the range $-2 < a < 2$, additional fixed points are given by $\cos\theta = (a/2)$; stable fixed points are denoted by continuous lines, while the unstable ones are denoted by dashed lines. Here, the red circles may be obtained from the exact analysis of the noisy dynamics (7.9) based on the Fokker-Planck and the path integral approach discussed in the text

the value of $\theta$ will remain pinned to zero, until the minimum of the potential $V(\theta)$ at $\theta = 0$ turns into a maximum. The value of $a$ at which this happens, obtained by solving $V''(0) = 0$, is given by $a = 2$. Beyond $a = 2$, the stable value of $\theta$ will change to the value at the new minima, given by $\theta = \pm\pi$. Concomitant with the aforementioned behavior, $\cos\theta$ versus $a$ will behave as shown in Fig. 7.3 for the case of increasing $a$. Following the above line of argument, one may easily obtain the behavior of $\cos\theta$ versus $a$ for the case when $a$ has a starting value greater than 2 and is adiabatically decreased to a value less than $-2$. The corresponding behavior is depicted in Fig. 7.3 for the case of decreasing $a$. Hysteretic behavior of $\cos\theta$ is clearly evident from the figure.

From the diagram Fig. 7.2, one may identify the following bifurcation points ($a = 2$, $\theta = 0$) and ($a = -2$, $\theta = \pm\pi$). It is then natural to wonder about the nature of the bifurcation at these points. Let us first consider the bifurcation point at ($a = 2, \theta = 0$). Here, as $a$ is decreased across $a = 2$, a line of unstable fixed points bifurcates into two lines of unstable fixed points and one line of stable fixed points. The new lines of unstable fixed points that emerge are symmetrically disposed about the line of stable fixed points. To understand the nature of the bifurcation, one expands Eq. (7.2) close to the bifurcation point and obtains upto the first two leading orders in $\theta$ that

$$\frac{\mathrm{d}\theta}{\mathrm{d}t} = (a-2)\theta + (8-a)\frac{\theta^3}{6}\,. \tag{7.8}$$

Then, comparing Eq. (7.8) in the limit $a \to 2^+$ with Eq. (5.33), one may see that here we have a subcritical pitchfork bifurcation. Proceeding in the same fashion, one may

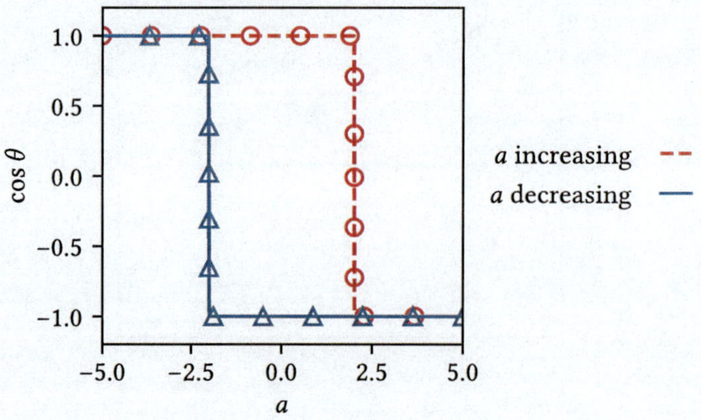


**Fig. 7.3** $\cos\theta$ **as a function of adiabatically-tuned** $a$**, showing hysteretic behavior.** The curves follow from the linear stability analysis of the noiseless dynamics (7.2) discussed in the text. On the other hand, the red circles and the blue triangles are obtained by numerically integrating the noisy dynamics (7.9) for the initial condition $\theta_0 = 0.75\pi$, and with $D = 10^{-5}$ and time step $\Delta t = 10^{-3}$, by using the stochastic version of the Euler method given in Appendix M.2; we first let the system reach the stationary state (reached at time $t = 10$) at a given value of $a < -2$, and then increase $a$ adiabatically to high values and back in a cycle. Here, the data correspond to one realization of the noisy dynamics

also check that the bifurcation that occurs with increasing $a$ at points $(a = -2, \theta = \pm\pi)$ is also a subcritical pitchfork bifurcation.

## 7.4 Noisy Dynamics

Our goal in this chapter is to show how the bifurcation behavior of a deterministic dynamical system may be obtained using the corresponding noisy dynamics. To proceed, we now introduce the noisy dynamics corresponding to the noiseless dynamics (7.2). Here, this is done by adding a Gaussian, white noise term $\eta(t)$ on the right hand side of Eq. (7.2) such that the corresponding noisy dynamics follows the Langevin equation:

$$\frac{\mathrm{d}\theta}{\mathrm{d}t} = a\sin\theta - \sin 2\theta + \eta(t)\,. \tag{7.9}$$

In Eq. (7.9), the Gaussian, white noise $\eta(t)$ satisfies

$$\langle\eta(t)\rangle = 0, \quad \langle\eta(t)\eta(t')\rangle = 2D\delta(t - t')\,, \tag{7.10}$$

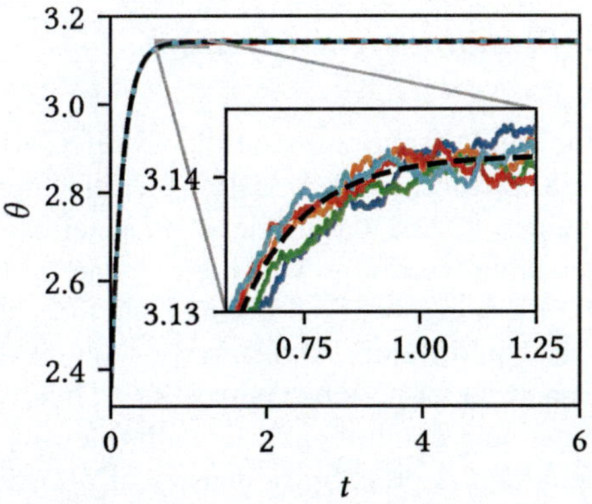


**Fig. 7.4 Dynamical trajectories of the noiseless and the noisy dynamics, Eqs.** (7.2) **and** (7.9)**, respectively, for the initial value $\theta_0 = 0.75\pi$.** The dashed line corresponds to the noiseless case, while the five continuous lines correspond to five independent realizations of the noisy dynamics. The trajectories are obtained by numerically integrating the corresponding equation of motion with integration time step $\Delta t = 10^{-3}$, by using the Euler method given in Appendix M.2. Here, we have chosen $a = 5$, $D = 10^{-5}$

where $D > 0$ sets the strength of the noise, and the angular brackets denote averaging over noise realizations. Note that by setting $D = 0$ one reduces the noisy dynamics to the noiseless one given in Eq. (7.2). Equation (7.9) may also be rewritten as

$$\frac{\mathrm{d}\theta}{\mathrm{d}t} = -V'(\theta) + \eta(t)\,, \tag{7.11}$$

which suggests that it corresponds to the noisy, overdamped dynamics of $\theta$ in the potential $V(\theta)$.

In Fig. 7.4, we show the dynamical trajectories $\theta(t)$ for the noiseless and the noisy dynamics obtained from Eqs. (7.2) and (7.9), respectively. From the figure, one may see that in the weak-noise limit ($D \to 0$), the trajectories obtained from the noisy dynamics occur as small fluctuations ($O(\sqrt{D})$) about the trajectory obtained from the noiseless dynamics. Owing to this observation, one may anticipate that it should be possible to extract the bifurcation behavior of the noiseless dynamics (7.2) from a suitable analysis of the noisy dynamics (7.9) with weak noise. As a straightforward numerical check of this expectation, one may perform numerical integration of the noisy dynamics (7.9) for small values of $D$. To this end, one may obtain the average value of $\cos\theta$ as a function of adiabatically-tuned $a$ and compare the results with that of the noiseless dynamics. Such a comparison is already shown in Fig. 7.3, where the lines and the points are obtained from the noiseless and the noisy dynamics, respectively, and it indeed shows a good match between the two results. In order to explain this match, we now turn to a detailed theoretical analysis of the noisy dynamics.

### 7.4.1 Phase-Space Probability Distribution

To locate dynamically the stable fixed points of the noiseless dynamics (7.2) for a given value of $a$, we need to run the dynamics for a sufficiently long time, while starting from an initial value of $\theta$, so that it relaxes to a stationary state. The latter would correspond to the stable fixed points of the dynamics. However, in case of the noisy dynamics (7.9), the system while evolving from the same initial condition follows different trajectories in different realizations of the noise $\eta(t)$. As a result, at all later times, one has a range of possible values of $\theta$. Therefore, to characterize the noisy dynamics, we define a conditional probability density $P(\theta, t|\theta_0, 0)$. The quantity $P(\theta, t|\theta_0, 0)$ denotes the probability density of observing the value $\theta$ at time $t$, given that the value $\theta_0$ was observed at the initial instant $t = 0$. We expect that studying the behavior of $P(\theta, t|\theta_0, 0)$ in the limit $t \to \infty$ and $D \to 0$ should allow recovery of the stable branches of the bifurcation diagram of the noiseless dynamics. Here, the function $P(\theta, t|\theta_0, 0)$ is $2\pi$-periodic in both $\theta$ and $\theta_0$, and obeys the following normalization condition:

$$\int_{-\pi}^{\pi} \mathrm{d}\theta \; P(\theta, t|\theta_0, 0) = 1 \; \forall \; \theta_0, t\,. \tag{7.12}$$

In the following, we discuss how one may obtain the functional form of the conditional probability density $P(\theta, t|\theta_0, 0)$ for a given value of $\theta_0$. To this end, we employ two different approaches, namely, the Fokker-Planck and the path integral approach.

#### 7.4.1.1 The Fokker-Planck Approach

The quantity $P(\theta, t|\theta_0, 0)$ evolves in time according to the Fokker-Planck equation corresponding to the Langevin equation (7.11). The Fokker-Planck equation is straightforwardly obtained as

$$\frac{\partial P(\theta, t|\theta_0, 0)}{\partial t} = -\frac{\partial}{\partial \theta}\left[-V'(\theta) P(\theta, t|\theta_0, 0)\right] + D\frac{\partial^2 P(\theta, t|\theta_0, 0)}{\partial \theta^2}\,, \tag{7.13}$$

with the initial condition

$$P(\theta, 0|\theta_0, 0) = \delta(\theta - \theta_0)\,. \tag{7.14}$$

The conditional probability density $P(\theta, t|\theta_0, 0)$, being $2\pi$-periodic in $\theta$, may be expanded in a Fourier series as

$$P(\theta, t|\theta_0, 0) = \sum_{n=-\infty}^{\infty} \widetilde{P}_n(t|\theta_0, 0)\, \mathrm{e}^{\mathrm{i}n\theta}\,. \tag{7.15}$$

As $P(\theta, t|\theta_0, 0)$ is a real quantity, we have $[\widetilde{P}_n(t|\theta_0, 0)]^\star = \widetilde{P}_{-n}(t|\theta_0, 0)$, where the star denotes complex conjugation. Substituting Eq. (7.15) in Eq. (7.13), we obtain the time evolution of the Fourier coefficients $\widetilde{P}_n(t|\theta_0, 0)$ as

$$\begin{aligned}\frac{\partial \widetilde{P}_n(t|\theta_0, 0)}{\partial t} &= -Dn^2\widetilde{P}_n(t|\theta_0, 0) + \frac{na}{2}\left[\widetilde{P}_{n+1}(t|\theta_0, 0) - \widetilde{P}_{n-1}(t|\theta_0, 0)\right] \\ &+ \frac{n}{2}\left[\widetilde{P}_{n-2}(t|\theta_0, 0) - \widetilde{P}_{n+2}(t|\theta_0, 0)\right], \end{aligned} \tag{7.16}$$

with the initial condition

$$\widetilde{P}_n(0|\theta_0, 0) = \frac{1}{2\pi}\,\mathrm{e}^{-in\theta_0}, \tag{7.17}$$

see Eq. (7.14).

In order to find $\widetilde{P}_n(t|\theta_0, 0)$ as a function of $t$ for any $n$ and a given value of $\theta_0$, one needs to solve the system of equations (7.16). This however is not an easy task as the system comprises coupled equations. Moreover, it is not closed and in fact involves an infinite hierarchy. Indeed, for a given value of $\theta_0$, to obtain $\widetilde{P}_n(t|\theta_0, 0)$ as a function of $t$, one requires the knowledge of $\widetilde{P}_{n+1}$ and $\widetilde{P}_{n+2}$, whose solution again requires knowing $\widetilde{P}_{n+3}$ and $\widetilde{P}_{n+4}$, and so on. However, for the initial condition (7.17), one may solve the system of equations by truncating it at a given value $n = n_{\max}$, i.e., by stipulating that $\widetilde{P}_n(t|\theta_0, 0) = 0$ for $n > n_{\max}$ and for all $t$. Here, one may choose $n_{\max}$ to be as large as possible, making sure that larger $n$ values do not contribute appreciably to the result, within the adopted level of numerical precision [1].

#### 7.4.1.2 The Path Integral Approach

We now discuss a complementary approach to obtain $P(\theta, t|\theta_0, 0)$ as a function of $t$, by invoking the Feynman-Kac path integral formalism of treating stochastic processes [5–8]. The path integral formalism of Feynman is reviewed in Appendix F. A benefit of using this approach is that it provides a closed-form expression for $P(\theta, t|\theta_0, 0)$, unlike the Fokker-Planck approach. We follow the procedure discussed in Ref. [9]. Consider a representation of the dynamics (7.11) in discrete times $t_i = i\Delta t$, with $i = 0, 1, 2, \ldots$, and $0 < \Delta t \ll 1$ being a small time step. Denoting $\theta_i \equiv \theta(t_i)$, the discrete-time dynamics is given by

$$\theta_i = \theta_{i-1} + \Delta t\;\overline{F}(\theta_i) + \Delta t\;\eta_i\,, \quad i \geq 1, \tag{7.18}$$

where we have

$$F(\theta_i) \equiv -V'(\theta_i)\,, \tag{7.19}$$

which for our model system (7.9) equals $F(\theta_i) = a\sin\theta_i - \sin 2\theta_i$, and $\overline{F}(\theta_i) \equiv (F(\theta_{i-1}) + F(\theta_i))/2$. In arriving at Eq. (7.18), we have used the Stratonovich rule, and not the Itô rule [10], in discretizing the dynamics (7.9). In Eq. (7.18), the Gaussian, white noise in the discretized form $\eta_i$ satisfies $\langle \eta_i \rangle = 0$; $\langle \eta_i \eta_j \rangle = (2D/\Delta t)\delta_{i,j}$. We discuss a proof of this claim in Appendix E. Defining $\xi_i \equiv \sqrt{\Delta t/2D}\,\eta_i$, so that

$$\langle \xi_i \rangle = 0\,; \qquad \langle \xi_i \xi_j \rangle = \frac{\Delta t}{2D} \langle \eta_i \eta_j \rangle = \delta_{i,j}\,, \tag{7.20}$$

let us rewrite Eq. (7.18) as

$$\theta_i = \theta_{i-1} + \Delta t\, \overline{F}(\theta_i) + \sqrt{2D\Delta t}\, \xi_i\,. \tag{7.21}$$

The joint probability distribution of occurrence of a given realization $\{\xi_i\}_{1 \le i \le N}$ (with $N$ being a positive integer) of the Gaussian, white noise is evidently given by

$$P[\{\xi_i\}] = \left(\frac{1}{2\pi}\right)^{N/2} \exp\left[-\frac{1}{2}\sum_{i=1}^{N} \xi_i^2\right]. \tag{7.22}$$

From the discrete-time dynamics (7.21) and the joint distribution (7.22), we obtain the probability of occurrence of a given phase trajectory $\{\theta_i\}_{0 \le i \le N} \equiv \{\theta_0, \theta_1, \theta_2, \ldots, \theta_{N-1}, \theta_N = \theta\}$ as

$$P[\{\theta_i\}] = \det(\mathcal{J}) \left(\frac{1}{2\pi}\right)^{N/2} \exp\left[-\frac{1}{2}\sum_{i=1}^{N} \frac{\left(\theta_i - \theta_{i-1} - \overline{F}(\theta_i)\Delta t\right)^2}{2D\Delta t}\right]. \tag{7.23}$$

Here, $\mathcal{J}$ is the Jacobian matrix for the transformation $\{\xi_i\} \to \{\theta_i\}$. The elements of $\mathcal{J}$ obtained from Eq. (7.21) are given by

$$\mathcal{J}_{1 \le i,j \le N} \equiv \frac{\partial \xi_i}{\partial \theta_j} = \frac{1}{\sqrt{2D\Delta t}} \left[\delta_{i,j}\left(1 - \frac{\Delta t\, F'(\theta_i)}{2}\right) + \delta_{i-1,j}\left(-1 - \frac{\Delta t\, F'(\theta_{i-1})}{2}\right)\right]. \tag{7.24}$$

For small $\Delta t$, we have

$$\det(\mathcal{J}) = \left(\frac{1}{2D\Delta t}\right)^{N/2} \prod_{i=1}^{N}\left(1 - \frac{\Delta t F'(\theta_i)}{2}\right) \approx \left(\frac{1}{2D\Delta t}\right)^{N/2} \exp\left[-\frac{\Delta t}{2}\sum_{i=1}^{N} F'(\theta_i)\right]. \tag{7.25}$$

On considering all possible trajectories, one gets the probability density that the phase starting at the value $\theta_0$ at time $t = 0$ evolves to the value $\theta$ at time $t = N\Delta t$ to be given by [9]

$$P(\theta, t|\theta_0, 0) = \left(\prod_{i=1}^{N-1} \int_{-\pi}^{\pi} d\theta_i\right) P[\{\theta_i\}]\,. \tag{7.26}$$

Using Eqs. (7.23) and (7.25), we obtain from Eq. (7.26) that

$$P(\theta, t|\theta_0, 0) = \left(\frac{1}{4\pi D\Delta t}\right)^{N/2} \left(\prod_{i=1}^{N-1} \int_{-\pi}^{\pi} \mathrm{d}\theta_i\right) \times \exp\left[-\Delta t \sum_{i=1}^{N} \left\{\frac{\left((\theta_i - \theta_{i-1})/\Delta t - \overline{F}(\theta_i)\right)^2}{4D} + \frac{F'(\theta_i)}{2}\right\}\right]. \tag{7.27}$$

In the limit of continuous time, i.e., $\Delta t \to 0,\ N \to \infty$, while keeping $N\Delta t = t$ fixed, on defining

$$\int_{\theta(0)=\theta_0}^{\theta(t)=\theta} \mathcal{D}[\theta(t)] \equiv \lim_{\substack{N\to\infty \\ \Delta t\to 0 \\ N\Delta t = t \text{ fixed}}} \left(\frac{1}{4\pi D\Delta t}\right)^{N/2} \left(\prod_{i=1}^{N-1} \int_{-\pi}^{\pi} \mathrm{d}\theta_i\right), \tag{7.28}$$

we get an exact expression for the corresponding probability density (7.27) to be given by the following path integral [9]:

$$P(\theta, t|\theta_0, 0) = \int_{\theta(0)=\theta_0}^{\theta(t)=\theta} \mathcal{D}[\theta(t)] \exp\left(-S[\theta(t)]\right). \tag{7.29}$$

Here, we have introduced the action as

$$S[\theta(t)] = \int_0^t \mathrm{d}t \left[\frac{[(\mathrm{d}\theta/\mathrm{d}t) - F(\theta)]^2}{4D} + \frac{F'(\theta)}{2}\right]. \tag{7.30}$$

Let us now invoke the Feynman-Kac formalism to identify the path integral on the right hand side of Eq. (7.29) with the propagator of a quantum mechanical evolution in (negative) imaginary time due to a quantum Hamiltonian $\hat{H}_\mathrm{q}$. We then obtain

$$P(\theta, t|\theta_0, 0) = \exp\left[\frac{1}{2D}\int_{\theta_0}^{\theta} \mathrm{d}\theta\, F(\theta)\right] G_\mathrm{q}(\theta, -\mathrm{i}t|\theta_0, 0) = \mathcal{F}(\theta, \theta_0)\, G_\mathrm{q}(\theta, -\mathrm{i}t|\theta_0, 0), \tag{7.31}$$

with

$$\mathcal{F}(\theta, \theta_0) \equiv \exp\left[\frac{1}{2D}\left\{a(\cos\theta_0 - \cos\theta) + \frac{\cos 2\theta - \cos 2\theta_0}{2}\right\}\right], \tag{7.32}$$
$$G_\mathrm{q}(\theta, -\mathrm{i}t|\theta_0, 0) \equiv \langle\theta|\exp(-\hat{H}_\mathrm{q}t)|\theta_0\rangle.$$

Here, the quantum Hamiltonian is

$$\hat{H}_{\mathrm{q}}(\theta) \equiv -\frac{1}{2m_{\mathrm{q}}}\frac{\partial^2}{\partial\theta^2} + V_{\mathrm{q}}(\theta)\,, \tag{7.33}$$

the mass in the equivalent quantum problem is

$$m_{\mathrm{q}} \equiv \frac{1}{2D}, \tag{7.34}$$

and the quantum potential is given by

$$V_{\mathrm{q}}(\theta) \equiv \frac{(F(\theta))^2}{4D} + \frac{F'(\theta)}{2} = \frac{(a\sin\theta - \sin 2\theta)^2}{4D} + \frac{a\cos\theta - 2\cos 2\theta}{2}\,. \tag{7.35}$$

Note that in the quantum propagator in Eq. (7.32), we have set the Planck's constant to unity.

When expressed in terms of the eigenvalues $E_n$ and the eigenfunctions $\Phi_n(\theta)$ of the Hamiltonian $H_{\mathrm{q}}(\theta)$, we have

$$G_{\mathrm{q}}(\theta, -\mathrm{i}t|\theta_0, 0) = \sum_n \Phi_n(\theta)\, \Phi_n^\star(\theta_0)\, \mathrm{e}^{-E_n t}\,. \tag{7.36}$$

We finally have [1]

$$P(\theta, t|\theta_0, 0) = \mathcal{F}(\theta, \theta_0) \sum_n \Phi_n(\theta)\, \Phi_n^\star(\theta_0)\, \mathrm{e}^{-E_n t}\,. \tag{7.37}$$

In the limit $t \to \infty$, one expects that only the eigenvalue equal to zero (provided it exists) will matter, so that we obtain

$$P(\theta, t \to \infty|\theta_0, 0) = \mathcal{F}(\theta, \theta_0)\, \Phi_0(\theta)\, \Phi_0^\star(\theta_0)\,. \tag{7.38}$$

Equations (7.37) and (7.38) are our exact expressions for the conditional probability. The form of the eigenvalues $E_n$ and eigenfunctions $\Phi_n$ depends on the potential $V(\theta)$.

We show in Fig. 7.5 a comparison of the conditional probability $P(\theta, t|\theta_0, 0)$ obtained from the Fokker-Planck and the path integral approach, for representative values of $a, D, t$ and $\theta_0$. In the Fokker-Planck approach, the truncation parameter has been taken to be $n_{\max} = 60$; we have made sure that higher values do not affect our results appreciably. In the path integral approach, we obtain the eigenvalues $E_n$ and the eigenfunctions $\Phi_n$ of the Hamiltonian (7.33) by discretizing $\theta$ over $[-\pi, \pi]$, expressing the Hamiltonian as a matrix and then solving numerically the corresponding eigenvalue equation using standard linear algebra packages. We have discussed in full detail in Appendix G the procedure of discretization of the Hamiltonian. In Fig. 7.5, we demonstrate an excellent agreement between the results obtained in the two approaches. It is evident from the figure that the probability is peaked at

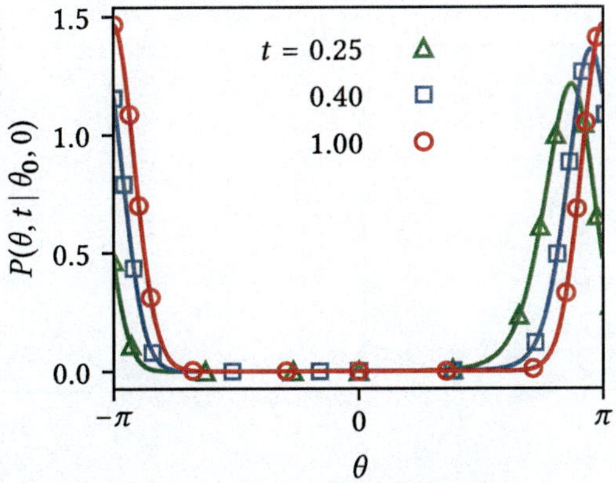


**Fig. 7.5 Comparison of the conditional probability distribution $P(\theta, t|\theta_0, 0)$ obtained from the Fokker-Planck and the path integral approach.** The lines are obtained by solving Eq. (7.16) with $n_{\max} = 60$, while the points are obtained by using Eq. (7.37). Here, we have chosen $a = 5.0$, $D = 0.5$, $\theta_0 = 0.5\pi$. As the dynamics evolves in time, the probability distribution gradually shows a peak at $\theta = \pm\pi$, the stable fixed point for $a > 2$, see Fig. 7.2

$\theta = \pm\pi$, the stable fixed point at the considered value of $a$; we expect that the density $P(\theta, t|\theta_0, 0)$ gets more sharply peaked as $D$ is decreased, thereby allowing to recover the stable fixed points of the noiseless dynamics from the noisy one.

## 7.5 Reconstruction of the Bifurcation Diagram Using the Noisy Dynamics

In this section, we discuss how to reconstruct the bifurcation diagram of the noiseless dynamics (7.2) using the conditional probability distribution for the noisy dynamics (7.9). We first discuss the case of small but finite noise, and then the case of infinitesimal noise.

### *7.5.1 Behavior of the Dynamics for Small but Finite Noise*

Here, we discuss the long-time behavior of the noisy dynamics (7.9) for small but finite noise. In order to proceed, let us obtain and analyze the behavior of the probability distribution $P(\theta, t|\theta_0, 0)$ in different $a$ regimes. From the bifurcation diagram Fig. 7.2, we see that there are three different regimes of $a$ across which the qualitative behavior of the dynamics changes abruptly. These regimes are the following: (I) $-2 < a < 2$, (II) $a > 2$ and (III) $a < -2$. Let us study these cases separately.

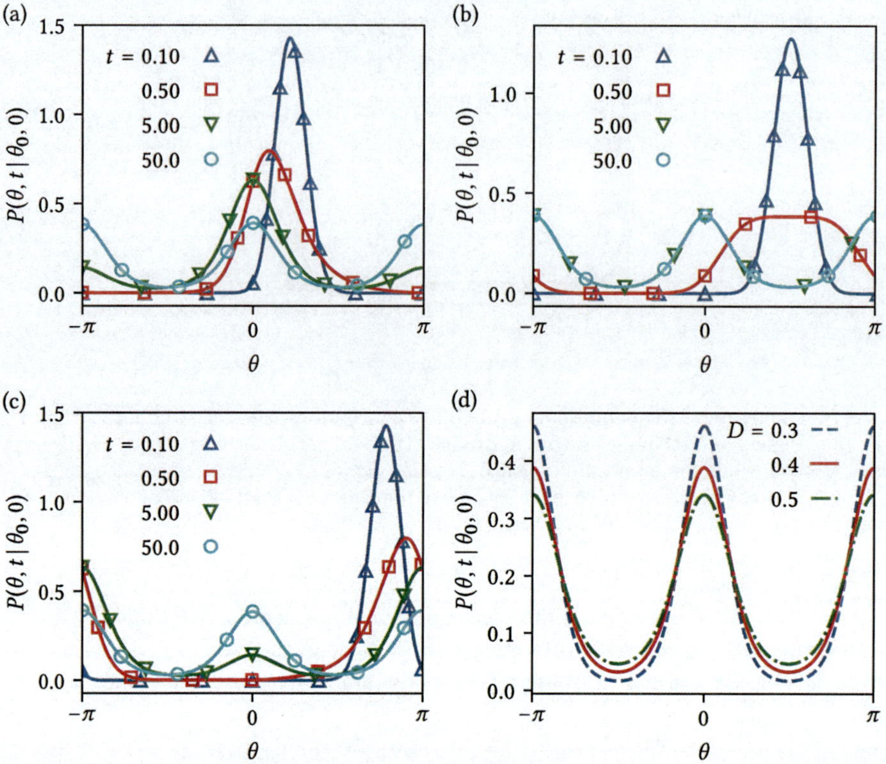


**Fig. 7.6 Conditional probability distribution** $P(\theta, t|\theta_0, 0)$ **versus** $\theta$ **for** $-2 < a < 2$. Here, we have $a = 0$. Evolution of $P(\theta, t|\theta_0, 0)$ with $D = 0.4$ for $\theta_0 = \pi/4,\ \pi/2$ and $3\pi/4$ are, respectively, shown in panels **a**, **b**, and **c**. All of these figures show that the initial peak at $\theta_0$ eventually shifts to $\theta = 0,\ \pm\pi$. Here, the lines are obtained by solving Eq. (7.16) with $n_{\max} = 60$, while the points are obtained by using Eq. (7.37). **d** $P(\theta, t|\theta_0, 0)$ for $\theta_0 = \pi/2$ and at long times, i.e., $t = 200$. We have checked that the probability distribution does not change for larger values of $t$, and that the form of the distribution is independent of $\theta_0$. We thus conclude that the probability distribution represents the stationary state. It is seen from the figure that the peaks of the distribution around $\theta = 0,\ \pm\pi$ become sharper as we decrease the noise strength $D$. Here, the lines are obtained by using Eq. (7.37)

**Case I:** $-2 < a < 2$

In this case, one may see from the bifurcation diagram Fig. 7.2 that the stable fixed points are $\theta = 0, \pm\pi$. In Fig. 7.6, we have shown the behavior of $P(\theta, t|\theta_0, 0)$ for $a = 0$. We have shown the time evolution of $P(\theta, t|\theta_0, 0)$ for different values of $\theta_0$, namely, $\pi/4$, $\pi/2$ and $3\pi/4$, in panels (a), (b) and (c), respectively. From these three panels, it is seen that the distribution is peaked around $\theta_0$ in the initial stage of the evolution. However, as time goes on, the peak of the distribution shifts and eventually splits into two peaks. In the long-time limit, i.e., $t \to \infty$, for all the initial conditions considered, the distribution peaks around the values $\theta = 0,\ \pm\pi$.

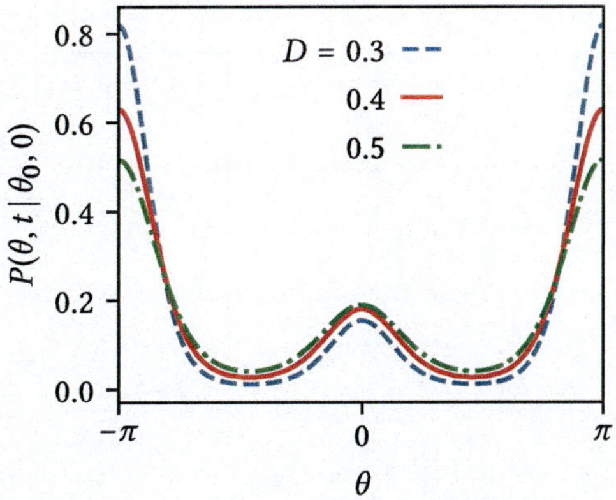


**Fig. 7.7** **$P(\theta, t|\theta_0, 0)$ versus $\theta$ for $a = 0.25$ in the stationary state for different values of $D$.** The stationary state is attained at long times, and is independent of $\theta_0$. The conditional probability distribution is peaked around $\theta = 0, \ \pm\pi$ at long times, and the peaks become sharper with decreasing $D$ (the latter fact may not be evident from the figure for the peak at $\theta = 0$, but one may check that it is actually the case by shifting the curves in the $y$-direction so that they all have the same $y$-value). However, unlike Fig. 7.6(d), the values of $P(\theta, t|\theta_0, 0)$ are not the same at $\theta = 0, \ \pm\pi$. The reason is that in this case, the potential $V(\theta)$ does not have the same value at these points, in contrast to the case with $a = 0$ that corresponds to Fig. 7.6d, see Fig. 7.1. Here, the lines are obtained by using Eq. (7.37)

It may be seen from Fig. 7.6b that unlike panels (a) and (c), the initial peak splits into two symmetric peaks before exhibiting any shift. The reason for this may be understood from the form of the potential $V(\theta)$ for $a = 0$ shown in Fig. 7.1. In this case, the minimum of the potential lies at $\theta = 0, \pm\pi$. Moreover, $V(\theta)$ has the same value at these points. In other words, $V(\theta)$ is symmetric about $\pi/2$ in the range $[0, \pi]$. This reflects in the dynamical evolution with $\theta_0 = \pi/2$ being symmetric about $\pi/2$.

In Fig. 7.6d, we have shown $P(\theta, t|\theta_0, 0)$ for $\theta_0 = \pi/2$ at long times, i.e., $t = 200$. We have checked that the probability distribution does not change for larger values of $t$, and that the form of the distribution is independent of $\theta_0$. We thus conclude that the probability distribution plotted in Fig. 7.6d represents the stationary state. It is seen from the figure that the peaks of the distribution around $\theta = 0, \pm\pi$ become sharper as we decrease the noise strength $D$. Now, one may wonder why the values of $P(\theta, t|\theta_0, 0)$ are the same at $\theta = 0, \pm\pi$. The reason is that the potential $V(\theta)$ for $a = 0$ has the same value at these points. To contrast this case, let us look at the stationary distribution for $a = 0.25$, see Fig. 7.7. In this case, it is clearly seen from the figure that heights of the peaks at $\theta = 0$ and $\theta = \pm\pi$ are different. This is because $V(\theta)$ at $a = 0.25$ does not have the same value at these points, as evident from Fig. 7.1. From the above discussions, we see that $P(\theta, t|\theta_0, 0)$ in the long-time limit is peaked around $\theta = 0, \pm\pi$, which are the stable fixed points for $-2 < a < 2$.

**Case II:** $a > 2$

For $a > 2$, we see from the bifurcation diagram Fig. 7.2 that there exist stable fixed points only at $\theta = \pm\pi$. For $a = 3$, the evolution of $P(\theta, t|\theta_0, 0)$ is shown in Fig. 7.8a

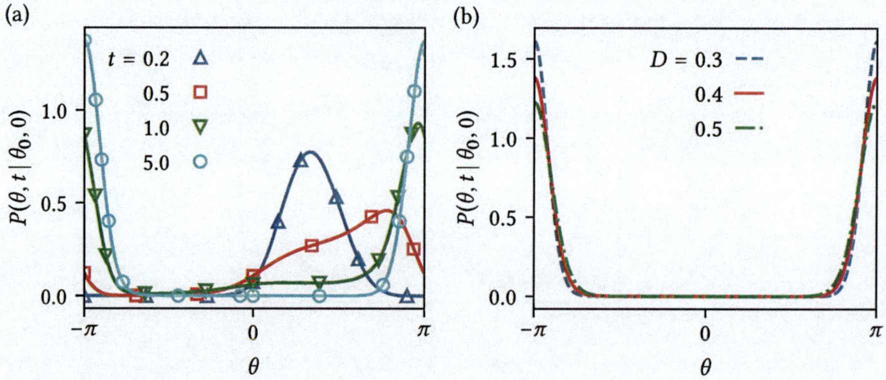


**Fig. 7.8 Conditional probability distribution** $P(\theta, t|\theta_0, 0)$ **versus** $\theta$ **for** $a > 2$. Here, we have $a = 3$. **a** The evolution of $P(\theta, t|\theta_0, 0)$ in time for $D = 0.4$, $\theta_0 = \pi/4$. The figure shows that $P(\theta, t|\theta_0, 0)$ is initially peaked around $\pi/4$. However, as time goes on, the peak shifts to $\pm\pi$. Here, the lines are obtained by solving Eq. (7.16) with $n_{\max} = 60$, while the points are obtained by using Eq. (7.37). **b** $P(\theta, t|\theta_0, 0)$ in the stationary state for three different values of the noise strength $D$. As the value of $D$ decreases, the peak around $\pm\pi$ becomes sharper. The stationary distribution clearly shows that $\pm\pi$ is the stable fixed point for $a > 2$, consistent with the bifurcation diagram, Fig. 7.2. Here, the lines are obtained by using Eq. (7.37)

with $\theta_0 = \pi/4$. From the figure, it is seen that initially, $P(\theta, t|\theta_0, 0)$ is peaked around $\pi/4$. During the evolution, the peak shifts to $\pm\pi$. As $t \to \infty$, the conditional probability no longer depends on $\theta_0$ and relaxes to a stationary state. In Fig. 7.8b, we have shown for $a = 3$ the stationary state for three different values of the noise strength $D$. The figure shows sharper peaks around $\pm\pi$ as $D$ is reduced. The long-time behavior of $P(\theta, t|\theta_0, 0)$ clearly suggests that there are stable fixed points at $\pm\pi$.

**Case III:** $a < -2$
For $a < -2$, there exists a stable fixed point only at $\theta = 0$ as seen from the bifurcation diagram Fig. 7.2. The conditional probability $P(\theta, t|\theta_0, t)$ in the stationary state is shown in Fig. 7.9 for three different values of the noise strength $D$. It is seen from the figure that $P(\theta, t|\theta_0, t)$ in the stationary state is peaked around $\theta = 0$. The peak becomes sharper as one reduces the value of $D$, thereby hinting at the existence of a stable fixed point at $\theta = 0$.

## 7.5.2 *Behavior of the Dynamics for Infinitesimal Noise: The Reconstruction*

### 7.5.2.1 A Suitable Observable

Let us now discuss the long-time behavior of the noisy dynamics (7.9) in the limit of infinitesimal noise. For given values of $\theta_0$ and $D$, one may calculate the average of

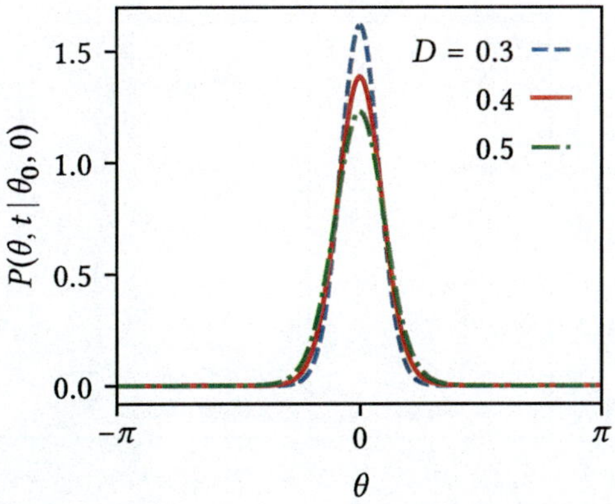


**Fig. 7.9** $P(\theta, t|\theta_0, 0)$ **versus** $\theta$ **in the stationary state for** $a < -2$ **and for different values of the noise strength** $D$**.** Here, we have $a = -3$. The figure shows that in the stationary state, the conditional probability distribution is peaked around zero, which is the stable fixed point for $a < -2$ as seen in the bifurcation diagram Fig. 7.2. It is also seen that the peak becomes sharper as the value of $D$ decreases. Here, the lines are obtained by using Eq. (7.37)

$\cos\theta$ at a given time $t$ and for different values of $a$ by using either the Fokker-Planck or the path integral result for the conditional probability density $P(\theta, t|\theta_0, 0)$, as [1]

$$\langle\cos\theta\rangle \equiv \langle\cos\theta\rangle(a, D, \theta_0, t) = \int_{-\pi}^{\pi} \mathrm{d}\theta \;\cos\theta \; P(\theta, t|\theta_0, 0)\,. \tag{7.39}$$

In Fig. 7.10, we show that the results obtained in the limit $D \to 0$ (specifically, for $D = 10^{-5}$) are in excellent agreement with the values of $\cos\theta$ obtained from numerical integration of the noiseless dynamics (7.2). This is also consistent with Fig. 7.4 showing that the noisy trajectories in the limit $D \to 0$ represent small fluctuations about the trajectory obtained from the noiseless dynamics.

#### 7.5.2.2 Reconstruction of the Stable and Unstable Branch

We now proceed to discuss how one may obtain the bifurcation behavior of the noiseless dynamics from the noisy one. We take $D = 10^{-5}$, and obtain for a given $\theta_0$ and a given time $t$ the behavior of $\langle\cos\theta\rangle$ versus $a$ by using Eq. (7.39). The results depicted in Fig. 7.11 show that the cross-over between the two limiting values of $\langle\cos\theta\rangle$, namely, $\langle\cos\theta\rangle = +1$ and $\langle\cos\theta\rangle = -1$, becomes steeper as $t$ increases. For larger $t$, we indeed have a sharp jump, as shown in Fig. 7.12. The same results are obtained as $t$ is increased further, and therefore Fig. 7.12 also characterizes the stationary behavior. In this case, it is further shown that the long-time values of $\cos\theta$ obtained in the noiseless dynamics lie on the curve for the noisy dynamics.

For a given $\theta_0$, let us now consider increasing $a$ from low to high values. One may see from Fig. 7.2 that this action implies moving along a straight line parallel

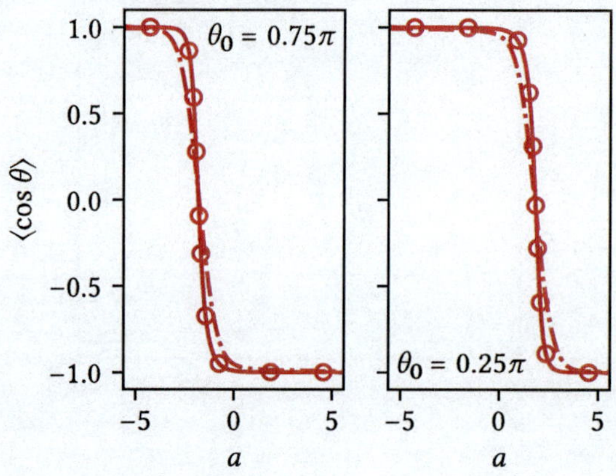


**Fig. 7.10** $\langle \cos\theta \rangle$ **versus** $a$ **at time** $t = 1$**, and for** $\theta_0 = 0.75\pi$ **(left panel) and** $\theta_0 = 0.25\pi$ **(right panel).** The lines are obtained by using Eq. (7.39): The continuous lines are for $D = 10^{-5}$, while the dash-dotted lines are for $D = 10^{-1}$. On the other hand, the circles are obtained by numerically integrating the noiseless dynamics (7.2) with integration time step $\Delta t = 10^{-3}$, by using the Euler method given in Appendix M.2. The plots show that the curves corresponding to the noiseless dynamics coincide with the noisy ones in the limit $D \to 0$

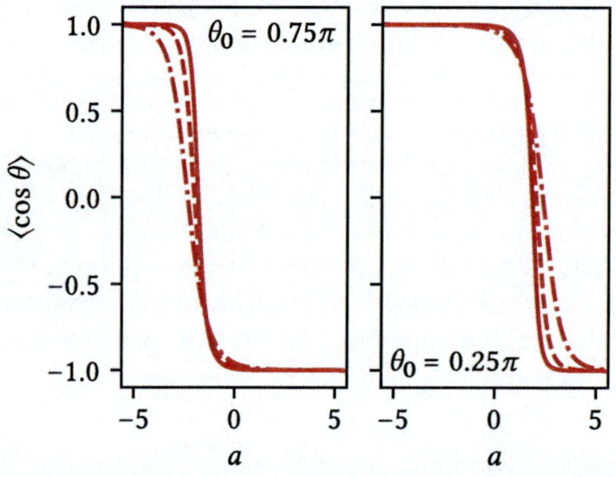


**Fig. 7.11** $\langle \cos\theta \rangle$ **versus** $a$ **for** $\theta_0 = 0.75\pi$ **(left panel) and** $\theta_0 = 0.25\pi$ **(right panel).** The lines are obtained by using Eq. (7.39). Here, we have chosen $D = 10^{-5}$. The dash-dotted, the dashed and the continuous line correspond respectively to times $t = 0.6, 0.8, 1.0$. The plots show that the crossover between values $\langle \cos\theta \rangle = +1$ and $\langle \cos\theta \rangle = -1$ with change of $a$ becomes steeper with the increase of $t$

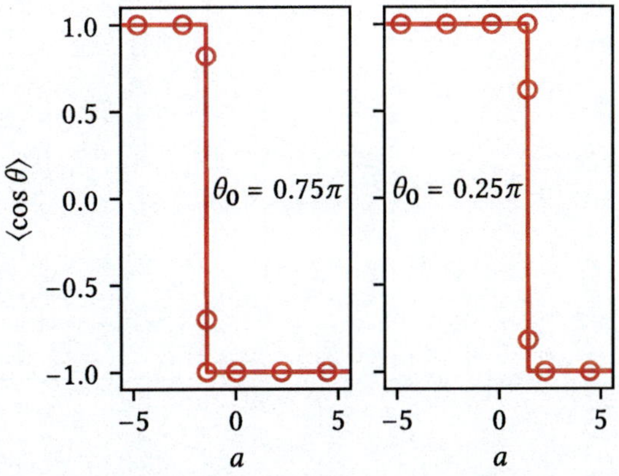


**Fig. 7.12** **$\langle \cos\theta \rangle$ versus $a$ obtained in the stationary state (time $t = 10$).** The left panel (respectively, the right panel) corresponds to the initial condition $\theta_0 = 0.75\pi$ (respectively, $\theta_0 = 0.25\pi$). Here, we have $D = 10^{-5}$. While the red circles involve using Eq. (7.39), the lines are obtained from numerical integration of the noiseless dynamics (7.2) by using the Euler method given in Appendix M.2 with integration time step $\Delta t = 10^{-3}$

to the $x$-axis while keeping a distance $\theta_0$ from it. As we increase $a$, the long-time value of $\cos\theta$ in the noiseless dynamics will be $+1$ so long as the straight line does not intersect the dashed curve in red lying in the region $-2 < a < 2$. Beyond the intersection point, the long-time value of $\cos\theta$ will switch to $-1$. One obtains the point of intersection by solving $\cos\theta_0 = a/2$. The point of intersection will act as a crossover point such that the stable value of $\cos\theta$ will be $+1$ and $-1$ for smaller and larger values of $a$, respectively. In view of Fig. 7.10 showing match between the noiseless and the noisy dynamics in the limit of weak noise, one may expect such a crossover behavior of $\langle \cos\theta \rangle$ versus $a$ at long times. This is indeed obtained by our exact results shown in Fig. 7.12. For a given $\theta_0$, obtaining the long-time behavior of $\langle \cos\theta \rangle$ versus $a$ as shown in Fig. 7.12, we find the stable fixed points corresponding to different values of $a$ [1]. One may repeat the procedure for different values of $\theta_0$ to reconstruct the stable branches of the bifurcation diagram Fig. 7.2. Consistent with the crossover behavior discussed above, one may check from Fig. 7.12 that the crossover point is obtained at the value of $a$ given by $a = 2\cos\theta_0 = \pm\sqrt{2}$ for the left and the right panel, respectively. By repeating plots as in Fig. 7.12 for different values of $\theta_0$, one finds the corresponding crossover points, which then allow to obtain the line of unstable fixed points in the range $-2 < a < 2$. As expected, the crossover points so obtained lie exactly on the unstable branch of the bifurcation diagram Fig. 7.2 for $-2 < a < 2$ [1]. Similarly, repeating plots as in Fig. 7.12 for $\theta_0 = 0$ and $\theta_0 = \pi$ one may obtain the crossover points $a = 2$ and $a = -2$, respectively. These points coincide with the bifurcation points in Fig. 7.2, thereby explaining the associated stability.

We may also obtain the unstable fixed points of the noiseless dynamics for $a > 2$ and $a < -2$ from the noisy dynamics. For example, in order to show that $\theta = 0$

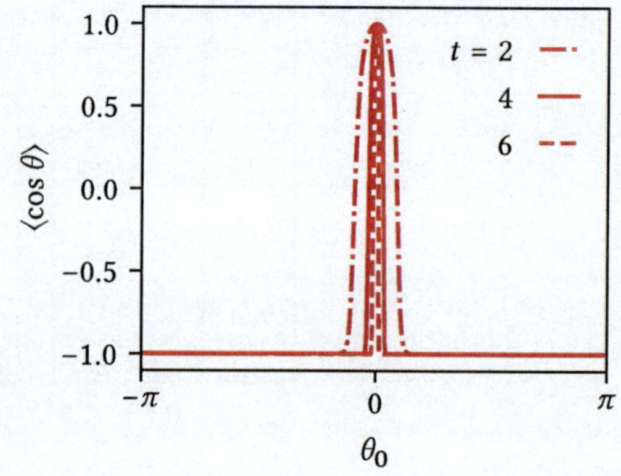


**Fig. 7.13** **$\langle \cos\theta \rangle$ versus $\theta_0$ obtained using Eq. (7.39) and for** $a = 2.5$. Here, we have $D = 10^{-5}$, and the different curves correspond to different times $t$

is unstable for a representative value $a = 2.5$, one may plot $\langle \cos\theta \rangle$, obtained from Eq. (7.39), as a function of $\theta_0$ for different times [1] as depicted in Fig. 7.13. One observes from the figure that as time increases, $\langle \cos\theta \rangle$ for increasing number of values of $\theta_0$ different from the specific value $\theta_0 = 0$ attains the value of $-1$. It is evident from the plot that in the long-time limit, only when $\theta_0 = 0$ does $\langle \cos\theta \rangle$ have the value of unity, which attains for all other values of $\theta_0$ the value of $-1$. This fact is fully consistent with the bifurcation diagram Fig. 7.2, and has been indicated in the figure by the red circle at $\theta = 0, a = 2.5$.

#### 7.5.2.3 How Weak Should the Noise Be

A reason why one could reconstruct the bifurcation diagram from the noisy dynamics is the choice of Gaussian noise for Langevin evolution, which ensures that typical trajectories for the noisy dynamics represent fluctuations of a given size around the noiseless ones, The size of the fluctuations is set by the variance $D$ of the Gaussian distribution for the noise. In the limit $D \to 0$, the typical trajectories for the noisy dynamics coincide with the noiseless one. Now, one may ask: numerically how small should $D$ be? The answer depends on whether we are studying the dynamics around a stable or an unstable fixed point. We have seen that a stable fixed point corresponds to a local minimum of the potential $V(\theta)$. As a result, for any reasonably small value of $D$ (smaller than a critical value $D_c^{(s)}$), almost all trajectories of the noisy dynamics are attracted towards the stable fixed point because of the existence of the minimum of the potential there. At long times, these trajectories settle into the stable fixed point. However, for an unstable fixed point, which corresponds to a local maximum of $V(\theta)$, the issue is a bit tricky. In this case, values of $D$ smaller than $D_c^{(s)}$ that ensured convergence of the noisy to the noiseless dynamics in the case of stable fixed points may prove to be "strong" enough that trajectories starting at the unstable fixed point are pushed away from it to settle into stable fixed points at long times. Such a scenario would invalidate the procedure of obtaining the unstable fixed points

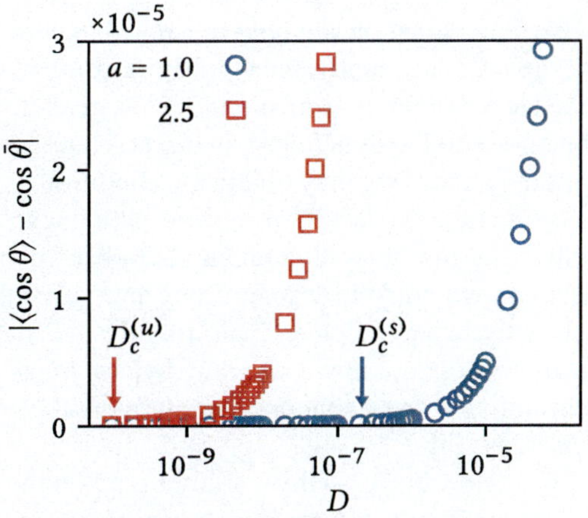


**Fig. 7.14** $|\langle\cos\theta\rangle - \cos\bar{\theta}|$ **versus** $D$ **for values of** $a$ **at which one has a fixed point at** $\theta = \bar{\theta}$**.** For $a = 1.0$ (respectively, $a = 2.5$), one has a stable (respectively, unstable) fixed point with $\bar{\theta} = 0$. Here, one obtains $\langle\cos\theta\rangle$ by using Eq. (7.39) and by choosing $\theta_0 = \bar{\theta}$ and $t = 6$. In the figure, we have indicated the approximate $D_c^{(s)}$ and $D_c^{(u)}$

of the noiseless dynamics from an analysis of the noisy one. In order to make the program successful, one requires to reduce further the noise strength, i.e., having a critical value $D_c^{(u)} < D_c^{(s)}$, and considering the convergence for values of $D < D_c^{(u)}$.

In order to illustrate the aforementioned conclusion, we show in Fig. 7.14 the outcome of the following numerical experiment. Referring to Fig. 7.2, let us choose a value of $a$ at which one has a stable fixed point $\bar{\theta}$, and another at which one has an unstable fixed point $\bar{\theta}$. Then, study the noisy dynamics for both cases with the corresponding initial condition $\theta_0 = \bar{\theta}$. One may expect that for $D$ small enough, i.e., for $D < D_c^{(s)}$ (respectively, $D < D_c^{(u)}$) in case of the stable (respectively, the unstable) fixed point, the long-time value of $\langle\cos\theta\rangle$ should coincide with the value of $\cos\bar{\theta}$. Figure 7.14 suggests that indeed we have $D_c^{(u)} < D_c^{(s)}$. We remark that the precision of match between the values of $\langle\cos\theta\rangle$ and $\cos\bar{\theta}$ depends on the precision employed in numerical evaluation of the quantities involved. Also, higher precision implies lower values of $D_c^{(s)}$ and $D_c^{(u)}$, which may be checked with numerics.

## 7.6 Conclusions and Perspectives

In this chapter, we discussed how one may obtain the bifurcation behavior of a nonlinear dynamical system by introducing noise into the dynamics and then studying the resulting noisy dynamics in the weak-noise limit. For a simple nonlinear model system, we showed that as a suitable quantity to capture the bifurcation behavior in the noisy dynamics one may consider the conditional probability to observe microscopic configurations at a given time while conditioned on observation of a given configuration at an earlier time. We analyzed the time evolution of the conditional probability by employing two different approaches, namely, the Fokker-Planck and the path integral approach. We showed that using the path integral approach one may obtain exact closed-form expressions for the conditional probability.

We remark that the analysis presented in Sect. 7.4 applies to any form of the potential $V(\theta)$. Consequently, the whole procedure of obtaining the bifurcation diagram of a noiseless dynamics corresponding to a given form of $V(\theta)$ by addition of Gaussian, white noise to the dynamics can be rather straightforwardly carried through. For the noisy dynamics, one may obtain the conditional probability distribution $P(\theta, t|\theta_0, 0)$ from either the Fokker-Planck or the path integral approach. As a remarkable feature of studying the noisy dynamics, one observes that the probability distribution for weak-enough noise is at long times naturally peaked around the stable fixed points of the noiseless dynamics. Exploiting this fact, once the fixed points of a given noiseless dynamics are known, we may bypass the need to perform a stability analysis in order to locate the stable ones by studying the corresponding noisy dynamics in the weak-noise limit.

Let us now discuss why a study based on the noisy dynamics is useful. In a real-world experiment, we measure various properties of a physical system that has an underlying dynamics. As an outcome of the experiment, we get some data on the behavior of the system under varying external conditions. Any modeling of the data should include effects of noise to account for measurement errors, although the actual underlying dynamics does not have this source of noise. We are usually interested in knowing how qualitatively different behavior observed in the data obtained with varying experimental parameters emerges from a bifurcation in the underlying dynamics. In experimental data, one typically observes signatures of only stable points of the dynamics, while the unstable points contribute mainly to short-time transients. As we have seen, a long-time analysis of the noisy dynamics automatically picks up the stable points of the actual dynamics, which allows one to infer the bifurcation behavior of the actual noiseless system from a study of the noisy dynamics.

As an application, one may use our method in studying the phenomenological neuron model proposed by Hindmarsh and Rose (the HR model [11]). Systems of neurons depending on various biophysical parameters show diverse dynamical behavior such as quiescence, spiking, bursting, etc. The HR model exhibits all of these behavior as the dynamical parameters are tuned. However, a complete analytical study of bifurcations in the model is full of challenges [12], and one has to rely mostly on numerical analysis or suitable approximations in specific parameter regimes. As a workaround, one may employ our procedure to see whether adding noise to the HR model and studying the corresponding noisy dynamics in the weak-noise limit allow to obtain the complete bifurcation diagram.

For the dynamical model studied in this chapter, we showed that the stable fixed points that exist for the deterministic dynamics may be recovered by adding Gaussian white noise to the system and then studying the resulting noisy dynamics in the limit in which the noise strength approaches zero. We believe that our work, therefore, is an illustration of stochastic stability. We may mention that we are not pretending to demonstrate that our approach in this chapter of obtaining the bifurcation diagram by adding noise is in any way simpler than the one based on application of the well-developed bifurcation theory [13]. Our motivation here is simply to show how for a model system one may add noise and efficiently employ the tools of statistical mechanics, and in particular, the Fokker-Planck and path integral methods, to unravel the bifurcation behavior of the corresponding deterministic dynamics. Let us add a word of caution here that adding noise to a deterministic dynamics may lead to a drastic change in the detailed properties of the resulting system. For example, in a recent reference [14], it is shown that the deterministic dynamics of a two-dimensional ordinary differential equation exhibiting Hopf bifurcation may with the addition of noise to the dynamics result in so-called random attractors.

## References


1. D. Das, S. Roy, S. Gupta, Commun. Nonlinear Sci. Numer. Simulat. **72**, 575 (2019). https://doi.org/10.1016/j.cnsns.2019.01.018
2. E. Ott, *Chaos in Dynamical Systems*, 2nd edn. (Cambridge University Press, Cambridge, 2002). https://doi.org/10.1017/CBO9780511803260
3. M. Cross, H. Greenside, *Pattern Formation and Dynamics in Nonequilibrium Systems* (Cambridge University Press, Cambridge, 2009). https://doi.org/10.1017/CBO9780511627200
4. T. Dauxois, M. Peyrard, *Physics of Solitons* (Cambridge University Press, Cambridge, 2006)
5. R.P. Feynman, A.R. Hibbs, D.F. Styer, *Quantum Mechanics and Path Integrals*, emended. (Dover Publications, Mineola, 2010)
6. L.S. Schulman, *Techniques and Applications of Path Integration* (Dover Publications, Mineola, 2005)
7. M. Kac, Trans. Am. Math. Soc. **65**(1), 1 (1949). https://doi.org/10.1090/S0002-9947-1949-0027960-X
8. M. Kac, in *Second Berkeley Symposium on Mathematical Statistics and Probability* (1951), pp. 189–215
9. E. Roldán, S. Gupta, Phys. Rev. E **96**(2), 022130 (2017). https://doi.org/10.1103/PhysRevE.96.022130
10. C.W. Gardiner, *Handbook of Stochastic Methods for Physics, Chemistry and the Natural Sciences* (Springer, Berlin, 1983)
11. J.L. Hindmarsh, R.M. Rose, Proc. R. Soc. Lond. B Biol. Sci. **221**(1222), 87 (1984). https://doi.org/10.1098/rspb.1984.0024
12. M. Storace, D. Linaro, E. de Lange, Chaos **18**(3), 033128 (2008). https://doi.org/10.1063/1.2975967
13. S.H. Strogatz, *Nonlinear Dynamics and Chaos* (CRC Press, Boca Raton, 2018). https://doi.org/10.1201/9780429492563
14. T.S. Doan, M. Engel, J.S.W. Lamb, M. Rasmussen, Nonlinearity **31**(10), 4567 (2018). https://doi.org/10.1088/1361-6544/aad208

# Chapter 8
# Relaxation Dynamics of Mean-Field Classical Spin Systems

**Abstract** In this chapter, we explore the facet of noise occurring in a finite-size system due to sampling of initial conditions and its interplay with that introduced through interaction with the environment. Within the ambit of a model long-range interacting system of classical Heisenberg spins, we pursue our study focusing specifically on the dynamical relaxation of the system to Boltzmann-Gibbs equilibrium. Considering the deterministic spin precessional dynamics, we reveal the full range of quasistationary behavior observed during the relaxation, whereby the system gets trapped in nonequilibrium states for times that diverge with the system size, and consequently exhibits a slow relaxation to equilibrium. By contrast, the corresponding stochastic dynamics, modeling the interaction of the system with the environment and constructed in the spirit of the stochastic Landau-Lifshitz-Gilbert equation, shows a fast relaxation to equilibrium on a size-independent timescale. In particular, no signature of quasistationarity is observed in this case when the noise is strong enough. We also observe similar fast relaxation in the Glauber Monte Carlo dynamics of the model. Our findings on the spin dynamics thus establish that the quasistationarity observed in deterministic dynamics is washed away by fluctuations induced through contact with the environment.

## 8.1 Introduction

Stochasticity is inherently present in any statistical system constituted by a finite number of interacting degrees of freedom. The finiteness of a system is known to induce fluctuations in both its static and dynamic properties. Stochasticity may arise in finite-size systems due to sampling of initial conditions and also due to interaction with the external environment.[1] In this chapter, we investigate how these two sources of stochasticity interplay in dictating the long-time behavior of the system, and in particular, in predicting the values of the macroscopic observables in the stationary

[1] Let us remark that stochasticity in the sense of unpredictability (due to sensitive dependence on the initial conditions) may also result from chaos in fully-deterministic dynamical systems even when the dynamics is low-dimensional.



D. Das and S. Gupta, *Facets of Noise*, Fundamental Theories of Physics 214,
https://doi.org/10.1007/978-3-031-45312-0_8

state. We explore these issues within the ambit of a model many-body interacting system comprising classical spins that are interacting with one another through an inter-particle potential that is long-ranged in nature. As we have discussed in Sect. 3.3, the inter-particle potential in long-range systems decays rather slowly as a function of the inter-particle separation $r$, specifically, as $1/r^\alpha$; $0 \leq \alpha \leq d$, with $d$ being the embedding spatial dimension of the system [1–5]. Here, the extreme limit of $\alpha = 0$ corresponds to a mean-field interaction. Such an interaction implies that the inter-particle potential does not depend on the distance. Consequently, in a mean-field system, all the particles interact with all other particles with equal strength.

Long-range interacting (LRI) systems are observed at all length scales, from atomic to astrophysical. They are known to exhibit a range of physical phenomena that appear counterintuitive when viewed from the perspective of short-range systems. As we have discussed in Sect. 3.3, a basic property that distinguishes LRI systems from short-range ones is the violation of additivity. We have seen in Sect. 4.5 that non-additivity results in unusual features such as inequivalent equilibrium ensembles and non-concave entropy function. It has however more striking consequence on dynamical properties, namely, that an isolated LRI system relaxes to equilibrium over a time scale that diverges with the number of degrees of freedom [3]. As a result, the larger the system is, the longer is the time that it takes to relax to equilibrium. Such a relaxation process results in the occurrence of slowly-evolving nonequilibrium states, making them directly accessible in experiments [3].

## 8.2 The Model

We consider the $N$ globally-coupled classical Heisenberg spins introduced in Eq. (3.99), and rewrite the Hamiltonian with $J = 1$ as

$$H = -\frac{1}{2N} \sum_{i,j=1}^{N} \mathbf{S}_i \cdot \mathbf{S}_j + D \sum_{i=1}^{N} S_{iz}^{2n} , \tag{8.1}$$

where $n > 0$ is an integer. We recall that the anisotropy term involving $D > 0$ in Eq. (8.1) lowers the energy by having the magnetization vector

$$\mathbf{m} \equiv \frac{\sum_{i=1}^{N} \mathbf{S}_i}{N} = (m_x, m_y, m_z) \tag{8.2}$$

pointing in the $xy$ plane. The length of the magnetization vector is given by

$$m \equiv \sqrt{m_x^2 + m_y^2 + m_z^2} . \tag{8.3}$$

We refer to the dynamics (8.1) for $n = 1$ and $n = 2$ as the quadratic and quartic model, respectively. The spin components in the Hamiltonian (8.1) satisfy

$$\{S_{i\alpha}, S_{j\gamma}\} = \delta_{i,j}\, \epsilon_{\alpha\gamma\delta} S_{j\delta}\,, \tag{8.4}$$

where $\{\cdot\,,\cdot\}$ denotes the Poisson bracket and the quantity $\epsilon_{\alpha\gamma\delta}$ is the Levi-Civita symbol.[2] As discussed in Sect. 3.8, the Poisson bracket for two functions $A, B$ of the spins is defined in terms of the canonical variables $\cos\theta_i$ and $\phi_i$ as [6] $\{A, B\} \equiv \sum_{i=1}^{N}(\partial A/\partial\phi_i\, \partial B/\partial(\cos\theta_i) - \partial A/\partial(\cos\theta_i)\, \partial B/\partial\phi_i)$, which may also be re-expressed as

$$\{A, B\} = \sum_{i=1}^{N} \mathbf{S}_i \cdot \frac{\partial A}{\partial \mathbf{S}_i} \times \frac{\partial B}{\partial \mathbf{S}_i}\,. \tag{8.5}$$

### 8.2.1 Spin Dynamics

In dimensionless times, the time evolution of individual spins of the system (8.1) is given by the set of coupled first-order differential equations

$$\dot{\mathbf{S}}_i \equiv \frac{\mathrm{d}\mathbf{S}_i}{\mathrm{d}t} = \{\mathbf{S}_i, H\}\,; \quad i = 1, 2, \ldots, N\,. \tag{8.6}$$

Using Eqs. (8.1), (8.4), and (8.6) one obtains the dynamical evolution of the spin components as

$$\dot{S}_{ix} = S_{iy}m_z - S_{iz}m_y - 2nDS_{iy}S_{iz}^{2n-1}\,, \tag{8.7}$$

$$\dot{S}_{iy} = S_{iz}m_x - S_{ix}m_z + 2nDS_{ix}S_{iz}^{2n-1}\,, \tag{8.8}$$

$$\dot{S}_{iz} = S_{ix}m_y - S_{iy}m_x\,. \tag{8.9}$$

The dynamics (8.6) conserves the length of each spin. This can be verified by taking the vector dot product of both sides of Eq. (8.6) with $\mathbf{S}_i$. Summing Eq. (8.9) over $i$, it is easily seen that $m_z$ is a constant of motion. In a similar manner, one may also check that in the absence of anisotropy, i.e., with $D = 0$, the total magnetization $m$ is a constant of motion. The dynamical evolution (8.6) also conserves the total energy of the system (8.1), and thus models the microcanonical dynamics of the latter. We remark that the dynamical setting of Eq. (8.6) is very different from that involving particles characterized by generalized coordinates and momenta and with time evolution governed by a Hamiltonian given by a sum of a kinetic and a potential energy contribution, e.g., that of the celebrated Hamiltonian mean-field (HMF) model [3]; see Appendix C. As a result, none of the results, static or dynamic, derived for the latter may be a priori expected to apply to the model (8.1).

---

[2] In this chapter, the Roman and the Greek indices are used to label the spins and the spin components, respectively.

Expressing the spin components in spherical polar coordinates $\theta_i \in [0, \pi]$ and $\phi_i \in [0, 2\pi)$, one may write

$$S_{ix} = \sin\theta_i \cos\phi_i\ , \quad S_{iy} = \sin\theta_i \sin\phi_i\ , \quad S_{iz} = \cos\theta_i\ , \tag{8.10}$$

and obtain from Eqs. (8.7)–(8.9) the time evolution of the variables $\theta_i$ and $\phi_i$ as

$$\dot{\theta}_i = m_x \sin\phi_i - m_y \cos\phi_i\ , \tag{8.11}$$

$$\dot{\phi}_i = m_x \cot\theta_i \cos\phi_i + m_y \cot\theta_i \sin\phi_i - m_z + 2nD\cos^{2n-1}\theta_i\ . \tag{8.12}$$

One may also interpret Eqs. (8.7)–(8.9) as the precessional dynamics of the spins in an effective magnetic field $\mathbf{h}_i^{\text{eff}} \equiv \mathbf{h}_i^{\text{eff}}(\{\mathbf{S}_i\})$, so that one has

$$\dot{\mathbf{S}}_i = \mathbf{S}_i \times \mathbf{h}_i^{\text{eff}}\ , \tag{8.13}$$

where $\mathbf{h}_i^{\text{eff}}$ denotes the effective field experienced by the $i$-th spin. One obtains the quantity $\mathbf{h}_i^{\text{eff}}$ from the Hamiltonian (8.1) as

$$\mathbf{h}_i^{\text{eff}} = -\frac{\partial H}{\partial \mathbf{S}_i} = \mathbf{m} + \mathbf{h}_i^{\text{aniso}}\ ; \quad \mathbf{h}_i^{\text{aniso}} \equiv (0, 0, -2nDS_{iz}^{2n-1})\ . \tag{8.14}$$

Equation (8.14) clearly suggests that the effective magnetic field has a global and a local contribution. The global contribution comes from the magnetization $\mathbf{m}$ set up in the system by the effect of all the spins, while the local one is due to the field $\mathbf{h}_i^{\text{aniso}}$ set up for each individual spin by the anisotropy term in the Hamiltonian (8.1).

Let us denote the energy density of the system, i.e., the energy per spin, by $\epsilon$ and the equilibrium magnetization by $m_{\text{eq}}$. Consider first the case of the quadratic model ($n = 1$). In this case, we have shown in Sect. 4.6 that the model in microcanonical equilibrium exhibits a magnetized ($m_{\text{eq}} \neq 0$) phase at low values of $\epsilon$ and a non-magnetized ($m_{\text{eq}} = 0$) phase at high values of $\epsilon$. Moreover, as $\epsilon$ increases, the model shows a continuous transition from the magnetized to the nonmagnetized phase at a critical value $\epsilon_c$. Within the microcanonical dynamics of the model starting from a class of nonmagnetized initial states, one finds there exists an additional threshold energy density $\epsilon^\star < \epsilon_c$, such that for $\epsilon^\star < \epsilon < \epsilon_c$, the relaxation to equilibrium magnetized state occurs over a time that scales superlinearly with $N$ [6, 7]. On the other hand, for $\epsilon < \epsilon^\star$, the microcanonical dynamics exhibits a fast relaxation over a time that scales as $\log N$. The mentioned particular class of initial states is the so-called waterbag (WB) state [6], in which the spins have $\phi$'s sampled independently and uniformly from the interval $[0, 2\pi)$, and $\theta$'s sampled independently and uniformly from an interval symmetric about $\pi/2$, i.e., from the interval $[\pi/2 - a, \pi/2 + a]$ with $0 < a < \pi/2$. In other words, the pair $(\theta_i, \phi_i)$ for every $i$ is sampled independently from the WB joint distribution $f_0(\theta, \phi)$ given by [6]

$$f_0(\theta,\phi) = \begin{cases} \dfrac{1}{2\pi}\dfrac{1}{2\sin a} & \text{if } \theta \in \left[\dfrac{\pi}{2} - a\,, \dfrac{\pi}{2} + a\right], \quad \phi \in [0, 2\pi)\,, \\ 0 & \text{otherwise}\,. \end{cases} \tag{8.15}$$

The aforementioned observations on the behavior of the quadratic model have been obtained by numerically integrating the equations of motion of the model and have been complimented by studying analytically a Vlasov-type equation for the single-spin phase space distribution in the thermodynamic limit $N \to \infty$ [6]. The latter distribution function gives the fraction of the total number of spins that have given $\theta$ and $\phi$ values. One may show that the distribution corresponding to the WB state is a stationary solution of the Vlasov equation. However, this Vlasov-stationary distribution is stable only for $\epsilon > \epsilon^\star$ and unstable for energies above $\epsilon < \epsilon^\star$. The eventual relaxation to equilibrium observed for $\epsilon^\star < \epsilon < \epsilon_c$ is due to the finite-size effects. A finite system of size $N$ induces statistical fluctuations into the dynamics by adding nonzero finite-$N$ corrections to the Vlasov equation that are at least of order greater than $1/N$ [8]. As a result, the nonmagnetized WB state, which in an infinite system is stationary and stable for $\epsilon^\star < \epsilon < \epsilon_c$, shows for finite $N$ a slow evolution towards the magnetized equilibrium state in the said energy regime, exemplifying the existence of the so-called quasistationary states (QSSs) [3].

Starting from the premises discussed above, we consider here the model (8.1) for general values of $n$, and present a detailed characterization of the relaxational dynamics and its ubiquity in the context of long-range spin models. We have already discussed the equilibrium properties of the model in Sect. 4.6, where we have obtained the critical energy density $\epsilon_c(n)$, see Eq. (4.144). We recall that the system is in a magnetized phase for $\epsilon < \epsilon_c$ and in a nonmagnetized phase for $\epsilon > \epsilon_c$. As discussed in Sect. 4.5, since the system (8.1) exhibits a continuous transition in canonical equilibrium, the microcanonical and canonical ensembles are equivalent in equilibrium for the system, although this fact is not guaranteed in general for LRI systems. Because of the equivalence, we may associate to every value of the conserved microcanonical energy density $\epsilon$ a temperature $T$ of the system in canonical equilibrium. The temperature $T$ is to be chosen in a way such that it guarantees that the average energy in canonical equilibrium equals $\epsilon$ in the limit $N \to \infty$. Then, the phase diagram of the model (8.1) in canonical equilibrium is given by a $D$ vs $T$ plot that corresponds to the $D$ vs $\epsilon$ phase diagram in microcanonical equilibrium shown in Fig. (4.7).

One may wonder whether the mentioned quasistationarity observed in the $n = 1$ model [6, 8] is due to the non-analytic behavior of the WB single-spin distribution at $\theta = \pi/2 \pm a$. To show that quasistationarity is rather generic to the model (8.1), in the following, we consider as initial states suitably smoothened versions of the WB state, the so-called Fermi-Dirac (FD) state, and study its evolution in time. We remark that the single-spin distribution for the FD state is a perfectly analytic function. Using a linear stability analysis of the FD state under the infinite-$N$ Vlasov dynamics, we find that there exists a threshold energy value $\epsilon^\star(n)$, such that the state is stationary but linearly unstable under the dynamics for energies $\epsilon < \epsilon^\star(n)$. We show for finite $N$ that the relaxation to equilibrium occurs as a two-step process: a relaxation from the initial FD state to a magnetized QSS, followed by a relaxation of the latter over a timescale $\sim N$ to the microcanonical equilibrium. The magnetized QSS has thus a lifetime $\sim N$. On the other hand, in the energy regime $\epsilon^\star(n) < \epsilon < \epsilon_c(n)$, the FD state is dynamically stable under the Vlasov evolution and exhibits for finite $N$ a relaxation towards equilibrium over a scale $\sim N^\alpha$. Here, the exponent $\alpha$ has an essential dependence on $n$. In this case, we conclude observing a nonmagnetized QSS with a lifetime $\sim N^\alpha$. For the quadratic model ($n = 1$), we obtain $\alpha = 3/2$ (as opposed to the value $\alpha = 2$ for the WB state reported in Ref. [7]), whereas for the quartic model ($n = 2$), one observes $\alpha = 1$.

To establish the aforementioned relaxation scenario, one may choose the magnetization as a preferable macroscopic observable and monitor its evolution over time. However, in the energy regime $\epsilon > \epsilon_c(n)$, it does not serve the purpose because both the initial FD and the final microcanonical equilibrium state are nonmagnetized. To overcome this issue, we identify suitable macroscopic observables (e.g., $\sum_{i=1}^N \cos^4\theta_i/N$ and $\sum_{i=1}^N \cos^2\theta_i/N$ for the quadratic and quartic model, respectively) and show that the relaxation to equilibrium occurs over an $N$-dependent timescale that is distinct from what was observed for magnetization relaxation for energies $\epsilon^\star(n) < \epsilon < \epsilon_c(n)$. Namely, the relaxation time scales as $N^2$ and $N^{3/2}$ for the quadratic model and the quartic model, respectively. As a result, one may conclude for energies $\epsilon > \epsilon_c(n)$ the existence of a nonmagnetized QSS with a lifetime that diverges with the system size.

## 8.3 Analysis of the Model with Deterministic Dynamics

Let us now discuss the dynamical evolution of the model (8.1) in the thermodynamic limit $N \to \infty$. Here, we invoke the corresponding Vlasov equation and study how the deterministic spin-precessional dynamics (8.13) relaxes to equilibrium while starting far from it.

### *8.3.1 Behavior in the Limit $N \to \infty$*

To characterize the relaxation process, we monitor the time evolution of the single-spin distribution function $P_0(\mathbf{S}, t)$ normalized as $\int \mathrm{d}\mathbf{S}\; P_0(\mathbf{S}, t) = 1\ \forall\ t$. The time evolution of $P_0(\mathbf{S}, t)$ is given by the Vlasov equation (see Appendix H.2 for a derivation)

$$\frac{\partial P_0(\mathbf{S}, t)}{\partial t} + \frac{\partial}{\partial \mathbf{S}} \cdot (\mathbf{S} \times \mathbf{h}^{\mathrm{eff},0}) P_0 = 0\,, \tag{8.16}$$

where one has

$$\mathbf{h}^{\mathrm{eff},0} \equiv \mathbf{h}^{\mathrm{eff},0}[P_0] = \mathbf{m}[P_0] + (0, 0, -2nDS_z^{2n-1})\,; \quad \mathbf{m}[P_0] \equiv \int \mathrm{d}\mathbf{S}\,\mathbf{S}\,P_0(\mathbf{S}, t)\,. \tag{8.17}$$

For later convenience, let us define the single-spin distribution $f(\theta, \phi, t)$, such that $f(\theta, \phi, t)\sin\theta \mathrm{d}\theta \mathrm{d}\phi$ denotes the probability to observe a spin at time $t$ with its angles between $[\theta, \theta + \mathrm{d}\theta]$ and $[\phi, \phi + \mathrm{d}\phi]$. The distribution $f(\theta, \phi, t)$ is related to $P_0(\mathbf{S}, t)$ as $f(\theta, \phi, t) = P_0(\mathbf{S}, t)$ and satisfies the normalization $\int_0^{2\pi} \mathrm{d}\phi \int_0^{\pi} \mathrm{d}\theta\, \sin\theta\, f(\theta, \phi, t) = 1\ \forall\, t$. In order to obtain the time evolution of $f$, one may express the second term on the left hand side of Eq. (8.16), namely, $(\mathbf{S} \times \mathbf{h}^{\mathrm{eff},0}) \cdot \partial P_0/\partial \mathbf{S}$, in spherical polar coordinates with unit vectors $(\hat{\mathbf{r}}, \hat{\boldsymbol{\theta}}, \hat{\boldsymbol{\phi}})$ and $r = \sqrt{S_x^2 + S_y^2 + S_z^2} = 1$. Then, one obtains

$$\begin{aligned}
&\Big[(S_y h_z^{\mathrm{eff}} - S_z h_y^{\mathrm{eff}})(\cos\theta\cos\phi\,\hat{\boldsymbol{\theta}} - \sin\phi\,\hat{\boldsymbol{\phi}}) + (S_z h_x^{\mathrm{eff}} - S_x h_z^{\mathrm{eff}})(\cos\theta\sin\phi\,\hat{\boldsymbol{\theta}} + \cos\phi\,\hat{\boldsymbol{\phi}}) \\
&\quad - (S_x h_y^{\mathrm{eff}} - S_y h_x^{\mathrm{eff}})\sin\theta\,\hat{\boldsymbol{\theta}}\Big] \cdot \Big[\hat{\boldsymbol{\theta}}\frac{\partial}{\partial\theta} + \frac{1}{\sin\theta}\hat{\boldsymbol{\phi}}\frac{\partial}{\partial\phi}\Big] f(\theta, \phi, t) \\
&= -\Big[m_y\cos\phi - m_x\sin\phi\Big]\frac{\partial f}{\partial\theta} \\
&\quad + \Big[m_x\cot\theta\cos\phi + m_y\cot\theta\sin\phi - m_z + (2n)D\cos^{2n-1}\theta\Big]\frac{\partial f}{\partial\phi}\,,
\end{aligned} \tag{8.18}$$

which when substituted in Eq. (8.16) yields the time evolution of $f(\theta, \phi, t)$:

$$\begin{aligned}
\frac{\partial f}{\partial t} &= \Big(m_y\cos\phi - m_x\sin\phi\Big)\frac{\partial f}{\partial\theta} \\
&\quad - \Big(m_x\cot\theta\cos\phi + m_y\cot\theta\sin\phi - m_z + (2n)D\cos^{2n-1}\theta\Big)\frac{\partial f}{\partial\phi}\,,
\end{aligned} \tag{8.19}$$

with

$$\begin{aligned}
(m_x, m_y, m_z) &\equiv (m_x, m_y, m_z)[f] \\
&= \int \sin\theta'\mathrm{d}\theta'\mathrm{d}\phi'\,\big(\sin\theta'\cos\phi',\ \sin\theta'\sin\phi',\ \cos\theta'\big)\,f(\theta', \phi', t)\,.
\end{aligned} \tag{8.20}$$

Note that Eq. (8.19) may also be obtained from Eq. (3.152) with the substitutions $p = \cos\theta$ and $\bar{f}(\phi, p, t) = f(\theta, \phi, t)$.

We now consider a far-from-equilibrium initial state $f_0(\theta, \phi)$ that is uniform in $\phi$ over $[0, 2\pi)$ and distributed in $\theta$ according to an arbitrary distribution $p(\theta)$ defined over a symmetric interval about $\theta = \pi/2$:

$$f_0(\theta, \phi) = \frac{A}{2\pi}\,p(\theta)\,, \tag{8.21}$$

with $p(\pi/2-\theta) = p(\pi/2+\theta)$. Note that $p(\theta)$ is normalized as $\int_0^{\pi} \mathrm{d}\theta\ \sin\theta\ p(\theta) = 1/A$. One may easily check that the state (8.21) is nonmagnetized (i.e., it gives $m_x = m_y = m_z = 0$) and therefore acts as a stationary solution of the Vlasov equation (8.19). The arbitrariness involved in the choice of the function $p(\theta)$ evidently shows that the Vlasov equation has a large number of stationary solutions. This is in fact a salient feature of the Vlasov equation. In general, the Vlasov equation admits an infinite number of stationary solutions that are also stable with respect to fluctuations [3].

Let us now study the Vlasov-stability of the stationary state (8.21) with respect to fluctuations. To this end, one may linearize the Vlasov equation (8.19) with respect to small fluctuations $F_1(\theta, \phi, t)$ by expanding $f(\theta, \phi, t)$ as [6]

$$f(\theta, \phi, t) = f_0 + \lambda F_1(\theta, \phi, t)\,, \tag{8.22}$$

where one has $|\lambda| \ll 1$. Substituting Eq. (8.22) in Eq. (8.19), we obtain the linearized Vlasov equation

$$\frac{\partial F_1}{\partial t} = \left(\widetilde{m}_y \cos\phi - \widetilde{m}_x \sin\phi\right)\frac{\partial f_0}{\partial\theta} - (2n)D\cos^{2n-1}\theta\frac{\partial F_1}{\partial\phi}\,, \tag{8.23}$$

where the quantities $\widetilde{m}_x$ and $\widetilde{m}_y$, linear in $F_1$, are given by

$$(\widetilde{m}_x, \widetilde{m}_y)[F_1] \equiv \int \mathrm{d}\theta\mathrm{d}\phi \sin\theta\ (\sin\theta\cos\phi, \sin\theta\sin\phi)\ F_1(\theta, \phi, t)\,. \tag{8.24}$$

As $F_1(\theta, \phi, t)$ is $2\pi$-periodic in $\phi$, one may implement the following Fourier expansion [6]:

$$F_1(\theta, \phi, t) = \sum_k \int \mathrm{d}\omega\ g_k(\theta, \omega)\ \mathrm{e}^{\mathrm{i}(k\phi+\omega t)}\,. \tag{8.25}$$

In the long-time limit, the linearized Vlasov dynamics is dominated by the Fourier mode with frequency $\omega$ having the smallest imaginary part. Effectively, then, one may then write $F_1(\theta, \phi, t) = \sum_k g_k(\theta, \omega)\ \mathrm{e}^{\mathrm{i}(k\phi+\omega t)}$ and obtain

$$\widetilde{m}_x = \pi\mathrm{e}^{\mathrm{i}\omega t}\left(I_+ + I_-\right), \quad \widetilde{m}_y = \mathrm{i}\pi\mathrm{e}^{\mathrm{i}\omega t}\left(I_+ - I_-\right), \tag{8.26}$$

with

$$I_\pm = \int \mathrm{d}\theta\ \sin^2\theta\ g_{\pm 1}(\theta, \omega)\,, \tag{8.27}$$

from which it follows that the relevant eigenmodes of Eq. (8.23) are those with $k = \pm 1$. Indeed, as follows from Eq. (8.23), modes $k \neq \pm 1$ only oscillate in time. This fact that only the long-wavelength (i.e., small-$k$) mode perturbations determine

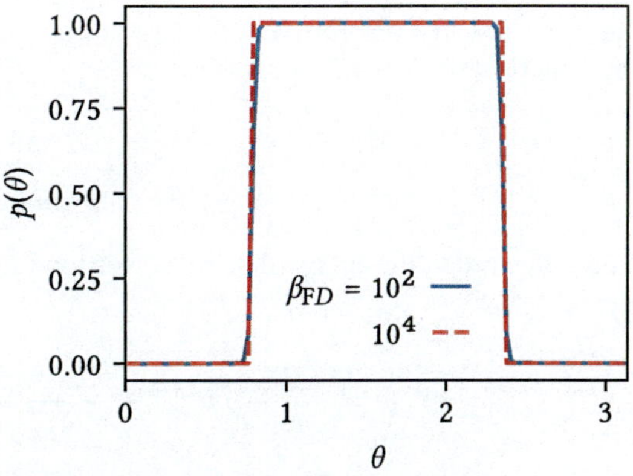


**Fig. 8.1** The distribution $p(\theta)$, corresponding to the Fermi-Dirac (FD) distribution (8.30), for two large values of $\beta_{\mathrm{FD}}$ and with $\mu = 0.5$

the stability of stationary states holds in general for long-range systems.[3] From Eqs. (8.23) and (8.26), it may be shown that the coefficients $g_{\pm 1}(\theta, \omega)$ satisfy

$$g_{\pm 1}(\theta, \omega) = \pi \frac{\partial f_0}{\partial \theta} \frac{I_\pm}{(2n) D \cos^{2n-1}\theta \pm \omega} . \tag{8.28}$$

Substituting the above equation into Eq. (8.27) and using the fact that $I_\pm \neq 0$, one may show that

$$\pi \int \mathrm{d}\theta \, \frac{\partial f_0}{\partial \theta} \frac{\sin^2\theta}{(2n) D \cos^{2n-1}\theta \pm \omega} = 1 . \tag{8.29}$$

We now consider a representative form of the function $f_0(\theta, \phi)$:

$$f_0(\theta, \phi) = \frac{A}{2\pi} \, p(\theta) ; \quad p(\theta) = \frac{1}{1 + \mathrm{e}^{\beta_{\mathrm{FD}}(\cos^2\theta - \mu)}} , \tag{8.30}$$

where $\mu \equiv \sin^2 a$ with $a > 0$ and $\beta_{\mathrm{FD}} > 0$ being real parameters, and $A$ is the normalization constant. Note that function $p(\theta)$ in the limit $\beta_{\mathrm{FD}} \to \infty$ yields a uniform distribution over the range $\theta \in [\pi/2 - a, \pi/2 + a]$, and hence the distribution (8.30) reduces to the WB state (8.15). On the other hand, for finite but large $\beta_{\mathrm{FD}}$, the distribution is smoothened around the boundaries at $\theta = \pi/2 \pm a$. We plot the distribution $p(\theta)$ for different values of $\beta_{\mathrm{FD}}$ and for $\mu = 0.5$ in Fig. 8.1, which evidently shows the similarity in the form of $p(\theta)$ to the Fermi-Dirac (FD) distribution. Henceforth, the distribution (8.30) will be referred to as the FD state. We remark that deriving analytical results for the FD distribution for general $\beta_{\mathrm{FD}}$ is nontrivial. However, for large $\beta_{\mathrm{FD}}$, one may obtain exact results, as detailed below.

---

[3] In this regard, one may refer to the phenomenon of Jeans instability in a prototypical long-range system, the gravitational systems [9].

As we show in Appendix H.1, for large values of $\beta_{\rm FD}$, one obtains correct to order $1/\beta_{\rm FD}^2$ that

$$A = \frac{1}{2\sqrt{\mu}}\left[1 + \frac{\pi^2}{24\beta_{\rm FD}^2\mu^2}\right]. \tag{8.31}$$

To the same order, the energy density corresponding to the state (8.30) becomes (see Appendix H.1)

$$\epsilon = \frac{D}{2n+1}\left[\mu^{2n/2} + \frac{(2n)^2\pi^2}{24\beta_{\rm FD}^2}\mu^{(2n-4)/2}\right]. \tag{8.32}$$

Substituting Eq. (8.30) in Eq. (8.29), one gets to order $1/\beta_{\rm FD}^2$ the equation (see Appendix H.1)

$$g(\mu)\mu^{-1/2} + \frac{\pi^2}{24\beta_{\rm FD}^2}\left[g(\mu)\mu^{-5/2} + 4g''(\mu)\mu^{-1/2}\right] = \frac{1}{nD}, \tag{8.33}$$

where one has

$$g(x) \equiv \frac{x^{(2n-1)/2} - x^{(2n+1)/2}}{(2n)^2D^2x^{2n-1} - \omega^2}, \tag{8.34}$$

while $\mu$ is to be obtained by solving Eq. (8.32). Equation (8.32) yields for $n = 1$ two possible values of $\mu$ given by

$$\mu = \left[\frac{3\epsilon}{D} - \frac{\pi^2 D}{18\beta_{\rm FD}^2\epsilon}\right] \text{ and } \frac{\pi^2 D}{18\beta_{\rm FD}^2\epsilon}, \tag{8.35}$$

and for $n = 2$ a single value given by

$$\mu = \sqrt{\frac{5\epsilon}{D} - \frac{2\pi^2}{3\beta_{\rm FD}^2}}. \tag{8.36}$$

Substituting Eq. (8.35) in Eq. (8.33) and keeping terms up to order $1/\beta_{\rm FD}^2$, one obtains for $n = 1$ two relations connecting $\omega$, $D$, $\beta_{\rm FD}$, and $\epsilon$. These two relations read as

$$\begin{aligned} &D\omega^4 - 3\omega^4\epsilon + D^2\omega^2\epsilon - 6D^2\omega\epsilon \\ &+ \frac{\pi^2 D}{18\beta_{\rm FD}^2\epsilon}\left(48D^4\epsilon - 8D^3\omega + 6D^2\omega\epsilon - 18D^2\omega^2\epsilon - D\omega^4\right) = \frac{2\pi^2D^3\omega^4}{3\beta_{\rm FD}^2\epsilon} - \omega^6, \end{aligned} \tag{8.37}$$

and

$$
\begin{aligned}
&D(\widetilde{\alpha}-\omega^2)\Big[(1-\widetilde{\gamma})(\widetilde{\alpha}-\omega^2)+\frac{\pi^2}{24\beta_{\rm FD}^2\widetilde{\gamma}^2}\Big(11\widetilde{\alpha}\widetilde{\gamma}-7\widetilde{\alpha}-3\widetilde{\gamma}\omega^2-\omega^2\Big)\Big]\\
&+\frac{\pi^2}{24\beta_{\rm FD}^2\widetilde{\gamma}^2}\Big[\widetilde{\alpha}\widetilde{\gamma}+3\widetilde{\alpha}-6\widetilde{\alpha}\widetilde{\gamma}\omega^2+6\widetilde{\alpha}\omega^2-3\widetilde{\gamma}\omega^4-\omega^4\Big]\\
&=\big(\widetilde{\alpha}-\omega^2\big)^3-\frac{2\pi^2D^2}{\beta_{\rm FD}^2\widetilde{\gamma}}\big(\widetilde{\alpha}-\omega^2\big)^2, \qquad (8.38)
\end{aligned}
$$

where we have $\widetilde{\gamma}\equiv 3\epsilon/D$ and $\widetilde{\alpha}\equiv 4D^2\widetilde{\gamma}$. Similarly, for $n=2$, substituting Eq. (8.36) in Eq. (8.33), we obtain to order $1/\beta_{\rm FD}^2$ that

$$
\begin{aligned}
&2D(\alpha-\omega^2)\Big[(\gamma-\gamma^2)(\alpha-\omega^2)-\frac{\pi^2}{24\beta_{\rm FD}^2\gamma}\Big(55\alpha-63\gamma\alpha+15\gamma\omega^2-7\omega^2\Big)\Big]\\
&+\frac{D\pi^2}{3\beta_{\rm FD}^2\gamma}\Big[-12D^2\gamma^4\alpha+60D^2\gamma^3\alpha-16\gamma\omega^2\alpha+16\omega^2\alpha-\frac{15\omega^4\gamma}{4}+\frac{3\omega^4}{4}\Big]\\
&=(\alpha-\omega^2)^2\Big[\alpha-\omega^2-\frac{3\pi^2\alpha}{\beta_{\rm FD}^2\gamma^2}\Big], \qquad (8.39)
\end{aligned}
$$

where we have $\gamma^2\equiv 5\epsilon/D$ and $\alpha\equiv 16D^2\gamma^3$. Equations (8.37), (8.38), and (8.39) are expected to be valid for all $\omega$, including $\omega=0$. However, Eq. (8.37) for $\omega=0$ leads to an inconsistent relation, namely, $8\pi^2D^5/(3\beta_{\rm FD}^2)=0$, and hence, may be discarded.

In the limit $\beta_{\rm FD}\to\infty$, when the FD state (8.30) reduces to the WB state (8.15), Eqs. (8.38) and (8.39) reduce to

$$
\omega^2=12D\epsilon+3\epsilon-D\,, \qquad (8.40)
$$

and

$$
\omega^2=2\sqrt{5\epsilon D}\left[40\epsilon-1+\sqrt{\frac{5\epsilon}{D}}\right], \qquad (8.41)
$$

respectively. Setting $\omega=0$ in Eqs. (8.40) and (8.41) yields the threshold value $\epsilon^\star(D,\beta_{\rm FD}\to\infty)$ of the energy density such that these equations give only real roots for the frequency $\omega$ for $\epsilon>\epsilon^\star$. One thus finds that $\epsilon^\star$ satisfies the following relations:

$$
\epsilon^\star=\frac{D}{3+12D} \quad \text{for } n=1\,, \qquad (8.42)
$$

and

$$
40\epsilon^\star+\sqrt{\frac{5\epsilon^\star}{D}}=1 \quad \text{for } n=2\,. \qquad (8.43)
$$

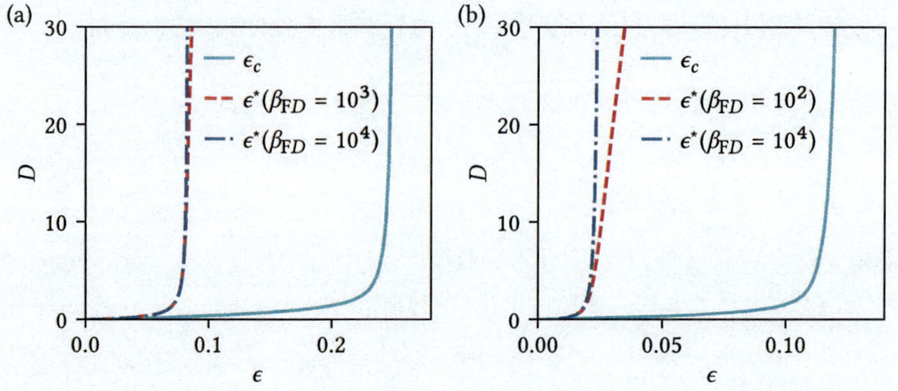


**Fig. 8.2 Phase diagram of the model** (8.1)**: a** $n = 1$, **b** $n = 2$. Both the diagrams show the equilibrium phase boundary $\epsilon_c$ and the Vlasov stability boundary $\epsilon^\star$ corresponding to the FD state (8.30) for two large values of $\beta_{\text{FD}}$. Both for $n = 1$ and $n = 2$, the result for the larger $\beta_{\text{FD}}$ value coincides with that obtained for the WB state (8.15)

Hence, we conclude that the WB state is linearly unstable under the Vlasov dynamics for $\epsilon < \epsilon^\star$, and is linearly stable for $\epsilon > \epsilon^\star$. When the WB state is linearly unstable (i.e., for $\epsilon < \epsilon^\star$), the perturbation $F_1(\theta, \phi, t)$ grows exponentially in time, and by setting $\omega^2 = -\Omega^2$ with real $\Omega$, we may write

$$F_1(\theta, \phi, t) \sim \mathrm{e}^{\pm \mathrm{i}\phi + \Omega t}\,. \tag{8.44}$$

Similarly, for finite but large $\beta_{\text{FD}}$, when Eqs. (8.38) and (8.39) remain valid, one may expect the existence of an energy threshold $\epsilon^\star(D, \beta_{\text{FD}}$ large) such that the FD state is linearly unstable and that the scaling (8.44) holds for $\epsilon < \epsilon^\star$, while the state is linearly stable for $\epsilon > \epsilon^\star$. In this case, one may obtain such an $\epsilon^\star$ by setting $\omega = 0$ in Eqs. (8.38) and (8.39). In doing so, we find that there exist more than one value of $\epsilon^\star$ for given $D$ and $\beta_{\text{FD}}$, and therefore we take for the physically meaningful $\epsilon^\star$ only the value that in the limit $\beta_{\text{FD}} \to \infty$ reduces to Eqs. (8.42) and (8.43). In Fig. 8.2, we show the results for $\epsilon^\star$ as a function of $D$ for both $n = 1$ and 2.

### 8.3.2 *Behavior for Finite N*

The Vlasov equation (8.16) describes the time evolution in an infinite system. In this section, we discuss the effects of large but finite system size $N$. Such a situation inevitably arises while studying the dynamics (8.13) numerically. In this case, the state of the system may be described by a discrete single-spin density function $P_{\text{d}}(\mathbf{S}, t)$, which to leading order in $N$ may be expanded as (see Appendix H.2)

$$P_{\mathrm{d}}(\mathbf{S},t) = P_0(\mathbf{S},t) + \frac{1}{\sqrt{N}}\delta P(\mathbf{S},t)\,. \tag{8.45}$$

For times $t \ll N$, the time evolution of $P_0$ follows Eq. (8.16), with that for $\delta P$ given by

$$\frac{\partial \delta P(\mathbf{S},t)}{\partial t} = -\frac{\partial}{\partial \mathbf{S}} \cdot \Big[(\mathbf{S} \times \delta\mathbf{h}^{\mathrm{eff}})P_0 + (\mathbf{S} \times \mathbf{h}^{\mathrm{eff},0})\delta P\Big], \tag{8.46}$$

where one has $\delta\mathbf{h}^{\mathrm{eff}} \equiv \delta\mathbf{h}^{\mathrm{eff}}[\delta P]$. Similar to Eq. (8.45), we may equivalently write

$$f_{\mathrm{d}}(\theta,\phi,t) = f(\theta,\phi,t) + \frac{1}{\sqrt{N}}\delta f(\theta,\phi,t)\,, \tag{8.47}$$

where the time evolution of $f$ for times $t \ll N$ is given by Eq. (8.19). As was done in obtaining Eq. (8.19), it may be shown that the time evolution of $\delta f(\theta,\phi,t) \equiv \delta P(\mathbf{S},t)$ is obtained from Eq. (8.46) as

$$\begin{aligned}\frac{\partial \delta f}{\partial t} &= \Big(m_y[\delta f]\cos\phi - m_x[\delta f]\sin\phi\Big)\frac{\partial f}{\partial \theta} + \Big(m_y[f]\cos\phi - m_x[f]\sin\phi\Big)\frac{\partial \delta f}{\partial \theta}\\ &- \Big(m_x[\delta f]\cot\theta\cos\phi + m_y[\delta f]\cot\theta\sin\phi - m_z[\delta f]\Big)\frac{\partial f}{\partial \phi}\\ &- \Big(m_x[f]\cot\theta\cos\phi + m_y[f]\cot\theta\sin\phi - m_z[f] + (2n)D\cos^{2n-1}\theta\Big)\frac{\partial \delta f}{\partial \phi}\,.\end{aligned} \tag{8.48}$$

If one chooses the initial state $f(\theta,\phi,t=0)$ to be $f_0(\theta,\phi)$ given by Eq. (8.21), then it follows from Eq. (8.19) that $f(\theta,\phi,t) = f(\theta,\phi,0)$ and Eq. (8.48) reduces to the same form as the linearized Vlasov equation (8.23):

$$\frac{\partial \delta f}{\partial t} = \Big(m_y[\delta f]\cos\phi - m_x[\delta f]\sin\phi\Big)\frac{\partial f}{\partial \theta} - (2n)D\cos^{2n-1}\theta\frac{\partial \delta f}{\partial \phi}\,. \tag{8.49}$$

Based on the previous analysis, one then concludes that for $\epsilon < \epsilon^\star$, when $f_0(\theta,\phi)$ is an unstable stationary solution of the Vlasov equation, Eq. (8.47) gives

$$(m_x, m_y, m_z)[f_{\mathrm{d}}] = \frac{1}{\sqrt{N}}(m_x, m_y, m_z)[\delta f]\,. \tag{8.50}$$

Observing the similarity in the forms of Eqs. (8.49) and (8.23), and considering the fact that Eq. (8.44) is a solution of Eq. (8.23), one thus obtains

$$m(t) \sim \frac{1}{\sqrt{N}}\mathrm{e}^{\Omega t}\,; \quad \epsilon < \epsilon^\star. \tag{8.51}$$

Equation (8.51) suggests that for $\epsilon < \epsilon^\star$ the relaxation time over which the magnetization acquires a value of $O(1)$ scales as $\log N$. On the other hand, for $\epsilon > \epsilon^\star$, when $f_0(\theta, \phi)$ acts as a stable-stationary solution of the Vlasov equation, the system would remain nonmagnetized for times $t \ll N$. In this case, the evolution at longer times is described by (see Appendix H.2):

$$\frac{\partial P_0(\mathbf{S}, t)}{\partial t} + \frac{\partial}{\partial \mathbf{S}} \cdot (\mathbf{S} \times \mathbf{h}^{\text{eff},0}) P_0 = -\frac{1}{N}\Big\langle \frac{\partial}{\partial \mathbf{S}} \cdot (\mathbf{S} \times \delta \mathbf{h}^{\text{eff}}) \delta P \Big\rangle. \tag{8.52}$$

For $\epsilon > \epsilon^\star$, the dynamics (8.52) comes into play only for longer times of order $N$, and therefore the initial nonmagnetized state would evolve only at these times. In a finite system, the state $f_0(\theta, \phi)$ thus manifests itself as a long-lived QSS that evolves very slowly, i.e., over a timescale that diverges with $N$.

### 8.3.3 *Numerical Results*

In support of our theoretical analysis of the preceding section, we now report on our numerical results for two representative values of $n$, namely, $n = 1, 2$. To integrate numerically the dynamics (8.13), we employ a fourth-order Runge-Kutta integration algorithm (discussed in Appendix M.1) with time step equal to 0.01. Our numerical results involve data averaging typically over several hundreds to thousand runs of the dynamics starting from different realizations of the FD state (8.30).

Let us first discuss the results for $n = 1$, for which we take $D = 5.0$. Then, using Eqs. (4.143) and (4.144), one obtains the equilibrium critical energy $\epsilon_c \approx 0.2381$. As the initial condition, we take the nonmagnetized FD state (8.30) with $\beta = 1000$, for which one has the stability threshold $\epsilon^\star \approx 0.0795$ as may be seen in the phase diagram 8.2. One may see from Fig. 8.3a that the system for energy $\epsilon < \epsilon^\star$ shows a fast relaxation out of the initial state on a timescale $\sim \log N$, which is more clearly seen in Fig. 8.3b. This observation is consistent with our theoretical prediction (8.51) and is further validated in Fig. 8.3c by showing the data collapse for $\sqrt{N} m(t)$ vs. $t$ for different values of $N$. In plotting Fig. 8.3c, the value of the growth rate $\Omega$ in Eq. (8.51) is obtained as the magnitude of imaginary part of the root of Eq. (8.38) for which the imaginary part is the largest in magnitude. In Fig. 8.4a, we show that the relaxation observed in Fig. 8.3 out of the initial FD state is not to equilibrium state but is to a magnetized QSS that has a lifetime that scales linearly with $N$. This latter fact on the time scale of the relaxation may be concluded on the basis of the plot in Fig. 8.4b. We thus summarize that for $\epsilon < \epsilon^\star$, the relaxation of the nonmagnetized FD state to equilibrium is a two-step process: (i) in the first step, the system relaxes over a timescale $\sim \log N$ to a magnetized QSS, (ii) in the second step, this QSS relaxes over a timescale $\sim N$ to the equilibrium state.

We now discuss the relaxation for energies $\epsilon^\star < \epsilon$, when the initial FD state is a stable stationary solution of the Vlasov equation. From Fig. 8.5a, one observes that consistent with our theoretical analysis, the initial FD state for $\epsilon^\star < \epsilon < \epsilon_c$ evolves

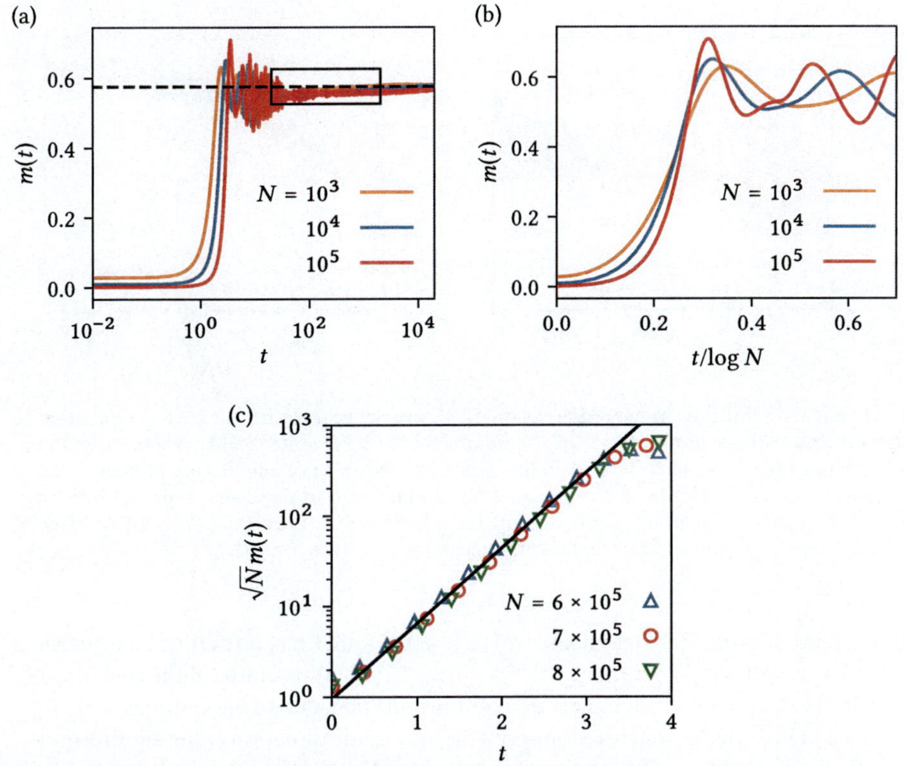


**Fig. 8.3 For the model (8.1) with $n = 1$ and $D = 5.0$ thus yielding $\epsilon_c \approx 0.2381$, the figure shows the relaxation under the deterministic dynamics (8.13) of an initial nonmagnetized FD state (8.30) with $\beta_{\text{FD}} = 1000$ (thus yielding stability threshold $\epsilon^\star \approx 0.0795$) for energy $\epsilon < \epsilon^\star$; here, we have chosen $\epsilon = 0.0212$.** One may observe a fast relaxation out of the initial FD state (panel (**a**)) over a time that scales with $N$ as $\log N$ (panel (**b**)). In panel (**a**), the dashed line represents the value of equilibrium magnetization at the studied energy value; here, the system sizes are indicated in the panel, and are given by $N = 10^3, 10^4, 10^5$ moving from left to right. The initial fast growth of the magnetization observed in **a** follows Eq. (8.51), as is evident from the data collapse for the scaled magnetization $\sqrt{N}m(t)$ as a function of $t$ shown in panel **c**. Here, the black line represents $e^{\Omega t}$, with $\Omega$ obtained as the magnitude of imaginary part of the root of Eq. (8.38) for which the imaginary part is the largest in magnitude; Here, we have $\Omega \approx 1.915$

to a nonmagnetized QSS that relaxes to equilibrium over a time which by virtue of the data presented in Fig. 8.5b may be concluded to be scaling with $N$ as $N^{3/2}$. One expects a QSS to show up also for energies $\epsilon > \epsilon_c$. However, in this energy regime, the magnetization is not the suitable quantity to monitor, since both the FD state and the equilibrium state are nonmagnetized. To tackle this issue, we choose $\langle \cos^4 \theta \rangle = (1/N) \sum_{i=1}^{N} \cos^4 \theta_i$ to monitor as a function of time.[4] From Fig. 8.6, one observes

[4] One may note that for the quadratic model ($n = 1$) with $\epsilon > \epsilon_c$, the quantity $\langle \cos^2 \theta \rangle$ is strictly a constant for infinite $N$. For finite $N$, it however shows fluctuations about this constant value.

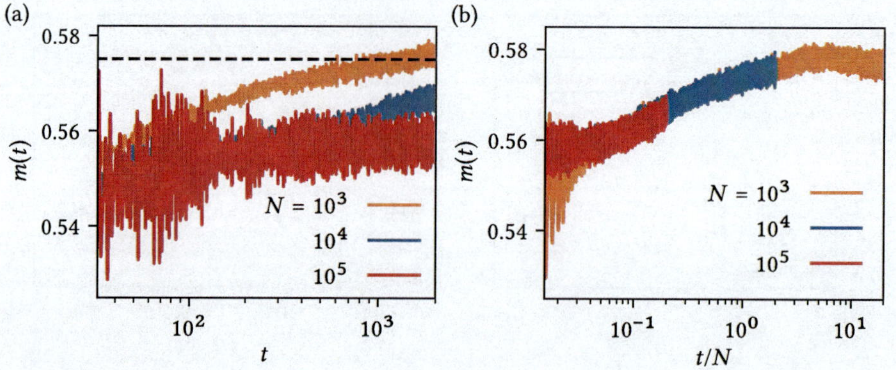


**Fig. 8.4 For the same parameter values as in Fig. 8.3, panel a shows for the case of relaxation under deterministic evolution (8.13) of the nonmagnetized FD state (8.30) a zoom in on the boxed part of Fig. 8.3a.** Here, the dashed line represents the value of equilibrium magnetization at the studied energy value; here, the system sizes are indicated in the panel, and are given by $N = 10^3, 10^4, 10^5$ moving from top to bottom along the line $t = 10^3$. The plot suggests the existence of a magnetized QSS with a lifetime that scales linearly with $N$ (see panel (**b**))

that indeed the initial FD state does evolve to a QSS that has a lifetime that scales quadratically with $N$. We remark that in all cases, when an initial nonmagnetized QSS relaxes eventually to a magnetized equilibrium state, due to the symmetry of the Hamiltonian (8.1) under spin rotation about the $z$-axis, the direction of the equilibrium magnetization may depend on the particular realization of the QSS under study. In the equilibrium state, the magnetization vector may even rotate in time, and only the application of an external field may lead it to take up a given orientation.

In order to show that the aforementioned relaxation behavior is generic to the spin model (8.1), we now present our numerical results for another value of $n$, namely, $n = 2$, in Figs. 8.7, 8.8, 8.9 and 8.10. From the figures, one may observe qualitatively similar behavior of the relaxation as that discussed above for $n = 1$. We remark that for energies $\epsilon > \epsilon_c$, in contrast to the quadratic model ($n = 1$), the quantity $\langle \cos^4 \theta \rangle$ for the quartic model ($n = 2$) is a constant in time. Therefore, to study the relaxation process for $\epsilon > \epsilon_c$ in this case we monitor $\langle \cos^2 \theta \rangle$ as a function of time, see Fig. 8.10. In the context of specific scaling relations of the QSSs, one may observe differences between the quadratic ($n = 1$) and the quartic ($n = 2$) model. In the quartic model, the nonmagnetized QSS occurring for $\epsilon^\star < \epsilon < \epsilon_c$ has a lifetime that scales as $N$, while the one occurring for $\epsilon > \epsilon_c$ has a lifetime that grows with $N$ as $N^{3/2}$.

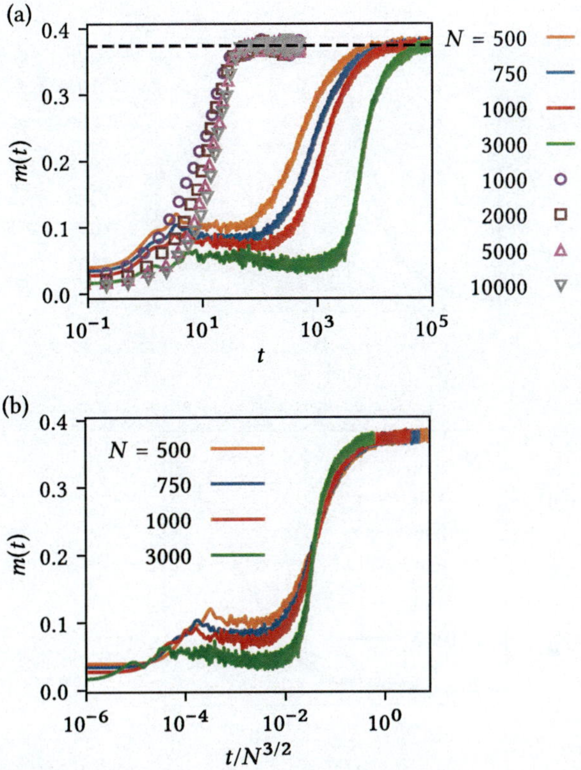


**Fig. 8.5 For the model** (8.1) **with** $n = 1$ **and** $D = 5.0$ **thus yielding** $\epsilon_c \approx 0.2381$**, the figure shows in lines the relaxation under the deterministic dynamics** (8.13) **of an initial nonmagnetized FD state** (8.30) **with** $\beta_{\text{FD}} = 1000$ **(thus yielding stability threshold** $\epsilon^\star \approx 0.0795$**) for energy** $\epsilon^\star < \epsilon < \epsilon_c$**; here, we have chosen** $\epsilon = 0.1473$**.** Here, the dashed line represents the value of equilibrium magnetization at the studied energy value; the system sizes are indicated in the panel, and are given by $N = 500, 750, 1000, 3000$ moving from left to right. One may observe the existence of a nonmagnetized QSS with a lifetime that diverges with the system size as $N^{3/2}$ (panel (**b**)). The points in panel **a** denote results based on numerical integration of the stochastic dynamics (8.53) for $\gamma = 0.05$ and at a value of temperature such that one obtains the same value of the equilibrium magnetization as that obtained at the value of energy chosen for the deterministic dynamics studied in **a**. The system sizes corresponding to the points in panel **a** are given by $N = 1000, 2000, 5000, 10000$ moving from left to right. The results imply a fast relaxation to equilibrium on a timescale that does not depend on $N$

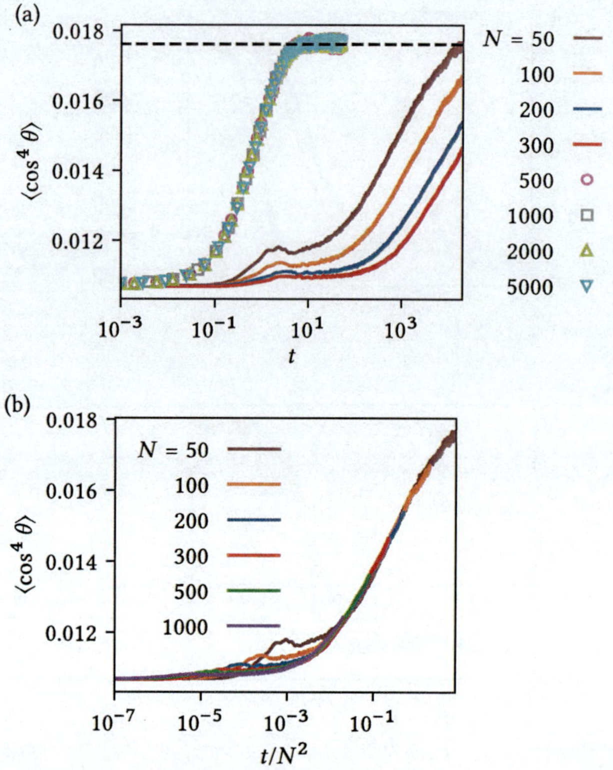


**Fig. 8.6 For the model (8.1) with $n = 1$ and $D = 5.0$ thus yielding $\epsilon_c \approx 0.2381$, the figure shows in lines the relaxation under the deterministic dynamics (8.13) of an initial nonmagnetized FD state (8.30) with $\beta_{\rm FD} = 1000$ (thus yielding stability threshold $\epsilon^\star \approx 0.0795$) for energy $\epsilon > \epsilon_c$; here, we have chosen $\epsilon = 0.3863$.** Here the dashed line denotes the equilibrium value of $\langle\cos^4\theta\rangle$ at the studied energy value, which may be obtained from the analysis in Sect. 4.6.1. The system sizes are indicated in the panel, and are given by $N = 50, 100, 200, 300$ moving from left to right. One may observe the existence of a nonmagnetized QSS with a lifetime that diverges with the system size as $N^2$ (panel (**b**)). The points in panel **a** represent results obtained from numerical integration of the stochastic dynamics (8.53) for $\gamma = 0.05$ and at a value of temperature for which one obtains the same value of equilibrium $\langle\cos^4\theta\rangle$ as that obtained at the value of energy chosen for the deterministic dynamics studied in **a**. The systems sizes corresponding to the points in panel **a** are $N = 500, 1000, 2000, 5000$. From the results, one may conclude a fast relaxation to equilibrium on a size-independent timescale, with no sign of quasistationarity

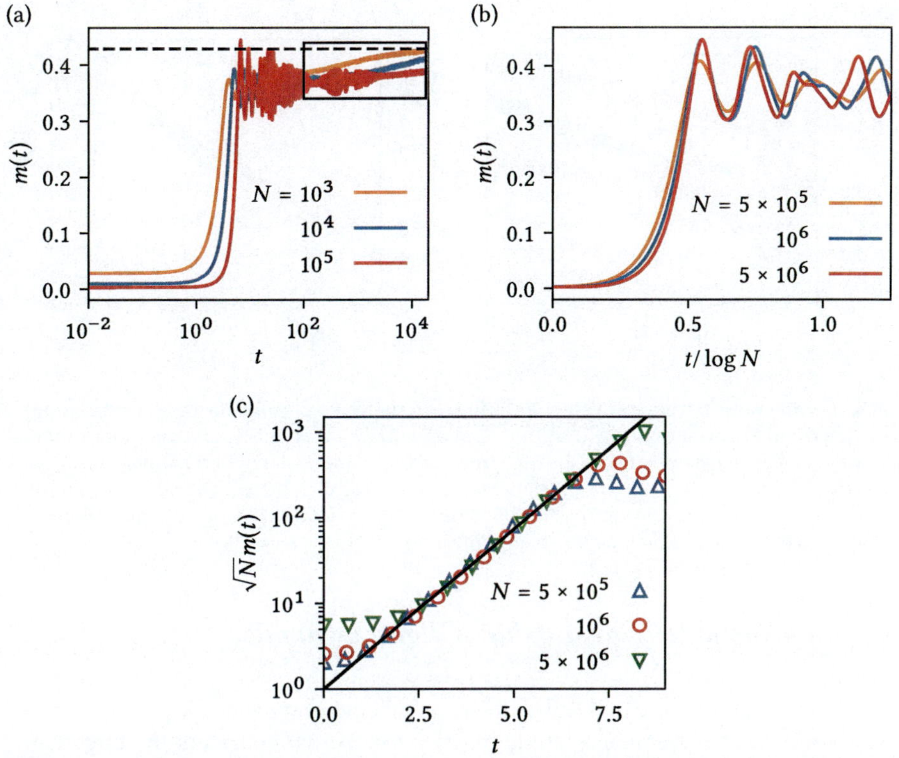


**Fig. 8.7 For the model (8.1) with $n = 2$ and $D = 15.0$ thus yielding $\epsilon_c \approx 0.1175$, the figure shows the relaxation under the deterministic dynamics (8.13) of an initial nonmagnetized FD state (8.30) with $\beta_{\mathrm{FD}} = 100$ (thus yielding stability threshold $\epsilon^\star \approx 0.0277$) for energy $\epsilon < \epsilon^\star$; here, we have chosen $\epsilon = 0.0111$.** One may observe a fast relaxation out of the initial FD state (panel (**a**)) over a time that scales with $N$ as $\log N$ (panel (**b**)). In panel **a**, the dashed line represents the value of equilibrium magnetization at the studied energy value; here, the system sizes are indicated in the panel, and are given by $N = 10^3, 10^4, 10^5$ moving from left to right. The initial fast growth of the magnetization observed in **a** follows Eq. (8.51), as is evident from the data collapse for the scaled magnetization $\sqrt{N}m(t)$ as a function of $t$ shown in panel **c**. Here, the black line represents $e^{\Omega t}$, with $\Omega$ obtained as the magnitude of imaginary part of the root of Eq. (8.39) for which the imaginary part is the largest in magnitude; Here, we have $\Omega \approx 0.859$

## 8.4 Analysis of the Model with Stochastic Dynamics

Let us now investigate the robustness of quasistationary states with respect to fluctuations induced through contact with external environment. For Hamiltonian particle dynamics (e.g., the HMF model), one may model the environment as a heat bath and invoke a scheme of coupling that allows for energy exchange between the system and the environment and a consequent stochastic Langevin evolution. In these cases, one observes a fast relaxation to equilibrium over a size-independent timescale in presence of strong-enough noise [10–12].

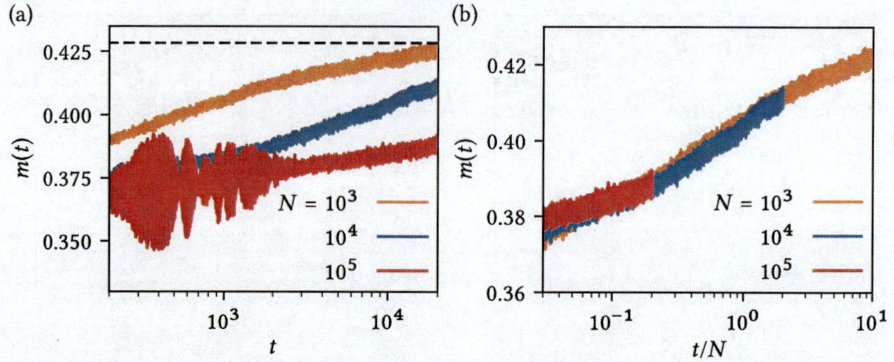


**Fig. 8.8 For the same parameter values as in Fig. 8.7, panel a shows for the case of relaxation under deterministic evolution (8.13) of the nonmagnetized FD state (8.30) a zoom in on the boxed part of Fig. 8.7a.** Here, the dashed line represents the value of equilibrium magnetization at the studied energy value; here, the system sizes are indicated in the panel, and are given by $N = 10^3, 10^4, 10^5$ moving from top to bottom along the line $t = 10^4$. The plot suggests the existence of a magnetized QSS with a lifetime that scales linearly with $N$ (see panel (**b**))

### *8.4.1 Stochastic Landau-Lifshitz-Gilbert Dynamics*

To study the effects of noise induced by the external environment on the system (8.1), we study a stochastic version of the dynamics (8.13) that considers the effective field $\mathbf{h}_i^{\text{eff}}$, to have an additional stochastic component due to interaction with the environment. In the spirit of the stochastic Landau-Lifshitz-Gilbert equation well known in studies of dynamical properties of magnetic systems [13], we write the resulting dynamics as

$$\dot{\mathbf{S}}_i = \mathbf{S}_i \times \left(\mathbf{h}_i^{\text{eff}} + \boldsymbol{\eta}_i(t)\right) - \gamma \mathbf{S}_i \times \left(\mathbf{S}_i \times \left(\mathbf{h}_i^{\text{eff}} + \boldsymbol{\eta}_i(t)\right)\right). \tag{8.53}$$

In the above equation, the second term on the right represents dissipation with the real parameter $\gamma > 0$ being the dissipation constant, and the vector $\boldsymbol{\eta}_i(t)$ is a Gaussian white noise with independent components that satisfy

$$\langle \eta_{i\mu}(t) \rangle = 0, \quad \langle \eta_{i\mu}(t)\eta_{j\nu}(t') \rangle = 2\delta_{i,j}\delta_{\mu,\nu}\mathcal{D}\delta(t - t'). \tag{8.54}$$

Here, the real constant $\mathcal{D} > 0$ sets the strength of the noise. The form of the stochastic dynamics (8.53) ensures that the length of each spin is conserved, as does the deterministic dynamics (8.13). We remark that Eq. (8.53) models the canonical dynamics for which the energy is not conserved during the dynamical evolution, while, as already mentioned earlier, Eq. (8.13) models the energy-conserving microcanonical dynamics.

Because of the noise in Eq. (8.53), the state of the system at a given time $t$, characterized by the set of values $\{\mathbf{S}_i(t)\}$, varies from one realization of the dynamics to

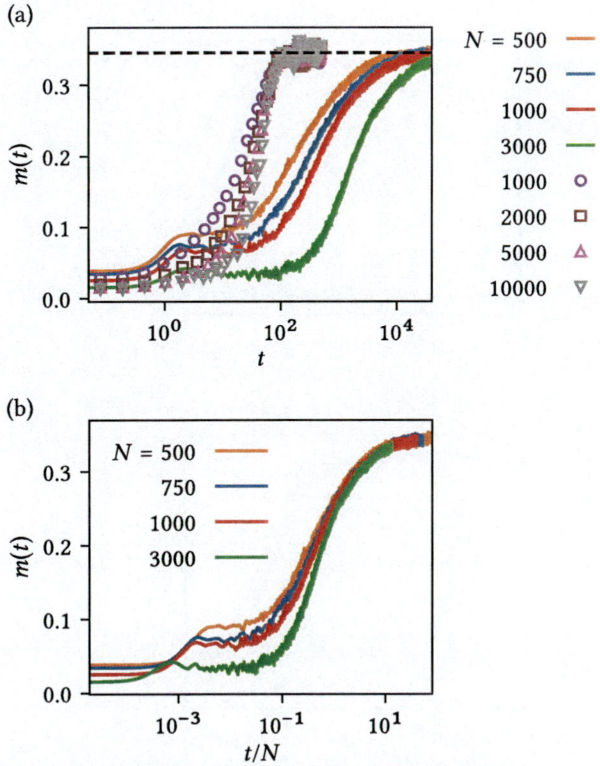


**Fig. 8.9 For the model** (8.1) **with** $n = 2$ **and** $D = 15.0$ **thus yielding** $\epsilon_c \approx 0.1175$**, the figure shows in lines the relaxation under the deterministic dynamics** (8.13) **of an initial nonmagnetized FD state** (8.30) **with** $\beta_{\text{FD}} = 100$ **(thus yielding stability threshold** $\epsilon^\star \approx 0.0277$**) for energy** $\epsilon^\star < \epsilon < \epsilon_c$**; here, we have chosen** $\epsilon = 0.0487$**.** Here, the dashed line represents the value of equilibrium magnetization at the studied energy value; the system sizes are indicated in the panel, and are given by $N = 500, 750, 1000, 3000$ moving from left to right. One may observe the existence of a nonmagnetized QSS with a lifetime that diverges with the system size as $N$ (panel (**b**)). The points in panel **a** denote results based on numerical integration of the stochastic dynamics (8.53) for $\gamma = 0.05$ and at a value of temperature such that one obtains the same value of the equilibrium magnetization as that obtained at the value of energy chosen for the deterministic dynamics studied in **a**. The system sizes corresponding to the points in panel **a** are given by $N = 1000, 2000, 5000, 10000$ moving from left to right. The results imply a fast relaxation to equilibrium on a timescale that does not depend on $N$

another, even when starting every time from the same initial condition. In Eq. (8.53), the noise and the dissipation term are incorporated in a way that it has the desirable feature of keeping the length of each spin to be unity at all times during the dynamical evolution. As the noise terms in Eq. (8.53) depend on the state of the system, itself stochastic in nature, the noise is said to be multiplicative in common parlance. As we argue later in Appendix H.2, requiring the dynamics (8.53) to relax at long times to canonical equilibrium at a given temperature $T$ fixes the constant $\mathcal{D}$ to be

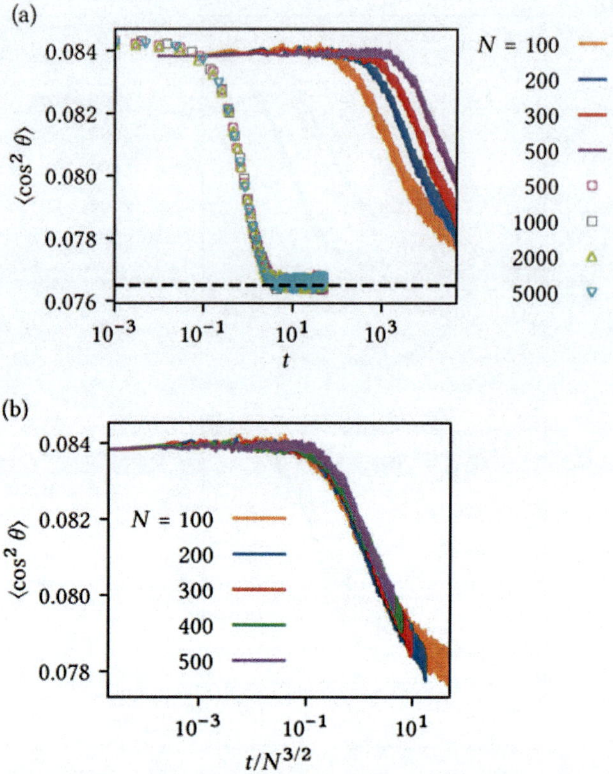


**Fig. 8.10 For the model (8.1) with $n = 2$ and $D = 15.0$ thus yielding $\epsilon_c \approx 0.1175$, the figure shows in lines the relaxation under the deterministic dynamics (8.13) of an initial nonmagnetized FD state (8.30) with $\beta_{\text{FD}} = 100$ (thus yielding stability threshold $\epsilon^\star = 0.0277$) for energy $\epsilon > \epsilon_c$; here, we have chosen $\epsilon = 0.1925$.** Here the dashed line denotes the equilibrium value of $\langle\cos^2\theta\rangle$ at the studied energy value, which may be obtained from the analysis in Sect. 4.6.1. The system sizes are indicated in the panel, and are given by $N = 100, 200, 300, 500$ moving from left to right. One may observe the existence of a nonmagnetized QSS with a lifetime that diverges with the system size as $N^{3/2}$ (panel (**b**)). The points in panel **a** represent results obtained from numerical integration of the stochastic dynamics (8.53) for $\gamma = 0.05$ and at a value of temperature for which one obtains the same value of equilibrium $\langle\cos^2\theta\rangle$ as that obtained at the value of energy chosen for the deterministic dynamics studied in **a**. The system sizes corresponding to the points in panel **a** are given by $N = 500, 1000, 2000, 5000$. From the results, one may conclude a fast relaxation to equilibrium on a size-independent timescale, with no sign of quasistationarity

$$\mathcal{D} = \frac{\gamma T}{1+\gamma^2}, \tag{8.55}$$

a choice we also consider in this work. In integrating numerically the dynamics (8.53), we follow a scheme detailed in Appendix M.3. In the following, we show that in presence of strong-enough noise, the system exhibits a fast relaxation to equilibrium on a size-independent timescale, with no existence of intermediate quasistationary states. A Monte Carlo scheme is also implemented as an alternative to the dynamics (8.53) to study the effects of environment-induced noise on the dynamics (8.6). Using the Monte Carlo scheme, one finds similar to the study of the dynamics (8.53) a fast relaxation to equilibrium on a size-independent timescale. These results reaffirm what has been observed earlier in particle dynamics of long-range systems, namely, that quasistationarity, observed in conservative dynamics, is completely washed away in presence of noise.

### 8.4.2 Behavior in the Limit $N \to \infty$

In the limit $N \to \infty$, the stochastic dynamics (8.53) may be studied by considering the time evolution of the single-spin distribution function $P_0(\mathbf{S}, t)$. The corresponding evolution equation derived in Appendix H.2 is given by

$$\begin{aligned}\frac{\partial P_0(\mathbf{S}, t)}{\partial t} &+ \frac{\partial}{\partial \mathbf{S}} \cdot (\mathbf{S} \times \mathbf{h}^{\mathrm{eff},0}) P_0(\mathbf{S}, t) \\ &= \gamma \frac{\partial}{\partial \mathbf{S}} \cdot \Big[(\mathbf{S} \times \mathbf{S} \times \mathbf{h}^{\mathrm{eff},0}) - \frac{1}{\beta}(\mathbf{S} \times (\mathbf{S} \times \frac{\partial}{\partial \mathbf{S}}))\Big] P_0(\mathbf{S}, t) . \qquad (8.56)\end{aligned}$$

It is worth noting that the state (8.30), or more generally, the state (8.21) is not a stationary solution of Eq. (8.56). From the structure of Eq. (8.56), one may conclude that the right hand side may be neglected for times $t \ll 1/\gamma$, and consequently, the time evolution is governed solely by the left hand side set to zero, which is nothing but the Vlasov equation (8.16). Consequently, the energy remains conserved for times $t \ll 1/\gamma$, and, based on the analysis presented in Sect. 8.3.1, the state (8.30) appears as an unstable stationary state for $\epsilon < \epsilon^\star$ and as a stable stationary state for $\epsilon > \epsilon^\star$. On the other hand, for times of order $1/\gamma$, one may not neglect the right hand side of Eq. (8.56), and hence, the energy would change over times of $O(1/\gamma)$ and the state (8.30) would evolve to relax to the stationary state of Eq. (8.56), which is nothing but the equilibrium state, see Appendix H.2.

### 8.4.3 Behavior for Finite N

For large but finite system, we need to add finite-$N$ corrections to Eq. (8.56). As shown in Appendix H.2, the time evolution of the single-spin distribution $P_0(\mathbf{S}, t)$ in this case is given by

$$\begin{aligned}
&\frac{\partial P_0(\mathbf{S},t)}{\partial t} + \frac{\partial}{\partial \mathbf{S}} \cdot (\mathbf{S} \times \mathbf{h}^{\mathrm{eff},0}) P_0 \\
&= \gamma \frac{\partial}{\partial \mathbf{S}} \cdot \Big[ (\mathbf{S} \times \mathbf{S} \times \mathbf{h}^{\mathrm{eff},0}) - (1/\beta)(\mathbf{S} \times (\mathbf{S} \times \frac{\partial}{\partial \mathbf{S}})) \Big] P_0 \\
&- \frac{1}{N} \Big\langle \frac{\partial}{\partial \mathbf{S}} \cdot \Big[ (\mathbf{S} \times \delta \mathbf{h}^{\mathrm{eff}}) \delta P - \gamma (\mathbf{S} \times \mathbf{S} \times \delta \mathbf{h}^{\mathrm{eff}}) \delta P \Big] \Big\rangle .
\end{aligned} \tag{8.57}$$

On the basis of our previous analysis, it may be concluded that for a given $N$, when the noise is strong enough that $1/\gamma \ll N$, the dynamics (8.57) would be dominated by the first term on the right hand side. Consequently, the FD state (8.30) would relax to the equilibrium state over times $t \sim 1/\gamma$, and no size-dependent relaxation and hence QSSs should be expected. However, in the opposite limit, i.e., for $1/\gamma \gg N$, the dynamics (8.57) would be dominated by the second term on the right hand side. As a result, the QSS observed for times $t \ll N$ would start evolving towards equilibrium over times of $O(N)$. Note that the relaxation would be further assisted by the noise after it comes into effect over times of order $1/\gamma$. We may then expect that for a given $N$, as the dissipation constant $\gamma$ is tuned from very small to very large values, we should see a cross-over behavior, from a size-dependent relaxation at small $\gamma$ to a size-independent one at large $\gamma$.

### *8.4.4 An Alternative Dynamics: A Monte Carlo Scheme*

As an alternative way of modeling the environment-induced noise, one may invoke a Monte Carlo update scheme of the spin values that guarantees that the long-time state of the system is the equilibrium state. In this scheme, one randomly selects spins and attempts to rotate them by a stipulated amount (which itself could be random) with a probability. The concerned probability depends on the change in the energy of the system as a result of the attempted update of the state of the system [14, 15]. Specifically, for the spin system (8.1), one may perform the Monte Carlo dynamics at a given temperature $T = 1/\beta$ by implementing the following steps [16]:

1. Start with a spin configuration in the nonmagnetized FD state.
2. Select a spin at random and attempt to change its direction at random, i.e., choose a value of $\theta$ uniformly in $[0, \pi]$ and a value of $\phi$ uniformly in $[0, 2\pi)$ and assign these values to the spin.
3. Compute $\Delta E$, the change in the energy of the system due to this attempted change of spin direction.
4. Accept the change if $\Delta E < 0$, i.e., the energy of the system is lowered by the change of spin direction.
5. On the other hand, if the energy increases due to the attempted change of the spin direction, i.e., $\Delta E > 0$, first compute the Boltzmann probability $p = \exp(-\beta \Delta E)$. Next, if a random number $r$ drawn uniformly from the range [0, 1]

satisfies $r < p$, accept the change in spin direction; otherwise, reject the attempted change and retain the previous spin configuration.

6. Measure time in units of Monte Carlo Steps (MCS), where one step corresponds to $N$ attempted changes in spin direction.
7. At the end of every MCS, compute the desired physical quantities such as the magnetization.

In practice, one repeats steps (2)–(5) to obtain values as a function of time of these physical quantities averaged over a sufficient number of independent configurations. We remark that unlike the deterministic dynamics (8.13), the above Monte Carlo scheme does not conserve energy.

### 8.4.5 Numerical Results

We now present the results obtained from numerical integration of the stochastic dynamics (8.53) using the midpoint algorithm discussed in Appendix M.3. We first discuss the numerical results for the quadratic ($n = 1$) model, for which we set $\gamma = 0.05$ and integration time step equal to $10^{-3}$. Similar to the case of the deterministic dynamics, here the numerical results involve data averaging of typically over several hundreds to thousand runs of the dynamics starting from different realizations of the FD state (8.30). Our primary focus is on comparing for a given energy $\epsilon$ the stochastic dynamics results with those observed for the deterministic dynamics. As we have already discussed, for our model in equilibrium, the microcanonical and canonical ensembles are equivalent. By virtue of this equivalence, we choose the value of temperature $T$ in the stochastic dynamics such that it ensures that one has in equilibrium the same value of magnetization as the one observed for the deterministic dynamics at energy $\epsilon$. One may make this specific choice of $T$ by using plots such as those in Fig. 4.8. It may be seen from Fig. 8.11a that under stochastic dynamics with $1/\gamma \ll N$, the initial FD state exhibits a fast relaxation to equilibrium on a size-independent timescale without any sign of quasistationarity during the relaxation process. From Fig. 8.11b, we see that at the chosen value of $T$, the average energy of the system in equilibrium coincides with the conserved energy of the deterministic dynamics, as one would expect. In Fig. 8.11c, we show for $N = 10^4$ the evolution of energy density $\epsilon(t)$ under the stochastic dynamics (8.53) for different values of the dissipation constant $\gamma$. Observing the scaling collapse of the data in the figure, one may conclude that the relaxation of the initial state occurs over the timescale $\sim 1/\gamma$, consistent with our theoretical analysis in Sect. 8.4.3. One also observes relaxation on a size-independent timescale for energies $\epsilon^\star < \epsilon < \epsilon_c$ (see Fig. 8.5a) and for energies $\epsilon > \epsilon_c$ (see Fig. 8.6a). One may note that in these cases too, we have $1/\gamma \ll N$. On the other hand, from Fig. 8.12, one may verify the expected cross-over in the relaxation behavior as the value of $\gamma$ is tuned from low to high values for a fixed $N$. The relaxation behavior observed of the quartic model ($n = 2$) is similar to that of the quadratic model ($n = 1$), as may be seen from Figs. 8.14, 8.9, and 8.10.

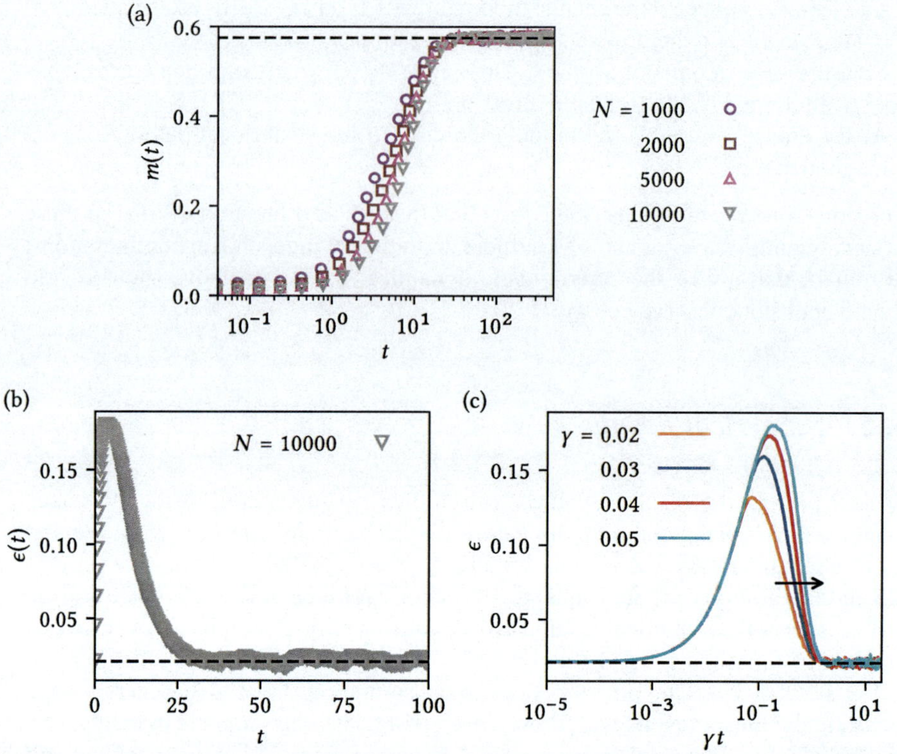


**Fig. 8.11 For the same parameter values as in Fig. 8.3, the points in panel a are obtained from numerical integration of the stochastic dynamics (8.53) for $\gamma = 0.05$ and at a value of temperature that ensures that one obtains the same value of the equilibrium magnetization as that obtained at the value of energy chosen for the deterministic dynamics studied in Fig. 8.3.** The results suggest a fast relaxation to equilibrium on a size-independent timescale, with no sign of quasistationarity. In panel **b**, we show for the parameter values of panel **a** that indeed at the studied value of temperature, the average energy of the system in equilibrium matches up to numerical accuracy the conserved value of energy (dashed line) chosen for the deterministic dynamics in Fig. 8.3. The figure in panel **c** shows for $N = 10000$ the evolution of energy under the stochastic dynamics (8.53) for four values of the dissipation parameter $\gamma$ (moving in the direction of the arrow, the values of $\gamma$ are given by 0.02, 0.03, 0.04, 0.05). Scaling collapse of the data suggests relaxation of the initial state over the timescale $\sim 1/\gamma$. Here the dashed line corresponds to the initial and the final energy value. Note that it is the final energy value that gets fixed in our numerical scheme by our choice of the temperature, while the choice of the initial energy is immaterial as it anyways is not conserved by the dynamics and will evolve to the allowed final value. In the plot of panel **c**, the initial energy value happens to have been chosen to equal the final allowed value

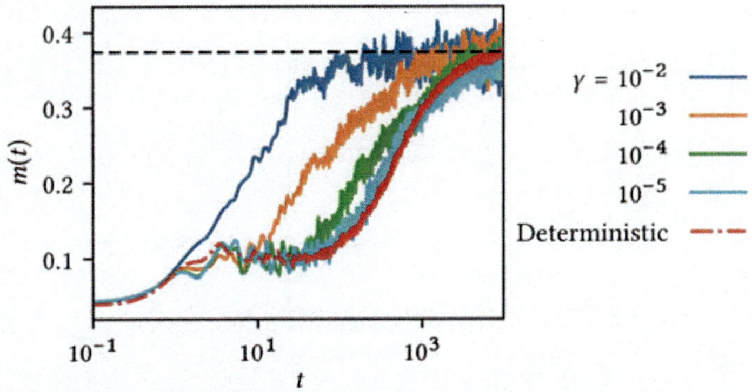


**Fig. 8.12 Considering the stochastic dynamics (8.53) with $n = 1$, $D = 5.0$, $N = 500$, four values of $\gamma$, and with (8.30) as the initial state with $\beta_{\text{FD}} = 1000$, initial energy $\epsilon = 0.1473$, the figure shows a cross-over in the relaxation behavior, from a fast to a slow one, as one tunes the parameter $\gamma$ from high to low values (moving from left to right, the values of $\gamma$ are $10^{-2}, 10^{-3}, 10^{-4}, 10^{-5}$).** Here, we keep the temperature fixed to a value such that one obtains the same value of the equilibrium magnetization as that obtained for the deterministic dynamics (8.13) with energy equal to $\epsilon$ whose results are also included in the plot. The observed behavior is consistent with the conclusion drawn in Sect. 8.4.3. Here, the dashed line represents the value of equilibrium magnetization at the studied temperature

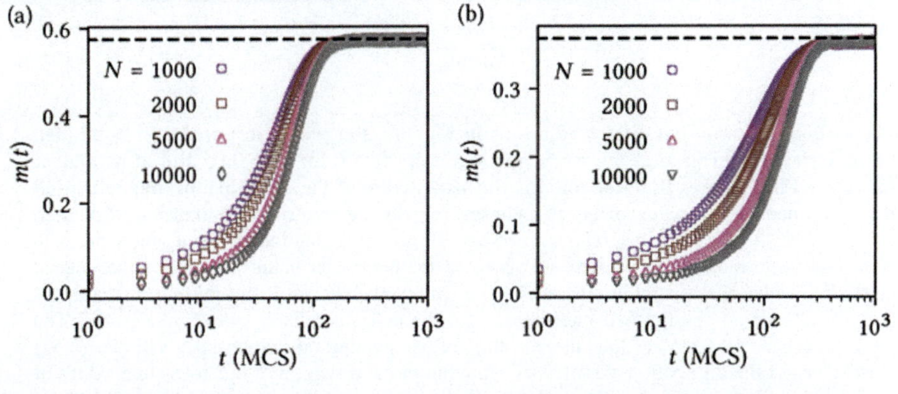


**Fig. 8.13 For the model (8.1) with $n = 1$ and $D = 5.0$, the figure shows the relaxation under Glauber Monte Carlo dynamics of an initial nonmagnetized FD state (8.30) with $\beta_{\text{FD}} = 1000$ for a the same temperature as in Fig. 8.11c and b the same temperature as in Fig. 8.5a.** Here the dashed line denotes the equilibrium magnetization value at the temperature at which the Monte Carlo dynamics is implemented. The figures suggest the absence of any quasistationary behavior and a fast relaxation on a size-independent timescale to equilibrium. We have observed similar fast relaxation also for temperatures corresponding to energies $\epsilon > \epsilon_c$ of the deterministic dynamics (data not shown here)

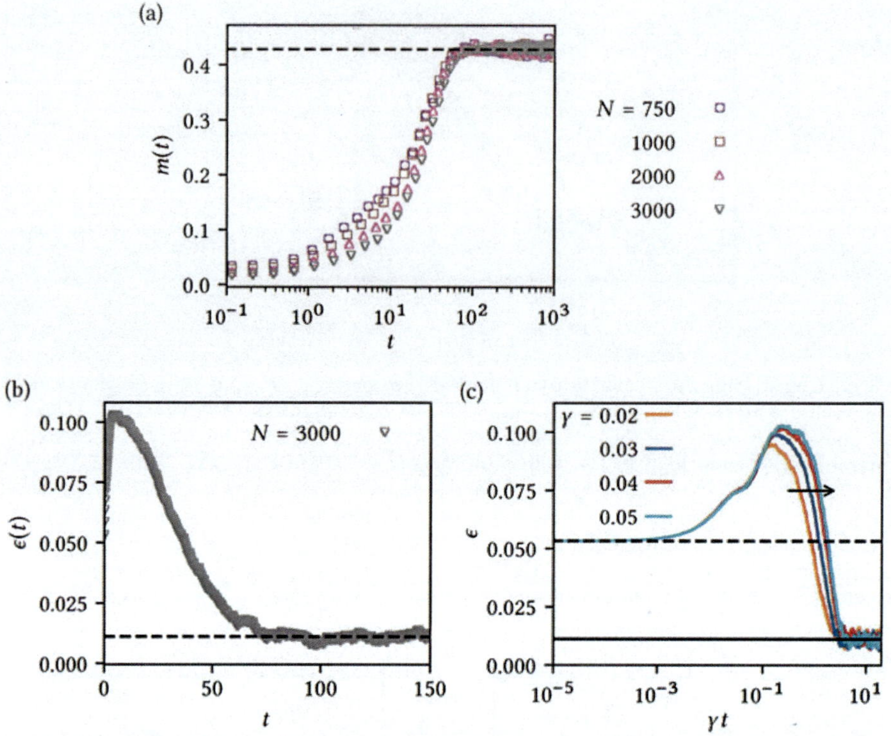


**Fig. 8.14 For the same parameter values as in Fig. 8.7, the points in panel a are obtained from numerical integration of the stochastic dynamics (8.53) for $\gamma = 0.05$ and at a value of temperature that ensures that one obtains the same value of the equilibrium magnetization as that obtained at the value of energy chosen for the deterministic dynamics studied in Fig. 8.7.** The results suggest a fast relaxation to equilibrium on a size-independent timescale, with no sign of quasistationarity. In panel **b**, we show for the parameter values of panel **a** that indeed at the studied value of temperature, the average energy of the system in equilibrium matches the conserved value of energy (dashed line) chosen for the deterministic dynamics in Fig. 8.7. The figure in panel **c** shows for $N = 3000$ the evolution of energy under the stochastic dynamics (8.53) for four values of the dissipation parameter $\gamma$ (moving in the direction of the arrow, the values of $\gamma$ are given by 0.02, 0.03, 0.04, 0.05). Scaling collapse of the data suggests relaxation of the initial state over the timescale $\sim 1/\gamma$. Here, the dashed line (respectively, the solid line) corresponds to initial (respectively, final) energy value

Let us now discuss the results obtained from the Glauber Monte Carlo simulation of the system (8.1) for $n = 1$ and 2. The results are shown in Figs. 8.13 and 8.15. Once again, a fast relaxation to equilibrium over a size-independent timescale may be seen from the figures. Here, one may explain the fast relaxation based on the fact that size-dependent relaxation is a feature of the energy-conserving Vlasov dynamics, as we have discussed in Sect. 8.3. We remark that within the stochastic dynamics (8.53), energy conservation is violated on the timescale $\sim 1/\gamma$, whereas within the Monte Carlo dynamics the same happens on the scale of one time unit. As a result, within

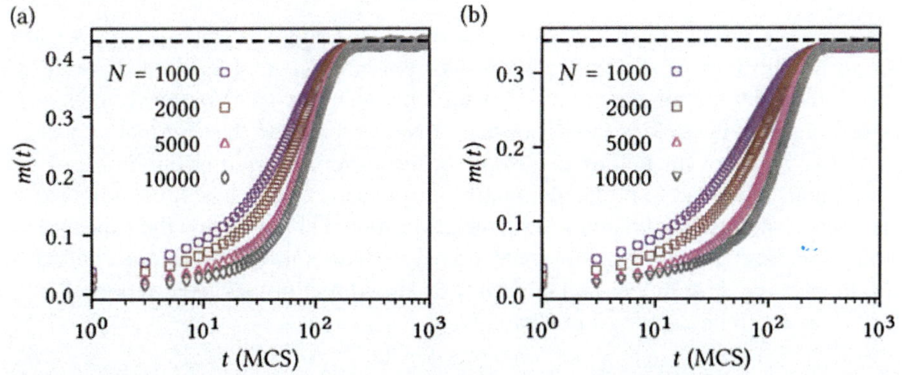


**Fig. 8.15 For the model** (8.1) **with** $n = 2$ **and** $D = 15.0$ **thus yielding** $\epsilon_c \approx 0.1175$**, the figure shows the relaxation under Glauber Monte Carlo dynamics of an initial nonmagnetized FD state** (8.30) **with** $\beta_{\mathrm{FD}} = 100$ **for a the same temperature as in Fig.** 8.14**c and b the same temperature as in Fig.** 8.9**a.** Here the dashed line denotes the equilibrium magnetization value at the temperature at which the Monte Carlo dynamics is implemented. The figures suggest the absence of any quasistationary behavior and a fast relaxation on a size-independent timescale to equilibrium. We have observed similar fast relaxation also for temperatures corresponding to energies $\epsilon > \epsilon_c$ of the deterministic dynamics (data not shown here)

the Monte Carlo dynamics, size-dependent relaxation will not be observed, while within the dynamics (8.53), fast relaxation requires choosing a strong-enough noise such that one has $1/\gamma \ll N$.

## 8.5 Conclusions and Perspectives

In this chapter, we have considered two sources of noise, namely, those arising from the finiteness of system size and those due to interaction with the external environment, Specifically, we have discussed the effects of noise arising from these two sources on the relaxation process of a model long-range interacting system of classical Heisenberg spins. The system under deterministic spin precessional dynamics and for a wide range of energy values exhibits a slow relaxation to equilibrium over a timescale that diverges with the system size. The corresponding stochastic dynamics that considers the interaction with the environment is constructed in the spirit of (i) the stochastic Landau-Lifshitz-Gilbert equation, and (ii) the Glauber Monte Carlo dynamics. We have shown that in both cases, the stochastic dynamics for strong enough noise shows a fast relaxation to equilibrium on a size-independent timescale, with no signature of quasistationarity. The results presented here thus establish unequivocally how quasistationarity observed in deterministic dynamics of long-range systems is washed away by fluctuations induced through the interactions with the environment.

One may be interested in studying how our long-range system (8.1) prepared either in equilibrium or in QSSs responds to an external field and what are the effects induced by the inherent slow relaxation process. For short-range systems in equilibrium, when the field is small, one has a linear response to the field that may be expressed in terms of fluctuation properties of the system in equilibrium. For long-range systems, studies of similar fluctuation-response relations have been pursued in the context of particle dynamics (e.g., the HMF model [17, 18]) and the existence of strange scaling of fluctuations in finite-size systems is known [19]. In the context of the long-range spin model (8.1), a study in this direction has been reported in Ref. [20], and will be discussed in Chap. 9.

## References

1. A. Campa, T. Dauxois, S. Ruffo, Phys. Rep. **480**(3–6), 57 (2009). https://doi.org/10.1016/j.physrep.2009.07.001
2. F. Bouchet, S. Gupta, D. Mukamel, Phys. A **389**(20), 4389 (2010). https://doi.org/10.1016/j.physa.2010.02.024
3. A. Campa, T. Dauxois, D. Fanelli, S. Ruffo, *Physics of Long-Range Interacting Systems*, 1st edn. (Oxford University Press, Oxford, 2014). https://doi.org/10.1093/acprof:oso/9780199581931.001.0001
4. Y. Levin, R. Pakter, F.B. Rizzato, T.N. Teles, F.P.C. Benetti, Phys. Rep. **535**(1), 1 (2014). https://doi.org/10.1016/j.physrep.2013.10.001
5. S. Gupta, S. Ruffo, Int. J. Mod. Phys. A **32**(09), 1741018 (2017). https://doi.org/10.1142/S0217751X17410184
6. S. Gupta, D. Mukamel, J. Stat. Mech: Theory Exp. **2011**(03), P03015 (2011). https://doi.org/10.1088/1742-5468/2011/03/P03015
7. C.R. Lourenço, T.M. Rocha Filho, Phys. Rev. E **92**(1), 012117 (2015). https://doi.org/10.1103/PhysRevE.92.012117
8. J. Barré, S. Gupta, J. Stat. Mech: Theory Exp. **2014**(2), P02017 (2014). https://doi.org/10.1088/1742-5468/2014/02/P02017
9. J. Binney, S. Tremaine, *Galactic Dynamics*, 2nd edn. (Princeton University Press, Princeton, 2011)
10. F. Baldovin, E. Orlandini, Phys. Rev. Lett. **96**(24), 240602 (2006). https://doi.org/10.1103/PhysRevLett.96.240602
11. F. Baldovin, E. Orlandini, Phys. Rev. Lett. **97**(10), 100601 (2006). https://doi.org/10.1103/PhysRevLett.97.100601
12. F. Baldovin, P.H. Chavanis, E. Orlandini, Phys. Rev. E **79**(1), 011102 (2009). https://doi.org/10.1103/PhysRevE.79.011102
13. M. Lakshmanan, Phil. Trans. R. Soc. A **369**(1939), 1280 (2011). https://doi.org/10.1098/rsta.2010.0319
14. R.J. Glauber, J. Math. Phys. **4**(2), 294 (1963). https://doi.org/10.1063/1.1703954
15. P.L. Krapivsky, S. Redner, E. Ben-Naim, *A Kinetic View of Statistical Physics* (Cambridge University Press, New York, 2010)
16. Z. Slanic, H. Gould, J. Tobochnik, Comput. Phys. **5**(6), 630 (1991). https://doi.org/10.1063/1.4823032
17. A. Patelli, S. Gupta, C. Nardini, S. Ruffo, Phys. Rev. E **85**(2), 021133 (2012). https://doi.org/10.1103/PhysRevE.85.021133

18. S. Ogawa, Y.Y. Yamaguchi, Phys. Rev. E **85**(6), 061115 (2012). https://doi.org/10.1103/PhysRevE.85.061115
19. Y.Y. Yamaguchi, Phys. Rev. E **94**(1), 012133 (2016). https://doi.org/10.1103/PhysRevE.94.012133
20. Y.Y. Yamaguchi, D. Das, S. Gupta, Phys. Rev. E **100**(3), 032131 (2019). https://doi.org/10.1103/PhysRevE.100.032131

# Chapter 9
# Critical Exponents in Mean-Field Classical Spin Systems

**Abstract** In this chapter, we study the facet of noise in the form of fluctuations due to the application of a small external field in the dynamics of the same mean-field classical spin system as discussed in the previous chapter. In the absence of the field, the system exhibits in the stationary state a second-order phase transition. Here, we analyze the corresponding phase-space evolution of the system following the Vlasov equation. Within the framework of the Vlasov dynamics, we obtain the values of the critical exponents describing the power-law behavior of response to a small external field. We show that the exponent values obtained from the Vlasov dynamics differ significantly from the ones obtained by analyzing a static analysis that makes no reference to the dynamics in the presence of the external field. This chapter illustrates that while extracting critical exponent values for mean-field systems, one has to be cautious against relying on a static approach and resort instead to a dynamic approach in which the actual evolution of the phase-space distribution is taken into account. This chapter is based on our published work, Ref. [1].

## 9.1 Introduction

Phase transitions in physical systems occur only in the thermodynamic limit. In contrast to first-order phase transitions, second-order transitions do not involve latent heat and are characterized by a power-law behavior of macroscopic variables close to the critical point. One may broadly classify second-order phase transitions in different systems into universality classes identified by different values of critical exponents describing the power-law behavior. As an example of a second-order transition, let us consider the ferromagnetic transition that occurs at a critical temperature $T_c$. In this case, the magnetization for temperatures $T \lesssim T_c$ shows a power-law dependence on the separation $(T_c - T)$, with the corresponding exponent being $\beta$. When an external field is applied, the magnetization increases as a function of the field strength. For an infinitesimal field, a linear growth in magnetization for $T \neq T_c$ implies a linear response and determines the zero-field susceptibility $\chi$. Close to the critical point $T_c$, the susceptibility diverges as a power law with the corresponding exponents



D. Das and S. Gupta, *Facets of Noise*, Fundamental Theories of Physics 214,
https://doi.org/10.1007/978-3-031-45312-0_9

denoted by $\gamma^+$ and $\gamma^-$ in the disordered ($T > T_c$) and the magnetized ($T < T_c$) phase, respectively. At the critical point $T = T_c$, the response to the external field becomes nonlinear, which is characterized by the critical exponent $\delta$. These critical exponents satisfy the scaling relation $\gamma^{\pm} = \beta(\delta - 1)$ [2–4].

As a representative class of systems exhibiting second-order phase transitions, one may consider the mean-field systems. When in thermal equilibrium and subject to a small external field, the values of the critical exponents may be obtained straightforwardly by assuming that the asymptotic distribution after the application of the field is also a thermal equilibrium state. Such an analysis does not consider the evolution of the initial thermal equilibrium state under the application of the field in order to obtain the asymptotic state, and therefore may be referred to as a static analysis. Using such a static analysis, one obtains the exponent values as $\beta = 1/2,\ \gamma^{\pm} = 1,\ \delta = 3$ [3]. However, if one follows the evolution of the initial state, one may obtain very different exponent values due to the mean-field nature of the dynamics. This is owing to the fact that the dynamics of a mean-field system in the thermodynamic limit is described by the so-called Vlasov equation. Now, for the Vlasov dynamics, there exists a vast number of stable stationary states, and thermal equilibrium is just one of them [5–8], see Sect. 3.7.3. Consequently, once the system is in a Vlasov-stable stationary state other than thermal equilibrium, it would not relax to thermal equilibrium. However, in contrast to this behavior observed in the thermodynamic limit, a large but finite system, despite being trapped initially in so-called quasistationary states (QSSs) identified as stable stationary solutions of the Vlasov equation, eventually relaxes to thermal equilibrium. In this case, finite-size effects allow the system to have a slow evolution of the QSSs towards thermal equilibrium over a timescale that diverges with the system size [9–11]. We have already discussed this system-size-dependent slow relaxation process in Chaps. 3 and 8.

A paradigmatic model of mean-field system of particles is the so-called Hamiltonian mean-field (HMF) model, which involves classical $XY$-spins evolving under a dynamics given by a Hamiltonian involving mean-field $XY$-interaction [12, 13]. We have provided a brief introduction to the model in Appendix C. A study of this model has shown that the critical exponents describing response of the system to a small external field and when obtained following the Vlasov evolution have the values $\gamma^+ = 1,\ \gamma^- = 1/4$ for a family of Vlasov-stable stationary initial states (including thermal equilibrium states) exhibiting continuous phase transitions with $\beta = 1/2$ [14]. At the critical point, the Vlasov dynamics yields $\delta = 3/2$. Note that these values of the critical exponents do not all match with those mentioned in the preceding paragraph that are obtained within the static analysis. More generally, for a class of Hamiltonian particle systems including the HMF model, the critical exponents within the Vlasov dynamics have been obtained as $\gamma^+ = 2\beta,\ \gamma^- = \beta/2,\ \delta = 3/2$ [15]. It is interesting to note that the critical exponents obtained within the two approaches satisfy the same scaling relation, namely, $\gamma^- = \beta(\delta - 1)$.

In the HMF model, the one-particle phase space is given by a cylinder, and the Poisson bracket between the spin components is taken to vanish identically. In this chapter, we consider the Heisenberg spin systems with mean-field interactions. Here, the single-spin phase space is given by the unit sphere, and the Poisson brackets between the spin components are strictly nonzero. We consider the time evolution of the spin components according to a Hamiltonian with a mean-field interaction and a local anisotropy, and seek for answers to several relevant questions. For example, we ask if the universality class for usual Hamiltonian systems defined on a cylinder (e.g., the HMF model) includes spin systems defined on the unit sphere, if the scaling relation $\gamma^- = \beta(\delta - 1)$ holds in case the latter falls in a different universality class, how the anisotropy affects the values of the critical exponents, etc.

The chapter is organized as follows. In Sect. 9.2, we discuss the dynamics of our model system and its characterization in terms of the Vlasov equation. In the following section, we recall the definition of various critical exponents of interest in a set up that involves response of an initial stationary state to a small external field. In the same section, we also obtain the exponent values using a static analysis that does not take into account the time evolution of the initial state under the application of the external field. Section 9.4 contains principal results of this chapter, in which we show how taking into account explicit evolution of the initial state according to the Vlasov dynamics leads to very different critical exponent values compared to the static analysis. These results are validated in numerical simulations of the dynamics reported in Sect. 9.5. The chapter ends with conclusions and perspectives in Sect. 9.6.

## 9.2 The Model

We consider the classical Heisenberg model (3.99) of $N$ spins in presence of a time-dependent external field $\mathbf{h}(t) \equiv (h_x(t), h_y(t), h_z(t))$:

$$H = -\frac{J}{2N} \sum_{i,j=1}^{N} \mathbf{S}_i \cdot \mathbf{S}_j + D \sum_{i=1}^{N} S_{iz}^{2n} - \mathbf{h}(t) \cdot \sum_{i=1}^{N} \mathbf{S}_i \,. \tag{9.1}$$

While the Kac's scaling of the ferromagnetic coupling $J > 0$ by the system size $N$ ensures that the energy of the system is an extensive quantity, the system continues to remain non-additive.[1] Let us recall that the second term with $D > 0$ on the right hand side of Eq. (9.1) accounts for local anisotropy and lowers the energy by having the magnetization vector pointing in the $xy$ plane. In this chapter, the model with exponent $2n$ in the anisotropy term is referred to as Model-$n$. We set $J = 1$ without loss of generality.

Here, we consider the external field $\mathbf{h}(t)$ to be absent for times $t < 0$, when the system is assumed to be existing in a reference state, e.g., a thermal equilibrium state. To measure the response of the reference state to an external field, we switch on at $t = 0$ a constant field such that one has

[1] Note that the system (9.1) without the last term on the right hand side is same as the one studied in Chap. 8 and introduced in Sect. 3.8, where these features of extensivity and non-additivity have been explained in detail; in presence of the last term, the discussions presented therein go through.

$$\mathbf{h}(t) = \Theta(t)\,\mathbf{h}\,, \tag{9.2}$$

where $\Theta(t)$ is the Heaviside step function, and $\mathbf{h}$ is a vector with a constant magnitude $h$. The singularity of the Heaviside function $\Theta(t)$ at $t = 0$ will have no effect on the values of the critical exponents obtained by studying the Vlasov dynamics, and one may replace it with a smooth function [16].

### 9.2.1 Spin Dynamics

In line with the discussions in Sect. 8.2.1, the evolution of individual spins of the system (9.1) in dimensionless time is given by

$$\dot{\mathbf{S}}_i \equiv \frac{\mathrm{d}\mathbf{S}_i}{\mathrm{d}t} = \{\mathbf{S}_i, H\}\,; \quad i = 1, 2, \ldots, N\,, \tag{9.3}$$

where $\{\cdot\,,\cdot\}$ denotes the Poisson bracket. The components $\mathbf{S}_i$ and $\mathbf{S}_j$ of two spins satisfy the Poisson bracket relations (8.4), namely,

$$\{S_{ix}, S_{jy}\} = \delta_{i,j} S_{iz}\,, \quad \{S_{iy}, S_{jz}\} = \delta_{i,j} S_{ix}\,, \quad \{S_{iz}, S_{jx}\} = \delta_{i,j} S_{iy}\,. \tag{9.4}$$

In our case, the time evolution of the spin components obtained using Eqs. (9.1), (9.3), and (9.4) is given by

$$\begin{aligned}
\dot{S}_{ix} &= S_{iy}(m_z + h_z) - S_{iz}(m_y + h_y) - 2nDS_{iy}S_{iz}^{2n-1}\,,\\
\dot{S}_{iy} &= S_{iz}(m_x + h_x) - S_{ix}(m_z + h_z) + 2nDS_{ix}S_{iz}^{2n-1}\,,\\
\dot{S}_{iz} &= S_{ix}(m_y + h_y) - S_{iy}(m_x + h_x)\,,
\end{aligned} \tag{9.5}$$

where the magnetization

$$\mathbf{m} \equiv \frac{1}{N}\sum_{i=1}^{N} \mathbf{S}_i = (m_x, m_y, m_z) \tag{9.6}$$

acts as the mean field governing the time evolution of the individual spins. Summing the last equation of (9.5) over $i$, one may see that $m_z$ is a constant of motion when the condition

$$m_x h_y - m_y h_x = 0 \tag{9.7}$$

holds. The length of each spin is a constant of motion, and so is the total energy of the system when the field $\mathbf{h}$ is time independent.

In spherical polar coordinates $\theta_i \in [0, \pi]$ and $\phi_i \in [0, 2\pi)$, one may write

$$S_{ix} = \sin\theta_i \cos\phi_i\,, \quad S_{iy} = \sin\theta_i \sin\phi_i\,, \quad S_{iz} = \cos\theta_i\,. \tag{9.8}$$

Then, the time evolution of the variables $\theta_i$ and $\phi_i$ is obtained from Eq. (9.5) as

$$\dot{\theta}_i = (m_x + h_x) \sin\phi_i - (m_y + h_y) \cos\phi_i \,, \tag{9.9}$$

$$\dot{\phi}_i = (m_x + h_x) \cot\theta_i \cos\phi_i + (m_y + h_y) \cot\theta_i \sin\phi_i - (m_z + h_z) + 2nD \cos^{2n-1}\theta_i \,. \tag{9.10}$$

In terms of the canonically-conjugate variables $\phi_i$ and

$$p_i \equiv \cos\theta_i \,, \tag{9.11}$$

the spin components are given by

$$S_{ix} = \sqrt{1 - p_i^2}\, \cos\phi_i \,, \quad S_{iy} = \sqrt{1 - p_i^2}\, \sin\phi_i \,, \quad S_{iz} = p_i \,. \tag{9.12}$$

For convenience, one may choose the dynamical variables of the $i$-th spin to be $\phi_i$ and $p_i$, and denote a volume element in the $(\phi_i, p_i)$ space by $\mathrm{d}\phi_i \, \mathrm{d}p_i$.

### 9.2.2 The Vlasov Dynamics

In the thermodynamic limit $N \to \infty$, the dynamics of the system (9.1) is described by the Vlasov equation, which may be obtained from Eq. (3.152) with the change of notation $\bar{f}(\phi, p, t) \to f(\phi, p, t)$:

$$\frac{\partial f}{\partial t} + \{f, H\} = 0 \,, \tag{9.13}$$

where $f(\phi, p, t)$ and $H(\phi, p, t)$ are the single-spin distribution function and the single-spin Hamiltonian, respectively. We recall that the quantity $f(\phi, p, t)$ denotes the probability density to find a spin $(\phi, p)$ at time $t$ in the single-spin $\mu \equiv (\phi, p)$ space and satisfies the normalization

$$\iint_{\mu} \mathrm{d}\phi \mathrm{d}p \; f(\phi, p, t) = 1 \; \forall \, t \,, \tag{9.14}$$

where the double integral is defined in the single-spin $\mu$ space and which for any function $X(\phi, p)$ is to be computed as

$$\iint_{\mu} \mathrm{d}\phi \mathrm{d}p \; X(\phi, p) \equiv \int_0^{2\pi} \mathrm{d}\phi \int_{-1}^{1} \mathrm{d}p \; X(\phi, p) \,. \tag{9.15}$$

The single-spin Hamiltonian in this case is given by

$$H(\phi, p, t) = Dp^{2n} - [\mathbf{m}(t) + \mathbf{h}(t)] \cdot \mathbf{S}, \tag{9.16}$$

with

$$\mathbf{S} \equiv \left(\sqrt{1-p^2}\,\cos\phi,\, \sqrt{1-p^2}\,\sin\phi,\; p\right). \tag{9.17}$$

The magnetization $\mathbf{m}(t) = (m_x(t), m_y(t), m_z(t))$ in Eq. (9.16) is given by

$$\mathbf{m}(t) = \iint_\mu \mathrm{d}\phi \mathrm{d}p\, \mathbf{S}\, f(\phi, p, t). \tag{9.18}$$

We remark that the single-spin Hamiltonian (9.16) depends on time $t$ through the magnetization $\mathbf{m}(t)$ and the external field $\mathbf{h}(t)$. Also, note that the function $f(\phi, p, t)$ is $2\pi$-periodic in $\phi$, i.e., we have

$$f(\phi + 2\pi, p, t) = f(\phi, p, t)\ \forall\ p, t. \tag{9.19}$$

From the single-spin Hamiltonian (9.16), one may obtain the single-spin Hamilton's equations of motion:

$$\dot{\phi} = \frac{\partial H}{\partial p} = \frac{p\left[(m_x + h_x)\cos\phi + (m_y + h_y)\sin\phi\right]}{\sqrt{1-p^2}} + 2nDp^{2n-1} - (m_z + h_z), \tag{9.20}$$

$$\dot{p} = -\frac{\partial H}{\partial \phi} = -(m_x + h_x)\sqrt{1-p^2}\,\sin\phi + (m_y + h_y)\sqrt{1-p^2}\,\cos\phi. \tag{9.21}$$

### 9.2.3 Casimir Invariants

Let us consider any functional $C[f](t)$ of the form

$$C[f](t) = \iint_\mu \mathrm{d}\phi \mathrm{d}p\, c(f(\phi, p, t)), \tag{9.22}$$

where $c$ is a function of the single-spin distribution $f(\phi, p, t)$. For any smooth function $c(f)$, the quantity $C[f](t)$ is a constant of motion, as we now show. Taking the derivative with respect to $t$ of both sides of Eq. (9.22), we get

$$\begin{aligned}\frac{\mathrm{d}t}{\mathrm{d}C}[f](t) &= \iint_{\mu} \mathrm{d}\phi \mathrm{d}p \, \frac{\mathrm{d}c}{\mathrm{d}f}\frac{\partial f}{\partial t} = \iint_{\mu} \mathrm{d}\phi \mathrm{d}p \, \frac{\mathrm{d}c}{\mathrm{d}f}\left[-\frac{\partial H}{\partial p}\frac{\partial f}{\partial \phi} + \frac{\partial H}{\partial \phi}\frac{\partial f}{\partial p}\right] \\ &= \iint_{\mu} \mathrm{d}\phi \mathrm{d}p \left[-\frac{\partial H}{\partial p}\frac{\partial c}{\partial \phi} + \frac{\partial H}{\partial \phi}\frac{\partial c}{\partial p}\right] \\ &= \iint_{\mu} \mathrm{d}\phi \mathrm{d}p \left[-\frac{\partial}{\partial \phi}\left\{\frac{\partial H}{\partial p}c\right\} + \frac{\partial}{\partial p}\left\{\frac{\partial H}{\partial \phi}c\right\}\right] \\ &= -\int_{-1}^{1} \mathrm{d}p \left[c(f(\phi, p, t))\frac{\partial H}{\partial p}\right]_{\phi=0}^{\phi=2\pi} + \int_{0}^{2\pi} \mathrm{d}\phi \left[c(f(\phi, p, t))\frac{\partial H}{\partial \phi}\right]_{p=-1}^{p=1} . \end{aligned} \tag{9.23}$$

Here, in the fourth step, we have used $\partial^2 H/\partial\phi\partial p = \partial^2 H/\partial p\partial\phi$, assuming that the Hamiltonian $H$ has continuous second-order partial derivatives at every point of the $\mu$-space; in obtaining the last step, we have performed integration by parts. Using Eqs. (9.19) and (9.20), one may see that the integrand in the first term on the right hand side of Eq. (9.23) vanishes. Similarly, using Eq. (9.21) that implies $\partial H/\partial\phi|_{p=1} = \partial H/\partial\phi|_{p=-1} = 0$, it is seen that the integrand in the second term on the right hand side of Eq. (9.23) also vanishes. Hence, we get

$$\frac{\mathrm{d}t}{\mathrm{d}C}[f](t) = 0 \, , \tag{9.24}$$

that is, the quantity $C[f]$ remains invariant in time. Such invariant quantities are called Casimir invariants, which hold even when the single-spin Hamiltonian depends on time. The Casimir invariants do not allow the relaxation of an initial state with $\mathbf{h} = \mathbf{0}$ to the thermal equilibrium state with $\mathbf{h} \neq \mathbf{0}$ when at least one of the Casimir invariants between the two states is not the same.

## 9.3 Setting and Definition of the Critical Exponents

For $t < 0$, the system (9.1) is considered to be existing in one of a family of stable stationary states of the Vlasov equation with external field $\mathbf{h}(t) = \mathbf{0}$. In order to study the critical exponents associated with the response of the system to an external field switched on for times $t \geq 0$, we restrict ourselves to a family of states that allows a second-order phase transition and consequently a critical point in the stationary state. Such a family of states is referred to as the reference state and is denoted by $f_0$. Any function $f_0$ of the form

$$f_0(\phi, p) = F(H_0(\phi, p)) = \frac{G\big(H_0(\phi, p)\big)}{\iint_{\mu} \mathrm{d}\phi \mathrm{d}p \, G\big(H_0(\phi, p)\big)} \, , \tag{9.25}$$

with $G$ an arbitrary function, implies that $\{f_0, H_0\} = 0$. Consequently, it is evident from Eq. (9.13) that it serves as a stationary solution of the Vlasov equation, and we have

$$H_0(\phi, p) = Dp^{2n} - \mathbf{m}_0 \cdot \mathbf{S}, \tag{9.26}$$

with

$$\mathbf{m}_0 = (m_{0x}, m_{0y}, m_{0z}) \tag{9.27}$$

satisfying the self-consistent equation

$$\mathbf{m}_0 = \iint_\mu \mathrm{d}\phi \mathrm{d}p\, \mathbf{S}\, f_0(\phi, p). \tag{9.28}$$

One may parametrize the family of functions $G$ by a parameter $T$, which in the case of thermal equilibrium coincides with the temperature[2]:

$$G(x) = \mathrm{e}^{-x/T}. \tag{9.29}$$

However, the analysis presented in the following applies to other family of functions $G$, such as the Fermi-Dirac-type family

$$G(x) = \frac{1}{\mathrm{e}^{(x-a)/b} + 1}. \tag{9.30}$$

In this case, the parameter $T$ may be identified with either of the two parameters $a$ and $b$.

Using the rotational symmetry of $H_0(\phi, p)$ on the $(S_x, S_y)$-plane, one may set $m_{0y} = 0$ without loss of generality. In addition, one may also assume that $m_{0z} = 0$, which solves the self-consistent equation for $m_{0z}$. Indeed, from the self-consistent equation, Eq. (9.28), one observes that

$$m_{0z} = \iint_\mu \mathrm{d}\phi \mathrm{d}p\, p\, f_0(\phi, p). \tag{9.31}$$

The right hand side of the above equation vanishes, since the Hamiltonian $H_0$, and consequently, $f_0(\phi, p)$, are even functions of $p$ once we set $m_{0y} = m_{0z} = 0$, see Eqs. (9.25) and (9.26). Denoting $m_{0x}$ by $m_0$, so that $\mathbf{m}_0 = (m_0, 0, 0)$, one obtains

$$H_0(\phi, p) = Dp^{2n} - m_0 \sqrt{1 - p^2}\, \cos\phi, \tag{9.32}$$

with the self-consistent Eq. (9.28) reducing to

[2] Here, we measure temperatures in units of the Boltzmann constant.

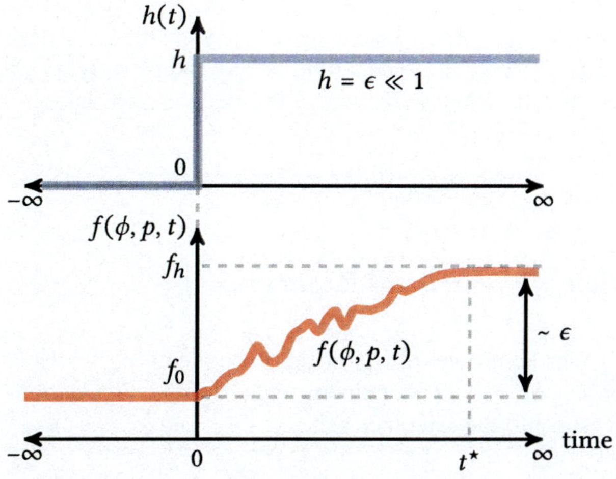


**Fig. 9.1 Schematic diagram for temporal evolution of the single-spin distribution** $f(\phi, p, t)$. Here, at $t = 0$, a small constant external field $h = \epsilon \ll 1$ is switched on. Because of this external field, the initial Vlasov-stationary state $f_0$ would asymptotically approach another Vlasov-stationary state $f_h$. As the field is small, we expect $(f_h - f_0) \sim \epsilon$. For times $t > t^\star$, the distribution $f(\phi, p, t)$ is almost close to $f_h$

$$m_0 = \iint_\mu \mathrm{d}\phi \mathrm{d}p \sqrt{1-p^2} \cos\phi \; f_0(\phi, p) \,. \tag{9.33}$$

A small constant external field $\mathbf{h} = (h, 0, 0)$ (such that $|h| = \epsilon \ll 1$) pointing in the direction of the reference magnetization $\mathbf{m}_0 = (m_0, 0, 0)$ is switched on at $t = 0$. Without loss of generality, we take $m_0$, and hence $h$, to be positive. Then, the single-spin Hamiltonian in presence of the external field becomes

$$H_h(\phi, p) = Dp^{2n} - (m_h + h)\sqrt{1-p^2}\cos\phi \,. \tag{9.34}$$

Consequently, for times $t > 0$, we have a time-dependent distribution $f(\phi, p, t)$ that evolves under the Vlasov dynamics (9.13) with the Hamiltonian $H_h$ and subject to the initial condition at $t = 0$ given by

$$f(\phi, p, t = 0) = f_0(\phi, p) \,. \tag{9.35}$$

In the limit $t \to \infty$, the distribution $f(\phi, p, t)$ approaches a stationary state $f_h(\phi, p)$, defined as

$$f_h(\phi, p) \equiv \lim_{t \to \infty} f(\phi, p, t) \,. \tag{9.36}$$

In Fig. 9.1, we have shown schematically the time evolution of the single-spin distribution $f(\phi, p, t)$ for given $\phi$, $p$.

Note that the distribution $f_h(\phi, p)$ is not necessarily the thermal equilibrium state proportional to $\exp(-H_h(\phi, p)/T)$, but could be any state satisfying $\{f_h, H_h\} = 0$, see Sect. 9.2.2. Within the Vlasov dynamics, the response to the external field is measured by

$$\delta m \equiv m_h - m_0 \,. \tag{9.37}$$

### 9.3.1 Definition of the Critical Exponents

Let us now recall the definitions of the critical exponents $\beta$, $\gamma^+$, $\gamma^-$, and $\delta$, see any standard reference on critical phenomena, e.g., Ref. [2]. One defines the critical exponent $\beta$ with respect to the reference state as

$$m_0(T) \propto (T_c - T)^{\beta}\,; \quad T \to T_c^- \,, \tag{9.38}$$

where $T_c$ is the critical temperature, and $m_0$ denotes the positive solution of the self-consistent Eq. (9.33).

The critical exponents $\gamma^{\pm}$ are defined in the regime of linear response to the external field. The response of the system $\delta m$ depends on both the temperature $T$ and the external field $h$, and one defines the susceptibility $\chi(T)$ as

$$\chi(T) \equiv \left. \frac{\partial(\delta m)}{\partial h} \right|_{h \to 0} . \tag{9.39}$$

At the critical point $T_c$, the susceptibility diverges as

$$\chi(T) \propto \begin{cases} (T - T_c)^{-\gamma^+}\,; & T \to T_c^+ \,, \\ (T_c - T)^{-\gamma^-}\,; & T \to T_c^- \,, \end{cases} \tag{9.40}$$

which defines the critical exponents $\gamma^{\pm} > 0$.

At the critical point $T_c$, we also have

$$\delta m \propto h^{1/\delta}; \quad T = T_c,\ h \to 0 \,, \tag{9.41}$$

which defines the critical exponent $\delta$. Using Eqs. (9.39) and (9.41), one obtains the susceptibility at the critical point $T_c$ as

$$\chi(T_c) \propto \frac{1}{\delta}\, h^{1/\delta - 1} \,. \tag{9.42}$$

It then follows that in order to be consistent with the behavior (9.40), one requires $\delta > 1$; note that $\delta = 1$ implies that $\chi(T)$ has a cusp at $T = T_c$.

### 9.3.2 *Critical Exponents from Static Analysis*

We now discuss how one may obtain the values of the critical exponents within the framework of a static analysis, without any reference to the dynamics of evolution after the application of the external field. The starting point is the equivalent of Eqs. (9.25), (9.26), and (9.28) in presence of a constant field $\mathbf{h} = (h, 0, 0)$:

$$f_h(\phi, p) = F(H_h(\phi, p)) = \frac{G(H_h(\phi, p))}{\iint_\mu \mathrm{d}\phi \mathrm{d}p\, G(H_h(\phi, p))}, \tag{9.43}$$

with

$$H_h(\phi, p) = Dp^{2n} - (\mathbf{m}_h + \mathbf{h}) \cdot \mathbf{S}, \tag{9.44}$$

and

$$\mathbf{m}_h = \iint_\mu \mathrm{d}\phi \mathrm{d}p\, \mathbf{S}\, f_h(\phi, p). \tag{9.45}$$

In our case, we have $\mathbf{m}_h = (m_h, 0, 0)$. Therefore, using Eqs. (9.43) and (9.45) is reduced to

$$m_h = \frac{\iint_\mu \mathrm{d}\phi \mathrm{d}p \sqrt{1-p^2}\, \cos\phi\, G\big(Dp^{2n} - (m_h + h)S_x\big)}{\iint_\mu \mathrm{d}\phi \mathrm{d}p\, G\big(Dp^{2n} - (m_h + h)S_x\big)}, \tag{9.46}$$

with $S_x = \sqrt{1-p^2}\cos\phi$. We assume $G$ to be a smooth function of its argument. Since we take $G$ to be parametrized by the parameter $T$, this means that $G$ is also smooth with respect to $T$.

In order to obtain the critical exponents, we now analyze Eq. (9.46) for small values of $(m_h + h)$. To this end, let us expand $G$ in a Taylor series around $Dp^{2n}$. Then, we have

$$\begin{aligned} G\big(&Dp^{2n} - (m_h + h)S_x\big) \\ &= G(Dp^{2n}) - (m_h + h)\sqrt{1-p^2}\, \cos\phi\, G'(Dp^{2n}) \\ &+ \frac{1}{2}(m_h + h)^2(1-p^2)\cos^2\phi\, G''(Dp^{2n}) \\ &- \frac{1}{6}(m_h + h)^3(1-p^2)^{3/2}\cos^3\phi\, G'''(Dp^{2n}) + \text{higher order terms.} \end{aligned} \tag{9.47}$$

Substituting the above Taylor series in Eq. (9.46) and then using

$$\int_0^{2\pi} \mathrm{d}\phi\, \cos^{2j+1}\phi = 0; \quad j = 0, 1, 2, \ldots, \tag{9.48}$$

we see that in the numerator on the right hand side of Eq. (9.46), only odd-order terms in $(m_h + h)$ survive. In the denominator, on the contrary, only even-order terms survive. Then, from the self-consistent Eq. (9.46), one obtains

$$
\begin{aligned}
m_h \approx \\
& \frac{-\pi(m_h+h)\int_{-1}^{1} \mathrm{d}p(1-p^2)\,G'(Dp^{2n}) - \frac{\pi}{8}(m_h+h)^3\int_{-1}^{1}\mathrm{d}p(1-p^2)^2\,G'''(Dp^{2n})}{2\pi\int_{-1}^{1}\mathrm{d}p\,G(Dp^{2n}) + \frac{\pi}{2}(m_h+h)^2\int_{-1}^{1}\mathrm{d}p\,(1-p^2)G''(Dp^{2n})} \\
&\approx \Big[-(m_h+h)\int_{-1}^{1}\mathrm{d}p\,(1-p^2)\,G'(Dp^{2n}) \\
&-\frac{1}{8}(m_h+h)^3\int_{-1}^{1}\mathrm{d}p\,(1-p^2)^2\,G'''(Dp^{2n})\Big] \\
&\times\frac{1}{2\int_{-1}^{1}\mathrm{d}p\,G(Dp^{2n})}\left[1-\frac{1}{4}(m_h+h)^2\,\frac{\int_{-1}^{1}\mathrm{d}p\,(1-p^2)\,G''(Dp^{2n})}{\int_{-1}^{1}\mathrm{d}p\,G(Dp^{2n})}\right]. \qquad (9.49)
\end{aligned}
$$

It may be seen that the right hand side of Eq. (9.49) has only odd-order terms in $(m_h + h)$. For brevity, omitting the argument from $G$ and its derivatives, Eq. (9.49) written up to order $(m_h + h)^3$ reads

$$
\begin{aligned}
(m_h+h)+&\frac{(m_h+h)}{\int_{-1}^{1}\mathrm{d}p\,G}\Bigg[\frac{1}{2}\int_{-1}^{1}\mathrm{d}p(1-p^2)G' + \frac{(m_h+h)^2}{16}\int_{-1}^{1}\mathrm{d}p(1-p^2)^2G''' \\
&-\frac{(m_h+h)^2\left(\int_{-1}^{1}\mathrm{d}p(1-p^2)G'\right)\left(\int_{-1}^{1}\mathrm{d}p(1-p^2)G''\right)}{8\int_{-1}^{1}\mathrm{d}pG}\Bigg] = h\,.
\end{aligned}
\qquad (9.50)
$$

Now, denoting the coefficients of $(m_h + h)$ and $(m_h + h)^3$ by $A(T)$ and $B(T)$, respectively, one may rewrite Eq. (9.50) as

$$
A(T)\,(m_h+h) + B(T)\,(m_h+h)^3 - h = 0\,, \qquad (9.51)
$$

with

$$
A(T) \equiv 1 + \frac{\int_{-1}^{1}\mathrm{d}p\,(1-p^2)\,G'(Dp^{2n})}{2\int_{-1}^{1}\mathrm{d}p\,G(Dp^{2n})}\,, \qquad (9.52)
$$

$$
\begin{aligned}
B(T) \equiv\ & \frac{\int_{-1}^{1}\mathrm{d}p(1-p^2)^2G'''(Dp^{2n})}{16\int_{-1}^{1}\mathrm{d}pG(Dp^{2n})} \\
&-\frac{\left(\int_{-1}^{1}\mathrm{d}p(1-p^2)G'(Dp^{2n})\right)\left(\int_{-1}^{1}\mathrm{d}p(1-p^2)G''(Dp^{2n})\right)}{8\left(\int_{-1}^{1}\mathrm{d}p\,G(Dp^{2n})\right)^2}\,. \qquad (9.53)
\end{aligned}
$$

For $h = 0$, Eq. (9.51) reads

$$A(T)\, m_0 + B(T)\, m_0^3 = 0\,, \tag{9.54}$$

from which one obtains the possible solutions

$$m_0 = 0\,, \text{ or, } m_0 = \pm\sqrt{-\frac{A(T)}{B(T)}}\,. \tag{9.55}$$

Assuming $B(T) > 0$, while $A(T)$ is positive (respectively, negative) for $T > T_c$ (respectively, $T < T_c$), one has the zero solution for $m_0$ for $T > T_c$ and the nonzero solution for $T < T_c$. Here, the critical point $T = T_c$ is where we have $A(T_c) = 0$. Consistent with our convention for the sign of $m_0$, we may choose the nonzero solution, giving the spontaneous magnetization for $T < T_c$, to be $m_0 = \sqrt{-A/B}$. The smoothness of $G$ with respect to the parameter $T$ allows $A(T)$ to be expanded in a Taylor series around $T_c$, giving $A(T) \propto (T - T_c)$ close to $T_c$, and this together with $m_0 = \sqrt{-A/B}$ gives $\beta = 1/2$. Note that in the foregoing, we did not use any specific form of the function $F$ in deriving the result $\beta = 1/2$, see Eq. (9.25). We therefore conclude that for our model, one has $\beta = 1/2$ quite generally for all choices of the family $F$ of the reference state.

The linear response is obtained by taking the derivative of Eq. (9.51) with respect to $h$, and then taking the limit $h \to 0$. Using the definition of susceptibility $\chi(T)$ given in Eq. (9.39), we obtain from Eq. (9.51) that

$$A(T)\Big(\chi(T) + 1\Big) + 3m_0^2 B(T)\,\Big(\chi(T) + 1\Big) = 1\,. \tag{9.56}$$

Now, using $m_0 = 0$ and $m_0 = \sqrt{-A/B}$, respectively, for $T > T_c$ and $T < T_c$, the susceptibility $\chi(T)$ is obtained from Eq. (9.56) as

$$\chi(T) = \begin{cases} (1 - A)/A\,; & T > T_c\,, \\ (1 + 2A)/(-2A)\,; & T < T_c\,. \end{cases} \tag{9.57}$$

From the behavior $A(T) \propto (T - T_c)$ around $T = T_c$, one obtains from Eq. (9.57) that

$$\lim_{T \to T_c^+} \chi(T) \propto \lim_{T \to T_c^+} \left[\frac{1 - (T - T_c)}{T - T_c}\right] \propto \frac{1}{T - T_c}\,, \tag{9.58}$$

$$\lim_{T \to T_c^-} \chi(T) \propto \lim_{T \to T_c^-} \left[-\frac{1 + 2(T - T_c)}{2(T - T_c)}\right] \propto \frac{1}{T_c - T}\,. \tag{9.59}$$

Hence, we get the critical exponents $\gamma^{\pm} = 1$.

At the critical point with $A(T_c) = 0$, the self-consistent equation (9.51) reduces to

$$B(T_c)\,(m_h + h)^3 = h\,. \tag{9.60}$$

For small values of $h$, since we have $\delta > 1$, the response $m_h \propto h^{1/\delta}$ (see Eq. (9.41)) is larger than $h$. Therefore, in the limit $h \to 0$, we have $(m_h + h) \to m_h$. Then, using Eq. (9.60), we obtain $m_h^3 \propto h$, implying $\delta = 3$.

To summarize, using a static analysis, with no reference to the dynamics, we obtain the critical exponents

$$\beta = \frac{1}{2}\,, \quad \gamma^{\pm} = 1\,, \quad \delta = 3\,, \tag{9.61}$$

irrespective of the value of the exponent $n$ characterizing the Hamiltonian (9.1). In the next section, we derive the values of the critical exponents within the Vlasov dynamics. We will obtain the response $\delta m$ within the Vlasov dynamics, and hence the exponents $\gamma^{\pm}$ and $\delta$ may very well take values different from the ones in Eq. (9.61).

## 9.4 Critical Exponents from the Vlasov Dynamics

Let us now proceed to obtain the values of the critical exponents by studying the Vlasov dynamics. As we have already mentioned, one has $\beta = 1/2$ quite generally for all choices of the family $F$ of the reference state. Keeping this in mind, we now focus on the computation of the critical exponents $\gamma^{\pm}$ and $\delta$ for a given value of the exponent $\beta$.

For $n = 0$, i.e., for Model-0, using Eq. (9.34), the single-spin Hamiltonian $H_h$ corresponding to the state $f_h$ may be written as

$$H_h = D - (m_h + h)S_x\,. \tag{9.62}$$

Note that the first term on the right hand side is a constant. Therefore, one can redefine the single-spin Hamiltonian $H_h$ as

$$H_h = -(m_h + h)S_x\,. \tag{9.63}$$

The equations of motion are obtained by using Eq. (9.5) with $\mathbf{m}_h = (m_h, 0, 0)$ as

$$\dot{S}_{ix} = 0, \quad \dot{S}_{iy} = S_{iz}(m_h + h), \quad \dot{S}_{iz} = -S_{iy}(m_h + h)\,. \tag{9.64}$$

Equation (9.64) evidently suggests that irrespective of the time dependence of the external field $h$, the spin component $S_{ix}$ $\forall\, i$ and consequently $m_x$ are constants of motion. Dynamically, this means that each spin rotates on a constant $S_x$ plane. The fact that $S_x$ is a constant of motion in both the cases with $h = 0$ and $h \neq 0$ implies that the reference state $f_0 = F(H_0) = F(-m_0 S_x)$ is stationary even after the external field is turned on, and we have $m_h = m_0$. Therefore, we may conclude that no response to the external field is observed within the Vlasov dynamics, and choose

$$\beta = \frac{1}{2}, \quad \gamma^{\pm} = 0, \quad \delta = 1. \tag{9.65}$$

One may argue for these values of $\gamma^{\pm}$ and $\delta$ in the following manner. As we see above, $m_h = m_0$, and consequently, no divergence of the susceptibility as $T \to T_c$ is obtained within the Vlasov dynamics. Hence, following the discussion in Sect. 9.3.1, we conclude that $\gamma^{\pm} = 0$ and $\delta = 1$. Note that the exponent values in Eq. (9.65) are quite different from the ones obtained within the static analysis, Eq. (9.61). For the case $G(x) = \exp(-x/T)$, Eq. (9.52) with $n = 0$ gives

$$\begin{aligned} A(T) = 1 + \frac{\int_{-1}^{1} \mathrm{d}p\,(1-p^2)\,G'(D)}{2\int_{-1}^{1} \mathrm{d}p\,G(D)} &= 1 + \frac{\int_{-1}^{1} \mathrm{d}p\,(1-p^2)\left(-\frac{1}{T}\,\mathrm{e}^{-D/T}\right)}{2\int_{-1}^{1} \mathrm{d}p\,\mathrm{e}^{-D/T}} \\ = 1 - \frac{1}{2T}\frac{\int_{-1}^{1} \mathrm{d}p\,(1-p^2)}{\int_{-1}^{1} \mathrm{d}p} &= 1 - \frac{1}{3T}. \end{aligned} \tag{9.66}$$

Then, solving $A(T_c) = 0$, we obtain in this case the critical temperature $T_c = 1/3$. As the Model-0 shows no response to the external field, it provides a simple and extreme example of dynamical suppression of response. However, in the following, we will see that the Model-$n$ with $n \geq 1$ does show nonzero response.

Let us now concentrate on computing the values of the critical exponents $\gamma^{\pm}$ and $\delta$ for Model-$n$ with $n \geq 1$. Starting from the state $f_0$ with the field turned on at $t = 0$, we need to obtain within the Vlasov dynamics the asymptotic state $f_h$ and hence the response $\delta m$. The single-spin Hamiltonian $H_h$ given in Eq. (9.34) depends on the conjugate pair $(\phi, p)$, with $\phi$ being the generalized coordinate and $p$ the generalized momentum. Now, the celebrated Liouville-Arnold theorem states that if there exist $k$ independent first integrals of motion in a Hamiltonian dynamical system with $k$ generalized coordinates, then the dynamics is integrable [17]. In our case, the Hamiltonian $H_h(\phi, p)$ has only one generalized coordinate. As $H_h(\phi, p)$ has no explicit dependence on time, we have $\mathrm{d}H_h/\mathrm{d}t = 0$, implying that the total energy $E$ is an integral of motion. Therefore, the dynamics governed by the Hamiltonian $H_h(\phi, p)$ is integrable. For an integrable system, there exists a canonical transformation from $(\phi, p)$ to a special set of variables called the 'angle-action' variables $(w, I)$, such that the transformed Hamiltonian depends only on the action variable $I$ [17]. Without inventing a new symbol, we continue to denote the transformed Hamiltonian as $H_h(I)$, while $H_h$ without any arguments denotes the Hamiltonian in $(\phi, p)$ coordinate system. Note that the functional form $F(x)$ is independent of the coordinate system we use for the Hamiltonian. In the transformed system, the Hamilton's equations of motion read

$$\dot{w} = \frac{\partial H_h(I)}{\partial I} = \frac{\mathrm{d}H_h(I)}{\mathrm{d}I} \equiv \Omega(I)\,, \tag{9.67}$$

$$\dot{I} = -\frac{\partial H_h(I)}{\partial w} = 0\,. \tag{9.68}$$

The time evolution of the angle-action variables is obtained from Eqs. (9.67) and (9.68) as

$$w(t) = w(0) + \Omega(I)t\,, \quad I(t) = I(0)\,. \tag{9.69}$$

By similar arguments, it may be seen that the dynamics described by the Hamiltonian $H_0$ in Eq. (9.32) is also integrable. Consequently, it has a similar angle-action description. We denote the angle-action variables for $H_0$ as $(w_0, I_0)$.

In terms of the angle-action variables $(w, I)$, one may write down the response formula

$$f_h(I) \equiv F(H_h(I)) = \langle f_0(\phi, p)\rangle_h \tag{9.70}$$

for Hamiltonian systems [15, 16]. The $\langle\cdot\rangle_h$ in Eq. (9.70) denotes the average over the angle variable $w$ corresponding to the Hamiltonian $H_h$ and which for any arbitrary function $A(\phi, p)$ is defined as

$$\langle A\rangle_h(I) \equiv \frac{1}{2\pi}\int_0^{2\pi} \mathrm{d}w\, A\left(\phi\left(w, I\right), p\left(w, I\right)\right)\,. \tag{9.71}$$

For a derivation of the response formula (9.70), see Appendix I. Equation (9.70) is to be understood as

$$f_h(I(\phi, p)) = \langle f_0(\phi, p)\rangle_h\,, \tag{9.72}$$

which implies that, in general, we may have

$$f_h(I(\phi, p)) \neq f_h(\phi, p)\,, \tag{9.73}$$

with $f_h(\phi, p)$ given by Eq. (9.43). Note that $f_h(I(\phi, p))$ is derived from an analysis of the Vlasov dynamics of the distribution function $f(\phi, p, t)$ under the action of the external field $h$ and with the initial condition $f(\phi, p, 0) = f_0(\phi, p)$, see Appendix I. On the other hand, $f_h(\phi, p)$ given by Eq. (9.43) is what one would naively expect to be the distribution at long times after the application of the field, namely, a distribution that has the same form as $f_0(\phi, p) = F(H_0(\phi, p))$, but with $H_0(\phi, p)$, the Hamiltonian in the absence of the field, replaced by $H_h(\phi, p)$, the Hamiltonian in its presence. Equation (9.73) is key to deriving in the following and eventually rationalizing the different values of the critical exponents obtained with a static *vis-à-vis* a dynamic analysis. Similar to the convention adopted for the transformed Hamiltonian, we will have in the following the quantity $f_h$ written without any arguments to be a function of $(\phi, p)$, while $f_h(I)$ is as denoted here a function of $I$.

Using the relation

$$\langle \varphi(I) \rangle_h = \varphi(I) \tag{9.74}$$

for any function $\varphi$, and the expansion of the single-spin Hamiltonian as

$$H_h = H_0 + \delta H , \quad \delta H = -(\delta m + h) S_x , \tag{9.75}$$

with $S_x = \sqrt{1 - p^2} \cos \phi$, one gets the expansion of $f_h$ from the response formula (9.72) as

$$\begin{aligned} f_h &= \langle f_0 \rangle_h = \langle F(H_0) \rangle_h = \langle F(H_h - \delta H) \rangle_h \\ &= \langle F(H_h) - F'(H_h)\, \delta H \rangle_h + O\big((\delta m + h)^2\big) \\ &= \langle F(H_h(I)) \rangle_h - \langle F'(H_h(I))\, \delta H \rangle_h + O\big((\delta m + h)^2\big) \\ &= F(H_h) - F'(H_h)\, \langle \delta H \rangle_h + O\big((\delta m + h)^2\big) . \end{aligned} \tag{9.76}$$

Using Eq. (9.75), one may reduce Eq. (9.76) to

$$\begin{aligned} f_h =& F(H_0) + F'(H_0)\, \delta H - F'(H_h)\, \langle \delta H \rangle_h + O\big((\delta m + h)^2\big) \\ =& f_0 - (\delta m + h)\, S_x\, F'(H_0) + (\delta m + h)\, \langle S_x \rangle_h\, F'(H_h) + O\big((\delta m + h)^2\big) \\ =& f_0 - (\delta m + h)\big[S_x - \langle S_x \rangle_0\big] F'(H_0) \\ & - (\delta m + h)\big[\langle S_x \rangle_0\, F'(H_0) - \langle S_x \rangle_h\, F'(H_h)\big] + O\big((\delta m + h)^2\big) . \end{aligned} \tag{9.77}$$

In the last step, we have added and subtracted the term $(\delta m + h)\langle S_x \rangle_0 F'(H_0)$, with $\langle \cdot \rangle_0$ defined as an average over the angle variable associated with the integrable Hamiltonian $H_0$.[3] Note that, for instance, we have $F'(x) = -F(x)/T$ for thermal equilibrium reference state (9.29).

We multiply Eq. (9.77) by $S_x = \sqrt{1 - p^2} \cos \phi$ and then integrate over $\phi$ and $p$ to obtain

$$\begin{aligned} \iint_\mu \mathrm{d}\phi \mathrm{d}p\, S_x (f_h - f_0) =& -(\delta m + h) \iint_\mu \mathrm{d}\phi \mathrm{d}p\, \big[S_x^2 - S_x \langle S_x \rangle_0\big] F'(H_0) \\ & - (\delta m + h) \iint_\mu \mathrm{d}\phi \mathrm{d}p \big[S_x \langle S_x \rangle_0 F'(H_0) - S_x \langle S_x \rangle_h F'(H_h)\big] + O\big((\delta m + h)^2\big) . \end{aligned} \tag{9.78}$$

The left hand side of the above equation by definition equals $\delta m$. Then, Eq. (9.78) may be reduced to

[3] In Appendix J, we discuss a method to compute the average $\langle \cdot \rangle_0$.

$$
\begin{aligned}
&(\delta m + h)\left[1 + \iint_{\mu} \mathrm{d}\phi \mathrm{d}p \, \left[S_x^2 - S_x \langle S_x \rangle_0\right] F'(H_0)\right] \\
&+ (\delta m + h) \iint_{\mu} \mathrm{d}\phi \mathrm{d}p \left[S_x \langle S_x \rangle_0 F'(H_0) - S_x \langle S_x \rangle_h F'(H_h)\right] - h = O\big((\delta m + h)^2\big) .
\end{aligned}
\tag{9.79}
$$

Let us compute one of the integrals occurring in Eq. (9.79) by changing the integration variables from $(\phi, p)$ to $(w, I)$. As they are related through a canonical transformation, we have $\mathrm{d}\phi \mathrm{d}p = \mathrm{d}w \mathrm{d}I$. Considering the last integral on the left hand side of Eq. (9.79), we have

$$
\begin{aligned}
\iint_{\mu} \mathrm{d}\phi \mathrm{d}p \; S_x \langle S_x \rangle_h F'(H_h) &= \iint \mathrm{d}w \mathrm{d}I \; S_x(\phi(w, I), p(w, I)) \; \langle S_x \rangle_h(I) \; F'(H_h(I)) \\
&= \int \mathrm{d}I \langle S_x \rangle_h(I) F'(H_h(I)) \int \mathrm{d}w S_x(\phi(w, I), p(w, I)) \\
&= 2\pi \int \mathrm{d}I \; \langle S_x \rangle_h^2(I) \; F'(H_h(I)) ,
\end{aligned}
\tag{9.80}
$$

where we have used Eq. (9.71) in the last step. Now, reintroducing an integral over the angle variable $w$, one obtains from Eq. (9.80) that

$$
\begin{aligned}
\iint_{\mu} \mathrm{d}\phi \mathrm{d}p \; S_x \langle S_x \rangle_h F'(H_h) &= \iint \mathrm{d}w \mathrm{d}I \; \langle S_x \rangle_h^2(I) \; F'(H_h(I)) \\
&= \iint_{\mu} \mathrm{d}\phi \mathrm{d}p \; \langle S_x \rangle_h^2 \; F'(H_h) .
\end{aligned}
\tag{9.81}
$$

By similar steps as above, one may show that

$$
\iint_{\mu} \mathrm{d}\phi \mathrm{d}p \; S_x \langle S_x \rangle_0 \; F'(H_0) = \iint_{\mu} \mathrm{d}\phi \mathrm{d}p \; \langle S_x \rangle_0^2 \; F'(H_0) . \tag{9.82}
$$

Substituting Eqs. (9.81) and (9.82) in Eq. (9.79), we get

$$
\begin{aligned}
&(\delta m + h)\left[1 + \iint_{\mu} \mathrm{d}\phi \mathrm{d}p \, \left[S_x^2 - \langle S_x \rangle_0^2\right] F'(H_0)\right] \\
&+ (\delta m + h) \iint_{\mu} \mathrm{d}\phi \mathrm{d}p \, \left[\langle S_x \rangle_0^2 \; F'(H_0) - \langle S_x \rangle_h^2 \; F'(H_h)\right] - h = O\big((\delta m + h)^2\big) .
\end{aligned}
\tag{9.83}
$$

The self-consistent Eq. (9.83) for the response $\delta m$ may be written as

$$
L(T) \; (\delta m + h) + \mathcal{N}(T) \; (\delta m + h) - h = O\big((\delta m + h)^2\big) , \tag{9.84}
$$

with

$$L(T) \equiv 1 + \iint_{\mu} \mathrm{d}\phi \mathrm{d}p \, \left[S_x^2 - \langle S_x \rangle_0^2\right] F'(H_0) \,, \tag{9.85}$$

$$\mathcal{N}(T) \equiv \iint_{\mu} \mathrm{d}\phi \mathrm{d}p \, \left[\langle S_x \rangle_0^2 \, F'(H_0) - \langle S_x \rangle_h^2 \, F'(H_h)\right] . \tag{9.86}$$

Let us focus on the second term in the integrand in Eq. (9.86). Using Eq. (9.75), one may write that

$$F'(H_h) = F'(H_0) + (\delta m + h) \, F''(H_0) + O\big((\delta m + h)^2\big) \,, \tag{9.87}$$

$$\langle S_x \rangle_h^2 = \langle S_x \rangle_0^2 + O\big((\delta m + h)^{\lambda}\big) \,; \quad \lambda > 0 \,. \tag{9.88}$$

Substituting Eqs. (9.87) and (9.88) in Eq. (9.86), we see that to leading order, we have

$$\mathcal{N}(T) = O\big((\delta m + h)^{\min(\lambda,1)}\big) \,. \tag{9.89}$$

Using Eqs. (9.89) and (9.84) may be written to order $(\delta m + h)$ as

$$L(T)(\delta m + h) + O\big((\delta m + h)^{\lambda+1}\big) - h = 0 \,, \tag{9.90}$$

where we have assumed $\lambda < 1$ (otherwise, we would have terms of $O((\delta m + h)^2)$ on both sides of Eq. (9.84)). Comparing the above equation with Eq. (9.54) and pursuing the analysis presented therein, we note that in the present case, the critical point $T_c$ is obtained by setting $h = 0$ in the above equation and then solving the equation

$$L(T_c) = 0 \,. \tag{9.91}$$

On the basis of the above, we conclude that provided $L(T) \neq 0$, the self-consistent Eq. (9.84) to leading order in $(\delta m + h)$ reads

$$L(T)(\delta m + h) = h \,. \tag{9.92}$$

Therefore, away from the critical point, i.e., for $L(T) \neq 0$, one obtains the linear response from Eq. (9.92) as

$$\delta m = \frac{1 - L}{L} h \,. \tag{9.93}$$

Then, within linear response, the divergence of the susceptibility,

$$\chi = \frac{1 - L}{L} \,, \tag{9.94}$$

is to be determined by the convergence behavior of $L$ as $T \to T_c^{\pm}$, thus enabling us to evaluate the critical exponents $\gamma^{\pm}$. On the other hand, at the critical point $T_c$, the self-consistent Eq. (9.84) to leading order in $(\delta m + h)$ reads

$$\mathcal{N}(T_c)(\delta m + h) = h\,. \tag{9.95}$$

In the following, we will see that consistent with Eq. (9.89) and our assumption $\lambda < 1$, one has to leading order that $\mathcal{N}(T_c) = O((\delta m + h)^{1/(n+1)})$. Equation (9.95) will help us to compute the critical exponent $\delta$.

### 9.4.1 Linear Response in the Disordered Phase

In the disordered phase ($T > T_c$), we have $m_0 = 0$ and the single-spin Hamiltonian $H_0$ given in Eq. (9.32) takes the form

$$H_0(p) = Dp^{2n}\,. \tag{9.96}$$

Equation (9.96) evidently shows that in this case, $H_0(p)$ is independent of $\phi$, suggesting that the angle variable $w_0$ for $H_0$ is nothing but $\phi$. Consequently, the variable $p$ becomes the action variable $I_0$. Then, we have

$$\langle S_x \rangle_0 = \left\langle \sqrt{1-p^2}\cos\phi \right\rangle_0 = \frac{1}{2\pi}\int_0^{2\pi} \mathrm{d}\phi\, \sqrt{1-p^2}\cos\phi = 0\,. \tag{9.97}$$

Substituting Eq. (9.97) in Eq. (9.85), we obtain

$$\begin{aligned} L(T) &= 1 + \iint_\mu \mathrm{d}\phi\,\mathrm{d}p\; S_x^2\; F'(H_0(p)) = 1 + \iint_\mu \mathrm{d}\phi\,\mathrm{d}p\; (1-p^2)\cos^2\phi\; F'(H_0(p)) \\ &= 1 + \pi \int_{-1}^{1} \mathrm{d}p\; (1-p^2)\; F'(H_0(p))\,, \end{aligned} \tag{9.98}$$

which suggests that $L(T)$ has no contribution from the dynamics induced by the external field, as there is no reference to the latter in Eq. (9.98). One may expand $L(T)$ in a Taylor series in $(T - T_c)$ around $T_c$ as

$$L(T) = L(T_c) + (T - T_c)L'(T_c) + O\big((T - T_c)^2\big)\,; \quad T \to T_c^{+}\,, \tag{9.99}$$

where we have assumed that $L(T)$ is analytic with respect to $T$. Now, using Eq. (9.91) in Eq. (9.99), one obtains the convergence of $L(T)$ as being proportional to $(T - T_c)$ as $T \to T_c^{+}$. Then, from Eq. (9.94), we see that

$$\chi \propto \frac{1}{(T - T_c)}; \quad T \to T_c^+, \tag{9.100}$$

from which one obtains

$$\gamma^+ = 1. \tag{9.101}$$

### 9.4.2 Linear Response in the Ordered Phase

In the ordered phase, for all temperatures $T < T_c$, there exists a nonzero magnetization $m_0(T)$. In the limit $T \to T_c^-$, one has $m_0 \to 0$. In order to find the value of the critical exponent $\gamma^-$ defined in Eq. (9.40), we need to know the behavior of $\chi(T)$ close to the critical point $T_c$. From Eq. (9.94), we see that $\chi(T)$ depends on $L(T)$. Therefore, let us first find the behavior of $L(T)$ near the critical point $T_c$. To this end, we employ Eqs. (9.94) and (9.40) to obtain

$$L = \frac{(T_c - T)^{\gamma^-}}{(T_c - T)^{\gamma^-} + a_n}; \quad T \to T_c^-, \tag{9.102}$$

where $a_n$ is a constant. Using Eq. (9.38), one may write from Eq. (9.102) that

$$L = \frac{m_0^{\gamma^-/\beta}}{m_0^{\gamma^-/\beta} + b_n} \sim m_0^{\gamma^-/\beta}; \quad T \to T_c^-, \tag{9.103}$$

where $b_n$ is a constant. We then numerically compute $L(T)$ from Eq. (9.85) by using the method discussed in Appendix J with the reference state $f_0$ given by the thermal equilibrium state at temperature $T$. In Fig. 9.2, we have shown the dependence of $L$ on $m_0$ for both Model-1 and Model-2. Here, the points are obtained by using Eq. (9.85). From the figure, we conclude that in the limit $m_0 \to 0$, the dependence of $L$ on $m_0$ is given by $m_0^{1/2}$ for Model-1, and by $m_0^{1/3}$ for Model-2. Then, for general $n \geq 1$ and in the limit $m_0 \to 0$ (that is, as $T \to T_c^-$), we postulate that

$$L \sim m_0^{1/(n+1)}; \quad T \to T_c^-. \tag{9.104}$$

Comparing Eqs. (9.103) and (9.104), we obtain

$$\gamma^- = \frac{\beta}{n+1}; \quad n \geq 1. \tag{9.105}$$

In order to check the validity of Eq. (9.105), we now compute $\chi$ by using Eq. (9.94) and show its dependence on $T_c - T$ close to the critical point $T_c$ in Fig. 9.3 for both Model-1 (panel (a)) and Model-2 (panel (b)). In both the panels, the black-dashed line corresponds to the behavior close to $T_c$. In the limit $T \to T_c^-$, by virtue of

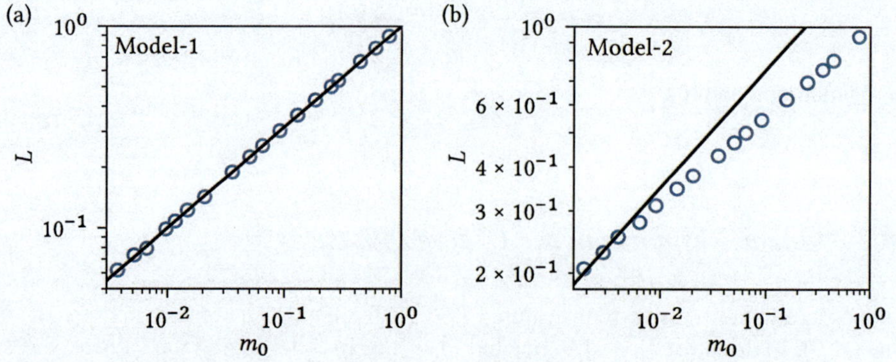


**Fig. 9.2 Dependence of $L$ on $m_0$ for $T < T_c$.** **a** Model-1. Here, we have $D = 5$. **b** Model-2. Here, we have $D = 15$. In both cases, the points are obtained by using Eq. (9.85) with the reference state $f_0$ given by the thermal equilibrium state at temperature $T$. The black lines show the behavior in the limit $m_0 \to 0$ and are given by the dependence $m_0^{1/2}$ for Model-1 and $m_0^{1/3}$ for Model-2

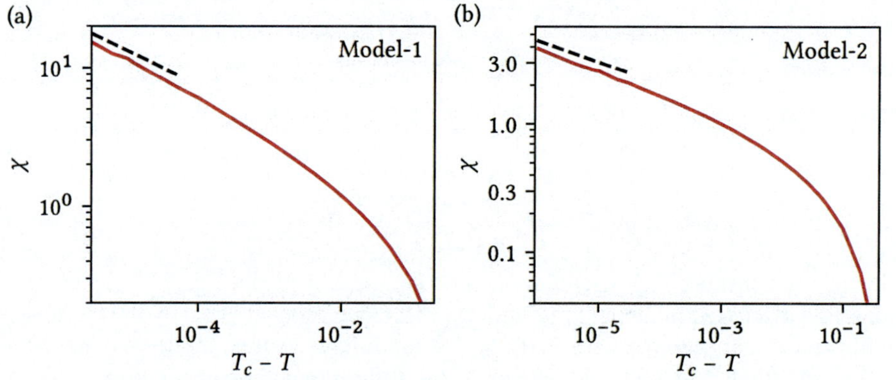


**Fig. 9.3 Susceptibility $\chi$ as a function of $T$ for $T < T_c$.** **a** Model-1. Here, we have $D = 5$. **b** Model-2. Here, we have $D = 15$. In both cases, the reference state is taken to be the thermal equilibrium state at temperature $T$. The continuous red line is obtained by using Eqs. (9.85) and (9.94). In both the panels, the black-dashed line corresponds to the behavior close to $T_c$, and by virtue of Eq. (9.40), yields the value of $\gamma^-$ to be $1/4$ for Model-1 and $1/6$ for Model-2

Eq. (9.40), one obtains the value of $\gamma^-$ to be $1/4$ for Model-1 and $1/6$ for Model-2, thus validating Eq. (9.105).

Let us now spend some time on the estimation of $L$ that will be helpful in the following. In the ordered phase, we divide $L$ into two parts, see Eq. (9.85):

$$L(T) = L_1(T) + L_2(T)\,, \tag{9.106}$$

with

$$L_1(T) \equiv 1 + \iint_{\mu} \mathrm{d}\phi \mathrm{d}p \; S_x^2 \; F'(H_0(\phi, p))\,, \tag{9.107}$$

and

$$L_2(T) \equiv -\iint_\mu \mathrm{d}\phi \mathrm{d}p \; \langle S_x \rangle_0^2 \; F'(H_0(\phi, p)) \,. \tag{9.108}$$

The behavior of $L_1(T)$ is obtained by following the same steps as in the disordered state discussed above (Sect. 9.4.1); one obtains

$$L_1(T) = O(T_c - T) \,; \quad T \to T_c^- \,. \tag{9.109}$$

The behavior of $L(T)$ may be obtained from Eq. (9.104) by substituting Eq. (9.38) in the latter. Then, one has

$$L(T) = O\big((T_c - T)^{\beta/(n+1)}\big) \,; \quad T \to T_c^- \,. \tag{9.110}$$

The decomposition (9.106) together with the behavior in Eqs. (9.109) and (9.110) and the fact that one has $\beta/(n+1) < 1$ for $n \geq 1$ and $\beta = 1/2$ suggest that the leading behavior of $L(T)$ as $T \to T_c^-$ has to come from the corresponding behavior of $L_2(T)$, and so we conclude that

$$L_2(T) = O\big((T_c - T)^{\beta/(n+1)}\big) = O\big((m_0)^{1/(n+1)}\big) \,; \quad T \to T_c^- \,. \tag{9.111}$$

### *9.4.3 Nonlinear Response at the Critical Point*

At the critical point $T = T_c$, we have $L(T_c) = 0$, and to leading order in $h$, one obtains from Eq. (9.84) that

$$\mathcal{N}(T_c) \; (\delta m + h) - h = 0 \,. \tag{9.112}$$

From Eq. (9.86), we obtain

$$\mathcal{N}(T_c) = -\iint_\mu \mathrm{d}\phi \mathrm{d}p \; \langle S_x \rangle_h^2 \; F'(H_h(\phi, p)) \,, \tag{9.113}$$

where we have used the fact that $m_0 = 0$ at $T = T_c$ gives $\langle S_x \rangle_0 = 0$. Note that the form of $\mathcal{N}(T_c)$ given in (9.113) reduces to the one for $L_2(T)$, Eq. (9.108), on replacing in the latter the reference state $f_0$ with the asymptotic state $f_h$ in performing the average over $S_x$. Then, using Eq. (9.111), we obtain an estimation of $\mathcal{N}(T_c)$ as

$$\mathcal{N}(T_c) = O\big((\delta m + h)^{1/(n+1)}\big) \,, \tag{9.114}$$

where we have replaced $m_0$ in Eq. (9.111) with $m_h + h = m_0 + \delta m + h = \delta m + h$. Substitution of this estimation (9.114) in Eq. (9.112) yields

$$(\delta m + h)^{(n+2)/(n+1)} \propto h\,, \tag{9.115}$$

and hence, that

$$\delta m \propto h^{(n+1)/(n+2)}\,. \tag{9.116}$$

Thus, the critical exponent $\delta$ is obtained as

$$\delta = \frac{n+2}{n+1}\,. \tag{9.117}$$

### 9.4.4 *Critical Exponents and the Scaling Relation*

The theoretically-predicted critical exponents, obtained within the Vlasov dynamics, are thus

$$\beta = \frac{1}{2}\,, \quad \gamma^+ = 1\,, \quad \gamma^- = \frac{\beta}{n+1}\,, \quad \delta = \frac{n+2}{n+1}\,; \quad n \geq 1\,, \tag{9.118}$$

and which satisfy the scaling relation

$$\gamma^- = \beta(\delta - 1)\,, \tag{9.119}$$

irrespective of the value of $\beta$. Note that the limit $n \to \infty$ eliminates from the Hamiltonian (9.32) the anisotropic term $Dp^{2n}$ as $|p| = |\cos\theta| < 1$; the resulting Hamiltonian is thus given by (9.63). Therefore, one may remark that the Model-0 corresponds to the limit $n \to \infty$. Equation (9.118) in this limit gives $\gamma^- = 0$ and $\delta = 1$, which is consistent with Eq. (9.65). We display the obtained critical exponents in Table 9.1.

**Table 9.1** Critical exponents of the spin model (9.1) obtained within a static analysis and within the Vlasov dynamics. The critical exponents $\gamma^\pm$ and $\delta$ in Model-0 have the same values under the two approaches. The scaling relation $\gamma^- = \beta(\delta - 1)$ holds for all cases

| | Model-$n$ | $\beta$ | $\gamma^+$ | $\gamma^-$ | $\delta$ |
|---|---|---|---|---|---|
| Static analysis | $n \geq 0$ | 1/2 | 1 | 1 | 3 |
| Vlasov dynamics | $n = 0$ | 1/2 | 0 | 0 | 1 |
| | $n \geq 1$ | 1/2 | 1 | $\frac{\beta}{n+1}$ | $\frac{n+2}{n+1}$ |

## 9.5 Numerical Results

We now present our numerical results on the critical exponents and check them with our theoretical predictions obtained in the preceding section. As a representative case, we consider thermal equilibrium states as the reference states, which are represented by Eq. (9.29) and which yield $\beta = 1/2$. However, we remark that the theoretical results obtained in Sect. 9.4 remain valid for other families of reference states. In order to do numerical checks, we perform numerical integration of the equations of motion (9.5) by using a fourth-order Runge-Kutta algorithm with time step $\delta t = 0.01$. As the theoretical results are obtained in the thermodynamic limit $N \to \infty$, while performing numerical checks, we take sufficiently large number of spins, namely, $N = 10^6$ or $10^7$. We will refer to $T$ as the temperature. However, note that $T$ is just a parameter that characterizes the reference state (9.29), and there is no thermal noise in the dynamics.

### 9.5.1 Evolution of Magnetization

Let us consider Model-1, and prepare the system in the thermal equilibrium state at a temperature $T > T_c$. Then, a constant field of strength $h = 0.01$ along the $x$-axis is turned on at $t = 0$. The resulting behavior of the magnetization $m_x$ for Model-1

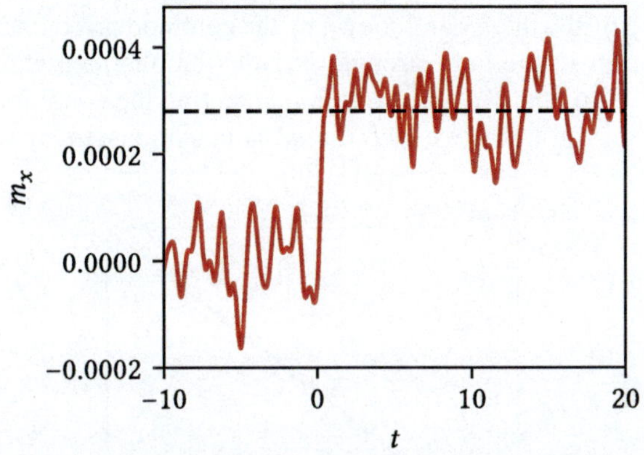


**Fig. 9.4 Model-1: Temporal evolution of magnetization $m_x$ for $D = 5$ and $N = 10^6$.** The initial state is the thermal equilibrium state at temperature $T (> T_c \approx 0.476) = 12.8$. A constant field of strength $h = 0.01$ along the $x$-axis is turned on at time $t = 0$. The dashed line gives the value of the magnetization induced by the field and given by Eq. (9.93) to be approximately 0.00028; the function $L(T)$ is computed numerically by using the method detailed in Appendix J. The data are obtained by numerically integrating the equations of motion (9.5) and averaging over 100 realizations of the dynamics

as a function of time is depicted in Fig. 9.4. The dashed line in the figure shows the value of the magnetization induced by the field, and is obtained by using Eq. (9.93).

The external field may induce periodic oscillations in the magnetization [18], but in our case, one may observe from Fig. 9.4 that the magnetization does not exhibit stable oscillations after the field is turned on. It rather exhibits due to finiteness of the number of spins only fluctuations about the theoretically-predicted value valid in the thermodynamic limit, as has also been observed in the HMF model [14].

### 9.5.2 *Critical Exponents*

For $N = 10^7$ spins, we now study numerically the response of the system (9.1) to a constant external field along the $x$-axis switched on at $t = 0$. We present the numerical results in Figs. 9.5 and 9.6 for Model-0, Figs. 9.7, 9.8 and 9.9 for Model-1, and in Figs. 9.10, 9.11 and 9.12 for Model-2. One may see from the figures that all our numerical results are consistent with the theoretical predictions given in Table 9.1. For Model-1, taking $D = 5$, using $G(x) = \exp(-x/T)$ in the expression (9.52) for the quantity $A(T)$, and then solving $A(T_c) = 0$, we obtain the critical temperature as $T_c \approx 0.476$. In a similar way, for Model-2, we take $D = 15$ and find $T_c \approx 0.47$. We remark that in Figs. 9.8 and 9.11, the theoretical results match with the numerical ones only for sufficiently small $h$. This is expected because the theoretical analysis is valid in the linear response regime obtained in the limit $h \to 0$. One may see that very close to $T_c$, the numerical results for finite $N$ depicted in Figs. 9.8 and 9.11 do not show the divergence predicted by the theory and shown by the continuous red lines. The reason for this is the finiteness of the field strength $h$. Note that the theoretical results are valid in the limits $h \to 0$ and $N \to \infty$, and implementing the condition $h > 1/\sqrt{N}$ ensures that the response dominates over finite-size fluctuations. In Figs. 9.8 and 9.11,

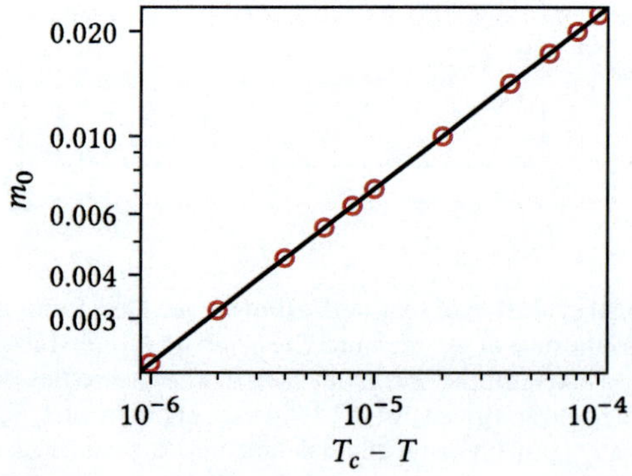


**Fig. 9.5 Model-0: Spontaneous magnetization $m_0$ for $T < T_c$.** Here, the points are obtained by solving the self-consistent equation (9.33) with thermal equilibrium as the reference state. The critical point is $T_c = 1/3$. The line corresponds to the behavior (9.38), with $\beta$ given by our theoretical analysis as $\beta = 1/2$, see Table 9.1

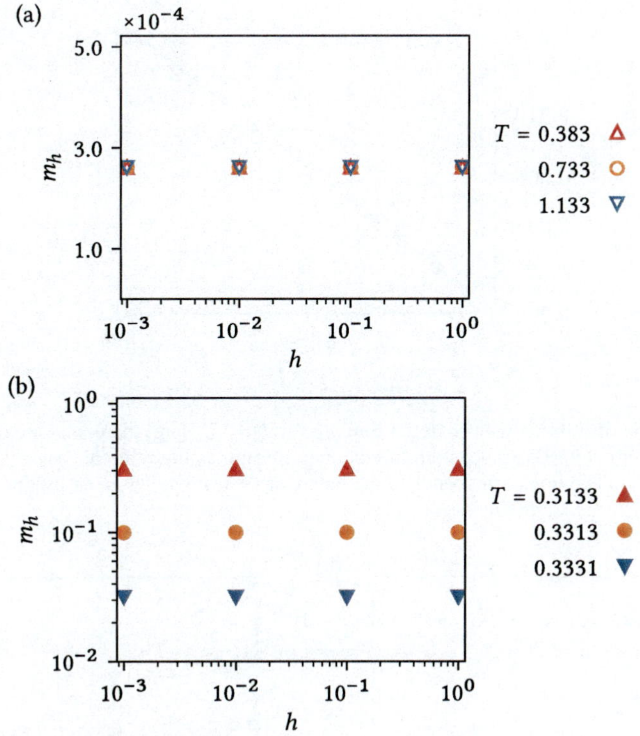


**Fig. 9.6 Model-0: Magnetization $m_h$ for (a) $T > T_c = 1/3$, (b) $T < T_c$.** The points are obtained by numerically integrating the equations of motion (9.64) with $N = 10^7$; we initiate the dynamics from the reference state $f_0$ given by the thermal equilibrium state at temperature $T$, turn on the external field at $t = 0$, let the system relax to the stationary state $f_h$. The data correspond to time average of the instantaneous magnetization in the stationary state $f_h$ over an interval of length 20, which is further averaged over 5 realizations of the aforementioned dynamical protocol. The magnetization $m_h$ is independent of $h$ in both cases, thus lending support to the behavior (9.40), with $\gamma^{\pm}$ given by our theoretical analysis as $\gamma^+ = \gamma^- = 0$, see Table 9.1

the value of $N$ is large enough that the condition $h > 1/\sqrt{N}$ is satisfied, although the limit $h \to 0$ is not quite achieved. We also remark that the convergence to the $h \to 0$ limit is slower for Model-2 than for Model-1.

## 9.6 Conclusions and Perspectives

In this chapter, we have studied the response to an external field in mean-field systems of classical Heisenberg spins exhibiting a second-order phase transition in the stationary state. In the thermodynamic limit, the dynamical evolution of such systems is described by the Vlasov equation involving the single-spin distribution function. For

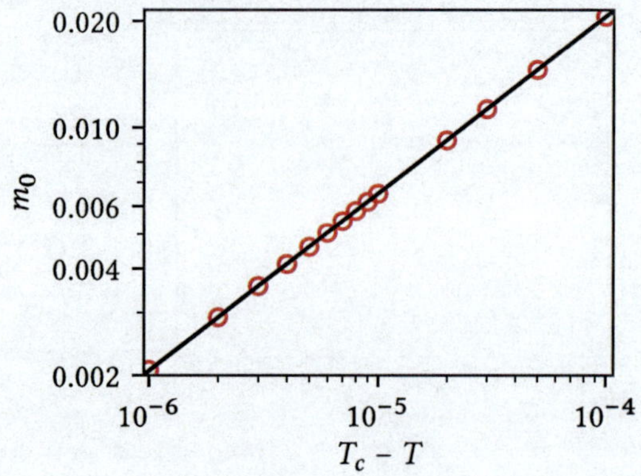


**Fig. 9.7 Model-1: Spontaneous magnetization $m_0$ for $D = 5$.** Here, the points are obtained by solving the self-consistent Eq. (9.33) with thermal equilibrium as the reference state. The critical point is $T_c \approx 0.476$. The line corresponds to the behavior (9.38), with $\beta$ given by our theoretical analysis as $\beta = 1/2$, see Table 9.1

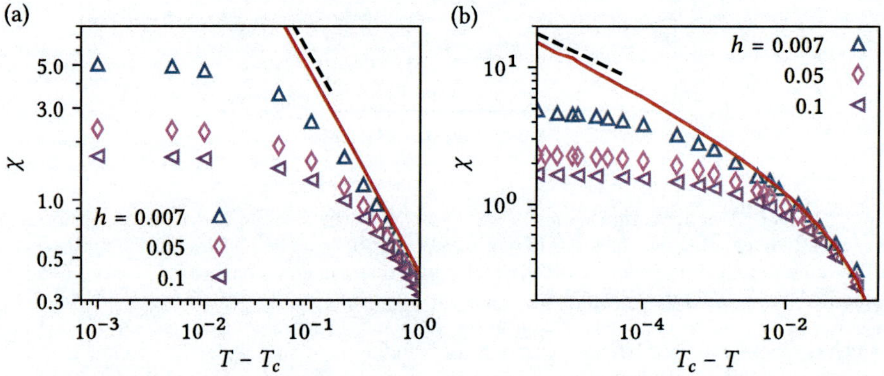


**Fig. 9.8 Model-1: Susceptibility $\chi(T)$ for $D = 5$.** Here, the points are obtained from values of magnetization $m_h$ that one gets by numerically integrating the equations of motion (9.5) with $N = 10^7$; we initiate the dynamics from the reference state $f_0$ given by the thermal equilibrium state at temperature $T$, turn on the external field at $t = 0$, let the system relax to the stationary state $f_h$. The data correspond to time average of the instantaneous magnetization in the stationary state $f_h$ over an interval of length 20, which is further averaged over 2 realizations of the aforementioned dynamical protocol. **a** The disordered phase at $T > T_c \approx 0.476$. **b** The ordered phase at $T < T_c$. In either case, the black dashed line corresponds to the behavior close to $T_c$, Eq. (9.40), with $\gamma^+ = 1$ and $\gamma^- = 1/4$ as given by our theoretical predictions, see Table 9.1. The red continuous lines in the figures are our theoretical result (9.94), with $L$ given by Eq. (9.85) computed numerically by using the method detailed in Appendix J

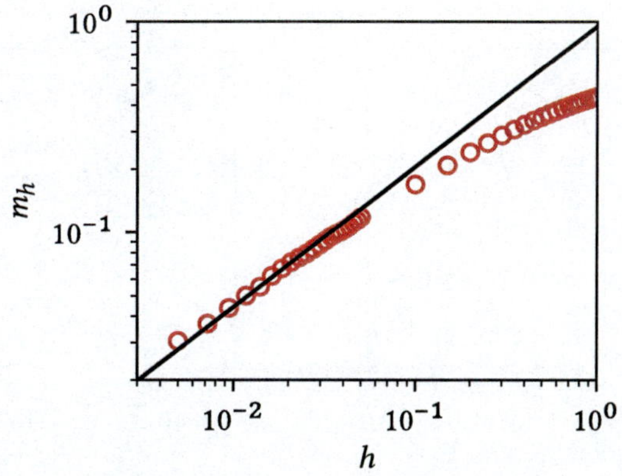


**Fig. 9.9 Model-1: Magnetization $m_h$ for $D = 5$ and at the critical point $T_c \approx 0.476$.** Here, the points are obtained by numerically integrating the equations of motion (9.5) with $N = 10^7$; we initiate the dynamics from the reference state $f_0$ given by the thermal equilibrium state at temperature $T$, turn on the external field at $t = 0$, let the system relax to the stationary state $f_h$. The data correspond to time average of the instantaneous magnetization in the stationary state $f_h$ over an interval of length 20, which is further averaged over 2 realizations of the aforementioned dynamical protocol. The black line corresponds to the behavior (9.41), with $\delta$ given by our theoretical analysis as $\delta = 3/2$, see Table 9.1. As expected, only for small $h$ does our theory match with numerical results

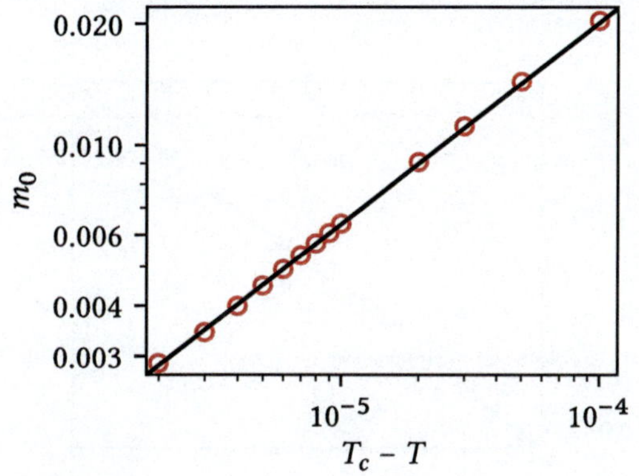


**Fig. 9.10 Model-2: Spontaneous magnetization $m_0$ for $D = 15$.** Here, the points are obtained by solving the self-consistent Eq. (9.33) with thermal equilibrium as the reference state. The critical point is $T_c \approx 0.47$. The line corresponds to the behavior (9.38), with $\beta = 1/2$, as predicted by our theory, see Table 9.1

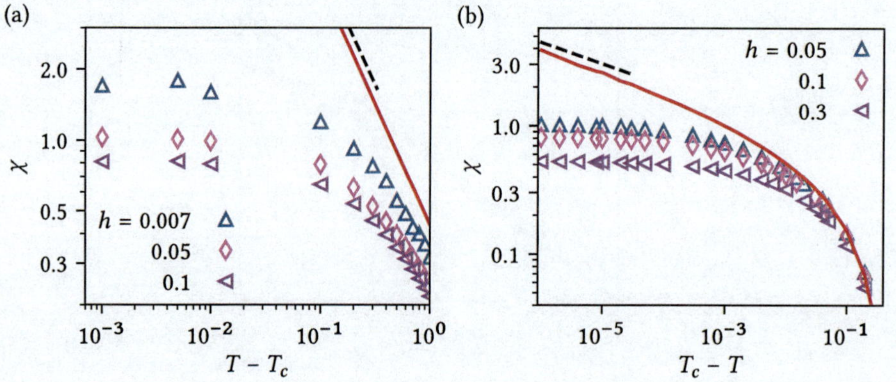


**Fig. 9.11 Model-2: Susceptibility** $\chi(T)$ **for** $D = 15$. Here, the points are obtained from values of magnetization $m_h$ that one gets by numerically integrating the equations of motion (9.5) with $N = 10^7$; we initiate the dynamics from the reference state $f_0$ given by the thermal equilibrium state at temperature $T$, turn on the external field at $t = 0$, let the system relax to the stationary state $f_h$. The data correspond to time average of the instantaneous magnetization in the stationary state $f_h$ over an interval of length 20, which is further averaged over 2 realizations of the aforementioned dynamical protocol. **a** The disordered phase at $T > T_c \approx 0.47$. **b** The ordered phase at $T < T_c$. In either case, the black dashed line corresponds to the behavior close to $T_c$, Eq. (9.40), with $\gamma^+ = 1$ and $\gamma^- = 1/6$ as given by our theoretical predictions, see Table 9.1. The red continuous lines in the figures are our theoretical result (9.94), with $L$ given by Eq. (9.85) computed numerically by using the method given in Appendix J

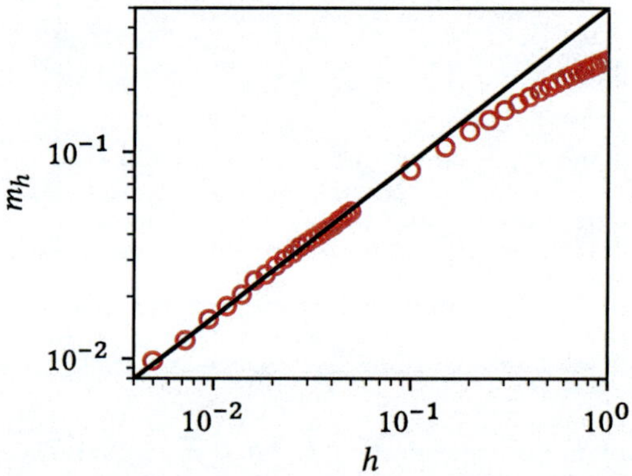


**Fig. 9.12 Model-2: Magnetization** $m_h$ **for** $D = 15$ **and at the critical point** $T_c \approx 0.47$. Here, the points are obtained by numerically integrating the equations of motion (9.5) with $N = 10^7$; we initiate the dynamics from the reference state $f_0$ given by the thermal equilibrium state at temperature $T$, turn on the external field at $t = 0$, let the system relax to the stationary state $f_h$. The data correspond to time average of the instantaneous magnetization in the stationary state $f_h$ over an interval of length 20, which is further averaged over 2 realizations of the aforementioned dynamical protocol. The black line corresponds to the behavior (9.41), with $\delta$ given by our theoretical analysis as $\delta = 4/3$, see Table 9.1. As expected, only for small $h$ does our theory match with numerical results

Vlasov-stationary states that allow a second-order phase transition, and when subject to a small external field, we have obtained the critical exponents characterizing the power-law behavior of response close to the critical point. We have shown that the exponent values obtained by studying the Vlasov dynamics are different from the ones obtained from a static analysis, with no reference to the dynamics of the initial state in the presence of the external field. Interestingly, we have found that both sets of the critical exponents satisfy the same scaling relation, which incidentally turns out to be the same as the one for mean-field Hamiltonian particle systems that are very different from the spin systems studied here. The discussions presented in this chapter hint at the universality of critical behavior for mean-field systems evolving under Vlasov dynamics. On the other hand, they also caution against relying on a static approach, with no reference to the dynamical evolution, while extracting critical exponent values for mean-field systems.

We remark that one can have a class of Vlasov-stationary states that for our model does not allow a second-order but rather a first-order phase transition. For example, we note that the HMF model, where the Vlasov-stationary thermal equilibrium state allows for a second-order phase transition, can also have other classes of Vlasov-stationary states that allow a first-order phase transition [19–22]. One may study the response of our model for other classes of Vlasov-stationary reference states that allow a first-order phase transition. In the context of our model, it would be interesting to investigate the validity of the Kubo fluctuation-dissipation theorem, which is valid for short-range systems in thermal equilibrium subject to small external fields. One may pursue investigations in this direction and compare the findings with the HMF model results [23]. Another interesting topic to study is the collective $1/f$ fluctuation due to the Casimir invariants [24].

## References


1. Y.Y. Yamaguchi, D. Das, S. Gupta, Phys. Rev. E **100**(3), 032131 (2019). https://doi.org/10.1103/PhysRevE.100.032131
2. M.E. Fisher, Rep. Prog. Phys. **30**(2), 615 (1967). https://doi.org/10.1088/0034-4885/30/2/306
3. H.E. Stanley, *Introduction to Phase Transitions and Critical Phenomena* (Oxford University Press, New York, 1987)
4. H. Nishimori, G. Ortiz, *Elements of Phase Transitions and Critical Phenomena* (Oxford University Press, Oxford, 2010). https://doi.org/10.1093/acprof:oso/9780199577224.001.0001
5. A. Campa, T. Dauxois, S. Ruffo, Phys. Rep. **480**(3–6), 57 (2009). https://doi.org/10.1016/j.physrep.2009.07.001
6. Y. Levin, R. Pakter, F.B. Rizzato, T.N. Teles, F.P.C. Benetti, Phys. Rep. **535**(1), 1 (2014). https://doi.org/10.1016/j.physrep.2013.10.001
7. A. Campa, T. Dauxois, D. Fanelli, S. Ruffo, *Physics of Long-Range Interacting Systems*, 1st edn. (Oxford University Press, Oxford, 2014). https://doi.org/10.1093/acprof:oso/9780199581931.001.0001
8. S. Gupta, S. Ruffo, Int. J. Mod. Phys. A **32**(09), 1741018 (2017). https://doi.org/10.1142/S0217751X17410184
9. Y.Y. Yamaguchi, J. Barré, F. Bouchet, T. Dauxois, S. Ruffo, Physica A **337**(1–2), 36 (2004). https://doi.org/10.1016/j.physa.2004.01.041

10. J. Barré, F. Bouchet, T. Dauxois, S. Ruffo, Y.Y. Yamaguchi, Physica A **365**(1), 177 (2006). https://doi.org/10.1016/j.physa.2006.01.005
11. J. Binney, S. Tremaine, *Galactic Dynamics*, 2nd edn. (Princeton University Press, Princeton, 2011)
12. S. Inagaki, T. Konishi, Publ. Astron. Soc. Jpn. **45**, 733 (1993)
13. M. Antoni, S. Ruffo, Phys. Rev. E **52**(3), 2361 (1995). https://doi.org/10.1103/PhysRevE.52.2361
14. S. Ogawa, A. Patelli, Y.Y. Yamaguchi, Phys. Rev. E **89**(3), 032131 (2014). https://doi.org/10.1103/PhysRevE.89.032131
15. S. Ogawa, Y.Y. Yamaguchi, Phys. Rev. E **91**(6), 062108 (2015). https://doi.org/10.1103/PhysRevE.91.062108
16. S. Ogawa, Y.Y. Yamaguchi, Phys. Rev. E **85**(6), 061115 (2012). https://doi.org/10.1103/PhysRevE.85.061115
17. R. Balescu, *Statistical Dynamics: Matter Out of Equilibrium*(Imperial College Press, London, 1997). https://doi.org/10.1142/p036
18. R. Pakter, Y. Levin, J. Stat. Phys. **150**(3), 531 (2013). https://doi.org/10.1007/s10955-012-0576-9
19. A. Antoniazzi, D. Fanelli, S. Ruffo, Y.Y. Yamaguchi, Phys. Rev. Lett. **99**(4), 040601 (2007). https://doi.org/10.1103/PhysRevLett.99.040601
20. R. Pakter, Y. Levin, Phys. Rev. Lett. **106**(20), 200603 (2011). https://doi.org/10.1103/PhysRevLett.106.200603
21. T.M. Rocha Filho, M.A. Amato, A. Figueiredo, Phys. Rev. E **85**(6), 062103 (2012). https://doi.org/10.1103/PhysRevE.85.062103
22. T.N. Teles, F.P.D.C. Benetti, R. Pakter, Y. Levin, Phys. Rev. Lett. **109**(23), 230601 (2012). https://doi.org/10.1103/PhysRevLett.109.230601
23. Y.Y. Yamaguchi, Phys. Rev. E **94**(1), 012133 (2016). https://doi.org/10.1103/PhysRevE.94.012133
24. Y.Y. Yamaguchi, K. Kaneko, Phys. Rev. E **98**(2), 020201 (2018). https://doi.org/10.1103/PhysRevE.98.020201

# Chapter 10
# Quantum Systems Subject to Random Projective Measurements

**Abstract** In contrast to the previous chapters, we deal in this chapter with quantum systems, and specifically, with the facet of noise in the sense of uncertainty associated with quantum measurements. This sort of noise is invariably of quantum origin, and we explore its manifestation in the set-up of a quantum system undergoing unitary evolution in time that is being repeatedly interrupted at random times with projective measurements collapsing the instantaneous state to the initial state of the system. Specifically, we ask: How does the survival probability, namely, the probability that an initial state survives even after a certain number $m$ of measurements, behave as a function of $m$? We pursue our studies in the context of two paradigmatic quantum systems, one, the quantum random walk evolving in discrete times, and the other, the tight-binding model evolving in continuous time, with both defined on a one-dimensional periodic lattice constituted by a finite number of sites. We unveil a host of interesting results on the survival probability that point to the curious nature of quantum measurement dynamics. This chapter is based on our published work, Ref. [1].

## 10.1 Introduction

Let us consider a quantum system that is described by a time-independent Hamiltonian $H$. Its state at any time $t$ is characterized by a state vector $|\psi(t)\rangle$, which is defined in the Hilbert space $\mathcal{H}_S$ of the system $(S)$. The state vector undergoes unitary evolution in time as $|\psi(t)\rangle = U(t, t_0)\,|\psi(t_0)\rangle\,;\ t > t_0$, where $U(t, t_0) \equiv \exp(-\mathrm{i}H(t-t_0))$ denotes the time-evolution operator[1]. Next, let us consider a series of instantaneous measurements performed on the system at random times. Following the measurement postulate of quantum mechanics [2], we may conclude that each measurement involves projecting the instantaneous state of the system onto the Hilbert space $\mathcal{H}_D \subset \mathcal{H}_S$ of the measuring device $(D)$. Starting at time $t = 0$ with a state vector with unit norm, each projective measurement reduces the norm, and

[1] In this chapter, we will take the Planck's constant to be unity.



D. Das and S. Gupta, *Facets of Noise*, Fundamental Theories of Physics 214,
https://doi.org/10.1007/978-3-031-45312-0_10

we may ask for its magnitude after a certain number $m \in \mathbb{Z}^+$ of measurements are performed on the system. The norm of the state vector at any time $t > 0$ has the physical interpretation of survival probability of the initial state at that time instant. In order to see this, consider a quantum particle undergoing motion in a spatial domain $\mathcal{D}$. Then, it follows that $|\psi(\mathbf{r},t)|^2 \mathrm{d}^3\mathbf{r}$ gives the probability of finding the particle between locations $\mathbf{r}$ and $\mathbf{r}+\mathrm{d}\mathbf{r}$ in $\mathcal{D}$ at time $t$, with $\psi(\mathbf{r},t) \equiv \langle \mathbf{r}|\,\psi(t)\rangle$. Since we have $\int_{\mathcal{D}} \mathrm{d}^3\mathbf{r}\ |\psi(\mathbf{r},t=0)|^2 = 1$, the particle is initially for sure within $\mathcal{D}$. Then, with $m$ instantaneous projective measurements being performed on the system at random times $t_1, t_2, t_3, \ldots, t_m;\ \ 0 < t_1 < t_2 < t_3 < \ldots < t_m$, the norm $\langle \psi(t_m)|\psi(t_m)\rangle = \int_{\mathcal{D}} \mathrm{d}^3\mathbf{r}\ |\psi(\mathbf{r},t_m)|^2$ of the state vector at time $t_m$ will give the probability $S_m$ that the particle is still inside $\mathcal{D}$ at the end of $m$ projective measurements. Put it differently, $S_m$ is the survival probability that the particle has survived in $\mathcal{D}$ up to time $t_m$. With the definition $t_0 = 0$, let us note that $S_0 = 1$. For latter purpose, we define the quantity

$$F_m \equiv S_{m-1} - S_m\,, \tag{10.1}$$

which stands for the first-detection probability, i.e., the probability that the particle gets detected by the measuring device at time $t_m$ *for the first time*. It is then pertinent to ask the question: How does $S_m$ vary from one realization of random times $\{t_1, t_2, \ldots, t_m\}$ to another? What is the dependence on $m$ of the averaged survival probability $\overline{S_m}$? How does the typical value of the survival probability, which is what one measures in a typical realization of the random times, depend on $m$?

In the aforementioned protocol, let us consider the successive time gaps $\tau_\alpha \equiv t_\alpha - t_{\alpha-1};\ \alpha = 1, 2, 3, \ldots, m$ as random variables sampled independently from a common distribution $p(\tau)$, i.e., the $\tau_\alpha$'s are independent and identically-distributed (i.i.d.) random variables. In this backdrop, let us be more specific regarding the framework of our study. Consider a generic state $|\psi(0)\rangle$ evolving for a random time $\tau_1$ according to the unitary operator $U_1 \equiv \exp(-\mathrm{i}H\tau_1)$. The evolved state $U_1\,|\psi(0)\rangle$ is subject to an instantaneous projective measurement according to a given projection operator $P$. Subsequent evolution may proceed with either the projected component $PU_1|\psi(0)\rangle$ of the state $U_1|\psi(0)\rangle$ or its leftover component given by $(\mathbb{I} - P)U_1|\psi(0)\rangle$, where $\mathbb{I}$ denotes the identity operator. We iterate $m$ times the aforementioned set of events, so we have two different schemes of time evolution of the system that involve one of the following repetitive sequence of event-pair:

- **Scheme 1**: A unitary evolution for a random time $\tau_\alpha$ according to the unitary operator $U_\alpha \equiv \exp(-\mathrm{i}H\tau_\alpha)$, followed by the action of the operator $P$. This scheme thus corresponds to subsequent evolution with the projected component after each measurement.

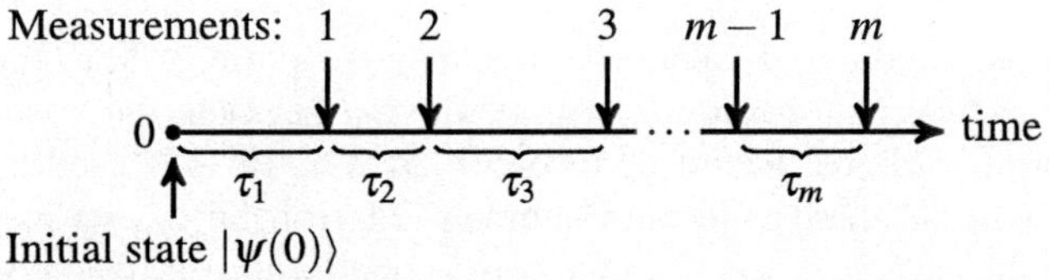


**Fig. 10.1 A typical time evolution of a quantum system subject to projective measurements at random times, as detailed in the text.** Starting from the state $|\psi(0)\rangle$, the evolution involves the following repetitive sequence of events: unitary evolution according to operator $U_\alpha$ for time $\tau_\alpha$, which is followed by a projective measurement (denoted by the down arrows). In **Scheme 1** (respectively, **Scheme 2**), subsequent evolution at the end of each measurement is implemented with the projected component of the instantaneous state (respectively, with the leftover component). The figure is taken from Ref. [1]

- **Scheme 2**: A unitary evolution for a random time $\tau_\alpha$ according to the unitary operator $U_\alpha$, followed by the action of the operator $\widetilde{P} \equiv \mathbb{I} - P$. This scheme thus corresponds to subsequent evolution with the leftover component after each measurement.

A typical time evolution for a given realization $\{\tau_\alpha\}_{1\le\alpha\le m}$ is shown schematically in Fig. 10.1. We take the projection operator as performing projection to the initial state:

$$P = |\psi(0)\rangle\langle\psi(0)|. \tag{10.2}$$

In the following, we use the notation $|\psi_\alpha^{(\mathrm{b})}\rangle$ (respectively, $|\psi_\alpha^{(\mathrm{a})}\rangle$) to denote the state of the system at the end of evolution for time $\tau_\alpha$ and just before (respectively, just after) the $\alpha$th projective measurement. It is then evident from the dynamical rules of evolution that one has $|\psi_\alpha^{(\mathrm{b})}\rangle = U_\alpha\,|\psi_{\alpha-1}^{(\mathrm{a})}\rangle$, while $|\psi_\alpha^{(\mathrm{a})}\rangle$ will be either $|\psi_\alpha^{(\mathrm{a})}\rangle = P\,|\psi_\alpha^{(\mathrm{b})}\rangle$ (**Scheme 1**) or $|\psi_\alpha^{(\mathrm{a})}\rangle = \widetilde{P}\,|\psi_\alpha^{(\mathrm{b})}\rangle$ (**Scheme 2**), and $\alpha = 1, 2, \ldots, m$. Let us define the random variable

$$S_m \equiv S_m(\{\tau_\alpha\}_{1\le\alpha\le m}) = \langle\psi_m^{(\mathrm{a})}|\psi_m^{(\mathrm{a})}\rangle \tag{10.3}$$

as the survival probability of the initial state after $m$ projective measurements and for the realization $\{\tau_\alpha\}_{1\le\alpha\le m}$. We may note that different values of $S_m$ may correspond to different total duration of evolution $\mathcal{T} \equiv \sum_{\alpha=1}^{m} \tau_\alpha$. Let us then ask: How does $\overline{S}_m$ depend on $m$? How does the dependence vary between the two schemes, namely, **Scheme 1** and **Scheme 2**? What is the essential role that is played by $p(\tau)$ in dictating the value of $\overline{S}_m$ in the two schemes?

In order to illustrate the two schemes of dynamical evolution, let us consider the representative example of a quantum particle moving between the sites of a one-dimensional periodic lattice. Referring to Fig. 10.2, wherein $n$ is used to label the lattice sites, the particle starts its journey at time $t = 0$ from site $n = 5$ in a lattice of 11 sites.[2] When the particle is on site $n$, the corresponding state is denoted by the ket $|n\rangle$ satisfying $\langle n|m\rangle = \delta_{n,m}$ and $\sum_n |n\rangle\langle n| = \mathbb{I}$. For Fig. 10.2, the initial state vector is given by $|\psi(0)\rangle = |5\rangle$. In the figure, the different panels show the site occupation probability $P_n(t) = |\langle n|\psi(t)\rangle|^2$, where $|\psi(t)\rangle$ gives the state vector at time $t$. Panel (a) depicts $P_n(t = 0) = \delta_{n,5}$. The first projective measurement is performed on the system at time $t_1 > 0$. Panel (b) shows $P_n(t)$ just before the first measurement and is obtained as the result of unitary evolution of the state in panel (a), while panels (c) and (d) show respectively results corresponding to the projected and the leftover state just after the first measurement, with the state being denoted by the generic symbol $|\psi_1^{(a)}\rangle$. The grey bars in these panels show the part of $P_n(t_1)$ projected out by the process of measurement. In **Scheme 1** (respectively, **Scheme 2**), subsequent unitary evolution proceeds with the state corresponding to panel (c) (respectively, panel (d)). If the second measurement happens at time $t_2 > t_1$, the state $|\psi_2^{(b)}\rangle$ is shown in panels (e) and (f) in the two schemes and is the result of unitary evolution of panels (c) and (d), respectively. Panels (g) and (h) show the results for the state $|\psi_2^{(a)}\rangle$ just after the second measurement and depict the projected and the leftover state, respectively, similar to panels (c) and (d). One goes on repeating the above sequence of pair of events: unitary evolution, subsequent evolution with either the projected state (**Scheme 1**) or the leftover state (**Scheme 2**). In terms of $|\psi_m^{(a)}\rangle$, the site occupation probability just after the $m$th measurement is given by

$$P_n(t = t_m^{(a)}) = |\langle n|\psi_m^{(a)}\rangle|^2, \tag{10.4}$$

with $t_m^{(a)}$ denoting the time instant just after the $m$th measurement. On the other hand, the survival probability in the two schemes is related to the respective $P_n(t)$ via

$$S_m = \sum_n P_n(t = t_m^{(a)}). \tag{10.5}$$

Thus, the survival probability is nothing but a measure of the site-averaged occupation probability of the particle in the two schemes.

[2] The time evolution shown in Fig. 10.2 follows the tight-binding model dynamics that we discuss later in Sect. 10.3.

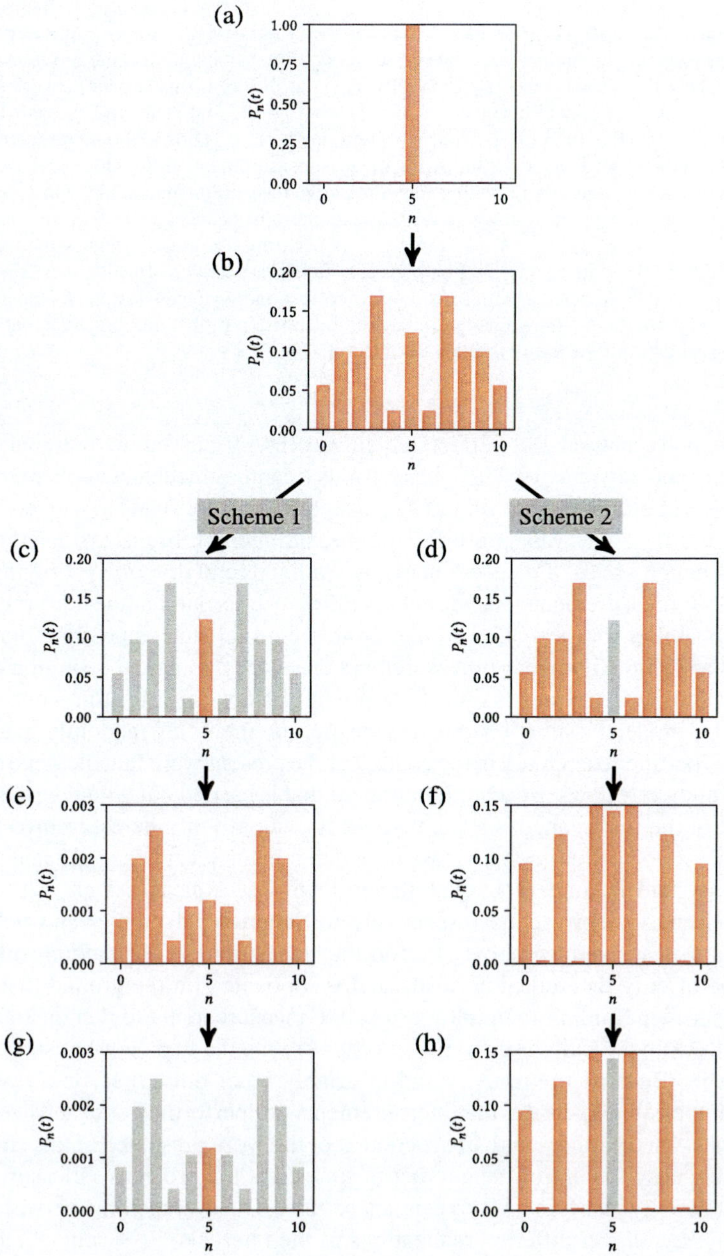
(a)
(b)
Scheme 1
Scheme 2
(c)
(d)
(e)
(f)
(g)
(h)
$P_n(t)$
$n$

◀**Fig. 10.2 Schematic representation of the dynamics studied under Scheme 1 and Scheme 2.** For a quantum particle moving between the sites of a one-dimensional lattice, the different panels in the figure show the site-occupation probability $P_n(t)$ at different times: panel **a** corresponds to the initial instant $t = 0$, panel **b** to the situation just before the first projective measurement is performed on the system at time $t = t_1$. Panels **c** and **d** denote $P_n(t)$ corresponding respectively to the projected and the leftover state just following the first measurement. In **Scheme 1** (respectively, **Scheme 2**), subsequent evolution following each projective measurement is implemented with the projected state (respectively, the leftover state). Correspondingly, panels **e** and **f** represent the state just before the second projective measurement that is performed at time $t_2$ and is the result of unitary evolution of the state represented by panels **c** and **d**, respectively. Finally, panels **g** and **h** denote respectively the projected and the leftover state corresponding to panels **e** and **f**, respectively. The grey bars in these panels show the part of the site-occupation probability projected out by the process of measurement. The figure is taken from Ref. [1]

The so-called quantum Zeno effect, which was first discussed in a seminal paper by Sudarshan and Misra in 1977 [3], is a curious quantum mechanical phenomenon that involves a quantum system subject to projective measurements to a given initial state at regular time intervals (stroboscopic measurements). In the extreme case of a frequent-enough series of measurements over a fixed total duration (implying that the successive measurements are infinitesimally close to one another in time), the survival probability for the system to remain in the initial state approaches unity in the limit of an infinite number of measurements, i.e., the dynamical evolution of the system gets completely frozen in time. Subsequently, stochastic quantum Zeno effect was studied to model the situation in which the measurements are randomly spaced in time [4]. It is pertinent to remark in the context of the present work that the Zeno effect involves evolution following each measurement that is carried out with the projected component of the instantaneous state, while it is evidently of interest to investigate the dynamics in the complimentary case in which it is the leftover component of the instantaneous state that undergoes evolution following each measurement. These two dynamical scenarios correspond respectively to **Scheme 1** dynamics and **Scheme 2** dynamics that are defined above. Performing measurements at random times is not an issue of only theoretical curiosity, but is motivated on the ground that after all, any experiment aiming to employ projective measurements to demonstrate the Zeno effect would typically use a timer in order to time the gap between successive measurements. Because the timer would invariably be of finite precision, it would be almost impossible to ensure that measurements are performed at exactly regular time intervals. On the other hand, in the context of the work presented in this chapter, it is relatively easy to control the number of times that the projective measurement is repeated, thereby justifying our dynamical set-ups. The average time between two intervals (averaged over different realizations of the intervals) is of course required to be finite in experiments.

In this chapter, we consider two representative quantum systems defined on a one-dimensional periodic lattice with a finite number of sites $N$, (i) the quantum random walk (QRW) model that evolves in discrete times [5], and (ii) the tight-binding model (TBM) that evolves in continuous time [6, 7]. Within the ambit of these two systems, we unveil a plethora of interesting and intriguing results, numerical as well as analytical, including universal features in the late-time behavior of both the average and the typical survival probability, all of which point to the peculiar nature of quantum measurement process. We note that experimentally, it is possible to implement a random or a stochastic sequence of measurement protocol [4, 8–11], and so we believe that our results are amenable to experimental realization.

The chapter is organized as follows. We study first the QRW, which is described in detail in Sect. 10.2.1. Here, we set out to compute analytically the site occupation probability, the results of which form a crucial input for the analytical and semi-analytical description of the survival probability that is pursued in Sect. 10.2.2. This latter section reports our main results on the survival probability of a generic initial state subject to instantaneous projective measurements to the initial state at random times. We report extensive numerical results that demonstrate that the average as well as the typical survival probability decays asymptotically as an exponential in $m$ in the case when the evolution following each projective measurement is carried on with the projected component of the instantaneous state. On the contrary, when the evolution proceeds with the leftover component, we find that the survival probability exhibits two characteristic $m$ values, namely, $m_1^\star(N) \sim N$ and $m_2^\star(N) \sim N^\delta$ with $\delta > 1$. We show that (i) for $m$ large and satisfying $m < m_1^\star(N)$, the decay of the survival probability is as $m^{-2}$ (correspondingly, $F_m$ decays as $m^{-3}$), (ii) for $m$ satisfying $m_1^\star(N) \ll m < m_2^\star(N)$, the decay is as $m^{-3/2}$ (correspondingly, $F_m$ decays as $m^{-5/2}$), while (iii) for $m \gg m_2^\star(N)$, the decay is as an exponential. These results are found to hold independently of the choice of the distribution of times between successive measurements, as we demonstrate by our results for a wide variety of distributions including exponential and power-law distributions as well as for the case of measurements at regular intervals. For the projected case, we support our numerical findings with explicit analytical calculations that employ the large deviation theory (LDT) well known in probability theory [12, 13], while for the leftover case, we provide a semi-analytical approach to reproduce correctly our numerical findings. Appendix L provides a primer on the LDT.

Next, we turn to the TBM and a discussion on its site occupation probability in Sect. 10.3.1. This we follow in Sect. 10.3.2 by a discussion of our numerical results for the survival probability for the case in which the evolution following each projective measurement is carried on with the projected component as well as for the case in which it is the leftover component of the instantaneous state that undergoes subsequent evolution until next measurement. Here, similar to the QRW case, we support our findings with analytical results that are derived using the LDT for the projected case and with semi-analytic calculations for the leftover case. Such calculations depend crucially on our results on site occupation probability presented in Sect. 10.3.1. Similar to the case of the QRW, we observe that for representative $p(\tau)$, the average as well as the typical survival probability decays as a function of $m$ as an exponential for the projected case and is characterized by the two scales $m_1^\star(N)$ and $m_2^\star(N)$ with associated behavior the same as for the leftover case in the case of the QRW. The results for the QRW and the TBM thus hint at the robustness of the same with respect to both discrete and continuous time evolution. The chapter ends with conclusions in Sect. 10.4. Appendix K collects some of the technical details of our analytical calculations of the most probable survival probability using the LDT.

In recent years, the TBM and related systems when subject to projective measurements with time evolution following **Scheme 2** of dynamics have been studied to address the issue of when does a quantum particle evolving under the dynamics arrive at a chosen set of sites [14–28]. A crucial difference with respect to our work is that, barring [28], these contributions almost exclusively implement measurements done at regular intervals of times, unlike the scheme considered in this work in which sequence of measurements at random times is implemented. Both the quantum random walk and the tight-binding model have been extensively studied in the literature, and the question of first-detection probability has also been studied in the past. However, to the best of our knowledge, the issue of first-detection probability when measurements are done not at regular but at random time intervals and the ensuing results that we report in this work have not been reported extensively in the literature. The fact that the survival probability $S_m$ exhibits in appropriate regimes the decay as $m^{-2}$ and as $m^{-3/2}$, that the first-detection probability $F_m$ decays in appropriate regimes as $m^{-3}$ and as $m^{-5/2}$ with oscillations at small $m$ have all been known in the literature only for the case of measurements at regular intervals (see, e.g., Refs. [14, 17]), and our main contribution as regards **Scheme 2** of the dynamics is to demonstrate that these results hold also when measurements are performed at random intervals, based on our analysis for the studied distributions of these random intervals.

## 10.2 A Discrete-Time Dynamics: The Quantum Random Walk Model

A quantum random walk (QRW) [5, 29, 30] model describes the quantum mechanical evolution in discrete times of a walker performing random walk on a lattice. Here, we consider the QRW model of a walker moving on a one-dimensional periodic lattice of $N$ sites, where the sites are labeled by the index $n \in \mathbb{Z}$. The Hilbert space $\mathcal{H}_P$ for the position of the walker is spanned by the orthonormal basis states $\{|n\rangle\}$ with $\langle m|n\rangle = \delta_{m,n}$. As we will discuss later in this section, the QRW dynamics also depends on the state of a "quantum" coin carried by the walker. The coin Hilbert space $\mathcal{H}_C$ is spanned by two orthonormal basis states $|\uparrow\rangle$ and $|\downarrow\rangle$ representing the head and the tail state of the coin, respectively. The Hilbert space $\mathcal{H}_S$ of the walker system is then given by a direct sum of the two Hilbert spaces $\mathcal{H}_P$ and $\mathcal{H}_C$, i.e., one has $\mathcal{H}_S = \mathcal{H}_P \oplus \mathcal{H}_C$. Let us denote the walker state at time $t$ by $|\psi(t)\rangle$, which is obtained as a linear combination of the basis states of the space $\mathcal{H}_S$ with time-dependent expansion coefficients. Note that the basis states of the space $\mathcal{H}_S$ are given by a direct product of those of the position and the coin space. One may also think the coin space as a spin space of a spin-1/2 particle, with the basis states being the eigenstates, the up (u) and the down (d) state, of the $z$-component of the spin operator. In such a case, the up and the down state correspond to the head and the tail state of the coin, respectively.

The discrete-time dynamics of the random walker is given by the following repetitive sequence of events: acting on the walker state, first by a unitary operator $C \otimes \mathbb{I}$, and then by a unitary operator $U$, with both operations constituting one time step of evolution. Here, the operator $U$ is given by

$$U \equiv |\uparrow\rangle\langle\uparrow| \otimes \sum_n |n+1\rangle\langle n| + |\downarrow\rangle\langle\downarrow| \otimes \sum_n |n-1\rangle\langle n| , \tag{10.6}$$

while the operator $C$ has the form

$$C \equiv \cos\theta |\uparrow\rangle\langle\uparrow| + \sin\theta |\uparrow\rangle\langle\downarrow| - \sin\theta |\downarrow\rangle\langle\uparrow| + \cos\theta |\downarrow\rangle\langle\downarrow| , \tag{10.7}$$

where $\theta \in [0, 2\pi)$ is a given parameter. In one time step of evolution, the operator $C$ generates a rotation in the coin space, while the operator $U$ generates a translation of the walker on the lattice, which is dependent on the state of the quantum coin.

We consider a representative class of initial conditions that corresponds to having the walker on a given site $n_0$ with an arbitrary superposition of the coin states. Such an initial condition is given by

$$|\psi(0)\rangle = \frac{1}{\sqrt{|a|^2 + |b|^2}} [a|\uparrow\rangle + b|\downarrow\rangle] \otimes |n_0\rangle , \tag{10.8}$$

with $a, b \in \mathbb{C}$. Following the rules of evolution of the random walker mentioned above, one may write down explicitly the state of the walker at the first and the second time step:

$$\begin{aligned}
|\psi(1)\rangle &= \frac{1}{\sqrt{|a|^2+|b|^2}}\Big(\big[a\cos\theta|\uparrow\rangle + b\sin\theta|\uparrow\rangle\big]\otimes|n_0+1\rangle \\
&\quad + \big[-a\sin\theta|\downarrow\rangle + b\cos\theta|\downarrow\rangle\big]\otimes|n_0-1\rangle\Big), \\
|\psi(2)\rangle &= \frac{1}{\sqrt{|a|^2+|b|^2}}\Big(\big[a\cos^2\theta|\uparrow\rangle + b\cos\theta\sin\theta|\uparrow\rangle\big]\otimes|n_0+2\rangle \\
&\quad + \big[-b\sin^2\theta|\downarrow\rangle - a\cos\theta\sin\theta|\downarrow\rangle\big]\otimes|n_0\rangle \\
&\quad + \big[-a\sin^2\theta|\uparrow\rangle + b\cos\theta\sin\theta|\uparrow\rangle\big]\otimes|n_0\rangle \\
&\quad + \big[-a\cos\theta\sin\theta|\downarrow\rangle + b\cos^2\theta|\downarrow\rangle\big]\otimes|n_0-2\rangle\Big).
\end{aligned} \tag{10.9}$$

Evidently, the state at time $t$ has the general structure

$$|\psi(t)\rangle = |\uparrow\rangle \otimes \sum_n \Psi_{\rm u}(n,t)|n\rangle + |\downarrow\rangle \otimes \sum_n \Psi_{\rm d}(n,t)|n\rangle, \tag{10.10}$$

where the sum extends over all the lattice sites. In Eq. (10.10), the quantities $\Psi_{\rm u}(n,t)$ and $\Psi_{\rm d}(n,t)$ denote the probability amplitude to find the walker on site $n$ at time $t$ with spin state up and down, respectively. Normalization of the state $|\psi(t)\rangle$ implies $\sum_n P_n(t) = 1\ \forall\, t$, where

$$P_n(t) \equiv |\Psi_{\rm u}(n,t)|^2 + |\Psi_{\rm d}(n,t)|^2 \tag{10.11}$$

denotes the site occupation probability for the walker to be on site $n$ at time $t$.

For $\theta = 0$ and $\theta = \pi$, we see from Eq. (10.9) that at any time step $t$, only the extreme possible sites to the left and to the right are going to be occupied. For $\theta = \pi/2$ and $\theta = 3\pi/2$, one may infer from Eq. (10.9) that the walker is going to occupy sites $(n_0 \pm 1)$ at odd time steps and only the initial site $n_0$ at even time steps. These scenarios occur due to the fact that the operator $C$ effectively implements no mixing of the spin states $|\uparrow\rangle$ and $|\downarrow\rangle$ for $\theta = 0$ and $\pi$; on the other hand, it generates a complete mixing implementing the transformations $|\uparrow\rangle \to -|\downarrow\rangle$ and $|\downarrow\rangle \to |\uparrow\rangle$ for $\theta = \pi/2$, and the transformations $|\uparrow\rangle \to |\downarrow\rangle$ and $|\downarrow\rangle \to -|\uparrow\rangle$ for $\theta = 3\pi/2$. Other values of $\theta$ allow partial mixing, yielding non-trivial results in which several sites are occupied at any time instant, see later.

### 10.2.1 *Site Occupation Probability*

We now proceed to compute analytically the site occupation probability $P_n(t)$ at time $t$ for the initial state (10.8). One may see that acting on the state (10.10) by $C$ followed by $U$ yields linear equations expressing $\Psi_{\rm u,d}(n, t+1)$ in terms of $\Psi_{\rm u,d}(n-1, t)$ and $\Psi_{\rm u,d}(n+1, t)$. Exploiting this fact, we may define a two-component vector [29]

$$|\Psi(n,t)\rangle \equiv \begin{bmatrix} \Psi_{\rm u}(n,t) \\ \Psi_{\rm d}(n,t) \end{bmatrix}, \tag{10.12}$$

which satisfies the matrix equation

$$|\Psi(n,t+1)\rangle = \begin{bmatrix} \cos\theta & \sin\theta \\ 0 & 0 \end{bmatrix} |\Psi(n-1,t)\rangle + \begin{bmatrix} 0 & 0 \\ -\sin\theta & \cos\theta \end{bmatrix} |\Psi(n+1,t)\rangle\,. \tag{10.13}$$

In order to solve Eq. (10.13), one uses the method of discrete Fourier transform [29] and writes $|\tilde{\Psi}(k,t)\rangle = \sum_n |\Psi(n,t)\rangle \exp(\mathrm{i}2\pi kn/N)$ and $|\Psi(n,t)\rangle = (1/N)\sum_k |\tilde{\Psi}(k,t)\rangle \exp(-\mathrm{i}2\pi kn/N)$, with $n, k \in [-(N-1)/2, (N-1)/2]$ if $N$ is odd, and $n, k \in [-N/2, N/2-1]$ if $N$ is even. Equation (10.13) then yields

$$|\tilde{\Psi}(k,t)\rangle = (M_k)^t\, |\tilde{\Psi}(k,0)\rangle\,, \tag{10.14}$$

where one has

$$M_k \equiv \begin{bmatrix} \mathrm{e}^{\mathrm{i}2\pi k/N}\cos\theta & \mathrm{e}^{\mathrm{i}2\pi k/N}\sin\theta \\ -\mathrm{e}^{-\mathrm{i}2\pi k/N}\sin\theta & \mathrm{e}^{-\mathrm{i}2\pi k/N}\cos\theta \end{bmatrix}. \tag{10.15}$$

The matrix $M_k$, when expressed in terms of its eigenvalues and orthonormal eigenvectors, is given by

$$M_k = \lambda_k^{(1)}\, |\phi_k^{(1)}\rangle\, \langle\phi_k^{(1)}| + \lambda_k^{(2)}\, |\phi_k^{(2)}\rangle\, \langle\phi_k^{(2)}|\,, \tag{10.16}$$

where we have

$$\begin{aligned} \lambda_k^{(1)} &= \mathrm{e}^{\mathrm{i}\omega_k}, \quad |\phi_k^{(1)}\rangle = \frac{1}{\sqrt{1+h_+^2(k)}} \begin{bmatrix} -\mathrm{i}\mathrm{e}^{\mathrm{i}2\pi k/N} h_+(k) \\ 1 \end{bmatrix}, \\ \lambda_k^{(2)} &= \mathrm{e}^{-\mathrm{i}\omega_k}, \quad |\phi_k^{(2)}\rangle = \frac{1}{\sqrt{1+h_-^2(k)}} \begin{bmatrix} -\mathrm{i}\mathrm{e}^{\mathrm{i}2\pi k/N} h_-(k) \\ 1 \end{bmatrix}, \end{aligned} \tag{10.17}$$

with $\cos\omega_k \equiv \cos(2\pi k/N)\cos\theta$, and $h_\pm(k) \equiv \cot\theta\sin(2\pi k/N) \pm \csc\theta\sin\omega_k$. Note that here we have $h_+(-k) = -h_-(k)$, and $h_+(k)h_-(k) = -1$. The initial condition corresponding to Eq. (10.8) is given by

$$|\tilde{\Psi}(k,0)\rangle = \frac{\mathrm{e}^{\mathrm{i}2\pi kn_0/N}}{\sqrt{|a|^2+|b|^2}} \begin{bmatrix} a \\ b \end{bmatrix} \quad \forall\, k\,. \tag{10.18}$$

Using these results, one then obtains $|\tilde{\Psi}(k,t)\rangle = (\lambda_k^{(1)})^t \langle\phi_k^{(1)}|\tilde{\Psi}(k,0)\rangle\,|\phi_k^{(1)}\rangle + (\lambda_k^{(2)})^t \langle\phi_k^{(2)}|\tilde{\Psi}(k,0)\rangle\,|\phi_k^{(2)}\rangle$. Finally, performing inverse Fourier transformation, we get for odd $N$ the result

$$\Psi_{\mathrm{u}}(n,t) = \sum_{k=-(N-1)/2}^{(N-1)/2} \mathcal{A}_{\mathrm{u}}(k,n,t)\,, \quad \Psi_{\mathrm{d}}(n,t) = \sum_{k=-(N-1)/2}^{(N-1)/2} \mathcal{A}_{\mathrm{d}}(k,n,t)\,, \tag{10.19}$$

while for even $N$, we obtain

$$\begin{aligned}\Psi_{\mathrm{u}}(n,t) &= (-1)^{n-n_0+t}\,\frac{\{a\cos(\theta t) + b\sin(\theta t)\}}{N\sqrt{|a|^2+|b|^2}} + \sum_{k=-N/2+1}^{N/2-1} \mathcal{A}_{\mathrm{u}}(k,n,t)\,,\\ \Psi_{\mathrm{d}}(n,t) &= (-1)^{n-n_0+t}\,\frac{\{-a\sin(\theta t) + b\cos(\theta t)\}}{N\sqrt{|a|^2+|b|^2}} + \sum_{k=-N/2+1}^{N/2-1} \mathcal{A}_{\mathrm{d}}(k,n,t)\,.\end{aligned} \tag{10.20}$$

The quantities $\mathcal{A}_{\mathrm{u}}(k,n,t)$ and $\mathcal{A}_{\mathrm{d}}(k,n,t)$ in Eqs. (10.19) and (10.20) are given by

$$\begin{aligned}\mathcal{A}_{\mathrm{u}}(k,n,t) &= \mathcal{N}(a,b,k)\big[a\cos\{2\pi k(n-n_0)/N + \omega_k t\}\\ &\qquad + bh_+(k)\sin\{2\pi k(n-n_0-1)/N + \omega_k t\}\big],\\ \mathcal{A}_{\mathrm{d}}(k,n,t) &= \mathcal{N}(a,b,k)\big[-ah_+(k)\sin\{-2\pi k(n-n_0+1)/N + \omega_k t\}\\ &\qquad + b\cos\{-2\pi k(n-n_0)/N + \omega_k t\}\big],\end{aligned} \tag{10.21}$$

with $\mathcal{N}(a,b,k) = 2/\left[N(1+h_+^2(k))\sqrt{|a|^2+|b|^2}\right]$. Using the above analysis and Eq. (10.11), one may study numerically the behavior of the site occupation probability in various cases of interest. Table 10.1 lists the possibilities for nonzero $P_n(t)$ for different choices of the initial location of the walker and for $N$ even and odd.

**Table 10.1 Site occupation probability $P(n,t)$ in the case of the QRW on a one-dimensional periodic lattice with $N$ sites.** Here, the initial state is given by Eq. (10.8)

| $N$ even | | $N$ odd | | | |
|---|---|---|---|---|---|
| Initial site even | Initial site odd | Initial site even | | Initial site odd | |
| For all $t$ | For all $t$ | $t<N$ | $t>N$ | $t<N$ | $t>N$ |
| **For even** $t$: nonzero $P(n,t)$ only for even $n$. **For odd** $t$: nonzero $P(n,t)$ only for odd $n$ | **For even** $t$: nonzero $P(n,t)$ only for odd $n$. **For odd** $t$: nonzero $P(n,t)$ only for even $n$ | **For even** $t$: nonzero $P(n,t)$ only for even $n$. **For odd** $t$: nonzero $P(n,t)$ only for odd $n$ | Nonzero $P(n,t)$ possible for all $n$ and $t$. | **For even** $t$: nonzero $P(n,t)$ only for odd $n$. **For odd** $t$: nonzero $P(n,t)$ only for even $n$ | Nonzero $P(n,t)$ possible for all $n$ and $t$. |

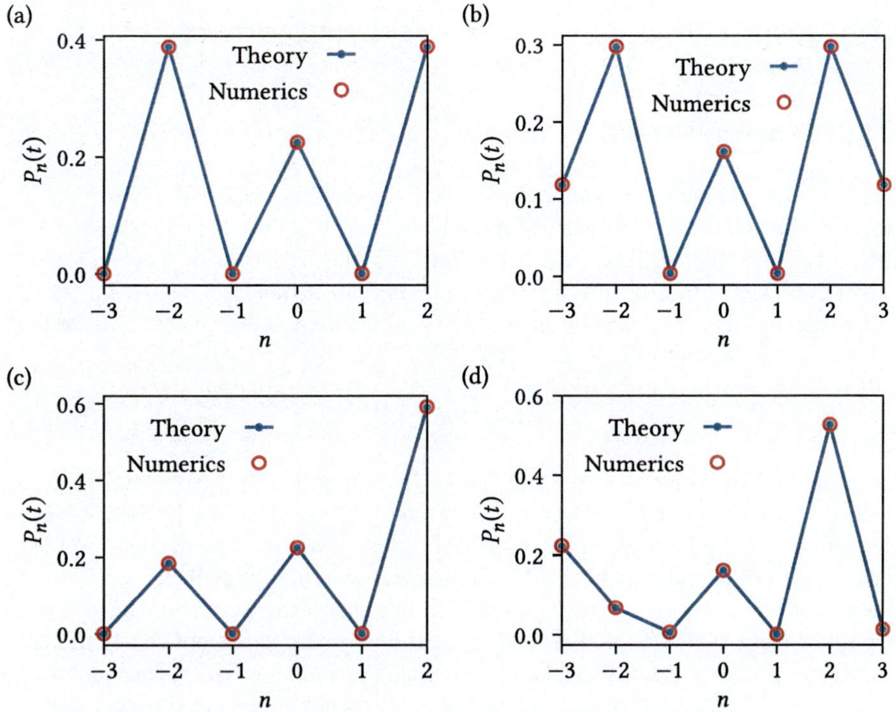


**Fig. 10.3 Site occupation probability $P_n(t)$ for the QRW on a one-dimensional periodic lattice.** Here, $N$ is the number of sites. The site occupation probability is plotted at time $t = 20$, while starting from the state (10.8) with $n_0 = 0$. The values of the parameters $a$ and $b$ defining the initial state are as follows: $a = 1$, $b = \mathrm{i}$ for panels **a** and **b**, and $a = b = 1$ for panels **c** and **d**. The value of $N$ is $N = 6$ for panels **a** and **c**, and $N = 7$ for panels **b** and **d**. The angle $\theta$ has a value in radian corresponding to 80°. Numerics correspond to bare evolution of the unitary dynamics of the QRW model, while analytical results are given by Eqs. (10.11), (10.19) and (10.20). The lines serve as a guide to the eye. The figure is taken from Ref. [1]

Note that the transformation $\theta \to \theta + \pi$ results in multiplying the operator $C$ and, consequently, the matrix $M_k$ by the factor of $-1$. Accordingly, the eigenvalues of $M_k$ get both multiplied by the factor of $-1$, although the corresponding eigenvectors remain unchanged. All of these however leave $|\Psi_\mathrm{u}(n,t)|^2$ and $|\Psi_\mathrm{d}(n,t)|^2$ and, consequently, the site occupation probability $P_n(t)$ unchanged. We thus conclude that the QRW is invariant under the transformation $\theta \to \theta + \pi$ and therefore restrict the values of $\theta$ to the range $[0, \pi]$. Figure. 10.3 shows a perfect match between the theoretical and numerical results for the site occupation probability $P_n(t)$.

## *10.2.2 Random Projective Measurements and Survival Probability*

### 10.2.2.1 Numerical Results

We present here numerical results on the survival probability $S_m$, and in particular, on the average and the typical survival probability, and also their dependence on the number of measurements $m$. We note in the context of the discussion on measurement schemes in Sect. 10.1 that the time intervals $\tau_\alpha$ for the case at hand are positive integers with underlying discrete distribution $p_\tau$.[3] Also, since the QRW evolves in discrete times, the unitary operator $U_\alpha$ is given by

$$U_\alpha \equiv [U(C \otimes \mathbb{I})]^{\tau_\alpha} . \tag{10.22}$$

Then, a single-step evolution of the QRW involves acting first by the operator $C \otimes \mathbb{I}$ and then by the operator $U$ (see Eqs. (10.6) and (10.7)). Since the QRW evolves in discrete times, the phrases "time" and "time step" have the same meaning in the context of the QRW and may be used interchangeably in the following.

As a representative case, we take the QRW lattice size $N$ to be even with $n_0 = 0$. It follows then from Table 10.1 that the site occupation probability for our choice of the initial state is nonzero on site 0 provided we consider even time steps. Consequently, we need to choose the i.i.d. random variables $\tau_\alpha$ as even numbers, as otherwise any projective measurement on the instantaneous state yields null result. For our purpose, we take two representative choices for the $\tau_\alpha$ distribution $p_\tau$: a discrete exponential distribution

$$\begin{aligned} &p_\tau = r(1-r)^{\tau/2-1}; \quad \tau = 2, 4, 6, \ldots, \quad 0 < r < 1; \\ &\langle \tau \rangle = \frac{2}{r}, \quad \mathrm{Var}[\tau] = \frac{4(1-r)}{r^2}, \end{aligned} \tag{10.23}$$

and a power-law distribution

$$\begin{aligned} &p_\tau = \frac{2^s}{\zeta(s)\tau^s}; \quad \tau = 2, 4, 6, \ldots, \quad s > 1; \\ &\langle \tau \rangle = \frac{2\zeta(s-1)}{\zeta(s)}, \ s > 2; \quad \mathrm{Var}[\tau] = \frac{4\left[\zeta(s)\zeta(s-2) - \zeta^2(s-1)\right]}{\zeta^2(s)}, \ s > 3; \end{aligned} \tag{10.24}$$

where $\zeta(s) \equiv \sum_{n=1}^{\infty} 1/n^s;\ s > 1$ is the Riemann zeta function. Here, $\langle \tau \rangle$ is the mean of the distribution $p_\tau$, while $\mathrm{Var}[\tau]$ is its variance. The aforementioned distributions satisfy the normalization $\sum_{\tau=2,4,6,\ldots}^{\infty} p_\tau = 1$. The exponential distribution (10.23) has all its moments, (and in particular, the mean) finite for all values of the parameter $r$. On the other hand, the power-law distribution (10.24) has a finite mean only for $s > 2$

[3] We will use the notation $p(\tau)$ (respectively, $p_\tau$) to denote the distribution for the case when $\tau$ is a continuous (respectively, a discrete) random variable.

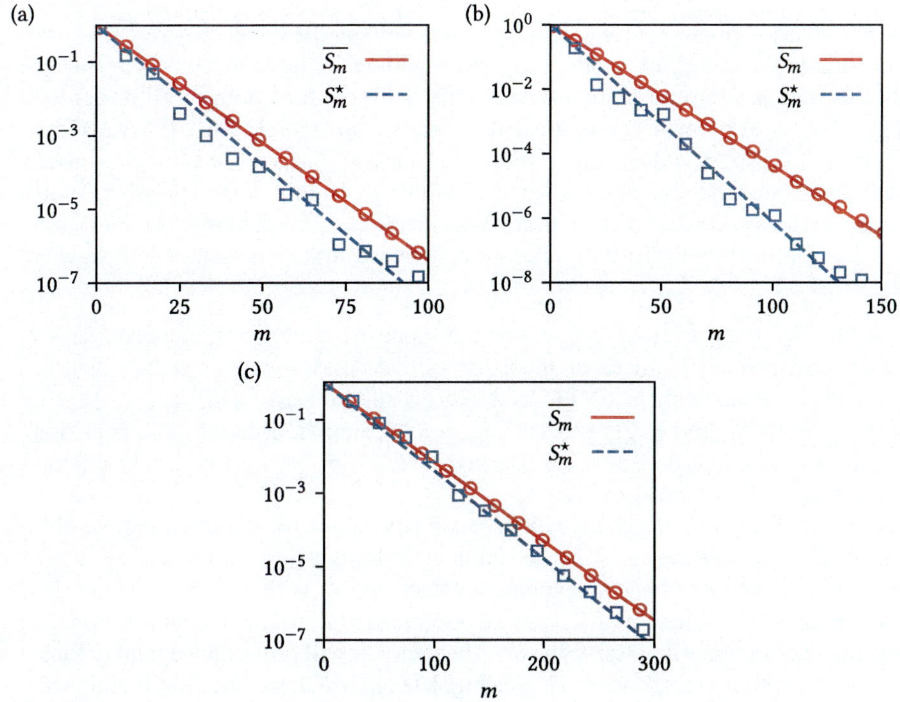


**Fig. 10.4 Average and typical survival probability for the QRW subject to instantaneous projective measurements at random times (Scheme 1).** The plots correspond to the initial state (10.8) with $n_0 = 0,\ a = 1,\ b = \mathrm{i}$, which is subject to repeated projective measurements at random times to the initial state and subsequent evolution with the projected component of the instantaneous state. Here, the time intervals $\tau_\alpha$ between two consecutive measurements are taken to be i.i.d. random variables sampled from the exponential distribution (10.23) with $r = 0.5$ (panel **a**) and from the power-law distribution (10.24) with $s = 2.5$ (panel **b**) and $s = 3.5$ (panel **c**). The angle $\theta$ characterizing the QRW evolution operator $C$ has the value in radian that corresponds to 80 degrees. The system size is $N = 500$. The points in the plots are based on results obtained from numerical implementation of the dynamics; the average survival probability $\overline{S_m}$ involves averaging over 3000 realizations of the set $\{\tau_\alpha\}_{1\le\alpha\le m}$, while the typical survival probability $S_m^\star$ corresponds to results obtained in a typical realization of the $\tau_\alpha$'s. The lines in the plots denote analytical results given by Eqs. (10.33), (10.34), and (10.32). The figure is taken from Ref. [1]

and a finite variance only when one has $s > 3$. Since any reasonable experimental set-up would necessarily allow measurements being performed at random time intervals that have a finite average, we will consider only values of $s$ that are larger than 2.

Figure 10.4 shows in case of the evolution under **Scheme 1** our numerical results on the average survival probability $\overline{S_m}$ averaged over typically hundreds of realizations $\{\tau_\alpha\}_{1\le\alpha\le m}$ and the survival probability $S_m^\star$ for a typical realization of the $\tau_\alpha$'s, with both plotted as a function of $m$, the number of measurements. In the figure, panel (a) has $\tau_\alpha$'s distributed according to the exponential distribution (10.23) with $r = 0.5$, while panels (b) and (c) stand for the power-law distribution (10.24) with

$s$ having values 2.5 and 3.5, respectively. The results in the panels suggest for both the quantities $\overline{S_m}$ and $S_m^\star$ an exponential decay with $m$ for large $m$.

Let us now present our results for **Scheme 2**. To contrast with those presented in Fig. 10.4 for **Scheme 1**, we will use the same initial state. In Fig. 10.5 we show our numerical results on the average survival probability $\overline{S_m}$ and the typical survival probability $S_m^\star$ for the exponential distribution (10.23) with $r = 0.5$ (panels (a) and (d)), and for the power-law distribution (10.24) with $s = 2.5$ (panels (b) and (e)) and $s = 3.5$ (panels (c) and (f)). Based on our results, we first summarize the behavior of $\overline{S_m}$.

1. We find that there are two characteristic $N$-dependent $m$ values, namely, $m_1^\star(N)$ and $m_2^\star(N)$, which characterize the behavior of $\overline{S_m}$. While $m_1^\star(N)$ scales linearly with $N$, the other scale $m_2^\star(N)$ is seen to grow superlinearly with $N$.
2. For $m < m_1^\star(N)$ and large, we have $\overline{S_m} \sim m^{-2}$, implying thereby the behavior of the average first-detection probability as $\overline{F_m} \sim m^{-3}$ (see Eq. (10.1) for the definition of $F_m$ in terms of $S_m$).
3. For $m$ satisfying $m_1^\star(N) \ll m < m_2^\star(N)$, we have $\overline{S_m} \sim m^{-3/2}$, implying thereby the behavior of the average first-detection probability as $\overline{F_m} \sim m^{-5/2}$.
4. For $m \gg m_2^\star(N)$, we have an exponential decay of $\overline{S_m}$ with $m$.
5. While the $m^{-3/2}$-behavior of $\overline{S_m}$ may be observed in Fig. 10.5, in the insets of panels (a) – (c), in order to observe the $m^{-2}$-behavior, it will prove convenient to look at the plot of the corresponding $\overline{F_m}$, see Fig. 10.6, in which the two aforementioned distinct behaviors of $m^{-3}$ and $m^{-5/2}$ may be clearly seen. The exponential decay of $\overline{S_m}$ with $m$ for $m \gg m_2^\star(N)$ may be observed from Fig. 10.7. Note that since $m_2^\star(N)$ grows superlinearly with $N$, in order to observe this exponential decay for the value of $N$ used in Fig. 10.5, one is required to obtain results for very large values of $m$, thereby requiring computation for a prohibitively-long time. Hence, we use in Fig. 10.7 a value of $N$ smaller than the one used in Fig. 10.5.
6. That the characteristic scale $m_1^\star(N)$ scales linearly with $N$ can be deduced from Fig. 10.8, which implies that

$$m_1^\star(N) \sim \frac{N}{\langle\tau\rangle}, \tag{10.25}$$

   where $\langle\tau\rangle$ is the average of the distribution $p_\tau$.
7. The superlinear scaling of $m_2^\star(N)$ with $N$ can be deduced from the insets of Fig. 10.7. We find:

$$m_2^\star(N) \sim N^\delta; \quad \delta = 3.0. \tag{10.26}$$

In the above backdrop, we mention here that if one studies an infinite system (i.e., the limit $N \to \infty$), the scale $m_1^\star(N)$ diverges, and hence, one would observe the large-$m$ behavior $\overline{S_m} \sim m^{-2}$ and $\overline{F_m} \sim m^{-3}$.

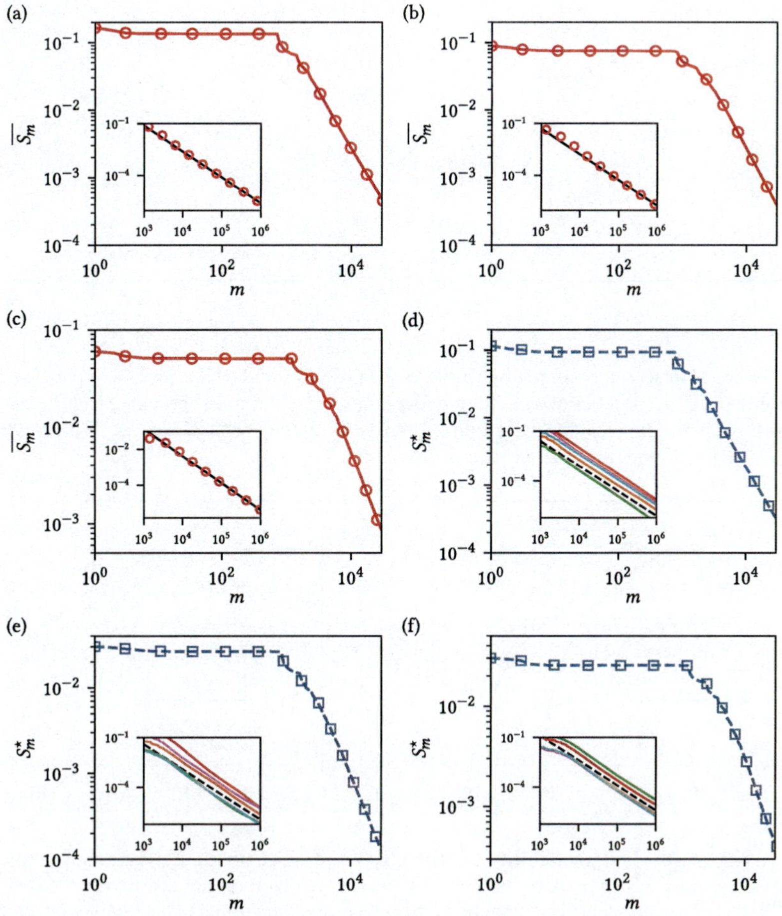

**Fig. 10.5 Average survival probability (panels a–c) and typical survival probability (panels d–f) for the QRW subject to instantaneous projective measurements at random times (Scheme 2).** The plots correspond to the initial state (10.8) with $n_0 = 0,\ a = 1,\ b = \mathrm{i}$, which is subject to repeated projective measurements at random times to the initial state and subsequent evolution with the leftover component of the instantaneous state after the measurement. Here, the time intervals $\tau_\alpha$ between two consecutive measurements are taken to be i.i.d. random variables sampled from the exponential distribution (10.23) with $r = 0.5$ (panels **a** and **d**), and from the power-law distribution (10.24) with $s = 2.5$ (panels **b** and **e**) and $s = 3.5$ (panels **c** and **f**). The angle $\theta$ characterizing the QRW evolution operator $C$ has the value in radian that corresponds to 80 degrees. The system size is $N = 500$. In the main plots, the points denote results obtained from numerical implementation of the dynamics; the average survival probability $\overline{S_m}$ involves averaging over 10 realizations (panel **a**), over 20 realizations (panel **b**), and over 10 realizations (panel **c**) of the set $\{\tau_\alpha\}_{1\le\alpha\le m}$, while the typical survival probability $S_m^\star$ corresponds to results obtained in a typical realization of the $\tau_\alpha$'s. The lines in the main plots are based on results obtained from the semi-analytical approach described in the text, see Sect. 10.2.2.2. In the insets in the upper row, while the points correspond to numerically-evaluated average survival probability, the line represents an $m^{-3/2}$ behavior. In the insets in the lower row, the continuous lines stand for numerically-evaluated survival probability for five typical realizations of $\{\tau_\alpha\}_{1\le\alpha\le m}$, while the dashed line represents an $m^{-3/2}$ behavior. It may be concluded from the insets in both the upper and the lower row that the average as well as the typical survival probability behaves at large $m$ as $m^{-3/2}$. The figure is taken from Ref. [1]

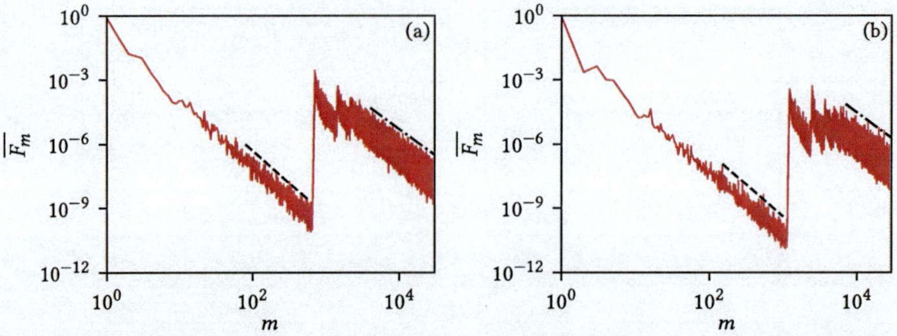


**Fig. 10.6 Average first-detection probability for the QRW model (Scheme 2).** For panels **a** and **b**, parameter values and number of averaging realizations are taken to be the same as in Fig. 10.5a and c, respectively. As shown in the plots, one observes two distinct behaviors $\sim m^{-3}$ (dashed line) and $\sim m^{-5/2}$ (dash-dotted line). The figure is taken from Ref. [1]

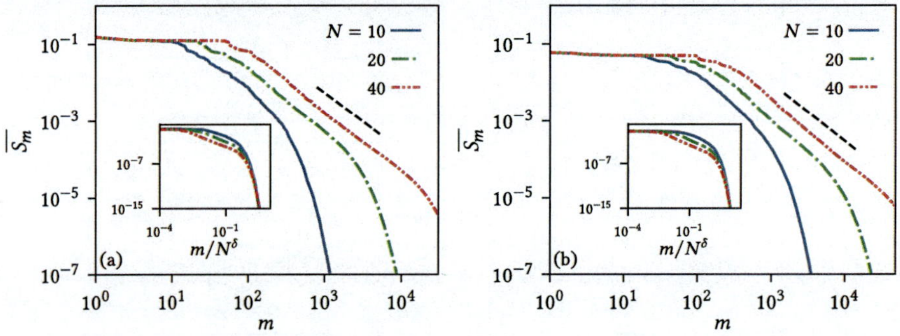


**Fig. 10.7 Average survival probability for the QRW model (Scheme 2).** For panels **a** and **b**, parameter values other than $N$ and the number of averaging realizations are taken to be the same as in Fig. 10.5a and c, respectively. The main plots depict a crossover from a $m^{-3/2}$-behavior (dashed line) to an exponential tail over the characteristic value $m_2^\star(N)$ of $m$; the collapse of the data shown in the insets is suggestive of the scaling (10.26). The figure is taken from Ref. [1]

One may wonder about how robust all of the aforementioned behavior is with respect to distributions $p_\tau$ other than the exponential distribution (10.23) and the power-law distribution (10.24) that we have considered until now. To address this issue, let us demonstrate that the same behavior is also observed for two other distributions, namely, the case corresponding to measurements performed at regular time interval $\tau_0$:

$$p_\tau = \delta_{\tau,\tau_0}; \quad \tau = 2, 4, 6, \ldots; \quad \tau_0 > 0; \quad \langle \tau \rangle = \tau_0, \quad \mathrm{Var}[\tau] = 0, \qquad (10.27)$$

and the Poisson distribution

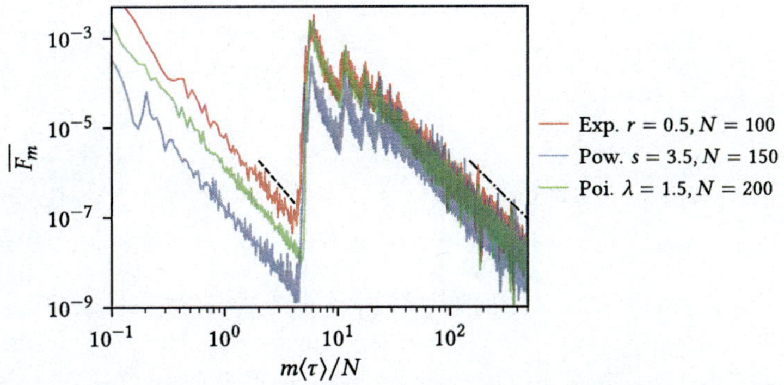


**Fig. 10.8 Average first-detection probability for the QRW model (Scheme 2) for the exponential distribution (10.23), the power-law distribution (10.24), and the Poisson distribution (10.28).** Here, the parameters $n_0$, $a$, $b$, and $\theta$ taken to have the same values as in Fig. 10.5. The number of averaging realizations in all cases is 50. The parameter $r$ for the exponential distribution and the parameter $s$ for the power-law distribution have the same values as the ones in Fig. 10.5, panels **a** and **c**, respectively. For the data plotted for the Poisson distribution (10.28), we have considered $\lambda = 1.5$. The plot suggests the scaling given by Eq. (10.25). The figure is taken from Ref. [1]

$$p_\tau = \frac{e^{-\lambda}\,\lambda^{\tau/2-1}}{(\tau/2-1)!}; \quad \tau = 2, 4, 6, \ldots; \quad \lambda > 0; \quad \langle\tau\rangle = 2(1+\lambda), \quad \mathrm{Var}[\tau] = 4\lambda\,. \tag{10.28}$$

In the latter, the parameter $\lambda$ may be seen to control the width of the distribution $p_\tau$, leading to a small (respectively, large) variance for small (respectively, large) $\lambda$. Results for the case of the delta-function distribution (10.27) are shown in Figs. 10.9 and 10.10. It may be observed from these figures and in appropriate regimes the $m^{-3/2}$ behavior of $\overline{S_m}$, the $m^{-3}$ behavior and the $m^{-5/2}$ behavior of $\overline{F_m}$, and the scaling of the crossover scale $m_1^\star(N)$ according to Eq. (10.25). Similar results are obtained for the case of the Poisson distribution (10.28), hence we do not present all the results here, excepting to show in Fig. 10.8 the scaling of $m_1^\star(N)$ following Eq. (10.25).

Thus far we have discussed the behavior of the average quantities, but even the typical survival probability $S_m^\star$ and the typical first-detection probability $F_m^\star$ may be shown to behave in a manner similar to their respective averages. Some representative plots are presented in Fig. 10.5, panels (d) – (f), and in Fig. 10.11.

Before closing this part, the last issue that we plan to discuss is the behavior of $\overline{S_m}$ and $\overline{F_m}$ for small $m \ll m_1^\star(N)$. These quantities show oscillations in such a regime, with the oscillations becoming more pronounced as the distribution $p_\tau$ becomes more narrow. In order to show a representative result, we show in Fig. 10.12 the quantity $\overline{F_m}$ versus $m$ for the case of the Poisson distribution (10.28) as one varies the parameter $\lambda$ that controls the narrowness of the distribution. It is evidently true from the plot that with decrease of $\lambda$, oscillations in the behavior of $\overline{F_m}$ become more pronounced.

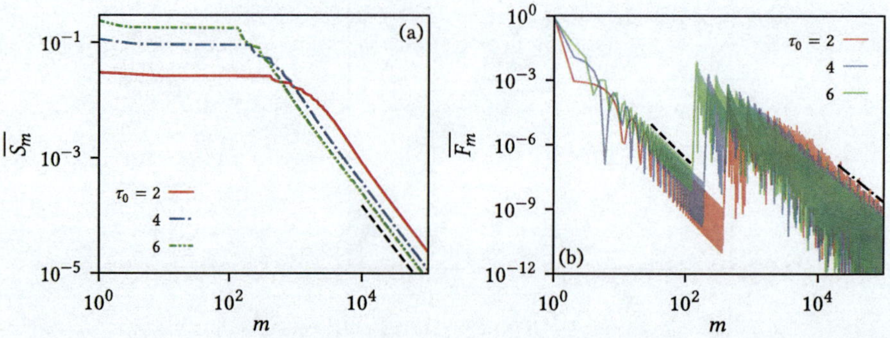


**Fig. 10.9 Average survival and first-detection probability for the QRW model (Scheme 2).** Here, the parameters $n_0$, $a$, $b$, and $\theta$ are taken to have the same values as in Fig. 10.5. The time intervals $\tau_\alpha$ between two consecutive measurements are taken to be i.i.d. random variables sampled from the distribution (10.27). The values of the parameter $\tau_0$ are shown in the figure. The system size is $N = 150$. In panel **a**, as shown in the plot, one observes a behavior $\sim m^{-3/2}$ (dashed line), while in panel **b**, two distinct behaviors $\sim m^{-3}$ (dashed line) and $\sim m^{-5/2}$ (dash-dotted line) may be observed. Note that since $\tau$ has only one allowed value, $\tau_0$, it is redundant to use the overbar over $S_m$ and $F_m$ to denote their average values. The figure is taken from Ref. [1]

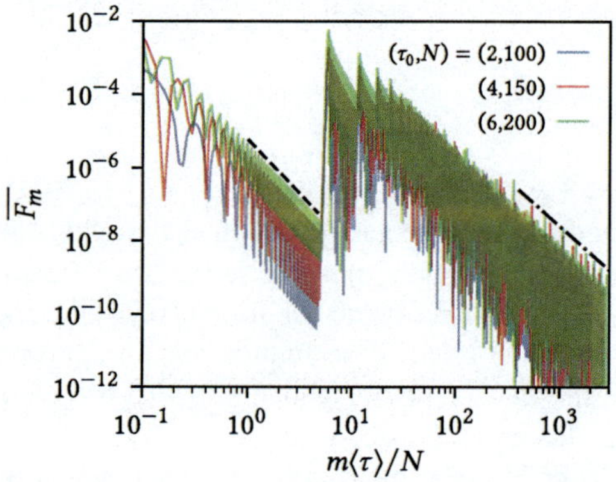


**Fig. 10.10 Average first-detection probability for the QRW model (Scheme 2).** Parameter values and other details are the same as those in Fig. 10.9. The plot suggests crossover from the $\sim m^{-3}$-behavior (dashed line) to the $\sim m^{-5/2}$-behavior (dash-dotted line) over the characteristic value $m_1^\star(N)$ of $m$ that follows the scaling (10.25). Note that since $\tau$ has only one allowed value, $\tau_0$, it is redundant to use the overbar over $F_m$ to denote its average. The figure is taken from Ref. [1]

This feature is seen to be generic across the different $p_\tau$'s that we considered (of course, for the delta-function distribution (10.27), the oscillations appear to be most striking).

We thus see a stark contrast between the obtained results on the survival probability under the two choices of the dynamics. It is our objective in the following to provide an analytical treatment of these results.

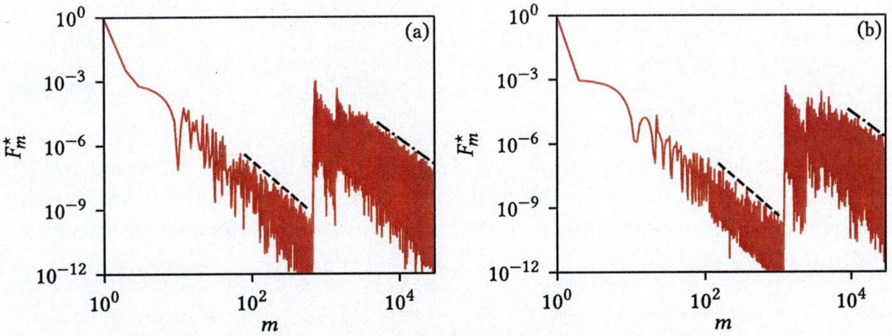


**Fig. 10.11 Typical first-detection probability for the QRW model (Scheme 2).** For panels **a** and **b**, parameter values are the same as those in Fig. 10.5a and c, respectively. As shown in the plots, one observes two distinct behaviors $\sim m^{-3}$ (dashed line) and $\sim m^{-5/2}$ (dash-dotted line). The figure is taken from Ref. [1]

**Fig. 10.12 Average first-detection probability for the QRW model (Scheme 2) for the Poisson distribution (10.28) with different values of $\lambda$.** Here, the parameters $n_0$, $a$, $b$, and $\theta$ have the same values as those in Fig. 10.5. The system size is $N = 150$, and the number of averaging realizations is 1000. The figure is taken from Ref. [1]

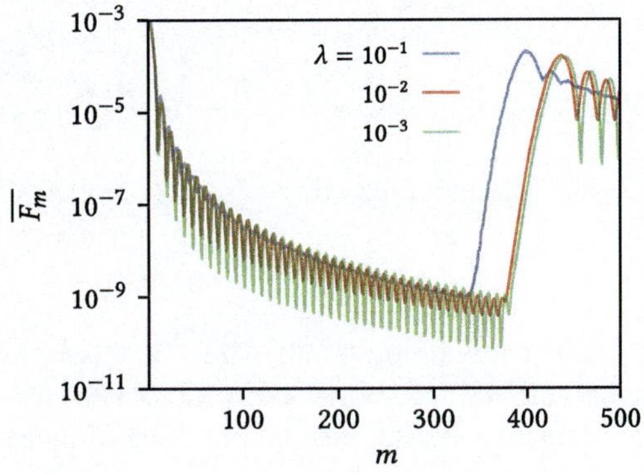


#### 10.2.2.2 Analytical Results

In **Scheme 1**, we obtain $|\psi_m^{(a)}\rangle$ for arbitrary $m$ as

$$|\psi_m^{(a)}\rangle = PU_m \dots PU_3PU_2PU_1 \, |\psi(0)\rangle \, . \tag{10.29}$$

From Eqs. (10.3), (10.2), and (10.29), we observe that $S_m$ may be expressed as

$$S_m = \prod_{\alpha=1}^{m} q(\tau_\alpha), \tag{10.30}$$

where the (quantum) probability $q(\tau_\alpha)$ is given by

$$q(\tau_\alpha) \equiv |\langle \psi(0)|U_\alpha|\psi(0)\rangle|^2 \tag{10.31}$$

$$= \frac{1}{|a|^2 + |b|^2} \left| a^* \Psi_{\mathrm{u}}(n_0, \tau_\alpha) + b^* \Psi_{\mathrm{d}}(n_0, \tau_\alpha) \right|^2, \tag{10.32}$$

with $\Psi_{\mathrm{u}}(n_0, \tau_\alpha)$ and $\Psi_{\mathrm{d}}(n_0, \tau_\alpha)$ given by Eqs. (10.19) and (10.20). We note that $q(\tau)$ is nothing but the probability for the walker to be found in the initial state $|\psi(0)\rangle$ after evolution for time $\tau$. Using Eq. (10.30), one may obtain the average survival probability as

$$\overline{S_m} = \prod_{i=1}^{m} \left( \sum_\tau p_\tau \, q(\tau) \right) = \exp\left( m \log \sum_\tau p_\tau \, q(\tau) \right). \tag{10.33}$$

As shown in Appendix K using the LDT, the most probable value of the survival probability $S_m^\star$ is given by

$$S_m^\star = \exp\left( m \sum_\tau p_\tau \, \log q(\tau) \right). \tag{10.34}$$

Using the Jensen's inequality $\overline{\exp(x)} \geq \exp(\overline{x})$, we arrive at

$$\overline{S_m} \geq S_m^\star, \tag{10.35}$$

with the equality holding only when there is no randomness in $\tau$, that is, when $\tau$ can take on only a single value: $p_\tau = \delta_{\tau,\tau_0}$, with $\tau_0 > 0$. On performing a large number $m$ of projective measurements, the value of the survival probability to remain in the initial state that one measures in a single experimental run equals $S_m^\star$ in the limit $m \to \infty$. On the other hand, averaging the survival probability over a large (ideally infinite) number of experimental runs yields the value $\overline{S_m}$.

In Fig. 10.4, we compare the theoretical results against those obtained in numerical implementation of the **Scheme 1** dynamics. In this figure, the continuous lines in red obtained from Eq. (10.33) depict the behavior of $\overline{S_m}$, while the blue-dashed lines obtained from Eq. (10.34) stand for the behavior of $S_m^\star$. We observe from the figure a very good match between theoretical and numerical results for the average survival probability, whereas the numerical results for the most probable survival probability fluctuate around the theoretical estimate, as one would expect.

In **Scheme 2**, we obtain $|\psi_m^{(\mathrm{a})}\rangle$ for arbitrary $m$ as

$$|\psi_m^{(\mathrm{a})}\rangle = \widetilde{P} U_m \ldots \widetilde{P} U_3 \widetilde{P} U_2 \widetilde{P} U_1 \, |\psi(0)\rangle. \tag{10.36}$$

In this scheme, to find an analytical closed form of $\overline{S_m}$ and $S_m^\star$ is non-trivial, as we explain below. As a result, we rely on a semi-analytical approach that involves implementation of the following four steps:

1. For a given choice of $a$, $b$, and $n_0$ that specifies the initial state $|\psi(0)\rangle$ in Eq. (10.8), we first obtain the vector $|\Psi(n, 0)\rangle$ by using Eqs. (10.10) and (10.12). Then, we implement the discrete Fourier transform given by $|\tilde{\Psi}(k, 0)\rangle = \sum_n |\Psi(n, 0)\rangle \, \mathrm{e}^{\mathrm{i}2\pi kn/N}$.
2. Subsequently, we use Eq. (10.14) to obtain $|\tilde{\Psi}(k, \tau_1)\rangle$ as the result of dynamical evolution for a random time $\tau_1$ sampled following either the exponential distribution (10.23) or the power-law distribution (10.24), and with $|\tilde{\Psi}(k, 0)\rangle$ as the initial condition. At the end of the evolution, one then has the set $\{|\tilde{\Psi}(k, \tau_1)\rangle\}$.
3. We next perform inverse discrete Fourier transform of the set $\{|\tilde{\Psi}(k, \tau_1)\rangle\}$ to obtain the set $\{|\Psi(n, \tau_1)\rangle\}$. In order to implement a projective measurement at the end of evolution for time $\tau_1$ and obtaining the corresponding leftover component of the state, we first obtain the state $|\psi(\tau_1)\rangle$ by using the obtained values of the elements of $\{|\Psi(n, \tau_1)\rangle\}$ in Eq. (10.10). We then compute the difference $|\psi(\tau_1)\rangle - |\psi(0)\rangle\langle\psi(0)|\psi(\tau_1)\rangle$, which yields the state $|\psi_1^{(a)}\rangle$.
4. Steps 1–3 are then applied in turn to the leftover component of the state corresponding to last projection; $m \geq 1$ number of repetitions would then generate the leftover component of the instantaneous state after $m$ projections, and this allows to obtain the survival probability $S_m$ for a given realization $\{\tau_\alpha\}_{1 \leq \alpha \leq m}$ of the dynamics.

The method is semi-analytical in that while the dynamical evolution in the Fourier space follows the exact solution (10.14), inverse Fourier transform to obtain $|\Psi(n, \tau)\rangle$ is done numerically. The semi-analytical results for the average and the typical survival probability are compared in Fig. 10.5 against numerical results; we observe a very good match.

Let us remark that in **Scheme 1**, each time a projective measurement is made, the dynamical evolution is seen to start afresh from a state that is just the initial state multiplied by a complex number (for example, the state for subsequent dynamical evolution after say the first measurement is $|\psi_1^{(a)}\rangle = \langle\psi(0)|\psi(\tau_1)\rangle|\psi(0)\rangle$). As a result, the survival probability $S_m$ is a product of the quantum probabilities $q(\tau_\alpha)$ over different times $\tau_\alpha$, see Eq. (10.30). This however is not the case in **Scheme 2**, and this defies the survival probability $S_m$ to be written as a product of probabilities for different $\tau_\alpha$'s. Consequently, an analytical estimate of the average and the typical survival probability for **Scheme 2** is not straightforward.

## 10.3 A Continuous-Time Dynamics: The Tight-Binding Model

We now discuss the tight-binding model (TBM) that involves quantum evolution of a particle on a one-dimensional lattice in continuous time. Similar to the case of the QRW, we consider here a periodic lattice of $N$ sites. The continuous-time dynamics in the TBM is generated by the Hamiltonian [6, 7]

$$H = -\gamma \sum_{j=0}^{N-1} (|j+1\rangle\langle j| + |j\rangle\langle j+1|)\,; \quad |N\rangle = |0\rangle, \tag{10.37}$$

where $\gamma > 0$ is a real parameter and the index $j$ refers to the lattice sites. Let us assume that the particle starts from site $n_0$ at time $t = 0$, i.e., one has $|\psi(0)\rangle = |n_0\rangle$. Then, according to the evolution equation, the state of the particle at time $t$ is given by $|\psi(t)\rangle = \exp(-\mathrm{i}Ht)|\psi(0)\rangle$, while the probability amplitude to find the particle on a site $n$ at time $t$ is given by $\psi_{n,n_0}(t) = \langle n|\psi(t)\rangle$, which satisfies the normalization $\sum_{n=0}^{N-1} |\psi_{n,n_0}(t)|^2 = 1\,\forall t$. For a small time interval $\Delta t$, one has the evolution equation $|\psi(t+\Delta t)\rangle = \exp(-\mathrm{i}H\Delta t)|\psi(t)\rangle$, from which we get

$$\psi_{n,n_0}(t+\Delta t) = \sum_{j=0}^{N-1} \langle n|\mathrm{e}^{-\mathrm{i}H\Delta t}|j\rangle\, \psi_{j,n_0}(t)\,, \tag{10.38}$$

where we have used the completeness relation $\sum_{j=0}^{N-1} |j\rangle\,\langle j| = \mathbb{I}$. In the limit $\Delta t \to 0$, expanding $\psi_{n,n_0}(t+\Delta t)$ in powers of $\Delta t$ and writing $\exp(-\mathrm{i}H\Delta t) \simeq \mathbb{I} - \mathrm{i}H\Delta t$, one obtains from Eq. (10.38) that

$$\frac{\partial \psi_{n,n_0}(t)}{\partial t} = -\mathrm{i}\sum_{j=0}^{N-1} \langle n|H|j\rangle\, \psi_{j,n_0}(t)\,. \tag{10.39}$$

From Eq. (10.37), we have $\langle j|H|k\rangle = -\gamma(\delta_{j,k-1} + \delta_{j,k+1})$, which when substituted in the above equation yields

$$\frac{\partial \psi_{n,n_0}(t)}{\partial t} = \mathrm{i}\gamma\, (\psi_{n-1,n_0}(t) + \psi_{n+1,n_0}(t))\,. \tag{10.40}$$

### 10.3.1 Site Occupation Probability

For the TBM model, the site occupation probability, i.e., the probability of finding the particle on site $n$ at time $t$ while starting from site $n_0$, is given by $P_n(t) = |\psi_{n,n_0}(t)|^2$. To obtain $P_n(t)$, one needs to solve Eq. (10.40) subject to the initial condition $\psi_{n,n_0}(0) = \delta_{n,n_0}$. To this end, we perform discrete Fourier transform of the set $\{\psi_{n,n_0}(t)\}_{0\le n\le N-1}$, given by the set $\{\widehat{\psi}_{q|n_0}(t)\}_{0\le q\le N-1}$, with $\widehat{\psi}_{q|n_0}(t) = \sum_{j=0}^{N-1} \psi_{j,n_0}(t)\exp(-\mathrm{i}2\pi jq/N)$; $\psi_{j,n_0}(t) = (1/N)\sum_{q=0}^{N-1} \widehat{\psi}_{q|n_0}(t)\exp(\mathrm{i}2\pi jq/N)$. Equation (10.40) may then be written as

$$\frac{\partial \widehat{\psi}_{q|n_0}(t)}{\partial t} = 2\mathrm{i}\gamma \cos\left(\frac{2\pi q}{N}\right) \widehat{\psi}_{q|n_0}(t)\,. \tag{10.41}$$

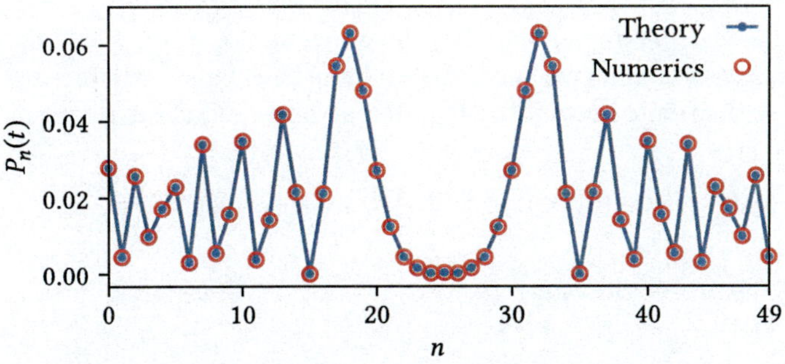


**Fig. 10.13 Site occupation probability $P_n(t)$ in the TBM on a one-dimensional periodic lattice.** Here, $N$ is the number of sites. The site occupation probability corresponds to time $t = 10$, while starting from initial site $n_0 = 0$. Here, we have $N = 50$, $\gamma = 1$. The numerical results correspond to bare evolution of the unitary dynamics of the model, while the analytical results are given by Eq. (10.43). The line serves as a guide to the eye. The figure is taken from Ref. [1]

Solving the above equation for the initial condition $\widehat{\psi}_{q|n_0}(0) = \exp(-\mathrm{i}2\pi n_0 q/N)$, we get $\widehat{\psi}_{q|n_0}(t) = \exp(\mathrm{i}2\gamma t \cos(2\pi q/N) - \mathrm{i}2\pi n_0 q/N)$, inverting which yields

$$\psi_{j,n_0}(t) = \frac{1}{N}\sum_{q=0}^{N-1} \mathrm{e}^{\mathrm{i}2\gamma t\cos(2\pi q/N)+\mathrm{i}2\pi q(j-n_0)/N}\,. \tag{10.42}$$

One may easily check that the solution in Eq. (10.42) satisfies the normalization $\sum_{j=0}^{N-1} |\psi_{j,n_0}(t)|^2 = 1$, as required. The site-occupation probability is then given by

$$P_n(t) = |\psi_{n,n_0}(t)|^2 = \left|\frac{1}{N}\sum_{q=0}^{N-1} \mathrm{e}^{\mathrm{i}2\gamma t\cos(2\pi q/N)+\mathrm{i}2\pi q(n-n_0)/N}\right|^2\,. \tag{10.43}$$

Figure 10.13 shows a perfect match between the theoretical and numerical results for the site occupation probability $P_n(t)$ in the TBM.

## 10.3.2 Random Projective Measurements and Survival Probability

### 10.3.2.1 Numerical Results

Let us now present numerical results on the average and the typical survival probability for the TBM that is subject to projective measurements at random times. Figure 10.1 shows a typical time evolution of the system. The initial state is $|\psi(0)\rangle = |n_0\rangle$, and we consider the projective measurement to be involving projection to the initial state $n_0$ implemented by the projection operator $P = |n_0\rangle\langle n_0|$. The random variable $\tau$ between two successive measurements is a continuous variable. We take $\tau$ to be distributed according to either an exponential distribution:

$$p(\tau) = re^{-r\tau}\,; \quad \tau \in [0, \infty),\ r > 0\,; \quad \langle\tau\rangle = \frac{1}{r}\,, \quad \mathrm{Var}[\tau] = \frac{1}{r^2}\,, \tag{10.44}$$

or a power-law distribution:

$$\begin{aligned} p(\tau) &= \frac{\alpha}{\tau_{\mathrm{ch}}\left(\frac{\tau}{\tau_{\mathrm{ch}}}\right)^{1+\alpha}}\,; \quad \tau \in [\tau_{\mathrm{ch}}, \infty)\,,\ \alpha > 0\,; \\ \langle\tau\rangle &= \frac{\tau_{\mathrm{ch}}\,\alpha}{\alpha - 1}\,,\ \alpha > 1\,; \quad \mathrm{Var}[\tau] = \frac{\tau_{\mathrm{ch}}^2\alpha}{(\alpha-2)(\alpha-1)^2}\,,\ \alpha > 2\,. \end{aligned} \tag{10.45}$$

The parameter $\tau_{\mathrm{ch}} > 0$ occurring in the power-law distribution (10.45) gives the lower cut-off scale, and we consider values of $\alpha$ larger than 1 so that the average time interval between two successive measurements remains finite. The exponential distribution (10.44) has a finite mean for all values of $r$.

Figure 10.14 shows in case of the evolution under **Scheme 1** our numerical results for the average survival probability $\overline{S_m}$ and the typical survival probability $S^\star_m$, both plotted with respect to $m$, the number of measurements. In the figure, panel (a) corresponds to the exponential distribution (10.44) with $r = 2$, while panels (b) and (c) are for the power-law distribution (10.45) with $\alpha$ being equal to 2.5 and 3.5, respectively. The panels suggest for both the quantities $\overline{S_m}$ and $S^\star_m$ an exponential decay with $m$, for large $m$.

As regards **Scheme 2** dynamics, the behavior of both the average and the typical value of the survival and the first-detection probability is indeed identical to what we reported for the QRW, see Sect. 10.2.2.1. The statement holds not only for the exponential distribution (10.44) and the power-law distribution (10.45), but also for the delta-function distribution corresponding to measurements performed at regular intervals:

$$p(\tau) = \delta(\tau - \tau_0); \quad \tau_0 > 0\,; \quad \langle\tau\rangle = \tau_0\,, \quad \mathrm{Var}[\tau] = 0\,, \tag{10.46}$$

and for the half-normal distribution:

$$\begin{aligned} p(\tau) &= \sqrt{\frac{2}{\pi\sigma^2}}\exp\left[-\frac{(\tau - \tau_{\mathrm{hn}})^2}{2\sigma^2}\right]; \quad \tau \in [\tau_{\mathrm{hn}}, \infty)\,,\ \sigma > 0\,; \\ \langle\tau\rangle &= \tau_{\mathrm{hn}} + \sigma\sqrt{\frac{2}{\pi}}\,, \quad \mathrm{Var}[\tau] = \sigma^2\left(1 - \frac{2}{\pi}\right), \end{aligned} \tag{10.47}$$

in which the quantity $\tau_{\mathrm{hn}}$ sets the lower limit of $\tau$. As shown in Figs. 10.15 and 10.16, we observe in $\overline{S_m}$ and $\overline{F_m}$ the occurrence of the $m^{-3/2}$ behavior in the former and of $m^{-3}$ and $m^{-5/2}$ behavior in the latter. The crossover between the $m^{-3}$ and the

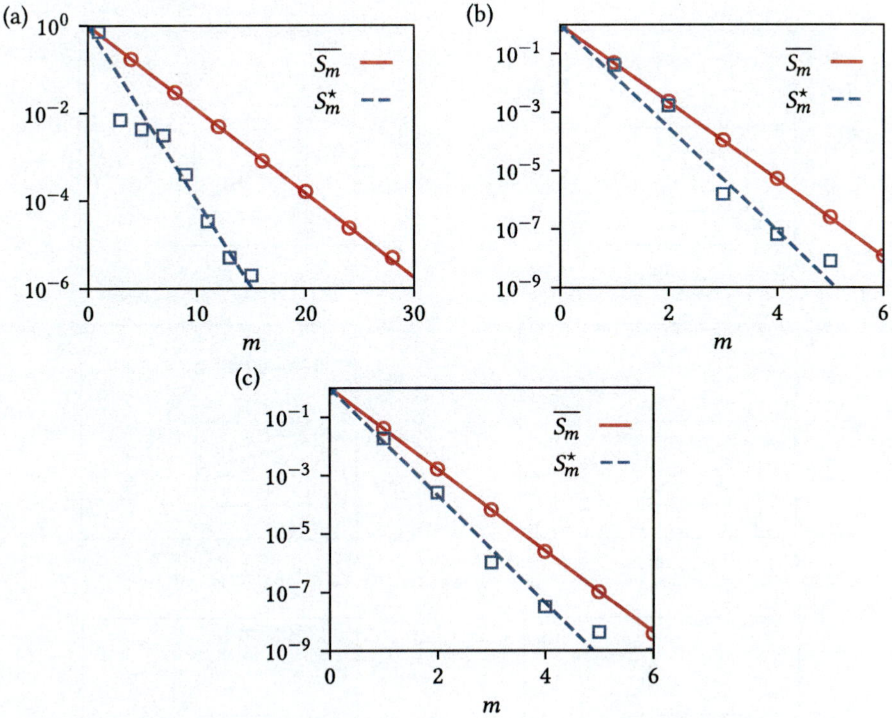


**Fig. 10.14 Average and typical survival probability for the TBM subject to instantaneous projective measurements at random times (Scheme 1).** The plots correspond to the initial state $|0\rangle$ (particle located on site $n_0 = 0$), which is subject to repeated projective measurements at random times to the initial state and subsequent evolution with the projected component of the instantaneous state. Here, the time intervals $\tau_\alpha$ between two consecutive measurements are taken to be i.i.d. random variables sampled from the exponential distribution (10.44) with $r = 2.0$ (panel **a**) and from the power-law distribution (10.45) with $\tau_{\text{ch}} = 1$ and $\alpha = 2.5$ (panel **b**) and $\alpha = 3.5$ (panel **c**). The system size is $N = 200$, while we have considered $\gamma = 1$. In the plots, the points are based on results obtained from numerical implementation of the dynamics; the average survival probability $\overline{S_m}$ involves averaging over $10^4$ realizations of the set $\{\tau_\alpha\}_{1\leq\alpha\leq m}$, while the typical survival probability $S_m^\star$ corresponds to results obtained in a typical realization of the $\tau_\alpha$'s. The lines in the plots denote analytical results given by Eqs. (10.48) and (10.49). The figure is taken from Ref. [1]

$m^{-5/2}$ behavior takes place over the characteristic scale $m_1^\star(N)$ satisfying the scaling (10.25), as implied by the results in Fig. 10.18. The existence of the characteristic value $m_2^\star(N)$ satisfying the scaling (10.26) may be observed in Fig. 10.17. The results for the delta-function distribution (10.46) are given in Figs. 10.19 and 10.20, while those for the typical survival probability in the case of the exponential distribution (10.44) and the power-law distribution (10.45) are given in Fig. 10.21. The existence of oscillations in the behavior of $\overline{F_m}$ at small $m < m_1^\star(N)$, which become more pronounced as the distribution $p(\tau)$ of the time interval between successive measurements becomes narrower, is shown for the case of the half-normal distribu-

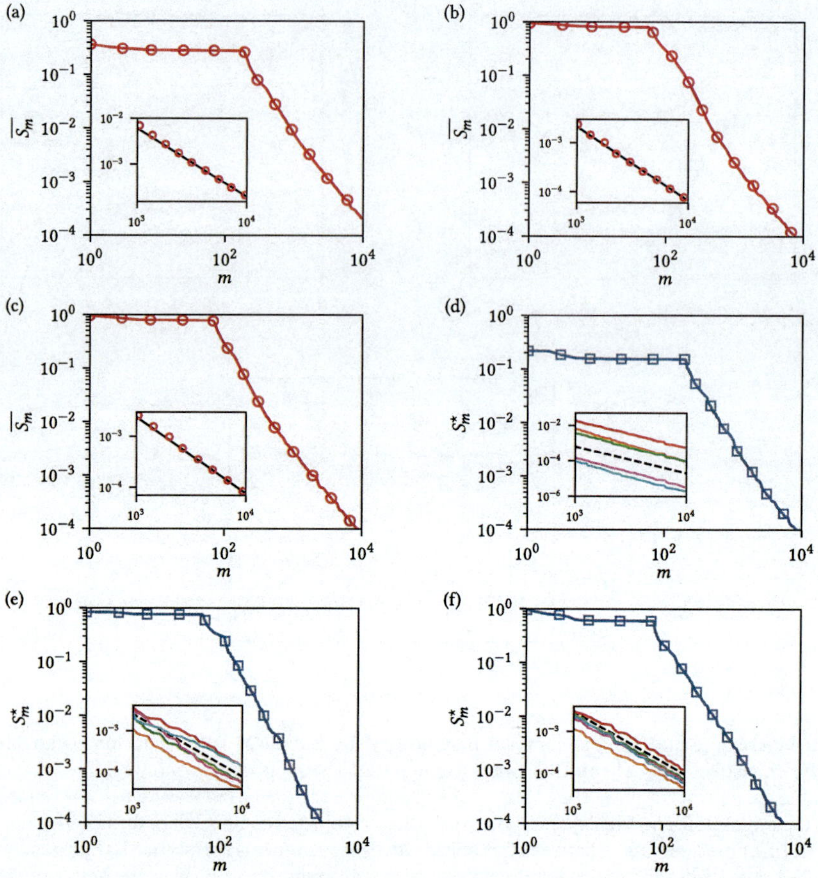

**Fig. 10.15 Average survival probability (panels a–c) and typical survival probability (panels d–f) for the TBM subject to instantaneous projective measurements at random times (Scheme 2).** The plots correspond to the initial state $|0\rangle$ (particle located on site $n_0 = 0$), which is subject to repeated projective measurements at random times to the initial state and subsequent evolution with the leftover component of the instantaneous state after the measurement. Here, the time intervals $\tau_\alpha$ between two consecutive measurements are taken to be i.i.d. random variables sampled from the exponential distribution (10.44) with $r = 2$ (panels **a** and **d**), and from the power-law distribution (10.45) with $\tau_{\mathrm{ch}} = 1$ and $\alpha = 2.5$ (panels **b** and **e**) and $\alpha = 3.5$ (panels **c** and **f**). The system size is $N = 200$, while we have taken $\gamma = 1$. In the main plots, the points are based on results obtained from numerical implementation of the dynamics; the average survival probability $\overline{S_m}$ involves averaging over 25 realizations of the set $\{\tau_\alpha\}_{1\leq\alpha\leq m}$, while the typical survival probability $S_m^\star$ corresponds to results obtained in a typical realization of the $\tau_\alpha$'s. The lines in the main plots denote results obtained from the semi-analytical approach described in the text, see Sect. 10.3.2.2. In the insets in the upper row, the points correspond to numerically-evaluated average survival probability, while the line on the other hand represents an $m^{-3/2}$ behavior. In the insets in the lower row, while the continuous lines correspond to numerically-evaluated survival probability for five typical realizations of $\{\tau_\alpha\}_{1\leq\alpha\leq m}$, the dashed line represents an $m^{-3/2}$ behavior. One may conclude from the insets in both the upper and the lower row that the average as well as the typical survival probability behaves at large $m$ as $m^{-3/2}$. The figure is taken from Ref. [1]

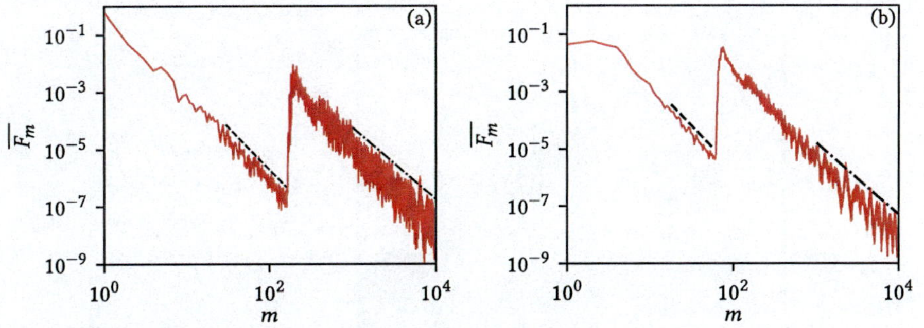


**Fig. 10.16 Average first-detection probability for the TBM (Scheme 2).** For panels **a** and **b**, parameter values and number of averaging realizations are the same as those in Fig. 10.15a and c, respectively. As shown in the plots, one observes two distinct behaviors $\sim m^{-3}$ (dashed line) and $\sim m^{-5/2}$ (dash-dotted line). The figure is taken from Ref. [1]

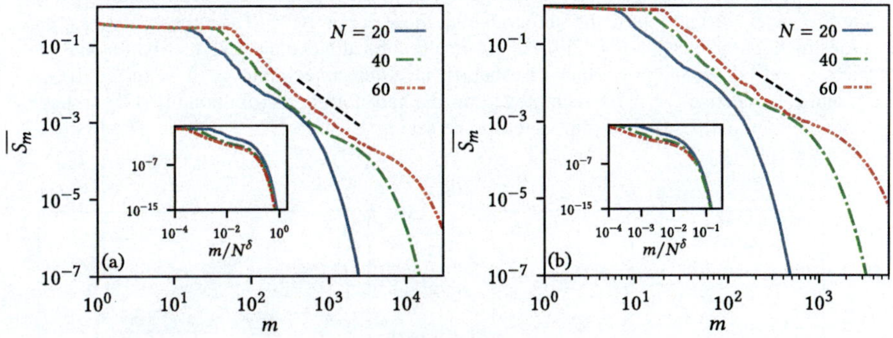


**Fig. 10.17 Average survival probability for the TBM (Scheme 2).** For panels **a** and **b**, parameter values other than $N$ and the number of averaging realizations are the same as those in Fig. 10.15a and c, respectively. The main plots depict a crossover from a $m^{-3/2}$-behavior (dashed line) to an exponential tail over the characteristic value $m_2^{\star}(N)$ of $m$; the collapse of the data shown in the insets is suggestive of the scaling (10.26). The figure is taken from Ref. [1]

tion (10.47) in Fig. 10.22. Let us remark that our results summarized above hold for finite $N$, while if one studies an infinite system (i.e., the limit $N \to \infty$), the scale $m_1^{\star}(N)$ diverges, and one observes the large-$m$ behavior $\overline{S_m} \sim m^{-2}$ and $\overline{F_m} \sim m^{-3}$.

On the basis of the foregoing, we remark that similar to the results for the QRW reported in Sect. 10.2.2, one observes a stark contrast in the behavior of the survival probability under **Scheme 1** and **Scheme 2** of the measurement dynamics in the case of the TBM.

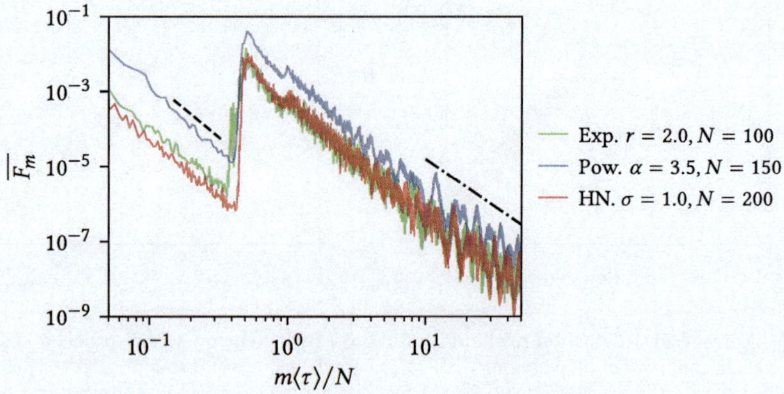


**Fig. 10.18 Average first-detection probability for the TBM (Scheme 2) for the exponential distribution (10.44), the power-law distribution (10.45), and the half-normal distribution (10.47).** Here, the parameters $n_0$ and $\gamma$ have the same values as those in Fig. 10.15. The number of averaging realizations in all cases is taken to be 50. The parameter $r$ for the exponential distribution and the parameters $\alpha$ and $\tau_{\text{ch}}$ for the power-law distribution have the same values as those in Fig. 10.15, panels **a** and **c**, respectively. For the data plotted for the half-normal distribution (10.47), we have taken $\tau_{\text{hn}} = 0.0$ and $\sigma = 1.0$. The plot suggests the scaling given by Eq. (10.25). The figure is taken from Ref. [1]

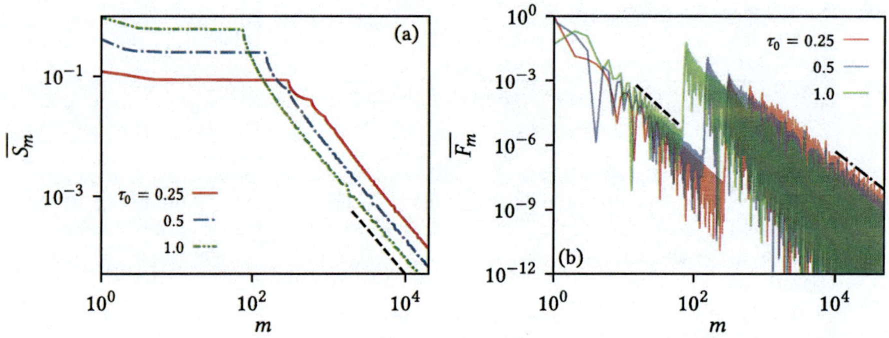


**Fig. 10.19 Average survival and first-detection probability for the TBM (Scheme 2).** Here, the parameters $n_0$ and $\gamma$ have the same values as those in Fig. 10.15. The time intervals $\tau_\alpha$ between two consecutive measurements are taken to be i.i.d. random variables sampled from the distribution (10.46). The values of the parameter $\tau_0$ are given in the figure. The system size is $N = 150$. In panel **a**, as shown in the plot, one observes a behavior $\sim m^{-3/2}$ (dashed line), while in panel **b**, two distinct behaviors $\sim m^{-3}$ (dashed line) and $\sim m^{-5/2}$ (dash-dotted line) may be observed. Note that since $\tau$ has only one allowed value, $\tau_0$, it is redundant to use the overbar over $S_m$ and $F_m$ to denote their average values. The figure is taken from Ref. [1]

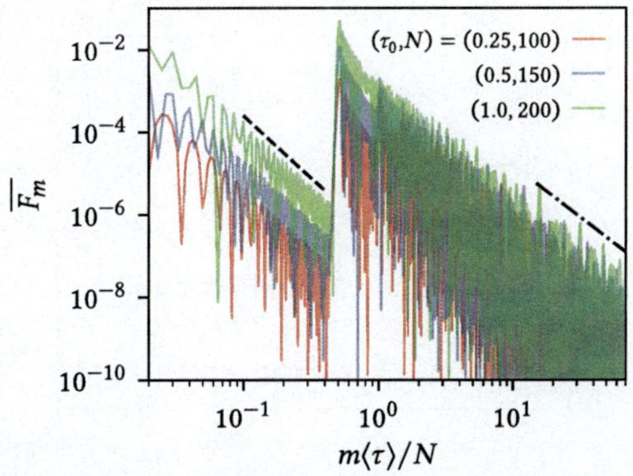


**Fig. 10.20 Average first-detection probability for the TBM (Scheme 2).** Parameter values and other details are the same as those in Fig. 10.19. The plot suggests the crossover from the $\sim m^{-3}$-behavior (dashed line) to the $\sim m^{-5/2}$-behavior (dash-dotted line) over the characteristic value $m_1^\star(N)$ of $m$ that follows the scaling (10.25). Note that since $\tau$ has only one allowed value, $\tau_0$, it is redundant to use the overbar over $F_m$ to denote its average. The figure is taken from Ref. [1]

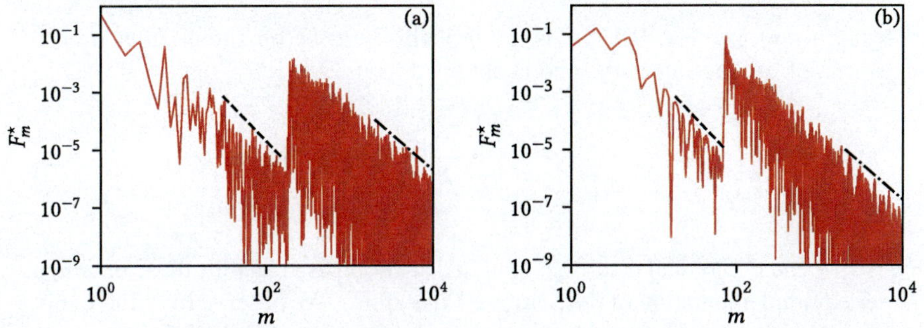


**Fig. 10.21 Typical first-detection probability for the TBM (Scheme 2).** For panels **a** and **b**, parameter values are the same as those in Fig. 10.15a and c, respectively. As shown in the plots, one observes two distinct behaviors $\sim m^{-3}$ (dashed line) and $\sim m^{-5/2}$ (dash-dotted line). The figure is taken from Ref. [1]

#### 10.3.2.2 Analytical Results

For the case of **Scheme 1**, on using the continuous-$\tau$ equivalent of Eqs. (10.33) and (10.34), we obtain the average and the typical survival probability for the TBM subject to projective measurements at random time intervals $\tau$ distributed according to the exponential and the power-law distribution, Eqs. (10.44) and (10.45), respectively. We get

$$\begin{aligned}\overline{S_m} &= \exp\left(m \log \int \mathrm{d}\tau \; p(\tau)q(\tau)\right), \\ S_m^\star &= \exp\left(m \int \mathrm{d}\tau \; p(\tau) \log q(\tau)\right).\end{aligned} \tag{10.48}$$

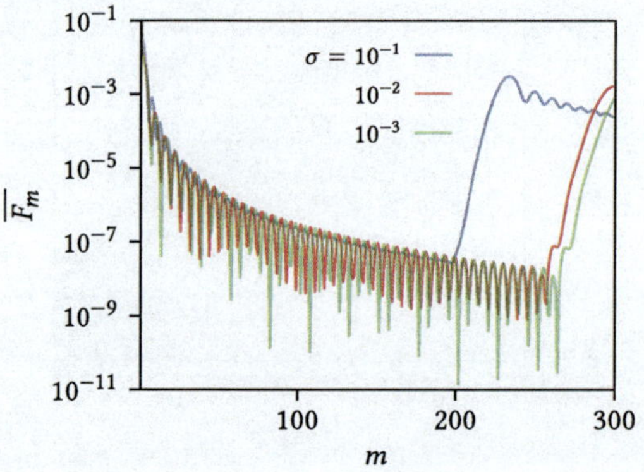


**Fig. 10.22 Average first-detection probability for the TBM (Scheme 2) for the half-normal distribution** (10.47) **with** $\tau_{\text{hn}} = 0.25$ **and different values of** $\sigma$. Here, the parameters $n_0$ and $\gamma$ have the same values as those in Fig. 10.15. The system size is $N = 150$, and the number of averaging realizations is 500. The figure is taken from Ref. [1]

In this case, $q(\tau)$, see Eq. (10.31), is the probability to be on site $n_0$ after time $\tau$ while starting from the same site, and is obtained from Eq. (10.43) as

$$q(\tau) = |\psi_{n_0,n_0}(\tau)|^2 = \left| \frac{1}{N} \sum_{q=0}^{N-1} e^{i2\gamma\tau \cos(2\pi q/N)} \right|^2. \tag{10.49}$$

In Fig. 10.14, the theoretical results so obtained are compared against those obtained in numerical implementation of the **Scheme 1** dynamics. We observe from the figure a very good match of the average and the typical survival probability results.

For the case $p(\tau) = \delta(\tau - \tau_0)$, we obtain $\overline{S_m} = S^\star_m = \exp(m \log q(\tau_0))$, and $\mathcal{T} = m\tau_0$. In the simultaneous limit $\tau_0 \to 0$, $m \to \infty$ with $\mathcal{T}$ kept constant (frequent measurements at close intervals), using Eq. (10.31) and the fact that $|\psi(0)\rangle$ is normalized to unity, it follows that $\overline{S_m} = S^\star_m = 1 - mO(\tau_0^2)$. This result thus implies that to leading order, there is no evolution of the initial state, which is an illustration of the quantum Zeno effect [3].

For **Scheme 2**, a semi-analytic approach to obtain $|\psi_m^{(a)}\rangle$, along the lines that we have employed for the QRW and have detailed in Sect. 10.2.2.2, involves the following steps:

1. For the initial state $|\psi(0)\rangle = |n_0\rangle$ so that $\psi_{n,n_0}(0) = \delta_{n,n_0}$, one has the discrete Fourier transform $\widehat{\psi}_{q|n_0}(0) = \exp(-i2\pi n_0 q/N)$ for $0 \le q \le N-1$.
2. Subsequently, $\widehat{\psi}_{q|n_0}(\tau_1)$, as the outcome of evolution according to (10.41) for a random time $\tau_1$ sampled following either the exponential distribution (10.44) or the power-law distribution (10.45) and with $\widehat{\psi}_{q|n_0}(0)$ as the initial condition, is given by

$$\widehat{\psi}_{q|n_0}(\tau_1) = \widehat{\psi}_{q|n_0}(0)\, e^{i2\gamma\tau_1 \cos(2\pi q/N)}. \tag{10.50}$$

3. Inverse discrete Fourier transform of the set $\{\widehat{\psi}_{q|n_0}(\tau_1)\}_{0\le q\le N-1}$ generates the set $\{\psi_{n,n_0}(\tau_1)\}_{0\le n\le N-1}$. The result of a projective measurement at the end of evolution for time $\tau_1$ to obtain the corresponding leftover component of the state is obtained from the set $\{\psi_{n,n_0}(\tau_1)\}_{0\le n\le N-1}$ with $\psi_{n_0,n_0}(\tau_1) = 0$.
4. We apply steps 1–3 to the leftover component of the state corresponding to last projection, in order to finally obtain the survival probability $S_m$ for a given realization $\{\tau_\alpha\}_{1\le\alpha\le m}$ of the dynamics.

Figure 10.15 depicts a very good agreement for both the average and the typical survival probability between the numerical results and those obtained based on the aforementioned semi-analytic approach.

## 10.4 Conclusions and Perspectives

In this chapter, we studied the issue of what happens when a quantum system undergoing unitary evolution in time is subject to repeated projective measurements to the initial state at random times. We considered two very distinct dynamical scenarios: **Scheme 1**, in which the evolution after every projective measurement continues with the projected component of the instantaneous state, and **Scheme 2**, in which the evolution after every projective measurement continues with the leftover component of the instantaneous state. We focused on a physical quantity of relevance: the survival probability of the initial state after a certain number $m$ of measurements are performed on the system. Based on results derived for two representative quantum systems defined on a one-dimensional periodic lattice with a finite number of sites $N$, namely, (i) the quantum random walk evolving in discrete times, and (ii) the tight-binding model evolving in continuous time, we showed that in **Scheme 1**, the average (averaged with respect to different realizations of the set of random time intervals $\{\tau_\alpha\}_{1\le\alpha\le m}$ between successive measurements) as well as the typical survival probability (obtained in a typical realization of the set $\{\tau_\alpha\}_{1\le\alpha\le m}$) decays as an exponential in $m$ for large $m$. By stark contrast, one obtains under **Scheme 2** that the behavior of the survival probability is characterized by two characteristic $m$ values: $m_1^\star(N) \sim N$ and $m_2^\star(N) \sim N^\delta$ with $\delta > 1$. These scales are such that (i) when $m$ is large and satisfies $m < m_1^\star(N)$, the decay of the survival probability is as $m^{-2}$, (ii) when $m$ satisfies $m_1^\star(N) \ll m < m_2^\star(N)$, the decay is as $m^{-3/2}$, while (iii) when $m$ satisfies $m \gg m_2^\star(N)$, the decay is as an exponential. These results are found to hold independently of the choice of the distribution of time intervals $\tau_\alpha$, a fact we demonstrated by considering a wide range of distributions including exponential and power-law distributions as well as for the case of measurements at regular intervals. It is of interest to extend our studies to the case of a many-body quantum system where additional dynamical timescales are expected to interplay with the average time between successive measurements in dictating a rich static and dynamical behavior.

## References


1. D. Das, S. Gupta, J. Stat. Mech. **2022**(3), 033212 (2022). https://doi.org/10.1088/1742-5468/ac5dc0
2. C. Cohen-Tannoudji, B. Diu, F. Laloë, *Quantum Mechanics, Volume 1: Basic Concepts, Tools, and Applications*, 2nd edn. (Wiley-VCH, Weinheim, 2020)
3. B. Misra, E.C.G. Sudarshan, J. Math. Phys. **18**(4), 756 (1977). https://doi.org/10.1063/1.523304
4. S. Gherardini, S. Gupta, F.S. Cataliotti, A. Smerzi, F. Caruso, S. Ruffo, New J. Phys. **18**(1), 013048 (2016). https://doi.org/10.1088/1367-2630/18/1/013048
5. Y. Aharonov, L. Davidovich, N. Zagury, Phys. Rev. A **48**(2), 1687 (1993). https://doi.org/10.1103/PhysRevA.48.1687
6. D.H. Dunlap, V.M. Kenkre, Phys. Rev. B **34**(6), 3625 (1986). https://doi.org/10.1103/PhysRevB.34.3625
7. D. Dunlap, V. Kenkre, Phys. Lett. A **127**(8–9), 438 (1988). https://doi.org/10.1016/0375-9601(88)90213-7
8. M.M. Müller, S. Gherardini, F. Caruso, Annalen der Physik **529**(9), 1600206 (2017). https://doi.org/10.1002/andp.201600206
9. S. Gherardini, C. Lovecchio, M.M. Müller, P. Lombardi, F. Caruso, F.S. Cataliotti, Quantum. Sci. Technol. **2**(1), 015007 (2017). https://doi.org/10.1088/2058-9565/aa5d00
10. S. Gherardini, L. Buffoni, M.M. Müller, F. Caruso, M. Campisi, A. Trombettoni, S. Ruffo, Phys. Rev. E **98**(3), 032108 (2018). https://doi.org/10.1103/PhysRevE.98.032108
11. H.V. Do, C. Lovecchio, I. Mastroserio, N. Fabbri, F.S. Cataliotti, S. Gherardini, M.M. Müller, N. Dalla Pozza, F. Caruso, New J. Phys. **21**(11), 113056 (2019). https://doi.org/10.1088/1367-2630/ab5740
12. R.S. Ellis, Entropy, Large Deviations, and Statistical Mechanics (Springer. Berlin, 2006). https://doi.org/10.1007/3-540-29060-5
13. H. Touchette, Phys. Rep. **478**(1–3), 1 (2009). https://doi.org/10.1016/j.physrep.2009.05.002
14. S. Dhar, S. Dasgupta, A. Dhar, J. Phys. A: Math. Theor. **48**(11), 115304 (2015). https://doi.org/10.1088/1751-8113/48/11/115304
15. S. Dhar, S. Dasgupta, A. Dhar, D. Sen, Phys. Rev. A **91**(6), 062115 (2015). https://doi.org/10.1103/PhysRevA.91.062115
16. H. Friedman, D.A. Kessler, E. Barkai, J. Phys. A: Math. Theor. **50**(4), 04LT01 (2017). https://doi.org/10.1088/1751-8121/aa5191
17. H. Friedman, D.A. Kessler, E. Barkai, Phys. Rev. E **95**(3), 032141 (2017). https://doi.org/10.1103/PhysRevE.95.032141
18. F. Thiel, E. Barkai, D.A. Kessler, Phys. Rev. Lett. **120**(4), 040502 (2018). https://doi.org/10.1103/PhysRevLett.120.040502
19. F. Thiel, I. Mualem, D.A. Kessler, E. Barkai, Quantum total detection probability from repeated measurements II. Exploiting symmetry (2019). https://doi.org/10.48550/arXiv.1909.02114
20. S. Lahiri, A. Dhar, Phys. Rev. A **99**(1), 012101 (2019). https://doi.org/10.1103/PhysRevA.99.012101
21. D. Meidan, E. Barkai, D.A. Kessler, J. Phys. A: Math. Theor. **52**(35), 354001 (2019). https://doi.org/10.1088/1751-8121/ab3305
22. R. Yin, K. Ziegler, F. Thiel, E. Barkai, Phys. Rev. Res. **1**(3), 033086 (2019). https://doi.org/10.1103/PhysRevResearch.1.033086
23. F. Thiel, I. Mualem, D. Meidan, E. Barkai, D.A. Kessler, Phys. Rev. Res. **2**(4), 043107 (2020). https://doi.org/10.1103/PhysRevResearch.2.043107
24. F. Thiel, I. Mualem, D.A. Kessler, E. Barkai, Phys. Rev. Res. **2**(2), 023392 (2020). https://doi.org/10.1103/PhysRevResearch.2.023392
25. V. Dubey, C. Bernardin, A. Dhar, Phys. Rev. A **103**(3), 032221 (2021). https://doi.org/10.1103/PhysRevA.103.032221
26. F. Thiel, I. Mualem, D. Kessler, E. Barkai, Entropy **23**(5), 595 (2021). https://doi.org/10.3390/e23050595

27. Q. Liu, K. Ziegler, D.A. Kessler, E. Barkai, Phys. Rev. Res. **4**(2), 023129 (2022). https://doi.org/10.1103/PhysRevResearch.4.023129
28. D.A. Kessler, E. Barkai, K. Ziegler, Phys. Rev. A **103**(2), 022222 (2021). https://doi.org/10.1103/PhysRevA.103.022222
29. A. Nayak, A. Vishwanath (2000). arXiv:quant-ph/0010117
30. J. Kempe, Contemp. Phys. **44**(4), 307 (2003). https://doi.org/10.1080/00107151031000110776

# Appendix A
# Left and Right Eigenvectors of a General Square Matrix

Consider a general square matrix $A$. Its right and left eigenvectors are defined respectively by [1]

$$A|\lambda_R\rangle = \lambda_R|\lambda_R\rangle\,, \tag{A.1}$$

$$\langle\lambda_L|A = \lambda_L\langle\lambda_L|\,. \tag{A.2}$$

The first equation rewrites as $(A - \lambda_R\mathbb{I})\,|\lambda_R\rangle = 0$, where $\mathbb{I}$ denotes the identity. Then, in order that one does not have the trivial solution $|\lambda_R\rangle = 0$, the inverse of the matrix $(A - \lambda_R\mathbb{I})$ must not exist. The latter fact implies that we need the eigenvalues $\lambda_R$ to satisfy $\text{Det}\,(A - \lambda_R\mathbb{I}) = 0$. Taking the transpose of the second equation above gives $A^T|\lambda_L\rangle = \lambda_L|\lambda_L\rangle$, so that following the same argument as in the foregoing, we see that the eigenvalues $\lambda_L$ satisfy $\text{Det}\left(A^T - \lambda_L\mathbb{I}\right) = 0 = \text{Det}\,(A - \lambda_L\mathbb{I})^T = \text{Det}\,(A - \lambda_L\mathbb{I})$. Here, we have used the fact that $\text{Det}\left(A^T\right) = \text{Det}(A)$. We thus conclude by comparing the two equations $\text{Det}\,(A - \lambda_R\mathbb{I}) = 0$ and $\text{Det}\,(A - \lambda_L\mathbb{I}) = 0$ that the left and right eigenvalues are the same: $\lambda_L = \lambda_R \equiv \lambda$. The left and right eigenvectors are however different in general, unless $A$ is a symmetric matrix: $A^T = A$. In that case, the first equation in Eq. (A.2) yields $A^T|\lambda_L\rangle = A|\lambda_L\rangle = \lambda|\lambda_L\rangle$ so that a comparison with the second equation shows that the right and the left eigenvector corresponding to the same eigenvalue are the same.

Consider next the diagonal matrix $D$ whose diagonal elements are the distinct eigenvalues of $A$. Next, define the matrix $\Lambda_R$ whose columns are the right eigenvectors of $A$. It then follows that

$$A\Lambda_R = \Lambda_R D\,. \tag{A.3}$$

In the same manner, it is checked on considering a matrix $\Lambda_L$ whose rows are the left eigenvectors of $A$ that we have

$$\Lambda_L A = D\Lambda_L\,. \tag{A.4}$$



D. Das and S. Gupta, *Facets of Noise*, Fundamental Theories of Physics 214,
https://doi.org/10.1007/978-3-031-45312-0

The last two equations yield respectively that

$$\Lambda_L A \Lambda_R = \Lambda_L \Lambda_R D\,, \tag{A.5}$$
$$\Lambda_L A \Lambda_R = D \Lambda_L \Lambda_R\,, \tag{A.6}$$

that is,

$$D\,(\Lambda_L \Lambda_R) = (\Lambda_L \Lambda_R)\,D\,. \tag{A.7}$$

In other words, the matrix $\Lambda_L \Lambda_R$ commutes with the diagonal matrix $D$. Now, the only matrices that commute with a diagonal matrix of distinct elements are themselves diagonal. Consequently, we conclude that $\Lambda_L \Lambda_R$ is diagonal. This implies that provided the eigenvalues of matrix $A$ are non-degenerate, each left eigenvector is orthogonal to all right eigenvectors except its corresponding one, and vice versa. We may invoke normalization by requiring that the product of the left and the right eigenvector for the same eigenvalue is unity for any matrix with non-degenerate eigenvalues. We thus have

$$\Lambda_L \Lambda_R = \mathbb{I} = \Lambda_R \Lambda_L\,. \tag{A.8}$$

It then follows by using Eq. (A.4) that $A = \Lambda_R D \Lambda_L$, that is,

$$A = \sum_\alpha \lambda^{(\alpha)} |\lambda_R^{(\alpha)}\rangle\langle\lambda_L^{(\alpha)}|\,, \tag{A.9}$$

where the index $\alpha$ labels the different eigenvalues.

## Reference

1 P. Dennery, A. Krzywicki, *Mathematics for Physicists* (Dover Publications, Mineola, 1996)

# Appendix B
# Stochastic Differential Equations: The Itô and the Stratonovich Approach

Let us consider the stochastic differential equation (SDE) [1]

$$\mathrm{d}x = f(x)\,\mathrm{d}t + g(x)\,\mathrm{d}W(t)\,, \tag{B.1}$$

where $W(t)$ is a Wiener process (standard Brownian process) characterized by the following properties [2]:

1. $W(0) = 0$.
2. $W(t)$ has independent increments: For $\tau \geq 0$, the random increment $[W(t+\tau) - W(t)]\;\forall\, t$ is independent of the past values $W(t')\;\forall\, t' \leq t$. For times $0 \leq t_0 < t_1 < \ldots < t_N$, all the increments $W(t_1) - W(t_0), \ldots, W(t_N) - W(t_{N-1})$ are independent of each other.
3. $W(t)$ has Gaussian increments: For $\tau > 0$, the increment $[W(t+\tau) - W(t)]\;\forall\, t$ is normally distributed with zero mean and variance proportional to $\tau$; equivalently, $W(t+\tau) - W(t) \sim \sqrt{\tau}\,\mathcal{N}(0,1)$, where $\mathcal{N}(0,1)$ denotes a normally-distributed random number with zero mean and unit variance, and the symbol $\sim$ means that the left hand side and the right hand side are equal in distribution. Here, for later convenience, we choose the variance of the increment $[W(t+\tau) - W(t)]$ to be $2D\tau$, where $D$ is a positive constant with suitable dimension determined from Eq. (B.1). In other words, we have $W(t+\tau) - W(t) \sim \sqrt{2D\tau}\,\mathcal{N}(0,1)$.
4. $W(t)$ is continuous in $t$, but is nowhere differentiable.

As the function $g$ on the right hand side of Eq. (B.1) depends on the variable $x$, the noise term $g(x)\,\mathrm{d}W(t)$ is called multiplicative noise, while for constant $g$, the noise is said to be additive. From Eq. (B.1), one may write

$$x(t+\tau) = x(t) + \int_t^{t+\tau} \mathrm{d}t'\; f\big(x(t')\big) + \int_t^{t+\tau} \mathrm{d}W(t')\; g\big(x(t')\big)\,. \tag{B.2}$$



D. Das and S. Gupta, *Facets of Noise*, Fundamental Theories of Physics 214,
https://doi.org/10.1007/978-3-031-45312-0

From the first and the third characteristic of the Wiener process given above, we obtain $W(t) = W(t) - W(0) \sim \sqrt{2Dt}\, \mathcal{N}(0, 1)$, which yields the mean and the variance of $W(t)$ as

$$\langle W(t) \rangle = 0\,, \tag{B.3}$$

$$\left\langle (W(t))^2 \right\rangle = 2Dt\,, \tag{B.4}$$

respectively. For two times $t_1$ and $t_2$ with $t_2 > t_1$, one may write

$$\begin{aligned}\left\langle W(t_1) W(t_2) \right\rangle &= \left\langle W(t_1) \left[ W(t_2) - W(t_1) + W(t_1) \right] \right\rangle \\ &= \left\langle W(t_1) \left[ W(t_2) - W(t_1) \right] \right\rangle + \left\langle (W(t_1))^2 \right\rangle \\ &= \left\langle \left[ W(t_1) - W(0) \right] \left[ W(t_2) - W(t_1) \right] \right\rangle + \left\langle (W(t_1))^2 \right\rangle . \end{aligned} \tag{B.5}$$

As $t_2 > t_1$, according to the second characteristic of $W(t)$, the increments $[W(t_1) - W(0)]$ and $[W(t_2) - W(t_1)]$ are independent of each other. Then, from Eq. (B.5), we get

$$\begin{aligned}\left\langle W(t_1) W(t_2) \right\rangle &= \left\langle \left[ W(t_1) - W(0) \right] \right\rangle \left\langle \left[ W(t_2) - W(t_1) \right] \right\rangle + \left\langle (W(t_1))^2 \right\rangle \\ &= \left\langle W(t_1) \right\rangle \left\langle \left[ W(t_2) - W(t_1) \right] \right\rangle + \left\langle (W(t_1))^2 \right\rangle \\ &= 2Dt_1\,, \end{aligned} \tag{B.6}$$

where we have used Eqs. (B.3) and (B.4). By using similar arguments, one obtains for $t_2 < t_1$ that $\left\langle W(t_1) W(t_2) \right\rangle = 2Dt_2$. Therefore, in general, we have

$$\left\langle W(t_1) W(t_2) \right\rangle = 2D\, \min(t_1, t_2)\,. \tag{B.7}$$

An integral of the form

$$A \equiv \int_0^{\tau} \mathrm{d}W(\tau')\, \Phi(W(\tau'), \tau') \tag{B.8}$$

is called a stochastic integral. Its evaluation in a spirit similar to what is done for usual "deterministic" integrals (e.g., the integral $\int_0^{\tau} \mathrm{d}\tau'\, \Phi(\tau')$) as a (discrete) Riemann sum depends crucially on the process of discretization of $W(\tau')$ over the interval $[0, \tau]$. The celebrated Itô and Stratonovich interpretations involve different forms of discretization, as [1]

$$\text{Itô rule:}\quad A_{\mathrm{I}} \equiv \lim_{\Delta \to 0} \sum_{i=0}^{N-1} \left[ W(\tau_{i+1}) - W(\tau_i) \right] \Phi(W(\tau_i), \tau_i)\,, \tag{B.9}$$

$$\text{Stratonovich rule:}\quad A_{\mathrm{S}} \equiv \lim_{\Delta\to 0}\sum_{i=0}^{N-1}\left[W(\tau_{i+1}) - W(\tau_i)\right] \times \Phi\left(\frac{W(\tau_i)+W(\tau_{i+1})}{2}, \frac{\tau_i+\tau_{i+1}}{2}\right), \tag{B.10}$$

where we have

$$\Delta = \max(\tau_{i+1}-\tau_i)\,;\ i \in [0, N-1]\,, \quad 0 = \tau_0 < \tau_1 < \ldots < \tau_N = \tau\,. \tag{B.11}$$

Note that in the Itô definition, $\Phi$ depends only on the initial point $\tau_i$ of the interval $[\tau_i, \tau_{i+1}]$. Contrarily, in the Stratonovich definition, $\Phi$ depends on both $\tau_i$ and $\tau_{i+1}$ symmetrically. It may be shown that the two sums (B.9) and (B.10), which may be interpreted as stochastic Riemann sums, would evaluate to give very different results even as $\Delta \to 0$; for an illustration, see, e.g., [3].

Using the Itô and Stratonovich prescriptions defined above, let us evaluate

$$B \equiv \left\langle \int_0^\tau \mathrm{d}W(\tau')\, W(\tau')\right\rangle, \tag{B.12}$$

which will be required in the following. From the Itô definition (B.9), one obtains

$$\begin{aligned} B_{\mathrm{I}} \equiv \left\langle \int_0^\tau \mathrm{d}W(\tau')\, W(\tau')\right\rangle_{\mathrm{I}} &= \left\langle \lim_{\Delta\to 0}\sum_{i=0}^{N-1}\left[W(\tau_{i+1}) - W(\tau_i)\right] W(\tau_i)\right\rangle \\ &= \lim_{\Delta\to 0}\sum_{i=0}^{N-1}\left[\langle W(\tau_{i+1})W(\tau_i)\rangle - \langle W(\tau_i)W(\tau_i)\rangle\right] \\ &= \lim_{\Delta\to 0}\sum_{i=0}^{N-1}\left[2D\tau_i - 2D\tau_i\right] \\ &= 0\,, \end{aligned} \tag{B.13}$$

where we have used Eq. (B.7). Similarly, using the Stratonovich definition (B.10), one obtains

$$\begin{aligned} B_{\mathrm{S}} \equiv \left\langle \int_0^\tau \mathrm{d}W(\tau')\, W(\tau')\right\rangle_{\mathrm{S}} &= \left\langle \lim_{\Delta\to 0}\sum_{i=0}^{N-1}\left[W(\tau_{i+1}) - W(\tau_i)\right]\left[\frac{W(\tau_i)+W(\tau_{i+1})}{2}\right]\right\rangle \\ &= \lim_{\Delta\to 0}\sum_{i=0}^{N-1}\frac{1}{2}\left[2D\tau_{i+1} - 2D\tau_i\right] \\ &= D\lim_{\Delta\to 0}\left[\tau_N - \tau_0\right] \\ &= D\tau\,. \end{aligned} \tag{B.14}$$

As discussed in the main text, the Fokker-Planck equation (FPE) corresponding to an SDE gives the time evolution of the probability density of the stochastic variable described by the latter. Here, as an alternative to what is presented in the main text as a general derivation of the FPE, we provide a simple derivation [4] of the FPE corresponding to the particular case of the SDE (B.1); our aim is to finally derive the form of the FPE for both the Itô and the Stratonovich prescription. Let $P(x, t)$ denote the probability density of observing the value $x$ at time $t$. Then, at a later time $t + \tau$ with $\tau > 0$ sufficiently small, the distribution function $P(x, t + \tau)$ may be written as

$$P(x, t + \tau) = \int_{-\infty}^{\infty} \mathrm{d}(\Delta x)\, P(x - \Delta x, t)\, \psi_\tau(x - \Delta x; \Delta x), \tag{B.15}$$

where $\psi_\tau(x; \Delta x)$ denotes the transition probability density that $x$ undergoes an increment of $\Delta x$ in time $\tau$. In writing Eq. (B.15), we have considered that the underlying stochastic dynamics (B.1) is Markovian, namely, the state of the system at time $t + \tau$ depends only on the state at time $t$, and is independent of the state at any time $t' < t$. As $\tau$ is small, it is physically meaningful to assume that $\Delta x$ is also a small quantity. Expanding $P(x - \Delta x, t)$ and $\psi_\tau(x - \Delta x; \Delta x)$ in Taylor series for small $\Delta x$, we obtain from Eq. (B.15) that

$$\begin{aligned}
P(x, t + \tau) = &\int_{-\infty}^{\infty} \mathrm{d}(\Delta x) \left[ P(x, t) - \Delta x\, \frac{\partial P(x, t)}{\partial x} + \frac{(\Delta x)^2}{2!} \frac{\partial^2 P(x, t)}{\partial x^2} - \cdots \right] \\
&\times \left[ \psi_\tau(x; \Delta x) - \Delta x\, \frac{\partial \psi_\tau(x; \Delta x)}{\partial x} + \frac{(\Delta x)^2}{2!} \frac{\partial^2 \psi_\tau(x; \Delta x)}{\partial x^2} - \cdots \right] \\
= &P(x, t) \int_{-\infty}^{\infty} \mathrm{d}(\Delta x)\, \psi_\tau(x; \Delta x) \\
&- \left[ \frac{\partial P(x, t)}{\partial x} + P(x, t) \frac{\partial}{\partial x} \right] \int_{-\infty}^{\infty} \mathrm{d}(\Delta x)\, \Delta x\, \psi_\tau(x; \Delta x) \\
&+ \frac{1}{2!} \left[ \frac{\partial^2 P(x, t)}{\partial x^2} + 2\, \frac{\partial P(x, t)}{\partial x} \frac{\partial}{\partial x} + P(x, t) \frac{\partial^2}{\partial x^2} \right] \\
&\times \int_{-\infty}^{\infty} \mathrm{d}(\Delta x)\, (\Delta x)^2\, \psi_\tau(x; \Delta x) - \cdots ,
\end{aligned} \tag{B.16}$$

where, although not shown explicitly, all the derivatives are to be evaluated at $x$. Using the identity $\int_{-\infty}^{\infty} \mathrm{d}(\Delta x)\, \psi_\tau(x; \Delta x) = 1$ (follows from the definition of $\psi_\tau(x; \Delta x)$ as a probability density), Eq. (B.16) is rewritten as

$$\begin{aligned}
P(x, t + \tau) - P(x, \tau) = &- \frac{\partial}{\partial x} \left[ P(x, t) \int_{-\infty}^{\infty} \mathrm{d}(\Delta x)\, \Delta x\, \psi_\tau(x; \Delta x) \right] \\
&+ \frac{1}{2!} \frac{\partial^2}{\partial x^2} \left[ P(x, t) \int_{-\infty}^{\infty} \mathrm{d}(\Delta x)\, (\Delta x)^2\, \psi_\tau(x; \Delta x) \right] - \cdots
\end{aligned} \tag{B.17}$$

Dividing both sides of the last equation by $\tau$ and then taking the limit $\tau \to 0$, one obtains

$$\frac{\partial P(x,t)}{\partial t} = \sum_{n=1}^{\infty} (-1)^n \frac{\partial^n x}{\partial^{\,n}} \Big[ D^{(n)}(x)\, P(x,t) \Big], \tag{B.18}$$

where the coefficients $D^{(n)}(x)$, known as the Kramers-Moyal expansion coefficients, are defined as [1]

$$D^{(n)}(x) \equiv \frac{1}{n!} \lim_{\tau \to 0} \frac{1}{\tau} \int_{-\infty}^{\infty} \mathrm{d}(\Delta x)\, (\Delta x)^n\, \psi_\tau(x; \Delta x)\,. \tag{B.19}$$

Note that the integral on the right hand side of Eq. (B.19) gives the average of the $n$-th power of the change $\Delta x$ that a given $x$ undergoes in the small time interval $\tau$. Consequently, we rewrite Eq. (B.19) as

$$D^{(n)}(x) = \frac{1}{n!} \lim_{\tau \to 0} \frac{1}{\tau} \Big\langle (\Delta x)^n \Big\rangle. \tag{B.20}$$

The first two Kramers-Moyal expansion coefficients $D^{(1)}(x)$ and $D^{(2)}(x)$ are known as the drift and the diffusion coefficient, respectively.

Now, in order to evaluate the quantities $\langle (\Delta x)^n \rangle$ and correspondingly, the Kramers-Moyal expansion coefficients $D^{(n)}(x)$, one has to use the SDE. In our case, for a given $x = x(t)$ at time $t$, the increment $\Delta x$ in time interval $\tau$ is given as

$$\Delta x = x(t+\tau) - x\,, \tag{B.21}$$

where $x(t+\tau)$ with $\tau > 0$ is the solution of Eq. (B.1) with $x(t)$ given to be equal to $x$. Then, one may rewrite Eq. (B.2) as

$$x(t+\tau) - x = \int_t^{t+\tau} \mathrm{d}t'\, f\big(x(t')\big) + \int_t^{t+\tau} \mathrm{d}W(t')\, g\big(x(t')\big)\,. \tag{B.22}$$

With the substitution $t' = t + \tau'$, Eq. (B.22) yields

$$x(t+\tau) - x = \int_0^{\tau} \mathrm{d}\tau'\, f\big(x(t+\tau')\big) + \int_0^{\tau} \mathrm{d}W(t+\tau')\, g\big(x(t+\tau')\big)\,. \tag{B.23}$$

Using iteratively the left hand side of the above equation on the right hand side to expand the functions $f$ and $g$ in Taylor series and then keeping only the leading terms, one obtains the first iteration from Eq. (B.23) as

$$\begin{aligned} x^{(1)}(t+\tau) - x &\approx \int_0^\tau \mathrm{d}\tau' \, f(x) + \int_0^\tau \mathrm{d}W(t+\tau')\, g(x) \\ &= f(x) \int_0^\tau \mathrm{d}\tau' + g(x) \int_0^\tau \mathrm{d}W(t+\tau') \\ &= \tau f(x) + (W(t+\tau) - W(t))\, g(x)\,. \end{aligned} \tag{B.24}$$

where in the last step, we have used

$$\begin{aligned} \int_0^\tau \mathrm{d}W(t+\tau') &= \lim_{\Delta\to 0} \sum_{i=0}^{N-1} \left[ W(t+\tau_{i+1}) - W(t+\tau_i) \right] \\ &= W(t+\tau_N) - W(t+\tau_0) \\ &= W(t+\tau) - W(t) \\ &\sim \sqrt{2D\tau}\, \mathcal{N}(0,1)\,, \end{aligned} \tag{B.25}$$

which holds for both the Itô and the Stratonovich prescription, see Eqs. (B.9), (B.10), and (B.11).

Now, the second term on the right hand side of Eq. (B.24) being of $O(\sqrt{\tau})$, while computing the first Kramers-Moyal expansion coefficient $D^{(1)}(x)$, the higher-order terms in the Taylor series expansions of the functions $f$ and $g$ may also contribute. So, we go for the second iteration of Eq. (B.23), which yields

$$\begin{aligned} x^{(2)}(t+\tau) - x \approx & \int_0^\tau \mathrm{d}\tau' \, f\Big(x + \int_0^{\tau'} \mathrm{d}\tau'' \, f\big(x(t+\tau'')\big) \\ & + \int_0^{\tau'} \mathrm{d}W(t+\tau'')\, g\big(x(t+\tau'')\big)\Big) \\ & + \int_0^\tau \mathrm{d}W(t+\tau')\, g\Big(x + \int_0^{\tau'} \mathrm{d}\tau''\, f\big(x(t+\tau'')\big) + \int_0^{\tau'} \mathrm{d}W(t+\tau'')\, g\big(x(t+\tau'')\big)\Big) \\ \approx & \int_0^\tau \mathrm{d}\tau' \Big[ f(x) + f'(x) \Big\{ \int_0^{\tau'} \mathrm{d}\tau''\, f\big(x(t+\tau'')\big) \\ & + \int_0^{\tau'} \mathrm{d}W(t+\tau'')\, g\big(x(t+\tau'')\big) \Big\} \Big] \\ & + \int_0^\tau \mathrm{d}W(t+\tau') \Big[ g(x) + g'(x) \Big\{ \int_0^{\tau'} \mathrm{d}\tau'' f\big(x(t+\tau'')\big) \\ & + \int_0^{\tau'} \mathrm{d}W(t+\tau'') g\big(x(t+\tau'')\big) \Big\} \Big] \\ \approx & \int_0^\tau \mathrm{d}\tau' \Big[ f(x) + f'(x) \Big\{ \int_0^{\tau'} \mathrm{d}\tau''\, f(x) + \int_0^{\tau'} \mathrm{d}W(t+\tau'')\, g(x) \Big\} \Big] \\ & + \int_0^\tau \mathrm{d}W(t+\tau') \Big[ g(x) + g'(x) \Big\{ \int_0^{\tau'} \mathrm{d}\tau''\, f(x) + \int_0^{\tau'} \mathrm{d}W(t+\tau'')\, g(x) \Big\} \Big]. \end{aligned} \tag{B.26}$$

We thus obtain

$$
\begin{aligned}
& x^{(2)}(t+\tau) - x \\
& \approx \int_0^\tau \mathrm{d}\tau' \left[ f(x) + f'(x)\, f(x)\, \tau' + f'(x)\, g(x)\, (W(t+\tau') - W(t)) \right] \\
& + \int_0^\tau \mathrm{d}W(t+\tau') \left[ g(x) + g'(x)\, f(x)\, \tau' + g'(x)\, g(x)\, (W(t+\tau') - W(t)) \right] \\
& = \tau f(x) + \frac{\tau^2}{2} f'(x) f(x) + f'(x)\, g(x) \int_0^\tau \mathrm{d}\tau' \, (W(t+\tau') - W(t)) \\
& + g(x)\, W(t+\tau) + g'(x) f(x) \int_0^\tau \mathrm{d}W(t+\tau')\, \tau' \\
& + g'(x) g(x) \int_0^\tau \mathrm{d}W(t+\tau')\, (W(t+\tau') - W(t)) \,, 
\end{aligned}
\tag{B.27}
$$

where in the last step we have used Eq. (B.25). Averaging Eq. (B.27) and keeping terms upto order $\tau$, we obtain

$$
\left\langle x^{(2)}(t+\tau) - x \right\rangle = \tau f(x) + g'(x)\, g(x) \left\langle \int_0^\tau \mathrm{d}W(t+\tau')\, (W(t+\tau') - W(t)) \right\rangle . \tag{B.28}
$$

In obtaining the last equation, we have used Eq. (B.3) and the fact that $\langle \int_0^\tau \mathrm{d}W(t+\tau')\, \tau' \rangle = 0$ for both the Itô and the Stratonovich prescription, see Eqs. (B.9) and (B.10).

Substituting Eq. (B.28) in Eq. (B.20), we get

$$
D^{(1)}(x) = f(x) + g'(x)\, g(x) \lim_{\tau \to 0} \frac{1}{\tau} \left\langle \int_0^\tau \mathrm{d}W(t+\tau')\, (W(t+\tau') - W(t)) \right\rangle . \tag{B.29}
$$

Equation (B.29) yields different results in the Itô and Stratonovich prescriptions. Proceeding as in Eqs. (B.13) and (B.14), one obtains

$$
\text{Itô rule:} \quad D_{\mathrm{I}}^{(1)}(x) \equiv f(x)\,, \tag{B.30}
$$

$$
\text{Stratonovich rule:} \quad D_{\mathrm{S}}^{(1)}(x) \equiv f(x) + D\, g'(x)\, g(x)\,. \tag{B.31}
$$

The second term on the right hand side of Eq. (B.31) is called the spurious or noise-induced drift, which does not occur in the Itô prescription. As mentioned earlier, the right hand side of Eq. (B.24) has two terms, which are of order $\sqrt{\tau}$ and $\tau$, respectively. It may be seen from the defining equation (B.20) that only the term linear in $\tau$ in the expression for $\langle (\Delta x)^2 \rangle$ has a non-vanishing contribution to the diffusion coefficient

$D^{(2)}(x)$, and all the higher-order terms in $\tau$ vanish in the limit $\tau \to 0$. Therefore, in order to obtain $D^{(2)}(x)$, it is sufficient to substitute Eq. (B.24) (instead of Eq. (B.27)) in Eq. (B.20). Then, we have

$$\begin{aligned} D^{(2)}(x) &= \frac{1}{2} \lim_{\tau \to 0} \frac{1}{\tau} \Big\langle \tau^2 \big(f(x)\big)^2 + 2\tau f(x) g(x) (W(t+\tau) - W(t)) \\ &+ \big(g(x)\big)^2 \big(W(t+\tau) - W(t)\big)^2 \Big\rangle \\ &= \frac{1}{2} \big(g(x)\big)^2 \lim_{\tau \to 0} \frac{1}{\tau} \Big\langle \big(W(t+\tau) - W(t)\big)^2 \Big\rangle \\ &= D \big(g(x)\big)^2 , \end{aligned} \tag{B.32}$$

where we have used Eqs. (B.3) and (B.4). Note that in evaluating $D^{(2)}(x)$, we do not have to deal with any stochastic integral of the form defined in Eq. (B.8). Therefore, Eq. (B.32) holds for both the Itô and the Stratonovich prescription.

For $n > 2$, the Kramers-Moyal expansion coefficients $D^{(n)}(x)$ vanish because in this case, the quantity $\langle (\Delta x)^n \rangle$ would have superlinear terms in $\tau$, and such terms would evidently not contribute to $D^{(n)}(x)$, see Eqs. (B.20) and (B.24). Using this result in Eq. (B.18), one finally obtains the FPE corresponding to the SDE (B.1):

$$\frac{\partial P(x,t)}{\partial t} = -\frac{\partial}{\partial x}\big[D^{(1)}(x) P(x,t)\big] + \frac{\partial^2}{\partial x^2}\big[D^{(2)}(x) P(x,t)\big] . \tag{B.33}$$

Then, substituting Eqs. (B.30), (B.31), and (B.32) in Eq. (B.33), the FPEs in the Itô and the Stratonovich prescription are obtained as

Itô FPE:

$$\frac{\partial P(x,t)}{\partial t} = -\frac{\partial}{\partial x}\Big[f(x) P(x,t)\Big] + D \frac{\partial^2}{\partial x^2}\Big[\big(g(x)\big)^2 P(x,t)\Big] , \tag{B.34}$$

Stratonovich FPE:

$$\begin{aligned} \frac{\partial P(x,t)}{\partial t} &= -\frac{\partial}{\partial x}\Big[\big\{f(x) + D g'(x) g(x)\big\} P(x,t)\Big] + D \frac{\partial^2}{\partial x^2}\Big[\big(g(x)\big)^2 P(x,t)\Big] \\ &= -\frac{\partial}{\partial x}\Big[f(x) P(x,t)\Big] + D \frac{\partial}{\partial x}\Big[g(x) \frac{\partial}{\partial x}\big\{g(x) P(x,t)\big\}\Big] . \end{aligned} \tag{B.35}$$

Note that for additive noise (i.e., when $g(x)$ is constant), both prescriptions lead to the same FPE.

It is worth mentioning that the SDE (B.1) is often written in physics literature (which is also what we have done elsewhere in the present monograph) as

$$\frac{dx}{dt} = f(x) + g(x)\,\eta(t) , \tag{B.36}$$

where $\eta(t)$ is a Gaussian, white noise satisfying

$$\langle \eta(t) \rangle = 0 \, ; \quad \langle \eta(t)\eta(t') \rangle = 2D\delta(t - t') \, , \tag{B.37}$$

and

$$\frac{\mathrm{d}W(t)}{\mathrm{d}t} = \eta(t) \, . \tag{B.38}$$

Note however that the left hand side of Eq. (B.38) is not mathematically defined, since a Wiener process $W(t)$ is not a differentiable function.

## References

1 H. Risken, *The Fokker-Planck Equation: Methods of Solution and Applications* (Springer, Berlin, 1996)
2 R. Durrett, *Probability: Theory and Examples*, 4th edn. (Cambridge University Press, Cambridge, 2010). https://doi.org/10.1017/CBO9780511779398
3 D.J. Higham., SIAM Rev. **43**(3), 525 (2001). https://doi.org/10.1137/S0036144500378302
4 S. Chandrasekhar, Rev. Mod. Phys. **15**(1), 1 (1943). https://doi.org/10.1103/RevModPhys.15.1

# Appendix C
# A Paradigmatic Long-Range System: The Hamiltonian Mean-Field Model

In the field of long-range interactions, the Hamiltonian mean-field (HMF) model has played a central role in elucidating many nontrivial features, static and dynamic, arising from long-range interactions, and has a cult status similar to the one enjoyed by the Ising model in statistical mechanics. Here, we provide a brief discussion of the HMF model. For a more extensive review of the model, see Ref. [1], on which this appendix is largely based. The model involves continuous degrees of freedom evolving under Hamiltonian dynamics, and may be taken to provide a simplified framework in which to study such diverse long-range systems as gravitational sheet models and free-electron lasers. The system comprises a bunch of $N$ point particles of mass $m > 0$ moving on a circle and interacting with each other via a mean-field interaction, which constitutes the extreme case of long-range interactions in which there is no decay at all of the inter-particle potential with distance between the particles (see Sect. 3.3). Denoting by $\theta_i \in [0, 2\pi)$ and $p_i \in (-\infty, \infty)$ the generalized coordinate and the momentum of the $i$-th particle, respectively, with $i = 1, 2, \ldots, N$, the Hamiltonian of the system is given by

$$H = \sum_{i=1}^{N} \frac{p_i^2}{2m} + \frac{K}{2N} \sum_{i,j=1}^{N} \left[1 - \cos\left(\theta_i - \theta_j\right)\right], \tag{C.1}$$

where $K > 0$ is the coupling constant. The HMF model may also be interpreted as involving classical $XY$-spins evolving under a dynamics given by a Hamiltonian that involves mean-field $XY$-interaction. Note that the Poisson bracket between the spin components is taken to vanish identically.

The Hamiltonian (C.1) is invariant under the O(2) symmetry group. As we show below, in thermal equilibrium and for energy densities (i.e., energy per particle) smaller than a critical value $\varepsilon_c \equiv 3K/4$, the symmetry gets spontaneously broken leading to a clustered state, whereby a macroscopic number of particles are clustered on the circle. The system consequently undergoes a continuous phase transition at $\varepsilon_c$, from a clustered phase for energy densities $\varepsilon < \varepsilon_c$ to an unclustered phase for



D. Das and S. Gupta, *Facets of Noise*, Fundamental Theories of Physics 214,
https://doi.org/10.1007/978-3-031-45312-0

$\varepsilon \geq \varepsilon_c$. The order parameter of clustering is provided by the quantity $r \equiv \sqrt{r_x^2 + r_y^2}$, defined as

$$(r_x, r_y) \equiv \frac{1}{N}\left(\sum_{i=1}^{N} \cos\theta_i, \sum_{i=1}^{N} \sin\theta_i\right) . \tag{C.2}$$

We note that $r$ is nothing but the magnetization in the $XY$-model.

We now proceed to derive the equilibrium properties of the model, in particular, when the system is in canonical equilibrium at temperature $T$. The canonical partition function, after performing the trivial Gaussian integration over the momenta, is given by

$$\begin{aligned} Z(\beta, N) = \exp\left(-\frac{N\beta K}{2}\right)\left(\frac{2m\pi}{\beta}\right)^{N/2} \\ \times \int \mathrm{d}\theta_1 \ldots \mathrm{d}\theta_N \exp\left\{\frac{\beta K}{2N}\left[\left(\sum_{i=1}^{N}\cos\theta_i\right)^2 + \left(\sum_{i=1}^{N}\sin\theta_i\right)^2\right]\right\} , \end{aligned} \tag{C.3}$$

where $\beta$ is as usual equal to $1/(k_B T)$. Using the Hubbard-Stratonovich transformation

$$\mathrm{e}^{bx^2} = \sqrt{\frac{b}{\pi}} \int_{-\infty}^{\infty} \mathrm{d}y\, \mathrm{e}^{-by^2 + 2bxy}; \quad b > 0 , \tag{C.4}$$

we obtain

$$\begin{aligned} Z(\beta, N) &= \exp\left(-\frac{N\beta K}{2}\right)\left(\frac{2m\pi}{\beta}\right)^{N/2} \int \mathrm{d}\theta_1 \ldots \mathrm{d}\theta_N \frac{\beta K}{2N\pi} \\ &\times \int \mathrm{d}x_1 \mathrm{d}x_2 \exp\left\{-\frac{\beta K(x_1^2 + x_2^2)}{2N} + \frac{\beta K}{N}\left(x_1 \sum_{i=1}^{N}\cos\theta_i + x_2 \sum_{i=1}^{N}\sin\theta_i\right)\right\} \\ &= \exp\left(-\frac{N\beta K}{2}\right)\left(\frac{2m\pi}{\beta}\right)^{N/2} \frac{\beta K}{2N\pi} \int \mathrm{d}x_1 \mathrm{d}x_2 \exp\left\{-\frac{\beta K(x_1^2 + x_2^2)}{2N}\right\} \\ &\times \left(\int \mathrm{d}\theta \exp\left\{\frac{\beta K}{N}(x_1\cos\theta + x_2\sin\theta)\right\}\right)^N . \end{aligned} \tag{C.5}$$

Rescaling the integration variables $x_1$ and $x_2$ by $N$ such that $x_1/N \to x_1$ and $x_2/N \to x_2$, we get from the above equation that

$$Z(\beta, N) = \exp\left(-\frac{N\beta K}{2}\right)\left(\frac{2m\pi}{\beta}\right)^{N/2} \frac{N\beta K}{2\pi} \int \mathrm{d}x_1 \mathrm{d}x_2 \, \exp\left\{-\frac{N\beta K(x_1^2 + x_2^2)}{2}\right\}$$
$$\times \left(\int \mathrm{d}\theta \, \exp\left\{\beta K(x_1 \cos\theta + x_2 \sin\theta)\right\}\right)^N, \quad \text{(C.6)}$$

which yields

$$= \exp\left(-\frac{N\beta K}{2}\right)\left(\frac{2m\pi}{\beta}\right)^{N/2} \frac{N\beta K}{2\pi}$$
$$\times \int \mathrm{d}x_1 \mathrm{d}x_2 \, \exp\left\{N\left[-\frac{\beta K(x_1^2 + x_2^2)}{2} + \ln(2\pi I_0(\beta K(x_1^2 + x_2^2)^{\frac{1}{2}}))\right]\right\}, \quad \text{(C.7)}$$

where $I_0(z)$ is the modified Bessel function of order zero, defined as

$$I_0(z) \equiv \frac{1}{2\pi}\int_0^{2\pi} \mathrm{d}\theta \; \exp(z\cos\theta) = \frac{1}{2\pi}\int_0^{2\pi} \mathrm{d}\theta \; \exp(z_1\cos\theta + z_2\sin\theta), \quad \text{(C.8)}$$

with $z \equiv \sqrt{z_1^2 + z_2^2}$. Expressed in terms of the polar coordinates in the $(x_1, x_2)$, we get

$$Z(\beta, N) = \exp\left(-\frac{N\beta K}{2}\right)\left(\frac{2m\pi}{\beta}\right)^{N/2}$$
$$\times N\beta K \int_0^{\infty} \mathrm{d}x \; x \exp\left\{N\left[-\frac{\beta K x^2}{2} + \ln(2\pi I_0(\beta K x))\right]\right\}. \quad \text{(C.9)}$$

Next, we consider the thermodynamic limit $N \to \infty$, when the integral in (C.9) can be computed by using the saddle point method involving the extremization problem of finding the particular value of $x$ that extremizes the function $\left[-\frac{\beta K x^2}{2} + \ln(2\pi I_0(\beta K x))\right]$. Thus, we are led to solving the equation

$$x = \frac{I_1(\beta K x)}{I_0(\beta K x)}, \quad \text{(C.10)}$$

where $I_1(z) = I_0'(z)$ is the modified Bessel function of order 1. In terms of the solution of this extremization problem, we obtain the rescaled free energy per particle as

$$\phi(\beta) \equiv \beta f(\beta) = -\lim_{N\to\infty} \frac{1}{N} \ln Z(\beta, N)$$
$$= \frac{\beta K}{2} - \frac{1}{2}\ln 2m\pi + \frac{1}{2}\ln\beta + \inf_{x\geq 0}\left[\frac{\beta K x^2}{2} - \ln(2\pi I_0(\beta K x))\right], \quad \text{(C.11)}$$

in which one has to choose the particular solution of (C.10) that minimizes the free energy (C.11). For $\beta K \leq 2$, Eq. (C.10) yields only one solution given by $x = r^* = 0$, while for $\beta K > 2$, we have an additional positive nonzero solution that monotonically increases with $\beta$, and approaches $r^* = 1$ for $\beta \to \infty$. The solution $r^* = 0$ of (C.10), valid for all values of $\beta$, is to be discarded for $\beta K > 2$, since it does not minimize the free energy. It may be shown that the value $r^*$ realizing the extremum in Eq. (C.11) is equal to the spontaneous magnetization in equilibrium. It may be noted that the spontaneous magnetization is defined only up to its modulus, while there is a continuous degeneracy in its direction. On the basis of the foregoing, it is evident that the HMF model displays in equilibrium a continuous phase transition at $\beta_c K = 2$ ($k_B T_c = K/2$). Using Eqs. (4.74) and (4.76), we obtain on taking the derivative of the rescaled free energy $\phi(\beta)$ in Eq. (C.11) with respect to $\beta$ that the energy per particle is given by

$$\varepsilon(\beta) = \frac{1}{2\beta} + \frac{K}{2} - \frac{K}{2}(r^*(\beta))^2 . \tag{C.12}$$

As already evident from the Hamiltonian, the lower bound of $\varepsilon$ is 0. At the critical temperature, we have $\varepsilon_c = 3K/4$. Since the HMF model has a continuous phase transition in the canonical ensemble, we may conclude on the basis of the discussion in Sect. 4.5 that microcanonical and canonical ensembles are equivalent for this model.

For the system (C.1), in order to study the dynamics of relaxation to equilibrium, one studies the Vlasov equation for the evolution of the single-particle phase space density $f_1(q, p, t)$ given by (see Sect. 3.7.3)

$$\frac{\partial f_1}{\partial t} + \frac{p}{m}\frac{\partial f_1}{\partial \theta} - \frac{\partial V[f_1](\theta, t)}{\partial \theta}\frac{\partial f_1}{\partial p} = 0 , \tag{C.13}$$

where the mean-field potential is given by

$$V[f_1](\theta, t) = K \int_0^{2\pi} \mathrm{d}\theta' \int_{-\infty}^{\infty} \mathrm{d}p \, [1 - \cos(\theta - \theta')] \, f_1(\theta', p, t) . \tag{C.14}$$

Note that the one-particle phase space $(\theta, p)$ for the HMF model is a cylinder on account of the fact that while $\theta$ is bounded to lie between 0 and $2\pi$, the momentum variable $p$ is unbounded. Now, it is easily seen that for distributions that are homogeneous with respect to $\theta$, the mean-field potential evaluates to a constant equal to $K$, implying that such distributions are stationary solutions of the Vlasov equation, i.e., $\partial f_1/\partial t = 0$ (see Eq. C.13). We denote such homogeneous stationary solutions by $f_1^{(0)}(p)$. Stationarity obviously does not guarantee stability, and so we now study the stability of such homogeneous distributions with respect to small perturbations, by linearizing the Vlasov equation (C.13) about $f_1^{(0)}(p)$. Skipping details which are amply detailed in standard references on long-range interactions, see Ref. [1], one obtains the result that the homogeneous distribution $f_1^{(0)}(p)$ is stable if and only if the quantity

$$I \equiv 1 + \pi K \int_{-\infty}^{\infty} \mathrm{d}p \, \frac{\partial f_1^{(0)}(p)/\partial p}{p/m} \tag{C.15}$$

is positive. We conclude that there can be an infinite number of Vlasov-stable stationary distributions depending on the form of $f_1^{(0)}(p)$. Below we briefly discuss some representative examples of $f_1^{(0)}(p)$.

- The first one is the Gaussian distribution:

$$f_1^{(0)}(p) = \frac{1}{2\pi}\sqrt{\frac{\beta}{2m\pi}} \exp\left(-\frac{\beta p^2}{2m}\right), \tag{C.16}$$

  which is expected at equilibrium. With the threshold condition (C.15), one recovers the result due to statistical mechanics reviewed above that the critical inverse temperature is $\beta_c K = 2$, and its associated critical stability threshold is $\varepsilon^\star = \varepsilon_c = 3K/4$. In other words, the distribution (C.16) is linearly stable for energies larger than $\varepsilon^\star$ and is unstable below.
- The next example is the so-called waterbag (WB) distribution:

$$f_1^{(0)}(p) = \frac{1}{2\pi}\frac{1}{2p_0}[\Theta(p + p_0) - \Theta(p - p_0)], \tag{C.17}$$

  which has often been used in the past to numerically demonstrate the out-of-equilibrium properties of the HMF model, and a version of which has been considered in Chap. 8. In the present case, such a distribution comprises momentum uniformly distributed in a given interval $[-p_0, p_0]$, where the parameter $p_0$ is related to the energy density as $p_0 = \sqrt{6m\varepsilon - 3mK}$. For such a distribution, one obtains the critical stability threshold $\varepsilon^\star = 7K/12$, and so the WB state is linearly stable for energies larger than $\varepsilon^\star$ and is unstable below.

We emphasize that the mentioned examples of Vlasov-stable stationary solutions are just two among infinitely many others, and there is no a priori reason to emphasize any one of them.

Now that we have learned about the existence of infinitely many Vlasov-stable stationary solutions of the HMF model, it follows then that when starting initially from one such solution in the stable regime (e.g., the water-bag distribution at energy density $\varepsilon > \varepsilon^\star = 7K/12$), and evolving under the Hamilton equations derived from the Hamiltonian (C.1):

$$\frac{\mathrm{d}\theta_i}{\mathrm{d}t} = \frac{p_i}{m}, \quad \frac{\mathrm{d}p_i}{\mathrm{d}t} = -Kr_x \sin\theta_i + Kr_y \cos\theta_i, \tag{C.18}$$

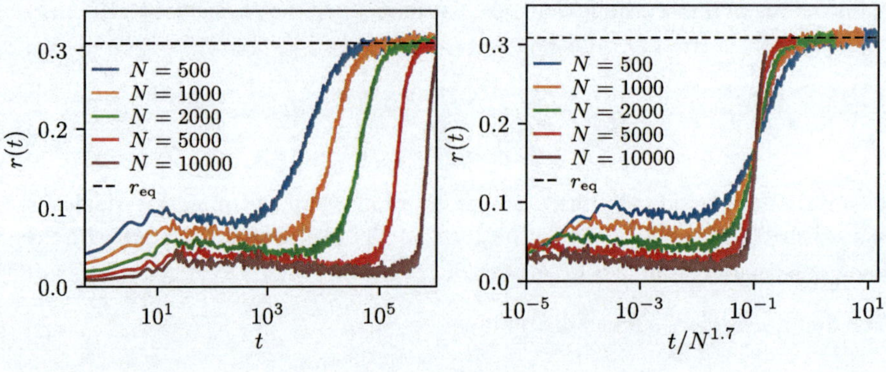


**Fig. C.1 Magnetization $r(t)$ as a function of time $t$ for the HMF model** (C.1). The initial condition corresponds to the waterbag distribution (C.17). Here, we have chosen the mass to be $m = 1$, the coupling constant to be $K = 1$, while the energy density is taken to be $\varepsilon = 0.69 > \varepsilon^\star$. The data are obtained from numerical integration of the equations of motion (C.18). In the plots, the BG equilibrium value of the magnetization shown by the dashed line is given by $r_{\text{eq}} \approx 0.31$. This value may be obtained as follows: we first solve Eq. (C.10) to obtain the canonical equilibrium magnetization $r^\star$ as a function of $\beta$ and then use Eq. (C.12) to obtain the average energy density $\varepsilon(\beta)$ as a function of $\beta$. Since the canonical and microcanonical ensembles are equivalent for the HMF model, we next identify $\varepsilon(\beta)$ with the conserved microcanonical energy density $\varepsilon$ and this way, we obtain $r^\star$ as a function of $\varepsilon$. Setting $\varepsilon = 0.69$ yields the quoted equilibrium value $r_{\text{eq}} \equiv r^\star \approx 0.31$. The panel to the left shows slow relaxation to equilibrium over a size-dependent time scale; the panel to the right suggests the scaling of the relaxation time with system size $N$ as $N^{1.7}$.

the magnetization in an infinite system remains zero at all times. However, when $N$ is finite, the resulting finite-$N$ effects drive the system away from the water-bag state, and through other stable stationary states. Such a slow quasi-stationary evolution across an infinite number of Vlasov-stable stationary states (note: stationary only in the limit $N \to \infty$) ends with the system in the Boltzmann-Gibbs (BG) equilibrium state, as shown in Fig. C.1. For the HMF model, it can be shown rigorously that Vlasov-stable homogeneous distributions do not evolve on timescales of order smaller or equal to $N$, and a scaling $\sim N^\gamma;\ \gamma > 0$, for the timescale of relaxation towards the BG equilibrium state has been observed in simulations. This slow relaxation to BG equilibrium depicted in Fig. C.1 is consistent with the general scenario of relaxation in long-range systems discussed in Sect. 3.7.3.

## Reference

1 A. Campa, T. Dauxois, D. Fanelli, S. Ruffo, *Physics of Long-Range Interacting Systems*, 1st edn. (Oxford University Press, Oxford, 2014). https://doi.org/10.1093/acprof:oso/9780199581931.001.0001

# Appendix D
# Stationary Distribution for a Noisy Nonuniform Oscillator in the Small-$\mathcal{K}$ Limit

The expression for the stationary distribution of the single noisy nonuniform oscillator is given in Eq. (6.71), which we reproduce here:

$$\mathcal{P}_{\rm st}(\theta, \omega) = \mathcal{A}(\omega)\, {\rm e}^{(-\mathcal{K}+\mathcal{K}\cos\theta+\omega\theta)/D}\mathcal{B}(\mathcal{K}), \tag{D.1}$$

where we have defined

$$\mathcal{B}(\mathcal{K}) \equiv 1 + \frac{({\rm e}^{-2\pi\omega/D} - 1)\int_0^\theta {\rm d}\theta'\, {\rm e}^{(-\omega\theta'-\mathcal{K}\cos\theta')/D}}{\int_0^{2\pi} {\rm d}\theta'\, {\rm e}^{(-\omega\theta'-\mathcal{K}\cos\theta')/D}}. \tag{D.2}$$

Let us note the integral

$$\int_0^\theta {\rm d}\theta' {\rm e}^{-\frac{\omega\theta'}{D}}\cos^2\theta' = \frac{4D^2 + 2\omega^2 - {\rm e}^{-\frac{\omega\theta}{D}}\left(4D^2 + \omega^2 + \omega^2\cos 2\theta - 2D\omega\sin 2\theta\right)}{2(4D\omega + \omega^3/D)}. \tag{D.3}$$

Expanding the term $\int_0^\theta {\rm d}\theta'\, {\rm e}^{(-\omega\theta'-\mathcal{K}\cos\theta')/D}$ on the right-hand side of Eq. (D.2) in orders of $\mathcal{K}$ and keeping terms up to $\mathcal{K}^2$, we have

$$\begin{aligned} I \equiv \int_0^\theta {\rm d}\theta'\, {\rm e}^{(-\omega\theta'-\mathcal{K}\cos\theta')/D} &= \int_0^\theta {\rm d}\theta' {\rm e}^{-\omega\theta'/D} - \frac{\mathcal{K}}{D}\int_0^\theta {\rm d}\theta' {\rm e}^{-\omega\theta'/D}\cos\theta' \\ &+ \frac{\mathcal{K}^2}{2D^2}\int_0^\theta {\rm d}\theta' {\rm e}^{-\omega\theta'/D}\cos^2\theta'. \end{aligned} \tag{D.4}$$

Substituting Eqs. (6.55), (6.56), and (D.3) in Eq. (D.4), one obtains



D. Das and S. Gupta, *Facets of Noise*, Fundamental Theories of Physics 214,
https://doi.org/10.1007/978-3-031-45312-0

$$I = \frac{D\left(1-e^{-\omega\theta/D}\right)}{\omega} - \frac{\mathcal{K}\left(\omega - \omega e^{-\omega\theta/D}\cos\theta + De^{-\omega\theta/D}\sin\theta\right)}{D^2+\omega^2} + \frac{\mathcal{K}^2\left[4D^2+2\omega^2 - e^{-\omega\theta/D}\left(4D^2+\omega^2+\omega^2\cos 2\theta - 2D\omega\sin 2\theta\right)\right]}{4D\,(4D^2\omega+\omega^3)}, \quad \text{(D.5)}$$

and consequently,

$$\int_0^{2\pi} d\theta' e^{(-\omega\theta'-\mathcal{K}\cos\theta')/D} = \left(1-e^{-2\pi\omega/D}\right)\left[\frac{D}{\omega} - \frac{\mathcal{K}\omega}{(D^2+\omega^2)} + \frac{\mathcal{K}^2(2D^2+\omega^2)}{2D(4D^2\omega+\omega^3)}\right]. \quad \text{(D.6)}$$

Using Eqs. (D.5) and (D.6), we obtain $\mathcal{B}(\mathcal{K})$ from Eq. (D.2) up to order $\mathcal{K}^2$ as

$$\mathcal{B}(\mathcal{K}) = e^{-\omega\theta/D}\left[1 + \frac{\mathcal{K}\omega(\omega - \omega\cos\theta + D\sin\theta)}{D(D^2+\omega^2)} + \frac{\mathcal{K}^2(-\omega^2+\omega^2\cos 2\theta - 2D\omega\sin 2\theta)}{4D^2(4D^2+\omega^2)} + \frac{\mathcal{K}^2(\omega^4 - \omega^4\cos\theta + D\omega^3\sin\theta)}{D^2(D^2+\omega^2)^2}\right]. \quad \text{(D.7)}$$

Substituting Eq. (D.7) in Eq. (D.1), we obtain up to order $\mathcal{K}^2$ that

$$\begin{aligned}\mathcal{P}_{\rm st}(\theta,\omega) = \mathcal{A}(\omega)\Bigg[1 &- \frac{\mathcal{K}(D - D\cos\theta - \omega\sin\theta)}{D^2+\omega^2} + \frac{\mathcal{K}^2}{2D^2(D^2+\omega^2)}\Big(D^2-\omega^2 \\ &+ D^2\cos^2\theta - \omega^2\cos^2\theta - 2D^2\cos\theta + 2\omega^2\cos\theta + D\omega\sin 2\theta - 2D\omega\sin\theta\Big) \\ &- \frac{\mathcal{K}^2\omega}{4D^2(D^2+\omega^2)^2(4D^2+\omega^2)}\Big(D^4\omega - 14D^2\omega^3 - 3\omega^5 + 16D^2\omega^3\cos\theta \\ &+ 4\omega^5\cos\theta - D^4\omega\cos 2\theta - 2D^2\omega^3\cos 2\theta - \omega^5\cos 2\theta - 16D^3\omega^2\sin\theta \\ &- 4D\omega^4\sin\theta + 2D^5\sin 2\theta + 4D^3\omega^2\sin 2\theta + 2D\omega^4\sin 2\theta\Big)\Bigg]. \quad \text{(D.8)}\end{aligned}$$

Invoking the normalization $\int_0^{2\pi} d\theta\, \mathcal{P}_{\rm st}(\theta,\omega) = 1$, one obtains from Eq. (D.8) that

$$2\pi\,\mathcal{A}(\omega)\left[1 - \frac{\mathcal{K}D}{D^2+\omega^2} + \frac{\mathcal{K}^2(6D^4+D^2\omega^2+\omega^4)}{2(D^2+\omega^2)^2(4D^2+\omega^2)}\right] = 1, \quad \text{(D.9)}$$

which up to order $\mathcal{K}^2$ yields

$$\mathcal{A}(\omega) = \frac{1}{2\pi}\left[1 + \frac{\mathcal{K}D}{D^2+\omega^2} + \frac{\mathcal{K}^2(2D^2-\omega^2)}{2(D^2+\omega^2)(4D^2+\omega^2)}\right]. \quad \text{(D.10)}$$

Substituting Eq. (D.10) in Eq. (D.8), we obtain the stationary distribution $\mathcal{P}_{\mathrm{st}}(\theta, \omega)$ up to order $\mathcal{K}^2$, as

$$\mathcal{P}_{\mathrm{st}}(\theta, \omega) = \frac{1}{2\pi} + \frac{\mathcal{K}(D\cos\theta + \omega\sin\theta)}{2\pi(D^2 + \omega^2)} + \frac{\mathcal{K}^2\{(2D^2 - \omega^2)\cos 2\theta + 3D\omega\sin 2\theta\}}{4\pi(D^2 + \omega^2)(4D^2 + \omega^2)}, \tag{D.11}$$

which is Eq. (6.99) of the main text.

# Appendix E
# Gaussian White Noise in Discrete Times

The Gaussian, white noise $\eta(t)$ in continuous time $t$ is characterized by

$$\langle \eta(t) \rangle = 0\,, \quad \langle \eta(t)\eta(t') \rangle = 2D\delta(t - t')\,, \tag{E.1}$$

where $D > 0$ is a constant. Here, angular brackets denote as usual averaging with respect to noise realizations. Let us discretize time in small steps of $\Delta t \to 0^+$, such that the $i$-th time step is denoted by $t_i = i\Delta t$; $i = 0, 1, 2, \ldots$. In this setting, let us choose two different times $t$ and $t'$, as

$$t = t_i = i\Delta t\,, \quad t' = t_j = j\Delta t\,; \quad i \neq j\,. \tag{E.2}$$

Suppose $f(t)$ is an arbitrary function of time. Then, using the definition of the Kronecker delta function, one may write

$$f_i \equiv f(t_i) = \sum_{j=-\infty}^{\infty} f_j\, \delta_{i,j} = \sum_{j=-\infty}^{\infty} \Delta t\, f_j\, \frac{\delta_{i,j}}{\Delta t}\,. \tag{E.3}$$

In the limit $\Delta t \to 0^+$, identifying

$$\sum_{j=-\infty}^{\infty} \Delta t \equiv \int_{-\infty}^{\infty} \mathrm{d}t', \tag{E.4}$$

and with $f_j = f(t')$, $f_i = f(t)$, we obtain from Eq. (E.3) that

$$f(t) = \int_{-\infty}^{\infty} \mathrm{d}t' \left[ \lim_{\Delta t \to 0} \frac{\delta_{i,j}}{\Delta t} \right] f(t')\,. \tag{E.5}$$


D. Das and S. Gupta, *Facets of Noise*, Fundamental Theories of Physics 214,
https://doi.org/10.1007/978-3-031-45312-0

From the definition of the Dirac delta function, one has

$$f(t) = \int_{-\infty}^{\infty} \mathrm{d}t' \, \delta(t - t') \, f(t') \, . \tag{E.6}$$

Hence, comparing Eqs. (E.5) and (E.6), we conclude that

$$\delta(t - t') = \lim_{\Delta t \to 0} \frac{\delta_{i,j}}{\Delta t} \, . \tag{E.7}$$

Using Eq. (E.7), and identifying $\eta_i = \eta(i \Delta t)$, one may write down from Eq. (E.1) and in the limit $\Delta t \to 0^+$ the properties of the Gaussian, white noise in discrete times as

$$\langle \eta_i \rangle = 0 \, , \quad \langle \eta_i \eta_j \rangle = \frac{2D}{\Delta t} \, \delta_{i,j} \, . \tag{E.8}$$

# Appendix F
# Review of Feynman's Path Integral Formalism

## F.1 The Propagator: Physical Meaning

The evolution of a quantum state is understood in terms of a time evolution operator. One may relate a quantum state $|\psi(t_0)\rangle$ at time $t_0$ to its evolved counterpart $|\psi(t)\rangle$ at a later time $t$ by

$$|\psi(t)\rangle = \widehat{U}(t, t_0)\,|\psi(t_0)\rangle\,; \qquad t > t_0\,, \tag{F.1}$$

where $\widehat{U}(t, t_0)$ is the time evolution operator satisfying

$$\widehat{U}(t_0, t_0) = \mathbb{I}\,. \tag{F.2}$$

Here, $\mathbb{I}$ is the identity operator.

One may obtain the form of operator $\widehat{U}(t, t_0)$ by substituting Eq. (F.1) into the time-dependent Schrödinger equation:

$$\mathrm{i}\hbar\frac{\partial}{\partial t}|\psi(t)\rangle = \widehat{H}|\psi(t)\rangle\,, \tag{F.3}$$

where $\widehat{H}$ is the Hamiltonian of the system. Then, one obtains

$$\frac{\partial}{\partial t}\widehat{U}(t, t_0) = -\frac{\mathrm{i}}{\hbar}\widehat{H}\widehat{U}(t, t_0)\,. \tag{F.4}$$

For a time-independent Hamiltonian, Eq. (F.4) may be solved by using the initial condition, Eq. (F.2). Then, one obtains

$$\widehat{U}(t, t_0) = \mathrm{e}^{-\frac{\mathrm{i}}{\hbar}\widehat{H}(t-t_0)}\,. \tag{F.5}$$



D. Das and S. Gupta, *Facets of Noise*, Fundamental Theories of Physics 214,
https://doi.org/10.1007/978-3-031-45312-0

Once we know the operator $\widehat{U}(t, t_0)$, we are able to predict the future of the quantum state. Therefore, the whole of quantum mechanics lies in obtaining $\widehat{U}(t, t_0)$. This time evolution operator is also called the propagator.

For an illustration of the above method, let us consider a quantum particle of mass $m$ moving in one dimension $x$ in presence of a time-independent potential. The state of the particle at time $t$ is denoted by $|\psi(t)\rangle$, so that $\langle x|\psi(t)\rangle = \psi(x, t)$ is the wave function. Then, $|\psi(x, t)|^2 \mathrm{d}x$ gives the probability to observe the particle between $x$ and $x + \mathrm{d}x$ at time $t$. The Hamiltonian operator for the particle is given by

$$\widehat{H} = \frac{\widehat{p}^2}{2m} + \widehat{V}\,, \tag{F.6}$$

where $\widehat{p}$ is the momentum operator, and $\widehat{V}$ is the potential energy operator. In order to understand the physical meaning of the propagator $\widehat{U}(t, t_0)$, one may think of the operator in some basis. For example, writing Eq. (F.1) in the position basis, one obtains

$$\langle x|\psi(t)\rangle = \langle x|\,\widehat{U}(t, t_0)\,|\psi(t_0)\rangle = \int_{-\infty}^{\infty} \mathrm{d}x_0\,\langle x|\mathrm{e}^{-\frac{\mathrm{i}}{\hbar}\widehat{H}(t-t_0)}|x_0\rangle\,\langle x_0|\psi(t_0)\rangle\,, \tag{F.7}$$

where we have used the completeness relation $\int_{-\infty}^{\infty} \mathrm{d}x_0\,|x_0\rangle\langle x_0| = \mathbb{I}$. Denoting

$$U(x, t\,; x_0, t_0) \equiv \langle x|\mathrm{e}^{-\frac{\mathrm{i}}{\hbar}\widehat{H}(t-t_0)}|x_0\rangle\,, \tag{F.8}$$

one obtains from Eq. (F.7) that

$$\psi(x, t) = \int_{-\infty}^{\infty} \mathrm{d}x_0\,U(x, t\,; x_0, t_0)\,\psi(x_0, t_0)\,. \tag{F.9}$$

From Eq. (F.9), it is seen that $U(x, t\,; x_0, t_0)$ denotes the transition probability amplitude for the particle to be at position $x$ at time $t > t_0$, given that the particle was at position $x_0$ at time $t_0$.

Let us now compute the propagator for a free particle in the position basis. In this case, the Hamiltonian operator is given by $\widehat{H} = \widehat{p}^2/(2m)$. To this end, we start with

$$U(x, t\,; x_0, t_0) = \langle x|\mathrm{e}^{-\frac{\mathrm{i}}{\hbar}\widehat{H}(t-t_0)}|x_0\rangle = \langle x|\mathrm{e}^{-\frac{\mathrm{i}}{2m\hbar}\widehat{p}^2(t-t_0)}|x_0\rangle\,. \tag{F.10}$$

Using the completeness relation of the momentum eigenstates $\int_{-\infty}^{\infty} \mathrm{d}p\,|p\rangle\,\langle p| = \mathbb{I}$, where $p$ denotes the eigenvalue of the momentum operator $\widehat{p}$ such that $\widehat{p}|p\rangle = p|p\rangle$, Eq. (F.10) is reduced to

$$U(x, t\,; x_0, t_0) = \int_{-\infty}^{\infty} \mathrm{d}p\,\mathrm{e}^{-\frac{\mathrm{i}}{2m\hbar}p^2(t-t_0)}\,\langle x|p\rangle\,\langle p|x_0\rangle\,. \tag{F.11}$$

Using the representation of the momentum operator in the position basis, we get for any state $|\Psi\rangle$ that

$$\langle x|\widehat{p}|\Psi\rangle = -\mathrm{i}\hbar\frac{\partial}{\partial x}\langle x|\Psi\rangle\,. \tag{F.12}$$

For $|\Psi\rangle = |p\rangle$, one gets from Eq. (F.12) that

$$-\,\mathrm{i}\hbar\frac{\partial}{\partial x}\langle x|p\rangle = p\,\langle x|p\rangle\,, \tag{F.13}$$

for which the solution is given by

$$\langle x|p\rangle = \mathcal{A}\,\mathrm{e}^{\frac{\mathrm{i}}{\hbar}px}\,, \tag{F.14}$$

where $\mathcal{A}$ is the normalization constant. One may find the value of $\mathcal{A}$ by using

$$\langle x|x'\rangle = \int_{-\infty}^{\infty}\mathrm{d}p\,\langle x|p\rangle\,\langle p|x'\rangle = |\mathcal{A}|^2\int_{-\infty}^{\infty}\mathrm{d}p\,\mathrm{e}^{\frac{\mathrm{i}}{\hbar}p(x-x')} = 2\pi\hbar|\mathcal{A}|^2\,\delta(x-x')\,. \tag{F.15}$$

The inner product $\langle x|x'\rangle$ is equal to $\delta(x-x')$. Then, choosing $\mathcal{A}$ to be a positive real constant by convention, we get from Eq. (F.15) that $\mathcal{A} = 1/\sqrt{2\pi\hbar}$. Hence, we have

$$\langle x|p\rangle = \frac{1}{\sqrt{2\pi\hbar}}\,\mathrm{e}^{\frac{\mathrm{i}}{\hbar}px};\qquad \langle p|x\rangle = \frac{1}{\sqrt{2\pi\hbar}}\,\mathrm{e}^{-\frac{\mathrm{i}}{\hbar}px}\,. \tag{F.16}$$

Substituting Eq. (F.16) in Eq. (F.11), we obtain

$$U(x,t\,;x_0,t_0) = \frac{1}{2\pi\hbar}\int_{-\infty}^{\infty}\mathrm{d}p\,\mathrm{e}^{-\frac{\mathrm{i}}{2m\hbar}p^2(t-t_0)}\,\mathrm{e}^{\frac{\mathrm{i}}{\hbar}p(x-x_0)}\,. \tag{F.17}$$

Now, using the result for Gaussian integral:

$$\int_{-\infty}^{\infty}\mathrm{d}p\,\mathrm{e}^{-\alpha p^2+\beta p+\gamma} = \sqrt{\frac{\pi}{\alpha}}\,\mathrm{e}^{\frac{\beta^2}{4\alpha}+\gamma}\,;\quad \mathrm{Re}[\alpha] > 0\,, \tag{F.18}$$

which may also be used for purely imaginary[1] $\alpha$, one obtains the propagator for a free particle from Eq. (F.17) as

$$U(x,t\,;x_0,t_0) = \sqrt{\frac{m}{2\pi\mathrm{i}\hbar(t-t_0)}}\,\exp\left[\frac{im(x-x_0)^2}{2\hbar(t-t_0)}\right]. \tag{F.19}$$

---

[1] In this case, one first multiplies the integrand on the left hand side of Eq. (F.18) by $\mathrm{e}^{-\epsilon p^2}$; $\epsilon > 0$ is real. Then, the limit $\epsilon \to 0^+$ is taken after the integration is performed.

## F.2 The Recipe for the Path Integral Formalism

We have seen in the preceding subsection that in order to obtain the propagator, we had to use the eigenvalues and the eigenstates of the Hamiltonian operator. There is an alternative formalism of quantum mechanics invented by Richard Phillips Feynman, namely, the path integral formalism, which allows to obtain the propagator differently [1]. To discuss the formalism, let us consider a system of $M$ particles evolving in three-dimensional space. The position of the $i$-th particle at time $t$ is specified by the vector $\mathbf{r}^{(i)}(t)$. In a $3M$-dimensional space, the 'position' of the whole system at time $t$ is given by the set of vectors $\{\mathbf{r}^{(i)}(t)\}_{1\leq i\leq M}$. Let us denote the positions at two times $t_a$ and $t_b(> t_a)$ by $\{\mathbf{r}_a^{(i)}\}$ and $\{\mathbf{r}_b^{(i)}\}$, respectively. In the path integral formalism, the recipe to find the propagator $U(\mathbf{r}_b^{(1)}, \mathbf{r}_b^{(2)}, \ldots, \mathbf{r}_b^{(M)}, t_b\,; \mathbf{r}_a^{(1)}, \mathbf{r}_a^{(2)}, \ldots, \mathbf{r}_a^{(M)}, t_a)$ is the following.

1. Consider all possible paths in the $(3M + 1)$-dimensional ($3M$ space dimensions and one time dimension) extended space that connect the points $(\{\mathbf{r}_a^{(i)}\}, t_a)$ and $(\{\mathbf{r}_b^{(i)}\}, t_b)$. A single path is defined by $3M$ time-dependent variables: $\{\mathbf{r}^{(i)}(t)\} \equiv \{r^{(i),(\alpha)}(t)\}_{1\leq i\leq M, 1\leq\alpha\leq 3}$. The $\alpha$ index denotes the $\alpha$-th component of the vector $\mathbf{r}^{(i)}(t)$.
2. Compute the action $S[\{\mathbf{r}^{(i)}(t)\}]$ for every path. For a given path $\{\mathbf{r}^{(i)}(t)\}$ connecting the points $(\{\mathbf{r}_a^{(i)}\}, t_a)$ and $(\{\mathbf{r}_b^{(i)}\}, t_b)$, the action $S[\{\mathbf{r}^{(i)}(t)\}]$ is defined by the time integral of the Lagrangian $\mathcal{L}(\{\mathbf{r}^{(i)}(t), \dot{\mathbf{r}}^{(i)}(t)\}\,; t)$ along the path. Here, the dot over $\mathbf{r}^{(i)}(t)$ denotes the derivative with respect to time $t$. In other words, one has

$$S[\{\mathbf{r}^{(i)}(t)\}] = \int_{t_a}^{t_b} \mathrm{d}t\ \mathcal{L}(\{\mathbf{r}^{(i)}(t), \dot{\mathbf{r}}^{(i)}(t)\}\,; t)\,. \tag{F.20}$$

3. The propagator is obtained as

$$\begin{aligned} &U(\mathbf{r}_b^{(1)}, \mathbf{r}_b^{(2)}, \ldots, \mathbf{r}_b^{(M)}, t_b\,; \mathbf{r}_a^{(1)}, \mathbf{r}_a^{(2)}, \ldots, \mathbf{r}_a^{(M)}, t_a) \\ &= \int_{\mathbf{r}^{(i)}(t_a)=\mathbf{r}_a^{(i)}}^{\mathbf{r}^{(i)}(t_b)=\mathbf{r}_b^{(i)}} \mathcal{D}^3[\mathbf{r}^{(1)}(t)]\,\mathcal{D}^3[\mathbf{r}^{(2)}(t)]\,\cdots\,\mathcal{D}^3[\mathbf{r}^{(M)}(t)]\,\mathrm{e}^{\frac{\mathrm{i}}{\hbar}S[\{\mathbf{r}^{(i)}(t)\}]}\,, \end{aligned} \tag{F.21}$$

where $\int_{\mathbf{r}^{(i)}(t_a)=\mathbf{r}_a^{(i)}}^{\mathbf{r}^{(i)}(t_b)=\mathbf{r}_b^{(i)}} \mathcal{D}^3[\mathbf{r}^{(1)}(t)]\,\mathcal{D}^3[\mathbf{r}^{(2)}(t)]\,\cdots\,\mathcal{D}^3[\mathbf{r}^{(M)}(t)]$ is a symbolic way to mean one has to integrate over all paths connecting $\{\mathbf{r}_a^{(i)}\}$ and $\{\mathbf{r}_b^{(i)}\}$ in the interval $t_b - t_a$. As seen in Eq. (F.21), the weight factor for each path is given by $\mathrm{e}^{\frac{\mathrm{i}}{\hbar}S[\{\mathbf{r}^{(i)}(t)\}]}$.

Let us use the above prescription to compute once again the free particle propagator in one dimension $x$. In this case, the position of the particle at time $t$ is specified by $x(t)$. In order to proceed, we discretize a given path $x(t)$ connecting $(x_a, t_a)$ and $(x_b, t_b)$ into $N$ fragments:

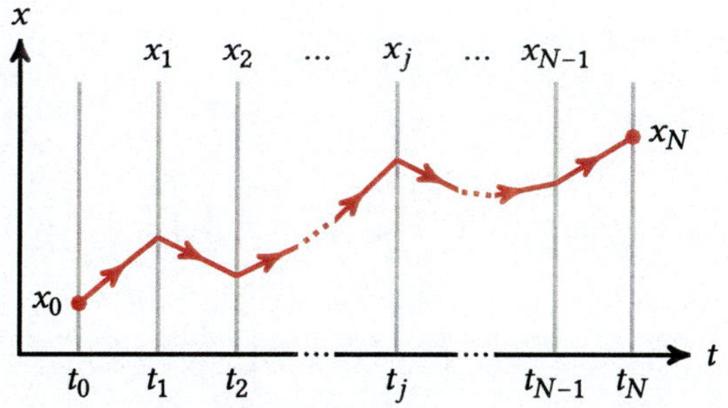


**Fig. F.1 Discretization of a path** $x(t)$. A path $x(t)$ is completely specified by the values of $x_0, x_1, x_2, \ldots, x_N$. To consider all possible paths that connect given points $x_0$ and $x_N$, one has to integrate over the remaining variables: $x_1, x_2, \ldots, x_{N-1}$.

$$x_0 \equiv x_a\,, \quad x_N \equiv x_b\,, \quad t_0 \equiv t_a\,, \quad t_N \equiv t_b\,, \tag{F.22}$$

$$t_j = t_0 + j\Delta t\,, \quad \Delta t = \frac{t_N - t_0}{N}\,; \quad j = 0, 1, 2, \ldots, N, \tag{F.23}$$

$$x_j = x(t_j)\,. \tag{F.24}$$

In Fig. F.1, we have shown this discretization schematically. In such a discretization, for given values of $x_0$ and $x_N$, every single path is specified by $N - 1$ variables $x_1, \ldots, x_{N-1}$. In order to sum over all paths, we need to integrate over all possible values of these $N - 1$ variables. Finally, to obtain the result in the continuum limit, one has to take the limit $N \to \infty$, $\Delta t \to 0$, while keeping $N\Delta t = t_N - t_0$ fixed and finite.

The action of a free particle is given by

$$S[x(t)] = \int_{t_0}^{t_N} \mathrm{d}t\ \mathcal{L}(x, \dot{x}; t) = \frac{m}{2}\int_{t_0}^{t_N} \mathrm{d}t\ \left(\dot{x}(t)\right)^2, \tag{F.25}$$

which in the discrete form is obtained as

$$S[x(t)] = \frac{m}{2}\sum_{j=0}^{N-1} \Delta t \left(\frac{x_{j+1} - x_j}{\Delta t}\right)^2. \tag{F.26}$$

In this discrete form, using Eq. (F.26), one may recast Eq. (F.21) as

$$U(x_N, t_N\,; x_0, t_0) = \lim_{\substack{N\to\infty \\ \Delta t\to 0}} A \left(\prod_{j=1}^{N-1}\int_{-\infty}^{\infty} \mathrm{d}x_j\right) \exp\left[\frac{im}{2\hbar}\sum_{j=0}^{N-1}\frac{(x_{j+1} - x_j)^2}{\Delta t}\right], \tag{F.27}$$

where $A$ is an overall normalization constant. In terms of a new set of variables

$$y_j \equiv \sqrt{\frac{m}{2\hbar\Delta t}}\, x_j\,, \tag{F.28}$$

Equation (F.27) is reduced to

$$
\begin{aligned}
U(y_N, t_N\,;\, y_0, t_0) \\
&= \lim_{\substack{N\to\infty \\ \Delta t\to 0}} A \left(\frac{2\hbar\Delta t}{m}\right)^{(N-1)/2} \left(\prod_{j=2}^{N-1} \int_{-\infty}^{\infty} \mathrm{d}y_j\right) \exp\left[-\sum_{j=2}^{N-1} \frac{(y_{j+1}-y_j)^2}{\mathrm{i}}\right] \\
&\quad \times \int_{-\infty}^{\infty} \mathrm{d}y_1 \exp\left[-\frac{(y_2-y_1)^2+(y_1-y_0)^2}{\mathrm{i}}\right]. \qquad \text{(F.29)}
\end{aligned}
$$

Using the Gaussian integral (F.18), one may compute the integral over $y_1$. Then, Eq. (F.29) reduces to

$$
\begin{aligned}
U(y_N, t_N\,;\, y_0, t_0) \\
&= \lim_{\substack{N\to\infty \\ \Delta t\to 0}} A \left(\frac{2\hbar\Delta t}{m}\right)^{(N-1)/2} \left(\prod_{j=2}^{N-1} \int_{-\infty}^{\infty} \mathrm{d}y_j\right) \exp\left[-\sum_{j=2}^{N-1} \frac{(y_{j+1}-y_j)^2}{\mathrm{i}}\right] \\
&\quad \times \sqrt{\frac{\mathrm{i}\pi}{2}} \exp\left[-\frac{(y_2-y_0)^2}{2\mathrm{i}}\right] \\
&= \lim_{\substack{N\to\infty \\ \Delta t\to 0}} A \left(\frac{2\hbar\Delta t}{m}\right)^{(N-1)/2} \left(\prod_{j=3}^{N-1} \int_{-\infty}^{\infty} \mathrm{d}y_j\right) \exp\left[-\sum_{j=3}^{N-1} \frac{(y_{j+1}-y_j)^2}{\mathrm{i}}\right] \\
&\quad \times \sqrt{\frac{\mathrm{i}\pi}{2}} \int_{-\infty}^{\infty} \mathrm{d}y_2 \exp\left[-\frac{(y_3-y_2)^2}{\mathrm{i}} - \frac{(y_2-y_0)^2}{2\mathrm{i}}\right]. \qquad \text{(F.30)}
\end{aligned}
$$

Using the Gaussian integral (F.18) again and computing the integral over $y_2$, one gets from Eq. (F.30) that

$$
\begin{aligned}
U(y_N, t_N\,;\, y_0, t_0) \\
&= \lim_{\substack{N\to\infty \\ \Delta t\to 0}} A \left(\frac{2\hbar\Delta t}{m}\right)^{(N-1)/2} \left(\prod_{j=3}^{N-1} \int_{-\infty}^{\infty} \mathrm{d}y_j\right) \exp\left[-\sum_{j=3}^{N-1} \frac{(y_{j+1}-y_j)^2}{\mathrm{i}}\right] \\
&\quad \times \sqrt{\frac{\mathrm{i}\pi}{2}} \sqrt{\frac{2\mathrm{i}\pi}{3}} \exp\left[-\frac{(y_3-y_0)^2}{3\mathrm{i}}\right]. \qquad \text{(F.31)}
\end{aligned}
$$

Similarly, carrying out all the remaining integrals over $y_3, y_4, \ldots y_{N-1}$, we obtain from Eq. (F.31) that

$$
U(y_N, t_N\,;\, y_0, t_0) = \lim_{\substack{N\to\infty \\ \Delta t\to 0}} A \left(\frac{2\mathrm{i}\pi\hbar\Delta t}{m}\right)^{N/2} \sqrt{\frac{m}{2\mathrm{i}\pi\hbar N\Delta t}} \exp\left[-\frac{(y_N-y_0)^2}{N\mathrm{i}}\right], \qquad \text{(F.32)}
$$

which when reverted to the variables $x_N,\ x_0$ by using Eq. (F.28) yields

$$U(x_N, t_N\,; x_0, t_0) = \lim_{\substack{N\to\infty \\ \Delta t\to 0}} A\left(\frac{2i\pi\hbar\Delta t}{m}\right)^{N/2}\sqrt{\frac{m}{2i\pi\hbar N\Delta t}}\exp\left[-\frac{m(x_N-x_0)^2}{2i\hbar N\Delta t}\right]. \tag{F.33}$$

Now, using the condition $N\Delta t = t_N - t_0 =$ fixed and finite, one obtains from Eq. (F.33) that

$$U(x_N, t_N; x_0, t_0) = \lim_{\substack{N\to\infty \\ \Delta t\to 0}} A\left(\frac{2i\pi\hbar\Delta t}{m}\right)^{N/2}\sqrt{\frac{m}{2i\pi\hbar(t_N-t_0)}}\exp\left[-\frac{m(x_N-x_0)^2}{2i\hbar(t_N-t_0)}\right]. \tag{F.34}$$

As $t_N \to t_0$, we know that $U(x_N, t_N\,; x_0, t_0) \to \delta(x_N - x_0)$. Then, using the following representation of the delta function:

$$\delta(x_N - x_0) = \lim_{\Delta\to 0}\frac{1}{\sqrt{\pi\Delta^2}}\exp\left[-\frac{(x_N-x_0)^2}{\Delta^2}\right], \tag{F.35}$$

and considering Eq. (F.34) in the limit $t_N \to t_0$, we obtain

$$A = \left(\frac{m}{2i\pi\hbar\Delta t}\right)^{N/2}. \tag{F.36}$$

Substituting the value of the normalization constant $A$ in Eq. (F.34), we obtain the free particle propagator as

$$U(x_N, t_N\,; x_0, t_0) = \sqrt{\frac{m}{2i\pi\hbar(t_N-t_0)}}\exp\left[\frac{im(x_N-x_0)^2}{2\hbar(t_N-t_0)}\right], \tag{F.37}$$

which with the identification $x_N = x,\ t_N = t$ is exactly the same as Eq. (F.19).

Conventionally, one absorbs the constant $A$ into the definition of the path integral. For a single particle in one dimension, the path integral is defined as

$$\int \mathcal{D}[x(t)] = \lim_{\substack{N\to\infty \\ \Delta t\to 0}}\left(\sqrt{\frac{m}{2i\pi\hbar\Delta t}}\right)^N\left(\prod_{j=1}^{N-1}\int_{-\infty}^{\infty}\mathrm{d}x_j\right). \tag{F.38}$$

In general, for $M$ particles each with mass $m$ moving in three dimensions, one defines the path integral as

$$\int \mathcal{D}^3[\mathbf{r}^{(1)}(t)]\,\mathcal{D}^3[\mathbf{r}^{(2)}(t)]\,\cdots\,\mathcal{D}^3[\mathbf{r}^{(M)}(t)]$$

$$= \lim_{\substack{N\to\infty\\ \Delta t\to 0}} \left(\sqrt{\frac{m}{2\mathrm{i}\pi\hbar\Delta t}}\right)^{3MN} \left(\prod_{i=1}^{M}\prod_{\alpha=1}^{3}\prod_{k=1}^{N-1}\int_{-\infty}^{\infty} \mathrm{d}r_k^{(i),(\alpha)}\right). \tag{F.39}$$

For a particle moving in one dimension under the influence of a potential $V(x)$, we have the Lagrangian $\mathcal{L}(x,\dot{x}) = m\dot{x}^2/2 - V(x)$. Then, using Eqs. (F.21) and (F.38), one writes the propagator for the particle as

$$U(x_N, t_N\,; x_0, t_0) = \langle x_N|\mathrm{e}^{-\frac{\mathrm{i}}{\hbar}\widehat{H}(t_N-t_0)}|x_0\rangle$$

$$= \lim_{\substack{N\to\infty\\ \Delta t\to 0}} \left(\sqrt{\frac{m}{2\mathrm{i}\pi\hbar\Delta t}}\right)^{N} \left(\prod_{j=1}^{N-1}\int_{-\infty}^{\infty} \mathrm{d}x_j\right)$$

$$\times \exp\left[\frac{\mathrm{i}}{\hbar}\sum_{j=0}^{N-1}\Delta t\left\{\frac{m}{2}\left(\frac{x_{j+1}-x_j}{\Delta t}\right)^2 - V(x_j)\right\}\right]. \tag{F.40}$$

Note that the Hamiltonian operator occurring in Eq. (F.40) is given by Eq. (F.6), which in position basis may be written as

$$\widehat{H} \equiv -\frac{\hbar^2}{2m}\frac{\partial^2}{\partial x^2} + V(x). \tag{F.41}$$

## F.3 The Feynman-Kac Formalism

In the Feynman-Kac formalism, one replaces the real time $t$ by imaginary time: $t \to -\mathrm{i}t$, and sets the reduced Planck's constant $\hbar$ to unity. Consequently, $\Delta t$ is replaced by $-\mathrm{i}\Delta t$, and the time-evolution operator $\mathrm{e}^{-\mathrm{i}\widehat{H}(t_N-t_0)/\hbar}$ is replaced by $\mathrm{e}^{-\widehat{H}(t_N-t_0)}$. With such replacements, Eq. (F.40) reduces to

$$U(x_N, -\mathrm{i}t_N\,; x_0, -\mathrm{i}t_0) \equiv \lim_{\substack{N\to\infty\\ \Delta t\to 0}} \left(\sqrt{\frac{m}{2\pi\,\Delta t}}\right)^{N} \left(\prod_{j=1}^{N-1}\int_{-\infty}^{\infty} \mathrm{d}x_j\right) \exp\left[-S[x(t)]\right], \tag{F.42}$$

where $S[x(t)]$ is the so-called Euclidean action in discrete times given by

$$S[x(t)] = \sum_{j=0}^{N-1}\Delta t\left\{\frac{m}{2}\left(\frac{x_{j+1}-x_j}{\Delta t}\right)^2 + V(x_j)\right\}. \tag{F.43}$$

In the continuous-time limit, one may write from Eq. (F.42) that

$$\langle x|\mathrm{e}^{-\widehat{H}(t-t_0)}|x_0\rangle = \int_{x(t_0)=x_0}^{x(t)=x} \mathcal{D}[x(t)]\ \exp\left(-S[x(t)]\right)\ , \tag{F.44}$$

with

$$\widehat{H} \equiv -\frac{1}{2m}\frac{\partial^2}{\partial x^2} + V(x)\,, \quad S[x(t)] = \int_{t_0}^{t} \mathrm{d}t\ \left[\frac{m}{2}\{\dot{x}(t)\}^2 + V(x(t))\right]. \tag{F.45}$$

### F.3.1 An Example: A Free Particle Diffusing on a Ring

The dynamical equation of a free particle diffusing on a ring is given in terms of its angle $\theta \in [0, 2\pi)$ as

$$\frac{\mathrm{d}\theta}{\mathrm{d}t} = \eta(t)\,, \tag{F.46}$$

where $\eta(t)$ is a Gaussian, white noise satisfying

$$\langle \eta(t)\rangle = 0, \quad \langle \eta(t)\eta(t')\rangle = 2D\,\delta(t-t')\,, \tag{F.47}$$

with $D > 0$, and angular brackets denoting as usual the average over noise realizations. For a given value $\theta_0$ at $t = 0$, the conditional probability distribution to observe a value around $\theta$ at time $t$, denoted by $P(\theta, t|\theta_0, 0)$, satisfies the following Fokker-Planck equation:

$$\frac{\partial P(\theta, t|\theta_0, 0)}{\partial t} = D\,\frac{\partial^2 P(\theta, t|\theta_0, 0)}{\partial \theta^2}\,, \tag{F.48}$$

with initial condition

$$P(\theta, 0|\theta_0, 0) = \delta(\theta - \theta_0)\,. \tag{F.49}$$

As $P(\theta, t|\theta_0, 0)$ is $2\pi$-periodic in $\theta$, we may expand it as a Fourier series in $\theta$:

$$P(\theta, t|\theta_0, 0) = \sum_{n=-\infty}^{\infty} \widetilde{P}_n(t|\theta_0, 0)\ \mathrm{e}^{in\theta}\,. \tag{F.50}$$

Substituting Eq. (F.50) in Eq. (F.48), one obtains the time evolution of the Fourier coefficients $\widetilde{P}_n$ as

$$\frac{\partial \widetilde{P}_n(t|\theta_0, 0)}{\partial t} = -n^2 D\ \widetilde{P}_n(t|\theta_0, 0)\,, \tag{F.51}$$

while Eq. (F.49) yields

$$\widetilde{P}_n(0|\theta_0, 0) = \frac{1}{2\pi}\,\mathrm{e}^{-in\theta_0}\,. \tag{F.52}$$

With the initial condition (F.52), one obtains the solution of Eq. (F.51) as

$$\widetilde{P}_n(t|\theta_0, 0) = \frac{1}{2\pi}\,\mathrm{e}^{-n^2Dt-in\theta_0}\,. \tag{F.53}$$

Then, from Eq. (F.50), the conditional probability density $P(\theta, t|\theta_0, 0)$ is obtained as

$$P(\theta, t|\theta_0, 0) = \frac{1}{2\pi}\sum_{n=-\infty}^{\infty}\mathrm{e}^{-n^2Dt+in(\theta-\theta_0)}\,. \tag{F.54}$$

In the long-time limit, i.e., $t \to \infty$, only the term $n = 0$ contributes to the sum on the right hand side of Eq. (F.54). Hence, one obtains the stationary distribution as

$$P(\theta, t \to \infty|\theta_0, 0) = \frac{1}{2\pi}\,. \tag{F.55}$$

Let us now employ the Feynman-Kac formalism to find $P(\theta, t|\theta_0, 0)$. Discretizing the dynamics (F.46), we obtain

$$\theta_i = \theta_{i-1} + \Delta t\;\eta_i\,, \tag{F.56}$$

with $\langle\eta_i\rangle = 0$; $\langle\eta_i\eta_j\rangle = (2D/\Delta t)\;\delta_{i,j}$ (see Appendix E). Here, we have $i, j = 1, 2, \ldots$, and $0 < \Delta t \ll 1$ is a small time step. Note that the dynamics (F.56) is the same as the dynamics (7.18) with $F(\theta_i) = 0$. Then, using Eqs. (7.29) and (7.30), one may write $P(\theta, t|\theta_0, 0)$ as

$$P(\theta, t|\theta_0, 0) = \int_{\theta(0)=\theta_0}^{\theta(t)=\theta}\mathcal{D}[\theta(t)]\;\exp\left(-S[\theta(t)]\right)\,, \tag{F.57}$$

with

$$S[\theta(t)] = \int_0^t \mathrm{d}t\;\frac{1}{4D}\left(\frac{\mathrm{d}\theta}{\mathrm{d}t}\right)^2\,. \tag{F.58}$$

Invoking the Feynman-Kac formalism to identify the path integral on the right hand side of Eq. (F.57) with the propagator of a quantum mechanical evolution in (negative) imaginary time due to a quantum Hamiltonian $\widehat{H}_\mathrm{q}$, we obtain

$$P(\theta, t|\theta_0, 0) = \langle\theta|\exp(-\widehat{H}_\mathrm{q}t)|\theta_0\rangle\,, \tag{F.59}$$

see Eq. (F.44), where the quantum Hamiltonian is given by

$$\widehat{H}_{\mathrm{q}}(\theta) \equiv -D \frac{\partial^2}{\partial \theta^2} . \tag{F.60}$$

In terms of the eigenvalues $E_n$ and the eigenfunctions $\Phi_n(\theta)$ of $\widehat{H}_{\mathrm{q}}(\theta)$, one has

$$P(\theta, t|\theta_0, 0) = \sum_n \Phi_n(\theta)\, \Phi_n^{\star}(\theta_0)\, \mathrm{e}^{-E_n t} . \tag{F.61}$$

The eigenvalues $E_n$ and the eigenfunctions $\Phi_n(\theta)$ for the Hamiltonian operator (F.60) are given by

$$E_n = n^2 D\, ; \quad \Phi_n(\theta) = \frac{1}{\sqrt{2\pi}}\, \mathrm{e}^{\mathrm{i}n\theta}\, ; \quad n = 0, \pm 1, \pm 2, \ldots . \tag{F.62}$$

Substituting Eq. (F.62) in Eq. (F.61), we get

$$P(\theta, t|\theta_0, 0) = \frac{1}{2\pi} \sum_{n=-\infty}^{\infty} \mathrm{e}^{-n^2 D t + \mathrm{i}n(\theta - \theta_0)} , \tag{F.63}$$

which is exactly the same as the result (F.54) obtained by solving the Fokker-Planck equation.

## Reference

1 R.P. Feynman, A.R. Hibbs, D.F. Styer, *Quantum Mechanics and Path Integrals*, emended edn. (Dover Publications, Mineola, 2010)

# Appendix G
# Discretization of One-Dimensional Hamiltonian

Any quantum Hamiltonian $\hat{H}$ of the form as in Eq. (7.33) may be written as

$$\hat{H} = -D \frac{\partial^2}{\partial\theta^2} + V(\theta)\,, \tag{G.1}$$

where we have dropped the subscript 'q' for simplicity of notation. We discretize the $\theta$-axis spanning between $-\pi$ and $\pi$ with a small lattice spacing $\Delta\theta$; $0 < \Delta\theta \ll 1$. Then, the coordinate of each 'lattice point' on the $\theta$-axis is given by $\theta_j \equiv j\Delta\theta$, where $j \in \mathbb{Z}$.

The Hamiltonian (G.1) acting on any function $F(\theta)$ yields

$$\hat{H}F(\theta) = -D \frac{\partial^2 F}{\partial\theta^2} + V(\theta)F(\theta)\,. \tag{G.2}$$

One computes $\hat{H}F(\theta)$ at every lattice point as

$$\hat{H}F\Big|_{\theta_j} = -D \left.\frac{\mathrm{d}^2 F}{\mathrm{d}\theta^2}\right|_{\theta_j} + V_j F_j\,, \tag{G.3}$$

where we have used $V_j \equiv V(\theta_j)$ and $F_j \equiv F(\theta_j)$. For small $\Delta\theta$, we use the method of finite difference to obtain

$$\left.\frac{\mathrm{d}^2 F}{\mathrm{d}\theta^2}\right|_{\theta_j} = \frac{F_{j+1} - 2F_j + F_{j-1}}{(\Delta\theta)^2}\,. \tag{G.4}$$

Substituting Eq. (G.4) in Eq. (G.3) with labeling of $\theta_j$ as $j$ on the left hand side, we obtain

$$\hat{H}F\Big|_j = -A\,F_{j-1} + (2A + V_j)\,F_j - AF_{j+1}\,, \tag{G.5}$$


D. Das and S. Gupta, *Facets of Noise*, Fundamental Theories of Physics 214,
https://doi.org/10.1007/978-3-031-45312-0

where $A \equiv D/(\Delta\theta)^2$. One may represent the action of the operator $\hat{H}$ on the function $F(\theta)$ in discrete space, i.e., Eq. (G.5) as the following matrix equation:

$$\hat{H}F\big|_j = \sum_i H_{ji}F_i = [\mathbf{H}\,\mathbf{F}]_j\,, \tag{G.6}$$

where the matrix $\mathbf{H}$ and the vector $\mathbf{F}$ are given by

$$\mathbf{H} = \begin{bmatrix} \ddots & -A & 0 & 0 & 0 \\ -A & V_{-1}+2A & -A & 0 & 0 \\ 0 & -A & V_0+2A & -A & 0 \\ 0 & 0 & -A & V_1+2A & -A \\ 0 & 0 & 0 & -A & \ddots \end{bmatrix}, \quad \mathbf{F} = \begin{bmatrix} \vdots \\ F_{-1} \\ F_0 \\ F_1 \\ \vdots \end{bmatrix}. \tag{G.7}$$

The tridiagonal matrix $\mathbf{H}$ is the discretized form of the one-dimensional Hamiltonian operator $\hat{H}$.

# Appendix H
# Derivation of a Few Equations Used in Chap. 8

## H.1 Derivation of Eqs. (8.31), (8.32), and (8.33)

Using the normalization of the Fermi-Dirac (FD) state (8.30), one obtains

$$
\begin{aligned}
1 &= A \int_0^{\pi} \mathrm{d}\theta \, \frac{\sin\theta}{1 + \mathrm{e}^{\beta_{\mathrm{FD}}(\cos^2\theta - \mu)}} \\
&= 2A \left[ \frac{1}{1 + \mathrm{e}^{\beta_{\mathrm{FD}}(1-\mu)}} - \int_0^1 \mathrm{d}x \, \sqrt{x} \, \frac{\partial f_{\mathrm{FD}}}{\partial x} \right],
\end{aligned}
\tag{H.1}
$$

where one has $f_{\mathrm{FD}}(x) \equiv 1/(1 + \mathrm{e}^{\beta_{\mathrm{FD}}(x-\mu)})$. The last equality of Eq. (H.1) is obtained by performing integration by parts. For large $\beta_{\mathrm{FD}}$, the first term on the right hand side of Eq. (H.1) may be neglected. Also, for large $\beta_{\mathrm{FD}}$ one has $\partial f_{\mathrm{FD}}(x)/\partial x = -\delta(x - \mu)$. Then, expanding the function $\sqrt{x}$ in a Taylor series about $\mu$ and substituting the result in Eq. (H.1) we obtain for $\beta_{\mathrm{FD}}$ that

$$
2A\Big[\sqrt{\mu}\,\bar{I}_0 + \frac{1}{2\beta_{\mathrm{FD}}\sqrt{\mu}}\bar{I}_1 - \frac{1}{8\beta_{\mathrm{FD}}^2\mu^{3/2}}\bar{I}_2\Big] = 1\,,
\tag{H.2}
$$

where we have

$$
\bar{I}_0 = \int_0^1 \mathrm{d}x \Big(-\frac{\partial}{\partial x} f_{\mathrm{FD}}(x)\Big) \approx \int_{-\infty}^{\infty} \mathrm{d}y \frac{\mathrm{e}^y}{(1+\mathrm{e}^y)^2} = 1\,,
\tag{H.3}
$$

$$
\bar{I}_1 = \int_0^1 \mathrm{d}x \; \beta_{\mathrm{FD}}(x-\mu)\Big(-\frac{\partial}{\partial x} f_{\mathrm{FD}}(x)\Big) \approx \int_{-\infty}^{\infty} \mathrm{d}y \frac{y\mathrm{e}^y}{(1+\mathrm{e}^y)^2} = 0\,,
\tag{H.4}
$$

$$
\bar{I}_2 = \int_0^1 \mathrm{d}x \; \beta_{\mathrm{FD}}^2(x-\mu)^2\Big(-\frac{\partial}{\partial x} f_{\mathrm{FD}}(x)\Big) \approx \int_{-\infty}^{\infty} \mathrm{d}y \frac{y^2\mathrm{e}^y}{(1+\mathrm{e}^y)^2} = \frac{\pi^2}{3}\,.
\tag{H.5}
$$



D. Das and S. Gupta, *Facets of Noise*, Fundamental Theories of Physics 214,
https://doi.org/10.1007/978-3-031-45312-0

In computing the integrals $\bar{I}_0, \bar{I}_1, \bar{I}_2$, we have considered the limit of large $\beta_{\text{FD}}$. Substituting Eqs. (H.3), (H.4), and (H.5) in Eq. (H.2), we obtain in the same limit the following result correct to order $1/\beta_{\text{FD}}^2$:

$$A = \frac{1}{2\sqrt{\mu}}\left[1 + \frac{\pi^2}{24\beta_{\text{FD}}^2\mu^2}\right], \tag{H.6}$$

which is Eq. (8.31) of the main text.

The energy density of the system (8.1) corresponding to the nonmagnetized FD state (8.30) is given by

$$\begin{aligned}\epsilon &= \frac{AD}{2\pi}\int_0^\pi \sin\theta \mathrm{d}\theta \mathrm{d}\phi \, \frac{\cos^{2n}\theta}{1 + \mathrm{e}^{\beta_{\text{FD}}(\cos^2\theta - \mu)}} \\ &= \frac{2AD}{2n+1}\left[\frac{1}{1 + \mathrm{e}^{\beta_{\text{FD}}(1-\mu)}} - \int_0^1 \mathrm{d}x \, x^{(2n+1)/2}\frac{\partial f_{\text{FD}}}{\partial x}\right],\end{aligned} \tag{H.7}$$

where once again we have used integration by parts to arrive at the last equality. For large $\beta_{\text{FD}}$, expanding the function $x^{(2n+1)/2}$ in a Taylor series about $x = \mu$, we obtain correct to order $1/\beta_{\text{FD}}^2$ that

$$\epsilon = \frac{2AD}{2n+1}\left[\mu^{(2n+1)/2}\bar{I}_0 + \frac{(2n+1)\mu^{(2n-1)/2}}{2\beta_{\text{FD}}}\bar{I}_1 + \frac{(2n+1)(2n-1)\mu^{(2n-3)/2}}{8\beta_{\text{FD}}^2}\bar{I}_2\right], \tag{H.8}$$

where $\bar{I}_0, \bar{I}_1, \bar{I}_2$ are given by Eq. (H.5). Substituting the values of $\bar{I}_0, \bar{I}_1, \bar{I}_2$ in the above equation, one gets for large $\beta_{\text{FD}}$ that

$$\epsilon = \frac{2AD}{2n+1}\left[\mu^{(2n+1)/2} + \frac{(2n+1)(2n-1)\pi^2\mu^{(2n-3)/2}}{24\beta_{\text{FD}}^2}\right], \tag{H.9}$$

correct to order $1/\beta_{\text{FD}}^2$. Finally, using Eq. (H.6) in Eq. (H.9), we obtain for large $\beta_{\text{FD}}$ that

$$\epsilon = \frac{D}{2n+1}\left[\mu^{2n/2} + \frac{(2n)^2\pi^2}{24\beta_{\text{FD}}^2}\mu^{(2n-4)/2}\right], \tag{H.10}$$

which is Eq. (8.32) of the main text and correct to order $1/\beta_{\text{FD}}^2$.

From Eqs. (8.29) and (8.30) of the main text, one obtains

$$1 = 2nDA\int_0^1 \mathrm{d}x \, g(x)\left(-\frac{\partial}{\partial x}f_{\text{FD}}\right); \quad g(x) = \frac{x^{(2n-1)/2} - x^{(2n+1)/2}}{(2n)^2D^2x^{2n-1} - \omega^2}. \tag{H.11}$$

For large $\beta_{\mathrm{FD}}$, using $\partial f_{\mathrm{FD}}(x)/\partial x = -\delta(x-\mu)$ and expanding $g(x)$ in a Taylor series about $x = \mu$, we get from Eq. (H.11) to order $1/\beta_{\mathrm{FD}}^2$ the result

$$2nDA\left[g(\mu) + \frac{g''(\mu)}{2\beta_{\mathrm{FD}}^2}\frac{\pi^2}{3}\right] = 1\,. \tag{H.12}$$

Using Eq. (H.6) in Eq. (H.12), we get to order $1/\beta_{\mathrm{FD}}^2$ the equation

$$g(\mu)\mu^{-1/2} + \frac{\pi^2}{24\beta_{\mathrm{FD}}^2}\left[g(\mu)\mu^{-5/2} + 4g''(\mu)\mu^{-1/2}\right] = \frac{1}{nD}\,, \tag{H.13}$$

where $\mu$ is to be obtained by solving Eq. (H.10). Equation (H.13) is Eq. (8.33) of the main text.

## H.2 Derivation of Eqs. (8.16) and (8.56)

In this appendix, we derive Eqs. (8.16) and (8.56) of the main text. We start by rewriting the stochastic equation of motion (8.53):

$$\dot{\mathbf{S}}_i = \mathbf{S}_i \times (\mathbf{h}_i^{\mathrm{eff}} + \boldsymbol{\eta}_i(t)) - \gamma \mathbf{S}_i \times (\mathbf{S}_i \times (\mathbf{h}_i^{\mathrm{eff}} + \boldsymbol{\eta}_i(t)))\,, \tag{H.14}$$

where $\boldsymbol{\eta}_i(t)$ is a Gaussian white noise with

$$\langle \eta_{i\alpha}(t)\rangle = 0\,; \quad \langle \eta_{i\alpha}(t)\eta_{j\beta}(t')\rangle = 2\mathcal{D}\delta_{ij}\delta_{\alpha\beta}\delta(t-t')\,, \tag{H.15}$$

and

$$\mathbf{h}_i^{\mathrm{eff}} = \mathbf{m} + \mathbf{h}_i^{\mathrm{aniso}};\ \mathbf{h}_i^{\mathrm{aniso}} = (0, 0, -2nDS_{iz}^{2n-1})\,. \tag{H.16}$$

In terms of components, Eq. (H.14) may be written as

$$\dot{S_i^\alpha} = f_i^\alpha(\mathbf{S}_i) + g_i^{\alpha\lambda}(\{\mathbf{S}_i\})\eta_i^\lambda\,, \tag{H.17}$$

where Einstein summation convention for repeated indices is assumed, and

$$f_i^\alpha(\{\mathbf{S}_i\}) = \epsilon_{\alpha\beta\lambda}S_i^\beta h_i^{\mathrm{eff},\lambda} - \gamma\epsilon_{\alpha\beta\lambda}\epsilon_{\lambda\sigma\rho}S_i^\beta S_i^\sigma h_i^{\mathrm{eff},\rho}\,, \tag{H.18}$$

$$g_i^{\alpha\beta}(\{\mathbf{S}_i\}) = \epsilon_{\alpha\lambda\beta}S_i^\lambda - \gamma\epsilon_{\alpha\lambda\rho}\epsilon_{\rho\sigma\beta}S_i^\lambda S_i^\sigma\,. \tag{H.19}$$

We now define the discrete single-spin density function $F_{\mathrm{d}}(\mathbf{S}, t)$ as

$$F_{\rm d}(\mathbf{S}, t) = \frac{1}{N} \sum_{i=1}^{N} \delta(\mathbf{S} - \mathbf{S}_i(t)) \,. \tag{H.20}$$

For a given noise realization $\{\boldsymbol{\eta}_i\}$, differentiating both sides of the above equation with respect to time, using Eq. (H.17) and the property $a\delta(a-b) = b\delta(a-b)$, we obtain the time evolution for $F_{\rm d}$ as

$$\frac{\partial}{\partial t} F_{\rm d}(\mathbf{S}, t) = -\frac{\partial}{\partial S^\alpha}\left[\left(f^\alpha + g^{\alpha\lambda}\eta^\lambda\right) F_{\rm d}(\mathbf{S}, t)\right], \tag{H.21}$$

where we have

$$f^\alpha(\mathbf{S}) = \epsilon_{\alpha\beta\lambda} S^\beta h^{{\rm eff},\lambda} - \gamma \epsilon_{\alpha\beta\lambda}\epsilon_{\lambda\sigma\rho} S^\beta S^\sigma h^{{\rm eff},\rho} \,; \tag{H.22}$$

$$\mathbf{h}^{\rm eff} \equiv \mathbf{h}^{\rm eff}[F_{\rm d}] = \mathbf{m}[F_{\rm d}] + (0, 0, -2nDS_z^{2n-1}); \quad \mathbf{m}[F_{\rm d}] \equiv \int {\rm d}\mathbf{S}\, \mathbf{S}\, F_{\rm d}(\mathbf{S}, t) \,, \tag{H.23}$$

$$g^{\alpha\beta}(\mathbf{S}) = \epsilon_{\alpha\lambda\beta} S^\lambda - \gamma \epsilon_{\alpha\lambda\rho}\epsilon_{\rho\sigma\beta} S^\lambda S^\sigma \,. \tag{H.24}$$

Now taking the average of Eq. (H.21) over the noise statistics (H.15), one gets for the averaged distribution $P_{\rm d}(\mathbf{S}, t)$ the equation [1, 2]

$$\frac{\partial P_{\rm d}(\mathbf{S}, t)}{\partial t} = -\frac{\partial}{\partial S^\alpha}\left[f^\alpha P_{\rm d} - \mathcal{D} g^{\alpha\beta} \frac{\partial}{\partial S^\lambda}(g^{\lambda\beta} P_{\rm d})\right]. \tag{H.25}$$

From Eq. (H.24), we obtain

$$\frac{\partial}{\partial S^\alpha} g^{\alpha\lambda} = \epsilon_{\alpha\beta\lambda}\delta_{\alpha\beta} - \gamma \delta_{\alpha\alpha} S^\lambda - \gamma S^\alpha \delta_{\alpha\lambda} = -4\gamma S^\lambda \,, \tag{H.26}$$

so that

$$g^{\alpha\beta} \frac{\partial}{\partial S^\lambda} g^{\lambda\beta} = -4\gamma\left(\epsilon_{\alpha\lambda\beta} S^\lambda - \gamma S^\alpha S^\beta + \gamma \delta_{\alpha\beta}\right) S^\beta = 0 \,, \tag{H.27}$$

where we have used the properties of the Levi-Civita symbol $\epsilon_{\alpha\lambda\beta}$. As a result, one may obtain from Eq. (H.25) that

$$\frac{\partial P_{\rm d}(\mathbf{S}, t)}{\partial t} = -\frac{\partial}{\partial S^\alpha}\left[f^\alpha - \mathcal{D} g^{\alpha\beta} g^{\lambda\beta} \frac{\partial}{\partial S^\lambda}\right] P_{\rm d} \,. \tag{H.28}$$

Note that from Eq. (H.24) we have

$$g^{\alpha\beta} g^{\lambda\beta} = (1 + \gamma^2)\epsilon_{\alpha\sigma\beta}\epsilon_{\lambda\rho\beta} S^\sigma S^\rho \,. \tag{H.29}$$

Consequently, the right hand side of Eq. (H.28) becomes

$$-\frac{\partial}{\partial S^{\alpha}}\Big[\Big(\epsilon_{\alpha\beta\lambda}S^{\beta}h^{\mathrm{eff},\lambda}-\gamma\epsilon_{\alpha\beta\lambda}\epsilon_{\lambda\sigma\rho}S^{\beta}S^{\sigma}h^{\mathrm{eff},\rho}$$
$$-\mathcal{D}(1+\gamma^{2})\epsilon_{\alpha\sigma\beta}\epsilon_{\lambda\rho\beta}S^{\sigma}S^{\rho}\frac{\partial}{\partial S^{\lambda}}\Big)P_{\mathrm{d}}\Big], \tag{H.30}$$

and hence Eq. (H.28) may be recast as

$$\frac{\partial P_{\mathrm{d}}(\mathbf{S},t)}{\partial t}=-\frac{\partial}{\partial \mathbf{S}}\cdot\Big[(\mathbf{S}\times\mathbf{h}^{\mathrm{eff}})P_{\mathrm{d}}-\gamma((\mathbf{S}\times\mathbf{S}\times\mathbf{h}^{\mathrm{eff}})P_{\mathrm{d}})$$
$$+\mathcal{D}(1+\gamma^{2})\left(\mathbf{S}\times\left(\mathbf{S}\times\frac{\partial}{\partial \mathbf{S}}\right)\right)P_{\mathrm{d}}\Big], \tag{H.31}$$

where note that $\partial/\partial\mathbf{S}\cdot(\mathbf{S}\times\mathbf{h}^{\mathrm{eff}})P_{\mathrm{d}}=(\mathbf{S}\times\mathbf{h}^{\mathrm{eff}})\cdot\partial P_{\mathrm{d}}/\partial\mathbf{S}$.

We now define the single-spin distribution function $P_0(\mathbf{S},t)$ as the average of $P_{\mathrm{d}}(\mathbf{S},t)$ with respect to a large number of initial conditions close to the same macroscopic state. Then, we may write to the leading order that

$$P_{\mathrm{d}}(\mathbf{S},t)=P_0(\mathbf{S},t)+\frac{1}{\sqrt{N}}\delta P(\mathbf{S},t)\,. \tag{H.32}$$

In the above equation, $\delta P$ denotes the deviation between $P_{\mathrm{d}}$ and $P_0$ (which is of order $N^0$) and has a zero average, namely, $\langle\delta P(\mathbf{S},t)\rangle=0$. Substituting Eq. (H.32) in Eq. (H.31), and using the expansion

$$\mathbf{h}^{\mathrm{eff}}=\mathbf{h}^{\mathrm{eff},0}[P_0]+\frac{1}{\sqrt{N}}\,\delta\mathbf{h}^{\mathrm{eff}}[\delta P]\,, \tag{H.33}$$

one obtains

$$\frac{\partial P_0(\mathbf{S},t)}{\partial t}+\frac{1}{\sqrt{N}}\frac{\partial\delta P(\mathbf{S},t)}{\partial t}$$
$$=-\frac{\partial}{\partial\mathbf{S}}\cdot\Big[(\mathbf{S}\times\mathbf{h}^{\mathrm{eff},0})-\gamma(\mathbf{S}\times\mathbf{S}\times\mathbf{h}^{\mathrm{eff},0})+\mathcal{D}(1+\gamma^{2})\left(\mathbf{S}\times\left(\mathbf{S}\times\frac{\partial}{\partial\mathbf{S}}\right)\right)\Big]P_0$$
$$-\frac{1}{\sqrt{N}}\frac{\partial}{\partial\mathbf{S}}\cdot\Big[(\mathbf{S}\times\delta\mathbf{h}^{\mathrm{eff}})P_0+(\mathbf{S}\times\mathbf{h}^{\mathrm{eff},0})\delta P-\gamma(\mathbf{S}\times\mathbf{S}\times\delta\mathbf{h}^{\mathrm{eff}})P_0$$
$$-\gamma(\mathbf{S}\times\mathbf{S}\times\mathbf{h}^{\mathrm{eff},0})\delta P+\mathcal{D}(1+\gamma^{2})(\mathbf{S}\times(\mathbf{S}\times\frac{\partial}{\partial\mathbf{S}}))\delta P\Big]$$
$$-\frac{1}{N}\frac{\partial}{\partial\mathbf{S}}\cdot\Big[(\mathbf{S}\times\delta\mathbf{h}^{\mathrm{eff}})\delta P-\gamma(\mathbf{S}\times\mathbf{S}\times\delta\mathbf{h}^{\mathrm{eff}})\delta P\Big]. \tag{H.34}$$

Taking the average of the above equation, and using $\langle\delta P\rangle=0$ that implies $\langle\delta\mathbf{h}^{\mathrm{eff}}\rangle=0$, we obtain

$$\frac{\partial P_0(\mathbf{S},t)}{\partial t}+\frac{\partial}{\partial \mathbf{S}}\cdot(\mathbf{S}\times\mathbf{h}^{\mathrm{eff},0})P_0$$
$$=-\frac{\partial}{\partial \mathbf{S}}\cdot\Big[-\gamma(\mathbf{S}\times\mathbf{S}\times\mathbf{h}^{\mathrm{eff},0})+\mathcal{D}(1+\gamma^2)\Big(\mathbf{S}\times\Big(\mathbf{S}\times\frac{\partial}{\partial \mathbf{S}}\Big)\Big)\Big]P_0$$
$$-\frac{1}{N}\Big\langle\frac{\partial}{\partial \mathbf{S}}\cdot\Big[(\mathbf{S}\times\delta\mathbf{h}^{\mathrm{eff}})\delta P-\gamma(\mathbf{S}\times\mathbf{S}\times\delta\mathbf{h}^{\mathrm{eff}})\delta P\Big]\Big\rangle. \tag{H.35}$$

In the limit $N\to\infty$, the last term on the right hand side of the above equation drops out. Now, requiring that the equilibrium state $P_0(\mathbf{S})\propto\exp(-\beta(-\mathbf{S}\cdot\mathbf{m}[P_0]+DS_z^{2n}))$ upto some normalization is the stationary solution of Eq. (H.35) in the limit $N\to\infty$, one gets

$$\frac{\partial}{\partial \mathbf{S}}\cdot\Big(\Big[(\mathbf{S}\times\mathbf{h}^{\mathrm{eff},0})-\gamma(\mathbf{S}\times\mathbf{S}\times\mathbf{h}^{\mathrm{eff},0})+\mathcal{D}(1+\gamma^2)\Big(\mathbf{S}\times\Big(\mathbf{S}\times\frac{\partial}{\partial \mathbf{S}}\Big)\Big)\Big]P_0(\mathbf{S})\Big)=0. \tag{H.36}$$

Using the following relations

$$\Big(\mathbf{S}\times(\mathbf{S}\times\frac{\partial P_0}{\partial \mathbf{S}})\Big)=\beta P_0\Big[\mathbf{S}\Big(\mathbf{S}\cdot\mathbf{m}[P_0]+\mathbf{S}\cdot\mathbf{h}^{\mathrm{aniso}}\Big)-\mathbf{m}[P_0]-\mathbf{h}^{\mathrm{aniso}}\Big], \tag{H.37}$$
$$\gamma(\mathbf{S}\times\mathbf{S}\times\mathbf{h}^{\mathrm{eff}})P_0=\gamma P_0\Big[\mathbf{S}\Big(\mathbf{S}\cdot\mathbf{m}[P_0]+\mathbf{S}\cdot\mathbf{h}^{\mathrm{aniso}}\Big)-\mathbf{m}[P_0]-\mathbf{h}^{\mathrm{aniso}}\Big], \tag{H.38}$$
$$\frac{\partial}{\partial \mathbf{S}}\cdot\Big[(\mathbf{S}\times\mathbf{h}^{\mathrm{eff}})P_0\Big]=(\mathbf{S}\times\mathbf{h}^{\mathrm{eff}})\cdot\frac{\partial P_0}{\partial \mathbf{S}}=\beta P_0(\mathbf{S}\times\mathbf{h}^{\mathrm{eff}})\cdot\mathbf{h}^{\mathrm{eff}}=0, \tag{H.39}$$

we see that Eq. (H.36) is satisfied provided

$$\mathcal{D}(1+\gamma^2)\beta=\gamma. \tag{H.40}$$

Considering that validity of Eq. (H.40), one may rewrite Eq. (H.35) as

$$\frac{\partial P_0(\mathbf{S},t)}{\partial t}+\frac{\partial}{\partial \mathbf{S}}\cdot(\mathbf{S}\times\mathbf{h}^{\mathrm{eff},0})P_0$$
$$=\gamma\frac{\partial}{\partial \mathbf{S}}\cdot\Big[(\mathbf{S}\times\mathbf{S}\times\mathbf{h}^{\mathrm{eff},0})-\frac{1}{\beta}\Big(\mathbf{S}\times\Big(\mathbf{S}\times\frac{\partial}{\partial \mathbf{S}}\Big)\Big)\Big]P_0$$
$$-\frac{1}{N}\Big\langle\frac{\partial}{\partial \mathbf{S}}\cdot\Big[(\mathbf{S}\times\delta\mathbf{h}^{\mathrm{eff}})\delta P-\gamma(\mathbf{S}\times\mathbf{S}\times\delta\mathbf{h}^{\mathrm{eff}})\delta P\Big]\Big\rangle, \tag{H.41}$$

which in the limit $N\to\infty$, yields Eq. (8.56) of the main text.

One may consider the case of the deterministic dynamics by setting $\gamma=0$. Then, in the limit $N\to\infty$, Eq. (H.41) reduces to the Vlasov equation (8.16) of the main text, namely,

$$\frac{\partial P_0(\mathbf{S},t)}{\partial t}+\frac{\partial}{\partial \mathbf{S}}\cdot(\mathbf{S}\times\mathbf{h}^{\mathrm{eff},0})P_0=0. \tag{H.42}$$

In other words, one may also say that in case of the deterministic dynamics, Eq. (H.42) describes for finite $N$ the time evolution for times $t \ll N$, with that for $\delta P$ obtained from Eq. (H.34) as

$$\frac{\partial \delta P(\mathbf{S}, t)}{\partial t} = -\frac{\partial}{\partial \mathbf{S}} \cdot \left[ (\mathbf{S} \times \delta \mathbf{h}^{\text{eff}}) P_0 + (\mathbf{S} \times \mathbf{h}^{\text{eff},0}) \delta P \right]. \tag{H.43}$$

The time evolution of the single-spin distribution in the deterministic dynamics for times of order $N$ is obtained from Eq. (H.41) as

$$\frac{\partial P_0(\mathbf{S}, t)}{\partial t} + \frac{\partial}{\partial \mathbf{S}} \cdot (\mathbf{S} \times \mathbf{h}^{\text{eff},0}) P_0 = -\frac{1}{N} \left\langle \frac{\partial}{\partial \mathbf{S}} \cdot (\mathbf{S} \times \delta \mathbf{h}^{\text{eff}}) \delta P \right\rangle. \tag{H.44}$$

## References

1 R. Kubo, J. Math. Phys. **4**(2), 174 (1963). https://doi.org/10.1063/1.1703941
2 M. Nishino, S. Miyashita, Phys. Rev. B **91**(13), 134411 (2015). https://doi.org/10.1103/PhysRevB.91.134411

# Appendix I
# The Response Formula for Mean-Field Classical Spin Systems

In this appendix, we briefly discuss the derivation of the response formula (9.70), which follows from Ref. [1]. The Vlasov equation (9.13), rewritten as

$$\frac{\partial f(\phi, p, t)}{\partial t} - \left\{H[f](\phi, p, t)\,,\ f(\phi, p, t)\right\} = 0\,, \tag{I.1}$$

is governed by the single-spin Hamiltonian (9.16):

$$H[f](\phi, p, t) = Dp^{2n} - \mathbf{S}\cdot\mathbf{m}[f](t) - \mathbf{S}\cdot\mathbf{h}(t)\,, \tag{I.2}$$

which depends on $f(\phi, p, t)$ through magnetization $\mathbf{m}[f](t) \equiv \mathbf{m}(t)$, see Eq. (9.18).

Now, $f(\phi, p, t)$ being a distribution function is bounded, i.e., $f(\phi, p, t)\ \forall\ t$ vanishes at the boundaries of $p$, and $f(\phi + 2\pi, p, t) = f(\phi, p, t)$. Moreover, it has a well-defined limiting value $f_h(\phi, p)$ given in Eq. (9.36). Then, one may show that the long-time average of $f(\phi, p, t)$ coincides with the $f_h(\phi, p)$, namely,

$$\lim_{t\to\infty} \frac{1}{t}\int_0^t \mathrm{d}t'\ f(\phi, p, t') = f_h(\phi, p)\,. \tag{I.3}$$

In order to prove Eq. (I.3), one uses Eq. (9.36), which implies for arbitrarily small $\nu > 0$ the existence of a time $t^\star > 0$ (see Fig. 9.1), such that

$$|f(\phi, p, t) - f_h(\phi, p)| < \nu \ \text{ for all } t > t^\star\,. \tag{I.4}$$

For $t > t^\star$, we start with


D. Das and S. Gupta, *Facets of Noise*, Fundamental Theories of Physics 214,
https://doi.org/10.1007/978-3-031-45312-0

$$\mathcal{A} \equiv \left| \frac{1}{t} \int_0^t \mathrm{d}t' \; f(\phi, p, t') - f_h(\phi, p) \right| = \left| \frac{1}{t} \int_0^t \mathrm{d}t' \; \left[ f(\phi, p, t') - f_h(\phi, p) \right] \right|$$
$$\leq \frac{1}{t} \int_0^t \mathrm{d}t' \left| f(\phi, p, t') - f_h(\phi, p) \right| . \tag{I.5}$$

In obtaining the last step, we have used the inequality $|z_1 + z_2| \leq |z_1| + |z_2|$ for two complex numbers $z_1$ and $z_2$. Separating the integral on the right hand side of the above equation in two parts, we obtain

$$\mathcal{A} \leq \frac{1}{t} \int_0^{t^\star} \mathrm{d}t' \left| f(\phi, p, t') - f_h(\phi, p) \right| + \frac{1}{t} \int_{t^\star}^t \mathrm{d}t' \left| f(\phi, p, t') - f_h(\phi, p) \right|$$
$$\leq \frac{1}{t} \left[ \int_0^{t^\star} \mathrm{d}t' | f(\phi, p, t') | + \int_0^{t^\star} \mathrm{d}t' | f_h(\phi, p) | + \int_{t^\star}^t \mathrm{d}t' \left| f(\phi, p, t') - f_h(\phi, p) \right| \right], \tag{I.6}$$

where in the last step we have used the inequality $|z_1 - z_2| \leq |z_1| + |z_2|$ for two complex numbers $z_1$ and $z_2$. Now, using Eq. (I.4), one may reduce Eq. (I.6) to

$$\left| \frac{1}{t} \int_0^t \mathrm{d}t' f(\phi, p, t') - f_h(\phi, p) \right| \leq \frac{t^\star \max_{t \in [0, t^\star]} | f(\phi, p, t) |}{t} + \frac{t^\star | f_h(\phi, p) |}{t}$$
$$+ \frac{(t - t^\star)\nu}{t} . \tag{I.7}$$

In the limit $t \to \infty$, Eq. (I.7) yields

$$\left| \lim_{t \to \infty} \frac{1}{t} \int_0^t \mathrm{d}t' \; f(\phi, p, t') - f_h(\phi, p) \right| \leq \nu . \tag{I.8}$$

As $\nu$ is arbitrarily small and positive, Eq. (I.8) leads to

$$\lim_{t \to \infty} \frac{1}{t} \int_0^t \mathrm{d}t' \; f(\phi, p, t') = f_h(\phi, p) , \tag{I.9}$$

which proves Eq. (I.3).

For $t > 0$, one may decompose $f(\phi, p, t)$ into the initial state $f_0(\phi, p)$ and an evolving perturbation $g_{\mathrm{E}}(\phi, p, t)$ as

$$f(\phi, p, t) = f_0(\phi, p) + \epsilon g_{\mathrm{E}}(\phi, p, t) . \tag{I.10}$$

As the external field is small, we assume that the quantity $f(\phi, p, t)$ remains in $O(\epsilon)$-neighborhood of the initial state $f_0(\phi, p)$, see Fig. 9.1. Then, $f(\phi, p, t)$ may also be decomposed into the asymptotically-stationary part $f_h(\phi, p)$ and a transient part $g_{\mathrm{T}}(\phi, p, t)$ as

$$f(\phi, p, t) = f_h(\phi, p) + \epsilon g_{\mathrm{T}}(\phi, p, t)\,, \tag{I.11}$$

with $g_{\mathrm{T}}(\phi, p, t) \to 0$ as $t \to \infty$. Substituting Eq. (I.11) in Eq. (I.2), one obtains the single-spin Hamiltonian as

$$H[f](\phi, p, t) = H_h + \epsilon K_{\mathrm{T}}\,, \tag{I.12}$$

with

$$H_h \equiv H[f_h](\phi, p) = Dp^{2n} - \mathbf{S}\cdot\mathbf{m}[f_h] - \mathbf{S}\cdot\mathbf{h}\,, \tag{I.13}$$

$$K_{\mathrm{T}} \equiv K_{\mathrm{T}}[g_{\mathrm{T}}](\phi, p, t) = -\,\mathbf{S}\cdot\mathbf{m}[g_T](t)\,, \tag{I.14}$$

where $H_h$ is the asymptotic part, characterizing the stationary ($t \to \infty$) state $f_h$, while $K_{\mathrm{T}}$ is the transient part. For our case with $\mathbf{h} = (h, 0, 0)$, the transient magnetization $\mathbf{m}[g_{\mathrm{T}}](t) = (m[g_{\mathrm{T}}](t), 0, 0)$ is obtained as

$$m[g_{\mathrm{T}}](t) = \iint_{\mu} \mathrm{d}\phi\,\mathrm{d}p\,\sqrt{1-p^2}\cos\phi\; g_{\mathrm{T}}(\phi, p, t)\,. \tag{I.15}$$

Substituting Eqs. (I.11) and (I.12) in Eq. (I.1), and keeping terms up to $O(\epsilon)$, the Vlasov equation (I.1) is rewritten as

$$\frac{\partial f(\phi, p, t)}{\partial t} = \left\{H_h\,,\ f(\phi, p, t)\right\} + \epsilon\left\{K_{\mathrm{T}}\,,\ f_h(\phi, p)\right\}. \tag{I.16}$$

The difference between $f_0(\phi, p)$ and $f_h(\phi, p)$ is of $O(\epsilon)$, see Fig. 9.1. As the second term on the right hand side of Eq. (I.16) is already suppressed by $\epsilon$, one may replace $f_h(\phi, p)$ with $f_0(\phi, p)$ in this term. With this slight modification, we write Eq. (I.16) as

$$\frac{\partial f(\phi, p, t)}{\partial t} = \mathcal{L}_h f(\phi, p, t) + \epsilon\left\{K_{\mathrm{T}}\,,\ f_0(\phi, p)\right\}, \tag{I.17}$$

where the operator $\mathcal{L}_h$ is defined as

$$\mathcal{L}_h(\cdot) = \left\{H_h\,, \cdot\right\}. \tag{I.18}$$

Equation (I.17) is solved using the initial condition $f(\phi, p, 0) = f_0(\phi(0), p(0))$. Then, the solution of Eq. (I.17) is given by

$$f(\phi, p, t) = \mathrm{e}^{t\mathcal{L}_h} f_0(\phi, p) + \epsilon \int_0^t \mathrm{d}s\; \mathrm{e}^{(t-s)\mathcal{L}_h}\left\{K_{\mathrm{T}}\,,\ f_0(\phi, p)\right\}. \tag{I.19}$$

Now, we assume that contribution from the second term on the right hand side, which contains the transient part having $K_{\mathrm{T}}$, is negligible; this is justified under some assumptions for Hamiltonian systems and may be related to the phenomenon of Landau damping, see Ref. [1] for details. With the aforementioned assumption, Eq. (I.19) reads

$$f(\phi, p, t) \approx \mathrm{e}^{t\mathcal{L}_h} f_0(\phi(0), p(0)) = f_0(\phi(t), p(t)) \,, \tag{I.20}$$

which represents temporal evolution of $f_0$ under Hamiltonian flow associated with the asymptotic Hamiltonian $H_h$. In Eq. (I.20), $(\phi(t),\ p(t))$ are obtained by solving the Hamilton's equations of motion corresponding to the Hamiltonian $H_h$ with the initial condition $(\phi(0), p(0))$. Then, using Eq. (I.3), one may obtain the asymptotically stationary distribution as

$$f_h(\phi, p) = \lim_{t\to\infty} \frac{1}{t} \int_0^t \mathrm{d}t' f(\phi, p, t') = \lim_{t\to\infty} \frac{1}{t} \left[ \int_0^{t^\star} + \int_{t^\star}^t \right] \mathrm{d}t' f(\phi, p, t') \,, \tag{I.21}$$

where $t^\star$ is defined by Eq. (I.4). As $t^\star$ is finite, the first term on the right hand side of Eq. (I.21) vanishes, and using Eq. (I.20), we get

$$f_h(\phi, p) = \lim_{t\to\infty} \frac{1}{t} \int_{t^\star}^t \mathrm{d}t' \; f_0(\phi(t'), p(t')) \,. \tag{I.22}$$

Let us recall from the discussions in Sect. 9.4 that the asymptotic Hamiltonian $H_h$ is integrable, and we may switch to the angle-action variables $(w, I)$, whose time evolution is given by

$$w(t) = w(0) + \Omega(I)t \,, \qquad I(t) = I(0) \,, \tag{I.23}$$

where we have $\Omega(I) = \mathrm{d}H_h(I)/\mathrm{d}I$. Then, in terms of the angle-action variables, Eq. (I.22) may be written as

$$f_h(\phi, p) = \lim_{t\to\infty} \frac{1}{t} \int_{t^\star}^t \mathrm{d}t' \; f_0\big(\phi(w(t'), I(t')), p(w(t'), I(t'))\big) \,. \tag{I.24}$$

As the angle variable $w(t)$ is periodic in time and the action variable $I(t)$ is a constant of motion, $f_0\big(\phi(w(t'), I(t')), p(w(t'), I(t'))\big)$ is periodic in time. Therefore, one may compute the time average in Eq. (I.24) over a single period $T_p$ on the iso-$I$ line. Thus, Eq. (I.24) reduces to

$$f_h(\phi, p) = \frac{1}{T_p} \int_{t^\star}^{t^\star + T_p} \mathrm{d}t' \; f_0\big(\phi(w(t'), I(t')), p(w(t'), I(t'))\big) \,. \tag{I.25}$$

Now, using Eq. (I.23), one may change the integration variable in Eq. (I.25) from $t'$ to $w$. In doing so, as the integrand in Eq. (I.25) is periodic, one may replace $t^\star$ in the limits of the integration with some $\bar{t} > t^\star$, such that $w(\bar{t}) = 0$ and $w(\bar{t} + T_p) = \Omega(I)T_p = 2\pi$. Finally, one obtains

$$f_h(\phi, p) = \frac{1}{T_p}\int_0^{\Omega T_p} \frac{\mathrm{d}w}{\Omega(I)} f_0\big(\phi(w, I), p(w, I)\big) = \frac{1}{2\pi}\int_0^{2\pi} \mathrm{d}w\, f_0\big(\phi(w, I), p(w, I)\big) . \quad \text{(I.26)}$$

From the right hand side of the above equation, it is evident that $f_h$ is a function of $I$ alone. Then, we have $f_h(I(\phi, p)) = \langle f_0(\phi, p)\rangle_h$, which is the response formula (9.70).

## Reference

1 S. Ogawa, Y.Y. Yamaguchi, Phys. Rev. E **89**(5), 052114 (2014). https://doi.org/10.1103/PhysRevE.89.052114

# Appendix J
# Mean-Field Classical Spin Systems: Computation of Averages Over Angle Variables

In this appendix, we discuss a method to compute the average $\langle\cdot\rangle_0$ over the angle variable associated with the single-spin Hamiltonian $H_0 = Dp^{2n} - m_0\sqrt{1-p^2}\cos\phi$, see Eq. (9.32). As discussed in Sect. 9.4, the Hamiltonian $H_0$ is integrable, and has an angle-action description such that the evolution of the angle-action variables $(w_0, I_0)$ is given by

$$w_0(t) = w_0(0) + \Omega(I_0)t\,, \quad I_0(t) = I_0(0)\,, \tag{J.1}$$

where one has $\Omega(I_0) = \mathrm{d}H_0(I_0)/\mathrm{d}I_0$. In this case, the dynamics takes place in the two-dimensional phase space $(w_0, I_0)$, and the phase-space trajectory is an iso-$I_0$ line. Along this line, $I_0$ is a constant, while $w_0$ changes periodically with frequency $\Omega(I_0)$ and time period $T_p^0 = 2\pi/\Omega(I_0)$. With the change of variable from $w_0$ to time $t$, one obtains the average of a quantity $B$ as

$$\langle B\rangle_0 = \frac{\int_0^{2\pi}\mathrm{d}w_0\; B(w_0, I_0)}{\int_0^{2\pi}\mathrm{d}w_0} = \frac{\int_0^{T_p^0}\mathrm{d}t\; B(w_0(t), I_0)\;\Omega(I_0)}{\int_0^{T_p^0}\mathrm{d}t\;\Omega(I_0)} = \frac{\int_0^{T_p^0}\mathrm{d}t\; B(w_0(t), I_0)}{\int_0^{T_p^0}\mathrm{d}t}\,. \tag{J.2}$$

The canonical equations of motion for the Hamiltonian $H_0$ are given by

$$\begin{aligned}\frac{\mathrm{d}\phi}{\mathrm{d}t} &= 2nDp^{2n-1} + m_0\frac{p}{\sqrt{1-p^2}}\cos\phi\,,\\ \frac{\mathrm{d}p}{\mathrm{d}t} &= -m_0\sqrt{1-p^2}\sin\phi\,.\end{aligned} \tag{J.3}$$

Then, the average $\langle\sqrt{1-p^2}\cos\phi\rangle_0$ may be computed as

$$\left\langle\sqrt{1-p^2}\cos\phi\right\rangle_0 = \frac{\int_0^{T_p^0}\mathrm{d}t\;\sqrt{1-p^2(t)}\;\cos\phi(t)}{\int_0^{T_p^0}\mathrm{d}t}\,. \tag{J.4}$$



D. Das and S. Gupta, *Facets of Noise*, Fundamental Theories of Physics 214,
https://doi.org/10.1007/978-3-031-45312-0

Note that the left hand side of Eq. (J.4) depends on the action $I_0$ only, with the initial condition $(\phi(0), p(0))$ determining the value of $I_0$. In order to evaluate the right hand side of Eq. (J.4), we need to integrate Eq. (J.3) with initial condition $(\phi(0), p(0))$ to obtain $(\phi(t), p(t))$. The initial condition $(\phi(0), p(0))$ corresponds to the reference state $f_0$. Note that in practice, when we do not have an a priori knowledge of $T_p^0$, the time integration on the right hand side of Eq. (J.4) is performed numerically over a long-enough duration of time.

# Appendix K
# The Most Probable Survival Probability Under Projective Measurements

In this appendix, we discuss the large deviation (LD) formalism to obtain the most probable value $S_m^\star$ of the survival probability as discussed in Chap. 10. Here, we follow the LD formalism of Ref. [1] and derive Eq. (10.34). For a brief introduction to the theory of large deviations, see Appendix L. To obtain $S_m^\star$, we focus on the case of $p_\tau$ being a $d$-dimensional Bernoulli distribution. In other words, we consider the situation in which the time interval $\tau$ between two successive measurements takes on $d$ possible discrete values $\tau^{(1)}, \tau^{(2)}, \ldots, \tau^{(d)}$, with corresponding probabilities $p^{(1)}, p^{(2)}, \ldots, p^{(d)}$ satisfying $\sum_{\alpha=1}^{d} p^{(\alpha)} = 1$. To invoke the LD formalism for the survival probability, let us consider the quantity

$$\mathcal{L}(\{\tau_\alpha\}_{1\le\alpha\le m}) \equiv \log\left(S_m(\{\tau_\alpha\}_{1\le\alpha\le m})\right) = \sum_{\alpha=1}^{d} n_\alpha \log q(\tau^{(\alpha)}), \tag{K.1}$$

where $n_\alpha$ denotes the number of times the value $\tau^{(\alpha)}$ occurs in the sequence $\{\tau_\alpha\}_{1\le\alpha\le m}$. The quantity $\mathcal{L}$ in Eq. (K.1) is a sum of i.i.d. random variables, and its probability distribution is given by (see Ref. [1])

$$\mathcal{P}(\mathcal{L}) = \frac{m!}{n_1'!\, n_2'! \,\ldots\, n_d'!} \prod_{\alpha=1}^{d} (p^{(\alpha)})^{n_\alpha'}, \tag{K.2}$$

where the quantities $n_\alpha'$ satisfy the constraints $\sum\limits_{\alpha=1}^{d} n_\alpha' = m$ and $\sum\limits_{\alpha=1}^{d} n_\alpha' \log q(\tau^{(\alpha)}) = \mathcal{L}$, implying that one has

$$m \log q\left(\tau^{(d)}\right) - \mathcal{L} = \sum_{\alpha=1}^{d-1} n_\alpha' \,\lambda\left(\tau^{(\alpha)}\right), \tag{K.3}$$



D. Das and S. Gupta, *Facets of Noise*, Fundamental Theories of Physics 214,
https://doi.org/10.1007/978-3-031-45312-0

with $\lambda(\tau^{(\alpha)}) \equiv \log q(\tau^{(d)}) - \log q(\tau^{(\alpha)})$. The solution of Eq. (K.3) is [1]

$$n'_\alpha = \frac{m \log q\left(\tau^{(d)}\right) - \mathcal{L}}{(d-1)\,\lambda\left(\tau^{(\alpha)}\right)}; \quad \alpha = 1, 2, \ldots, d-1\,, \tag{K.4}$$

and $n'_d = m - \sum_{\alpha=1}^{d-1} n'_\alpha$. Using this solution in Eq. (K.2), it may be shown that in the limit $m \to \infty$, the distribution $\mathcal{P}(\mathcal{L})$ has the following LD form [1]

$$\mathcal{P}(\mathcal{L}) \to \mathcal{P}(\mathcal{L}/m) \approx \exp\left(-mI\left(\mathcal{L}/m\right)\right)\,, \tag{K.5}$$

with

$$I\,(x) \equiv \sum_{\alpha=1}^{d} f(\tau^{(\alpha)}) \log\left(\frac{f(\tau^{(\alpha)})}{p^{(\alpha)}}\right), \tag{K.6}$$

$$f(\tau^{(\alpha)}) \equiv \frac{\log q(\tau^{(d)}) - x}{(d-1)\lambda(\tau^{(\alpha)})}; \quad \alpha = 1, \ldots, (d-1)\,, \tag{K.7}$$

$$f(\tau^{(d)}) \equiv 1 - \sum_{\alpha=1}^{d-1} f(\tau^{(\alpha)})\,. \tag{K.8}$$

From Eq. (K.5), it follows that the minimum of the function $I\,(\mathcal{L}/m)$ corresponds to the most probable value $\mathcal{L}^\star$ of $\mathcal{L}$ as $m \to \infty$. From Eq. (K.6), the condition $\partial I\,(\mathcal{L}/m)\,/\partial \log q(\tau^{(\alpha)})|_{\mathcal{L}=\mathcal{L}^\star} = 0$ gives, on performing a series of algebraic manipulations, that [1]

$$\mathcal{L}^\star = m \sum_{\alpha=1}^{d} p^{(\alpha)} \log q(\tau^{(\alpha)})\,. \tag{K.9}$$

Using Eqs. (K.1) and (K.5), one may obtain an LD form for the distribution of the survival probability $S_m$ as [1]

$$\mathcal{P}(S_m) \approx \exp\left(-m\ J\,(S_m)\right), \tag{K.10}$$

with $J\,(S_m) \equiv \min_{\mathcal{L}:\mathcal{L}=\log S_m} I\,(\mathcal{L}/m)$. The value $S^\star_m$ that minimizes the function $J\,(S_m)$ is the most probable value of the survival probability in the limit $m \to \infty$; one gets [1]

$$S^\star_m = \exp\left(m \sum_{\alpha=1}^{d} p^{(\alpha)} \log q(\tau^{(\alpha)})\right), \tag{K.11}$$

which rewritten suitably yields Eq. (10.34) of the main text.

## Reference

1 S. Gherardini, S. Gupta, F.S. Cataliotti, A. Smerzi, F. Caruso, S. Ruffo, New J. Phys. **18**(1), 013048 (2016). https://doi.org/10.1088/1367-2630/18/1/013048

# Appendix L
# A Primer on the Theory of Large Deviations

In this appendix, we provide a very brief summary of the theory of large deviations as applied to the paradigmatic example of tossing of a fair coin. For a detailed exposition, the reader is referred to Ref [1]. Let us consider $N$ independent trials of tossing a fair coin. As the coin is fair, the probability of getting heads in each of the trials is $1/2$ and so is the probability of getting tails. Then, the probability of observing $M$ number of heads in such $N$ trials is given by the binomial distribution

$$P(M, N) = \frac{1}{2^N} \binom{N}{M} = \frac{1}{2^N} \frac{N!}{M!(N-M)!}. \tag{L.1}$$

Note that for a given value of $N$, the random variable $M$ can take any integer value in the range $[0, N]$. From Eq. (L.1), one obtains the mean $\langle M \rangle$ and the variance $\sigma_M^2$ of the random variable $M$ as

$$\langle M \rangle = \sum_{M=0}^{N} M\, P(M, N) = \frac{N}{2}, \quad \sigma_M^2 = \langle M^2 \rangle - \langle M \rangle^2 = \frac{N}{4}, \tag{L.2}$$

respectively.

One may express the random variable $M$ as

$$M = \sum_{i=1}^{N} \Delta_i, \tag{L.3}$$

where $\Delta_i$ is a random variable defined as

$$\Delta_i = \begin{cases} 1 & \text{if the outcome of the } i\text{th toss is heads,} \\ 0 & \text{if the outcome of the } i\text{th toss is tails.} \end{cases} \tag{L.4}$$



D. Das and S. Gupta, *Facets of Noise*, Fundamental Theories of Physics 214,
https://doi.org/10.1007/978-3-031-45312-0

As the tosses are independent of each other, the $\Delta_i$'s are independent random variables. Moreover, they all can take only two possible values 1 or 0 with equal probabilities, i.e., they are identically distributed. The mean and the variance of $\Delta_i$ are respectively given by

$$\langle \Delta_i \rangle = \frac{1}{2}, \quad \sigma^2_{\Delta_i} = \langle \Delta_i^2 \rangle - \langle \Delta_i \rangle^2 = \frac{1}{4}, \tag{L.5}$$

for all $i$. Equation (L.3) suggests that the random variable $M$ is a sum of $N$ independent and identically-distributed (i.i.d) random variables $\Delta_i$.

Now, let us consider the case of $M$, $N$ large, such that $x \equiv M/N$ is finite. Using the Stirling's approximation

$$N! \approx \sqrt{2\pi N}\, \mathrm{e}^{N \log N - N}, \tag{L.6}$$

we obtain from Eq. (L.1) that

$$\mathcal{P}(x, N) \equiv \lim_{\substack{M, N \to \infty \\ x \equiv M/N \sim \text{finite}}} P(M, N) = \sqrt{\frac{1}{2\pi N x(1-x)}}\, \mathrm{e}^{-NI(x)}, \tag{L.7}$$

where we have

$$I(x) \equiv x \log x + (1-x)\log(1-x) + \log 2\,. \tag{L.8}$$

Let us now compute the most probable value of $x$. In order to proceed, denoting the derivative with respect to $x$ by prime, we obtain

$$\mathcal{P}'(x, N) = \frac{\mathrm{e}^{-NI(x)}\left[(2x-1)(x-x^2)^{-3/2} - 2N(x-x^2)^{-1/2} I'(x)\right]}{2\sqrt{2\pi N}}, \tag{L.9}$$

with

$$I'(x) = \log x - \log(1-x)\,. \tag{L.10}$$

Setting $\mathcal{P}'(x, N) = 0$ in Eq. (L.9), one obtains $x = 1/2$. Hence, $\mathcal{P}(x, N)$ has an extremum at $x = 1/2$, i.e., for $M = N/2$. Taking derivative with respect to $x$ on both sides of Eq. (L.9), we get

$$\mathcal{P}''(x, N) = \frac{(x-x^2)^{-1/2}\, \mathrm{e}^{-NI(x)}}{\sqrt{2\pi N}} \Big[(x-x^2)^{-1} + 3\,(2x-1)^2 (x-x^2)^{-3/2} \\ - N(2x-1)(x-x^2)^{-1} I'(x) + N^2 \{I'(x)\}^2 - N I''(x)\Big], \tag{L.11}$$

where we have

$$I''(x) = \frac{1}{x} + \frac{1}{1-x}\,. \tag{L.12}$$

Noting that $I(1/2) = I'(1/2) = 0$ and $I''(1/2) = 4$, we obtain from Eq. (L.11) that

$$\mathcal{P}''(x = 1/2, N) = \frac{8}{\sqrt{2\pi N}} - \frac{8N}{\sqrt{2\pi N}} = -\sqrt{\frac{2}{\pi N}}\,4(N-1) < 0\,. \tag{L.13}$$

Similarly, with a few straightforward steps, one gets

$$\mathcal{P}'''(x = 1/2, N) = 0\,, \quad \mathcal{P}''''(x = 1/2, N) = \sqrt{\frac{2}{\pi N}}\,16\,(3N^2 - 8N + 9)\,. \tag{L.14}$$

From Eq. (L.13), we conclude that $\mathcal{P}(x, N)$ is maximum at $x = 1/2$ and the maximum value is given by

$$\mathcal{P}^{\max}(N) \equiv \mathcal{P}(x = 1/2, N) = \sqrt{\frac{2}{\pi N}}\,. \tag{L.15}$$

The behavior of $\mathcal{P}(x, N)$ around the point $x = 1/2$ is obtained by expanding it in a Taylor series. To this end, writing

$$x = \frac{1}{2} + \nu\,; \quad |\nu| \ll 1\,, \tag{L.16}$$

and keeping terms upto order $\nu^4$, we obtain

$$\begin{aligned}\mathcal{P}(1/2 + \nu, N) &= \mathcal{P}(1/2, N) + \nu\,\mathcal{P}'(1/2, N) + \frac{\nu^2}{2!}\,\mathcal{P}''(1/2, N) \\ &\quad + \frac{\nu^3}{3!}\,\mathcal{P}'''(1/2, N) + \frac{\nu^4}{4!}\,\mathcal{P}''''(1/2, N)\,.\end{aligned} \tag{L.17}$$

Using Eqs. (L.13), (L.14), (L.15), and $\mathcal{P}'(1/2, N) = 0$ in Eq. (L.17), we obtain for large $N$ that

$$\mathcal{P}(1/2 + \nu, N) = \sqrt{\frac{2}{\pi N}}\,\left[1 - 2N\nu^2 + 2N^2\nu^4\right]. \tag{L.18}$$

For $\nu \ll 1/\sqrt{N}$ (implying $(M - N/2) \ll \sqrt{N}$), Eq. (L.18) yields

$$\mathcal{P}(1/2 + \nu, N) = \sqrt{\frac{2}{\pi N}}\,(1 - 2N\nu^2) \approx \sqrt{\frac{2}{\pi N}}\,\mathrm{e}^{-2N\nu^2}\,, \tag{L.19}$$

which when written in terms of $x$ using Eq. (L.16) and then in terms of $M$ using $x = M/N$ yields the Gaussian form

$$P(M,N) = \frac{1}{\sqrt{2\pi(N/4)}} \exp\left[-\frac{\left(M - N/2\right)^2}{2(N/4)}\right]; \quad (M - N/2) \ll \sqrt{N}. \quad \text{(L.20)}$$

Thus, when the deviation of $M$ from the mean $N/2$ is of order $\sqrt{N}$, $P(M,N)$ is approximately a Gaussian.

We remark that the Gaussian form (L.20) may also be obtained by using the Central Limit Theorem (CLT) discussed in Sect. 1.8. Note that in line with Eq. (1.59), here one may write from Eq. (L.3) that

$$x = \frac{M}{N} = \frac{1}{N}\sum_{i=1}^{N} \Delta_i. \quad \text{(L.21)}$$

We denote the mean and the variance of $\Delta_i$ by $\mu$ and $\sigma^2$, respectively, and write from Eq. (L.5) that $\mu = \langle \Delta_i \rangle = 1/2$ and $\sigma^2 = \sigma^2_{\Delta_i} = 1/4$. According to the CLT, in the limit $N \to \infty$, the distribution $P_Y(y)$ of the scaled random variable $y \equiv (x - \mu)/(\sigma/\sqrt{N}) = 2\sqrt{N}(x - 1/2)$ is a Gaussian with zero mean and unit variance. Therefore, one may write

$$P_Y(y) = \frac{1}{\sqrt{2\pi}} \exp(-\frac{y^2}{2}), \quad \text{(L.22)}$$

from which one may obtain the distribution of $x$ using $y = 2\sqrt{N}(x - 1/2)$. This distribution of $x$ when written in terms of $M$ using $x = M/N$ yields the Gaussian form (L.20). In the following, we denote $P(M,N)$ in Eq. (L.20) as $P_{\text{CLT}}(M,N)$. In Fig. L.1, we have shown $P_{\text{CLT}}(M,N)$ and $\mathcal{P}(x = M/N, N)$ as a function of $M$ for $N = 10^4$. One may see from the figure that these two distributions become identical near the mean $N/2$ shown by the gray dashed-dotted line. However, when the deviation of $M$ from the mean $N/2$ is sufficiently larger than $\sqrt{N}$, they do not agree.

Equation (L.7) may be put in a form commonly known as the large deviation form:

$$\mathcal{P}(x,N) \asymp \mathrm{e}^{-NI(x)}, \quad \text{(L.23)}$$

where the symbol $\asymp$ is to be interpreted as representing 'asymptotically' equal, i.e.,

$$\lim_{N\to\infty}\left[-\frac{1}{N}\log\left(\mathcal{P}(x,N)\right)\right] = I(x). \quad \text{(L.24)}$$

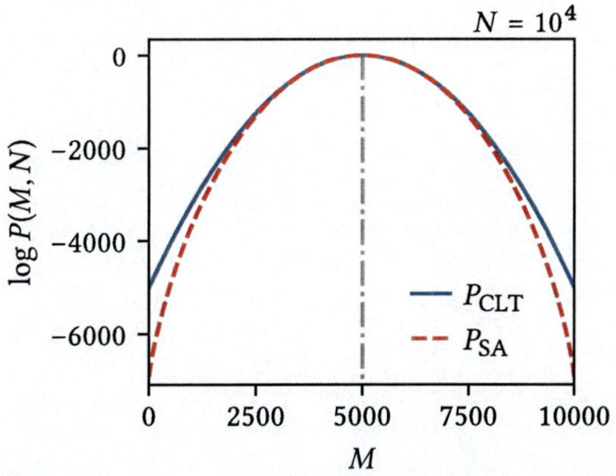


**Fig. L.1 Behavior of** $\log P_{\text{CLT}}(M, N)$ **and** $\log \mathcal{P}(x = M/N, N)$ **versus** $M$ **for** $N = 10^4$. The blue line is obtained by using the Gaussian form given in Eq. (L.20); Here, the fluctuations about the mean are of $O(\sqrt{N})$. The dashed line in red is obtained from Eq. (L.7) with $x = M/N$. The grey dashed-dotted line shows the average value $\langle M \rangle = N/2 = 5000$. It is seen from the figure that near the peak the distributions $P_{\text{CLT}}(M, N)$ and $\mathcal{P}(x = M/N, N)$ become indistinguishable. However, for large deviations from the mean (i.e., when deviation of $M$ from the mean is much larger than $\sqrt{N}$), one has $P_{\text{CLT}}(M, N) \gg \mathcal{P}(x = M/N, N)$

From the worked-out example here, we see that the theory of large deviations deals with fluctuations that are much large with respect to the mean than those captured by the CLT. In other words, the theory deals with fluctuations that have a very small (exponentially small) probability of occurrence, see Eq. (L.23) and Fig. L.1.

## Reference

1 H. Touchette, Phys. Rep. **478**(1–3), 1 (2009). https://doi.org/10.1016/j.physrep.2009.05.002

# Appendix M
# Numerical Integration Algorithms

## M.1 Fourth-Order Runge-Kutta Method

We consider an initial value problem specified by a first-order ordinary differential equation with the general form

$$\frac{\mathrm{d}y}{\mathrm{d}t} = f(t, y); \quad y(0) = y_0 . \tag{M.1}$$

Here, $y(t)$ is an unknown function of time. Our job is to approximate $y(t)$ using the initial value $y_0$ and the differential equation in Eq. (M.1). The rate of change of $y$ given by $\mathrm{d}y/\mathrm{d}t$ depends on $t$ and $y$ itself.

In order to integrate Eq. (M.1) numerically, we first discretize the dynamics, i.e., we choose a time step size $0 < \Delta t \ll 1$, such that the $n$-th time step is given by $t_n = n\Delta t;\ n = 0, 1, 2, \ldots$. It then follows that $y(t_0)$ is given by the initial value $y_0$. In the fourth-order Runge-Kutta algorithm, one update from $t_n$ to $t_n + \Delta t$ involves four intermediate updates. The various steps are as follows:

$$k_1 = f(t_n,\ y(t_n)) , \tag{M.2}$$

$$k_2 = f\left(t_n + \frac{\Delta t}{2},\ y(t_n) + \frac{\Delta t}{2}\, k_1\right) , \tag{M.3}$$

$$k_3 = f\left(t_n + \frac{\Delta t}{2},\ y(t_n) + \frac{\Delta t}{2}\, k_2\right) , \tag{M.4}$$

$$k_4 = f(t_n + \Delta t,\ y(t_n) + \Delta t\, k_3) , \tag{M.5}$$

$$y(t_n + \Delta t) = y(t_n) + \frac{\Delta t}{6}(k_1 + 2k_2 + 2k_3 + k_4) . \tag{M.6}$$

Given $y(t_n)$, Eq. (M.6) yields the value of $y$ at the next instant $t_{n+1}$, namely, the value $y(t_{n+1})$. Following this, we repeat the steps (M.2)–(M.6) to obtain the value of $y$



D. Das and S. Gupta, *Facets of Noise*, Fundamental Theories of Physics 214,
https://doi.org/10.1007/978-3-031-45312-0

at the next time step $t_{n+2}$, and so on. It may be shown [1] that the error per step is of the order of $(\Delta t)^5$, while the accumulated error is of order $(\Delta t)^4$. Let us note in passing that an $n$-th order ordinary differential equation may always be reduced to $n$ coupled first-order ordinary differential equations in $n$ variables, so that one may use the aforementioned algorithm to numerically integrate the $n$-th order equation.

## M.2 Euler Method for Stochastic Differential Equations

Suppose we have an initial value problem specified by a first-order stochastic differential equation with the general form

$$\frac{\mathrm{d}y}{\mathrm{d}t} = f(t, y) + \eta(t) ; \quad y(0) = y_0 , \tag{M.7}$$

where $\eta(t)$ is a Gaussian, white noise satisfying

$$\langle \eta(t) \rangle = 0 ; \quad \langle \eta(t)\eta(t') \rangle = 2D\delta(t - t') , \tag{M.8}$$

with $D > 0$. Here, the angular bracket denotes as usual averaging over noise realizations. Integrating Eq. (M.7) over a small time step $\Delta t$ between $t$ and $t + \Delta t$, we obtain to leading order in $\Delta t$ that

$$y(t + \Delta t) = y(t) + f(t, y(t))\Delta t + \zeta(t) ; \quad \zeta(t) \equiv \int_t^{t+\Delta t} \mathrm{d}t' \eta(t') . \tag{M.9}$$

As $\eta(t)$ is sampled from a Gaussian distribution, the new quantity $\zeta(t)$ is also a Gaussian random variable. Using Eq. (M.8), one finds that

$$\langle \zeta(t) \rangle = 0 , \tag{M.10}$$

$$\langle \zeta^2(t) \rangle = \int_t^{t+\Delta t} \mathrm{d}t' \int_t^{t+\Delta t} \mathrm{d}t'' \, \langle \eta(t)\eta(t'') \rangle = 2D\Delta t . \tag{M.11}$$

Equations (M.10) and (M.11) suggest that the Gaussian-distributed random variable $\zeta(t)$ has zero mean, while its variance is given by $2D\Delta t$. Consequently, one may rewrite Eq. (M.9) as

$$y(t + \Delta t) = y(t) + f(t, y(t)) \, \Delta t + \sqrt{2D\Delta t} \, \xi(t) , \tag{M.12}$$

where $\xi(t)$ is a Gaussian-distributed random number with zero mean and unit variance.

In order to integrate Eq. (M.7) numerically, we first discretize the dynamics, i.e., we choose a time step size $0 < \Delta t \ll 1$, such that the $n$-th time step is given by $t_n = n\Delta t$; $n = 0, 1, 2, \ldots$. It then follows that $y(t_0)$ is given by the initial value $y_0$.

In accordance with Eq. (M.12), given $y(t_n)$, its value is updated by the following rule [2]:

$$y(t_n + \Delta t) = y(t_n) + f(t_n, y(t_n))\, \Delta t + \sqrt{2D\Delta t}\, \xi_n\,, \tag{M.13}$$

where $\xi_n$ is a random variable sampled from a Gaussian distribution with zero mean and unit variance. Given $y(t_n)$, Eq. (M.13) yields the value of $y$ at the next instant $t_{n+1}$, namely, the value $y(t_{n+1})$. The rule (M.13) may then be applied to obtain the value of $y$ at the next time step $t_{n+2}$, and so on. Note that during every update, one has to choose $\xi_n$ independently.

In the absence of noise $\eta(t)$ in the dynamics (M.7), one may set $D = 0$ in the aforementioned algorithm to obtain what is known as the Euler method of numerical integration. It may be shown [1] that the error per step in the Euler method is of the order of $(\Delta t)^2$, while the accumulated error is of order $\Delta t$.

## M.3 Midpoint Method for Stochastic Landau-Lifshitz-Gilbert Equation

In this section, we summarize a method [3] to numerically integrate the dynamics (8.53) for given values of $\gamma$, $T$ and $N$. To integrate the dynamics over a time interval of duration $\mathcal{T}$, we first choose a time step size $0 < \Delta t \ll 1$, and set $t_n = n\Delta t$ as the $n$-th time step of the dynamics, with $n = 0, 1, 2, ..., \mathcal{N}_t$, and $\mathcal{N}_t = \mathcal{T}/\Delta t$. One step of the update scheme from $t_n$ to $t_{n+1} = t_n + \Delta t$ involves the following updates of the dynamical variables for $i = 1, 2, \ldots, N$ and $\mu, \nu, \ldots = x, y, z$:

$$S_i^\mu(t_n + \Delta t) = S_i^\mu(t_n) + F_i^\mu(\{\mathbf{S}_i(t_n + \Delta t/2)\})\, \Delta t + g_i^\mu(\{\mathbf{S}_i(t_n + \Delta t/2)\})\,, \tag{M.14}$$

$$S_i^\mu(t_n + \Delta t/2) = S_i^\mu(t_n) + F_i^\mu(\{\mathbf{S}_i(t_n)\})\, \Delta t/2 + g_i^\mu(\{\mathbf{S}_i(t_n)\})\,, \tag{M.15}$$

$$\begin{aligned} F_i^\mu(\{\mathbf{S}_i(t_n)\}) &= \epsilon_{\mu\nu\delta}\, S_i^\nu(t_n)\, (h_i^{\mathrm{eff}})^\delta(\{\mathbf{S}_i\}(t_n)) \\ &\quad - \gamma\epsilon_{\mu\nu\delta}\, S_i^\nu(t_n)\, \epsilon_{\delta\eta\zeta}\, S_i^\eta(t_n)\, (h_i^{\mathrm{eff}})^\zeta(\{\mathbf{S}_i\}(t_n))\,, \end{aligned} \tag{M.16}$$

$$g_i^\mu(\{\mathbf{S}_i(t_n)\}) = \epsilon_{\mu\nu\delta}\, S_i^\nu(t_n)\, \Delta W_i^\delta - \gamma\epsilon_{\mu\nu\delta}\, S_i^\nu \epsilon_{\delta\eta\zeta}\, S_i^\eta\, \Delta W_i^\zeta\,, \tag{M.17}$$

where Einstein summation convention is implied. Here, $\Delta W_i^\nu$ is a Gaussian distributed random number with zero mean and variance equal to $2\gamma T$.

## References

1 W.H. Press, S.H. Teukolsky, W.T. Vetterling, B.P. Flannery, *Numerical Recipes: The Art of Scientific Computing*, 3rd edn. (Cambridge University Press, Cambridge, 2007)
2 P.E. Kloeden, E. Platen, *Numerical Solution of Stochastic Differential Equations* (Springer, Berlin, 1992). https://doi.org/10.1007/978-3-662-12616-5
3 M. Nishino, S. Miyashita, Phys. Rev. B **91**(13), 134411 (2015). https://doi.org/10.1103/PhysRevB.91.134411